S0-AKY-027

Springer Series in

CLUSTER PHYSICS

Springer
Berlin
Heidelberg
New York
Hong Kong
London
Milan
Paris
Tokyo

Springer Series in

CLUSTER PHYSICS

The intent of the Springer Series in Cluster Physics is to provide systematic information on developments in this rapidly expanding field of physics. In comprehensive books prepared by leading scholars, the current state-of-the-art in theory and experiment in cluster physics is presented.

Mesoscopic Materials and Clusters
Their Physical and Chemical Properties
Editors: T. Arai, K. Mihama, K. Yamamoto and S. Sugano

Cluster Beam Synthesis of Nanostructured Materials
By P. Milani and S. Iannotta

Theory of Atomic and Molecular Clusters
With a Glimpse at Experiments
Editor: J. Jellinek

Metal Clusters at Surfaces
Structure, Quantum Properties, Physical Chemistry
Editor: K.-H. Meiwes-Broer

Clusters and Nanomaterials
Theory and Experiment
Editors: Y. Kawazoe, T. Kondow and K. Ohno

Quantum Phenomena in Clusters and Nanostructures
By S.N. Khanna and A.W. Castleman, Jr.

Water in Confining Geometries
Editors: V. Buch and J.P. Devlin

Series homepage – http://www.springer.de/phys/books/cluster-physics/

V. Buch J. P. Devlin (Eds.)

Water in Confining Geometries

With 185 Figures
Including 25 in Color

Springer

Professor Dr. Victoria Buch
Fritz Haber Research Center for Molecular Dynamics
and Department of Physical Chemistry
The Hebrew University
Jerusalem, 91904, Israel

Professor Dr. J. Paul Devlin
Department of Chemistry
Oklahoma State University
Stillwater, OK 74078, USA

ISSN 1437-0395
ISBN 3-540-00411-4 Springer-Verlag Berlin Heidelberg New York

Library of Congress Cataloging-in-Publication Data: Water in confining geometries/ Victoria Buch, J. Paul Devlin (eds.) p. cm. – (Springer series in cluster physics, ISSN 1437-0395) Includes bibliographical references and index. ISBN 3-540-00411-4 (alk. paper) 1. Water. 2. Ice. 3. Interfaces (Physical sciences) 4. Surfaces (Physics) I. Buch, Victoria, 1955–. II. Devlin, J. Paul, 1935– . III. Series. QC147.W38 2003 530.4'17–dc21 2003042483

Springer-Verlag Berlin Heidelberg New York
a member of BertelsmannSpringer Science+Business Media GmbH

http://www.springer.de

Data conversion by LE-TeX, Leipzig
Cover concept: eStudio Calamar Steinen
Cover production: *design & production* GmbH, Heidelberg
Printed on acid-free paper SPIN: 10852116 57/3141/ba - 5 4 3 2 1 0

Preface

The evolution of the physical/chemical sciences towards understanding the behavior of matter at the molecular level has been accompanied by a rapid increase in studies of the properties and functioning of ***confined water***; that is, water in small clusters and nanoparticles or confined to solid/liquid thin films, surfaces and interfaces. These studies represent a convergence of interests and methodologies. That is, much emerging science, both basic and applied, depends on an understanding of confined water for significant advances; and the technical ability to gain that understanding has evolved only during the past decade or two.

Firm concepts of the behavior of water in a variety of confining geometries are basic to advances in molecular biology, weather phenomena, atmospheric chemistry, interstellar and interplanetary physics and chemistry; as well as to the complete understanding of properties of macroscopic amounts of water and water-solute systems. In recognition of the growing importance of studies of confined water, a Telluride (Colorado) workshop was convened in August of 2000. This was an exceptionally strong 5-day conference with numerous informative talks by leading scientists on both basic and applied aspects of the subject. Lively discussions left the participants spent.

That success of the Telluride conference became the motivation for preparation of this book on “Water in Confining Geometries”. Though the subject is too broad for full coverage, an effort has been made to present the current state of knowledge at the fundamental level while also examining important examples in which the properties of confined water are critical to the functioning of a particular system. A desired breadth of coverage has been achieved by joining writings of several conference participants with chapters prepared by other scholars. The chapters are presented to convey an understanding of the state of knowledge within particular areas with emphasis on advances in which the research groups of the authors have participated. In this way the book offers an introduction to the fundamentals of “water in confining geometries”, a basis for recognition of the role of confined water in specific areas of science, and a description of current research methods and results.

Stillwater and Jerusalem,
December 2002

J. Paul Devlin
Victoria Buch

Preface

Contents

List of Contributors

Patrick Ayotte
Environmental Molecular
Sciences Laboratory,
Pacific Northwest
National Laboratory,
Richland, WA 99352,
USA

Steve Baldelli
Department of Chemistry,
University of Houston,
Houston, TX 77204,
USA

Raul A. Baragiola
Laboratory for Atomic
and Surface Physics,
Engineering Physics,
University of Virginia,
Charlottesville, VA 22904,
USA

Lawrence S. Bartell
Department of Chemistry,
University of Michigan,
Ann Arbor, MI 48109,
USA

Victoria Buch
Fritz Haber Institute
for Molecular Dynamics,
The Hebrew University,
Jerusalem, 91904,
Israel

Udo Buck
Max-Planck-Institut
für Strömungsforschung,
Bunsenstr. 10,
D-37073 Göttingen,
Germany

Robert Bukowski
Department of Physics
and Astronomy,
University of Delaware,
Newark, DE 19716,
USA

Will Cantrell
Department of Physics,
Michigan Technological University,
Houghton, MI 49931-1295,
USA

A. W. Castleman, Jr.
Departments of Chemistry
and Physics,
The Pennsylvania State University,
University Park,
Pennsylvania 16802,
USA

Tsun-Mei Chang
Department of Chemistry,
University of Wisconsin, Parkside,
900 Wood Road, Box 2000,
Kenosha, WI 53141-2000,
USA

Yaroslav G. Chushak
Department of Chemistry,
University of Michigan,
Ann Arbor, MI 48109,
USA

Liem X. Dang
Environmental Molecular
Sciences Laboratory,
Pacific Northwest
National Laboratory,
Richland, WA 99352,
USA

John. L. Daschbach
Environmental Molecular
Sciences Laboratory,
Pacific Northwest
National Laboratory,
Richland, WA 99352,
USA

J. Paul Devlin
Department of Chemistry,
Oklahoma State University,
Stillwater, OK 74078,
USA

Zdenek Dohnálek
Environmental Molecular
Sciences Laboratory,
Pacific Northwest
National Laboratory,
Richland, WA 99352,
USA

George E. Ewing
Department of Chemistry,
Indiana University,
Bloomington, IN 47405,
USA

Michelle Foster
Department of Chemistry,
University of Massachusetts,
Boston, MA 02125,
USA

Bruce C. Garrett
Environmental Molecular
Sciences Laboratory,
Pacific Northwest
National Laboratory,
902 Battelle Boulevard,
PO Box 9999,
Richland, WA 99352,
USA

J. J. Gilligan
Section on Metabolic Analysis
and Mass Spectrometry,
Laboratory of Cellular
and Molecular Biophysics,
National Institute of Child Health
and Human Development,
National Institutes of Health,
Bethesda, MD 20892,
USA

Andrew B. Horn
Department of Chemistry,
University of Manchester,
Oxford Road, Manchester M13 9PL,
England

Bogumil Jeziorski
Department of Chemistry,
University of Warsaw,
Pasteura 1,
02-093 Warsaw,
Poland

Kenneth D. Jordan
Department of Chemistry
and Center for Molecular
and Materials Simulations,
University of Pittsburgh,
Pittsburgh, PA 15260,
USA

Pavel Jungwirth
J. Heyrovsk'y Institute
of Physical Chemistry,
Academy of Sciences
of the Czech Republic
and Center for Complex
Molecular Systems
and Biomolecules,
Dolejvskova 3,
18223 Prague 8,
Czech Republic

K. Karapetian
Department of Chemistry
and Center for Molecular
and Materials Simulations,
University of Pittsburgh,
Pittsburgh, PA 15260,
USA

Shawn M. Kathmann
Environmental Molecular
Sciences Laboratory,
Pacific Northwest
National Laboratory,
902 Battelle Boulevard,
PO Box 9999,
Richland, WA 99352,
USA

Bruce D. Kay
Environmental Molecular
Sciences Laboratory,
Pacific Northwest
National Laboratory,
Richland, WA 99352,
USA

Greg A. Kimmel
Environmental Molecular
Sciences Laboratory,
Pacific Northwest
National Laboratory,
Richland, WA 99352,
USA

Kenichiro Koga
Department of Chemistry,
Fukuoka University of Education,
Japan

William I-Feng Kuo
Department of Chemistry,
University of California-Irvine,
Irvine, CA 92697,
USA

Ali Razmara
Department of Chemistry,
University of California-Irvine,
Irvine, CA 92697,
USA

Vlad Sadtchenko
Department of Chemistry,
George Washington University,
Washington, DC 20052,
USA

Gregory K. Schenter
Environmental Molecular
Sciences Laboratory,
Pacific Northwest
National Laboratory,
902 Battelle Boulevard,
PO Box 9999,
Richland, WA 99352,
USA

Cheryl Schnitzer
Department of Chemistry,
Stonehill College,
North Easton, MA 02357,
USA

Mary Jane Shultz
Department of Chemistry,
Pearson Laboratory,
Tufts University,
Medford, MA 02155,
USA

Danielle Simonelli
Intel Corporation,
Ronler acres,
2501 NW 229th Street,
Hillsboro, OR 97124,
USA

R. Scott Smith
Environmental Molecular
Sciences Laboratory,
Pacific Northwest
National Laboratory,
Richland, WA 99352,
USA

John R. Sodeau
Department of Chemistry,
University College Cork,
Cork,
Ireland

Christof Steinbach
Max-Planck-Institut
für Strömungsforschung,
Bunsenstr. 10,
D-37073 Göttingen,
Germany

Krzysztof Szalewicz
Department of Physics
and Astronomy,
University of Delaware,
Newark, DE 19716
and
Department of Chemistry,
Princeton University,
Princeton, NJ 08544,
USA

Hideki Tanaka
Department of Chemistry,
Faculty of Science,
Okayama University Okayama,
Japan

Fu-Ming Tao
Department of Chemistry
and Biochemistry,
California State University,
Fullerton, CA 92834,
USA

Mounir Tarek
Department of Chemistry,
University of California-Irvine,
Irvine, CA 92697,
USA

Glenn Teeter
Environmental Molecular
Sciences Laboratory,
Pacific Northwest
National Laboratory,
Richland, WA 99352,
USA

Douglas J. Tobias
Department of Chemistry,
University of California-Irvine,
Irvine, CA 92697,
USA

Introduction

J. P. Devlin and Victoria Buch

Studies of confined water are prompted by varied considerations. Most fundamental is that, despite the high disorder/low structural symmetry common to confined water, the basic structural and dynamic properties are often amenable to detailed study by modern computational and experimental techniques. As a result, the science is marked by molecular perceptions generated in these basic studies. However, lacking a flexible water potential that fully accounts for many-body effects important to the hydrogen-bonded network structure, no universally consistent fundamental description of water systems exists. The information derived from studies of various forms of confined water has the potential to alter the situation. Also, as the size of small water systems is expanded, observations of the evolution of molecular-level properties can help connect the dots leading to properties of macroscopic systems. In some instances water within important structures, such as globular proteins, cloud nuclei, the quasi-liquid layer of warm ice, icy interstellar particles and polar ice fogs, naturally assumes a confined form; in which case, studies of appropriate model systems advance both basic and applied objectives. As is common throughout science, advanced understanding of idealized model systems can create the most direct path to enhanced knowledge of the more complex systems. Within this monograph of monographs, the juxtaposition of reviews of several independent studies, by scientists working with an array of theoretical, computational and experimental methods, each applied to confined water but for a great variety of conditions, makes apparent the progression towards full understanding of this most vital of substances as well as the interplay between the basic and applied science of confined water.

A broad range of subjects is addressed within this lively science. The monograph begins with a focused look at the evolution towards a water potential that faithfully describes fundamental properties of pure water in the bulk phase as well as in the smallest of water clusters (Ch.1). This evolution has confirmed that non-additive forces have a profound effect on the hydrogen-bonded structure of liquid water. A theoretical analysis follows of the difficult basic question of the thermochemical properties of small clusters of pure water; and the optimum definition of a water cluster for purposes of nucleation of liquid water (Ch.2). Here, also, the necessity, that the potential represent polarizable water, i.e., with non-additive forces, is emphasized.

Important questions, about these nuclei or, more generally, cold small water clusters, resemble ones typically asked about individual molecules; except that bonding *between* molecules assumes a dominant role. Whether of water only, or with molecules of other substances included, the hydrogen bond enjoys a prominence within water clusters that, for molecules, is reserved for the chemical bond. Though intermolecular interactions are naturally highlighted, the internal response of the component molecules is also important. For example, the response of solute molecules to a stepwise increase in water content of a cluster, can reveal the essence of the solvation process while displaying the transition towards a condensed phase (Ch.4,5). For the smallest of clusters $(H_2O)_n$, (with n less than ~8) the details of structure, vibrational dynamics and structural transitions are discovered through *ab initio* computations (Ch.4,5), and an outline of the multidimensional intermolecular potential may be generated. For similarly sized through much larger clusters, classical molecular dynamics (MD) simulations, using water-water intermolecular potentials fitted to empirical data and often calibrated using *ab initio* parameters, can give useful estimates of the relative stability of isomeric structures and permit the assignment of infrared bands to specific cluster or cluster-surface vibrational modes (Ch.3,17).

When the water "cluster" is in contact with other substances, the focus switches to the effect of the interfacial interactions on the properties of the confined water. The topic of interfacial water is nicely introduced by an example of small water clusters attached to a well-defined surface (Ch.6). The results, using the Dang-Chang polarizable water potential with a polarizable model of graphite, show, for example, that hexameric water clusters are perturbed only marginally by a graphite single-crystal surface; but the relative-stability sequence of the isomeric forms is scrambled for fundamental reasons that are laid out in detail. By contrast, water severely confined within the quasi-one-dimensional space of carbon nanotubes, as simulated using MD with classical intermolecular potentials or as examined in *ab initio* computations, is characterized by novel network structures that are determined by the tube diameter (Ch.7). Similarly, water bilayers within a slit pore (i.e., a quasi-two-dimensional space) access several remarkable network structures (liquid, amorphous and crystalline) with variation of temperature and pressure (Ch.7). Both the one- and two-dimensional results are reminiscent of the richness of the phase diagram of water in three-dimensional space.

Water at biological interfaces typically experiences an environment that is both chemically and structurally rich. Modeling of the structure and dynamics of systems such as the interface with globular proteins reveals the nature of specific water interactions with distributed sites of varied H-bonding ability, while also accounting for the convoluted nature of the interface (Ch.9). By comparison, the interfacial structure is more regular for water films on extended flat insulator surfaces; and, through the use of spectroscopic and simulation methods, the water ad-layers have been characterized for a variety of ionic insulators (Ch.8). Still, the manner by which the interplay between substrate lattice structure and local surface morphology influences nucleation of hexagonal ice from the adsorbed water films remains largely mysterious. This is a continuing area of investigation with applications to natural systems that range from weather phenomena to the suppression of ice formation in biological structures.

Interfaces between water and a second liquid substance also occur commonly in chemistry, biochemistry and nature. Details of molecular-level interfacial structure, as well as free energy profiles and mechanisms of mass transport across the interface, have best been probed using MD simulations applied to liquids for which the necessary intermolecular potentials are well defined (Ch.10). The use of a polarizable potential for water is identified as a criterion for sensible results at the water-CCl_4 interface. Sum-frequency generation (SFG) spectroscopy, which is uniquely sensitive to the liquid-liquid interface because of a structural asymmetry selection rule, holds particular promise for experimental evaluation of these liquid-liquid interfacial properties.

Applications of the SFG technique have revealed interesting properties of the water-vapor interface of pure water and, more particularly, for water containing a variety of electrolytes (Ch.11). Chemistry at surfaces of water-rich aerosols, liquid and solid, is pervasive in the earth's atmosphere and elsewhere in nature (e.g., at the ocean-air interface); so the new insights to the distribution and orientation of ions near the interface, and the consequences for the interfacial water, have immediate application within atmospheric and environmental science. The SFG results parallel nicely a detailed computation of the distribution of ions within microaerosols of aqueous sea salt. In a study that extends the understanding of the influence of such aerosols on other atmospheric components, such as ozone in the polar troposphere, MD simulations revealed that anions with large polarizabilities readily occupy water surface sites (Ch.12). This was shown to occur to such an extent that the bromide ion actually functions as a surfactant. Similarly, insight to important stratospheric chemical and photochemical processes has been obtained by laboratory spectroscopic observations of interaction of adsorbates (HCl, N_2O_5, $ClONO_2$, etc.) with thin ice films (Ch.13). Photochemical reactions, along with changes induced by high-energy particles, are increasingly recognized as influential to the behavior of atmospheric systems (Ch.5,13,15). This influence undoubtedly extends to the chemistry and physics of icy particles in interstellar space (Ch.13,15). Together, the growing interest in atmospheric and space science foretells a parallel increase in efforts to mimic radiation effects in laboratory studies.

Though nearly all water ice on earth is crystalline, most ice in the universe is thought to be amorphous (in the form of an accretion onto dust particles in the cold regions of the dense interstellar clouds). Further, amorphous solid water (ASW) is a useful model of liquid water, the study of which is facilitated by the relative absence of thermal motion. For such reasons, the properties of thin films of ASW are of both fundamental and astrophysical interest. However, the nature of this solid version of confined water has been shown to be extremely sensitive to preparatory methods; the sensitivity made apparent through the use of controlled molecular beams in ultra-high vacuum environments (Ch.14,15). One outcome is the need for serious evaluation of much of the ASW data in the literature (Ch.15). With the basic nature of ASW more nearly circumscribed (Ch.14,15), its value in laboratory studies of extraterrestrial ice is being extended through irradiation and chemical treatments designed to mimic the impact of natural environments (Ch.13,15).

Cold-water clusters experience a size range in which the interior (core) structure closely resembles that of ASW. This range, which has not been thoroughly studied, extends from ~50 to a size of 200 - 400 water molecules (Ch.3,17). Larger water particles favor a crystalline cubic ice form; so $(H_2O)_{300}$ represents an approximate demarcation line between large water clusters and ice nanocrystals. Electron-diffraction patterns show that nanocrystals of cubic ice can nucleate to a remarkably low temperature and with an astounding rate from droplets formed in a supersonic expansion beam (Ch.16); although retaining an element of disorder. The disordered nature of the surface of ice nanocrystals has been established in spectroscopic (Ch.17) and simulation (Ch.16,17) studies; along with effects of that disorder on H-bond chemistry at the ice surface (Ch.17).

In addition to the strikingly large ice nucleation rate from cold nanodroplets, which serves to underscore the lack of understanding of the ubiquitous transformation of water to ice, surprises abound in results of studies of confined water. For example: the companion bond of water molecules with a free O-H group within a small water cluster can be significantly stronger than the H-bonds within ice, with stretch frequencies below 3150 cm^{-1} for n = 7-9 in $(H_2O)_n$ (Ch.3); water molecules of the magic-number dodecahedral cage of $(H_2O)_{21}H^+$ can be replaced by methanol molecules to from a mixed clathrate-like structure (Ch.5); the liquid-water surface is severely altered in a sulfuric-acid solution of less than 10 mole % (Ch.11); a barium fluoride surface does not provide exceptional nucleation sites for ice despite an excellent match of lattice parameters (Ch.8); amorphous ice is not microporous when prepared by vapor deposition from a molecular beam *normal* to a substrate surface (Ch.14); the number of "reactive" 3-coordination sites on the ice surface is greater for an oxygen-ordered surface than for the disordered surface of ice nanocrystals (Ch.17). Numerous other examples are contained within this monograph, suggesting that further surprises are waiting to be discovered in ongoing and future studies of confined water.

Part I

Water Clusters

1 Nature of Many-Body Forces in Water Clusters and Bulk

Krzysztof Szalewicz, Robert Bukowski, and Bogumil Jeziorski

1.1 Introduction

All the properties of water clusters and of bulk water can in principle be predicted by solving the Schrödinger equation for the motion of nuclei on the potential energy surface of a given system. This level of theory does neglect several physical effects such as nonadiabatic couplings of electronic and nuclear motions or relativistic effects, but these effects would contribute much below the current uncertainties of both measurements and theory for systems of this size. Several properties of bulk water can be described reasonably well by solving classical rather than quantum equations of motion, a significant further simplification. However, accuracy of all predictions depends critically on the accuracy of the potential energy surface. This surface can be decomposed into intramonomer contributions, i.e., the potentials within single water molecules, the pair potentials, and the so-called nonadditive potentials. Since derivatives of potential energy surfaces define forces, one may alternatively use the term "force fields" equivalently with "potentials". This chapter will be devoted to many-body potentials of water molecules with emphasis on elucidating the physical origins of interactions.

Let us start from a precise definition of the potential energy surface and its components for a system consisting of N molecules. This surface, denoted by $E_{tot}(\boldsymbol{Q}_1, ..., \boldsymbol{Q}_N)$, where $\boldsymbol{Q}_i = (\boldsymbol{R}_i, \boldsymbol{\omega}_i, \boldsymbol{\xi}_i)$ stands for the set of all coordinates needed to specify the spatial position $\boldsymbol{R}_i$, orientation $\boldsymbol{\omega}_i$, and the internal geometry $\boldsymbol{\xi}_i$ of the ith monomer, is defined as an eigenvalue of the clamped-nuclei electronic Hamiltonian for a given configuration of nuclei. It is convenient to separate this quantity into the sum of internal energies $E_i(\boldsymbol{\xi}_i)$ of the monomers and the interaction energy $E_{\text{int}}(\boldsymbol{Q}_1, ..., \boldsymbol{Q}_N)$:

$$E_{tot}(\boldsymbol{Q}_1, ...\boldsymbol{Q}_N) = E_{int}(\boldsymbol{Q}_1, ...\boldsymbol{Q}_N) + \sum_i E_i(\boldsymbol{\xi}_i). \tag{1.1}$$

The interaction energy of monomers can be expressed as a sum of terms involving interactions of 2,3,....monomers

$$E_{int} = E_{int}[2, N] + E_{int}[3, N] + ... + E_{int}[N, N], \tag{1.2}$$

where K-body contributions to the N-mer energy, $E_{\text{int}}[K, N]$, can be written as the following sums

$$E_{int}[2, N] = \sum_{i<j} E_{int}(\boldsymbol{Q}_i, \boldsymbol{Q}_j)[2, 2], \tag{1.3}$$

$$E_{int}[3, N] = \sum_{i<j<k} E_{int}(\boldsymbol{Q}_i, \boldsymbol{Q}_j, \boldsymbol{Q}_k)[3, 3], \tag{1.4}$$

etc. The two-body or *pairwise-additive* interaction energies $E_{\text{int}}[2, 2]$ are just the dimer interaction energies defined by (1.1) for $N = 2$. Their sum, $E_{\text{int}}[2, N]$, is the (pairwise) additive component of the interaction energy of an N-mer. The higher-body terms, i.e., the *nonadditive* contributions to the N-mer interaction energy, are defined recursively. For example, the three-body contribution to a trimer interaction energy, $E_{\text{int}}[3, 3]$, is the difference between the total interaction energy of a given trimer and the sum of all pair energies. Each K-body contribution $E_{\text{int}}(\boldsymbol{Q}_{i_1}, \boldsymbol{Q}_{i_2}, ..., \boldsymbol{Q}_{i_K})[K, K]$ is formally defined in terms of the energy of the K-member cluster (with monomers placed at $\boldsymbol{Q}_{i_1}, \boldsymbol{Q}_{i_2}, ..., \boldsymbol{Q}_{i_K}$) and of the energies of all its sub-clusters. However, in the perturbative approach discussed in Sec. 1.2, this contribution can be obtained directly, without the knowledge of interaction energies of clusters containing less than K monomers. This perturbation approach also shows that the $E_{\text{int}}[K, K]$ contributions with different K have different physical interpretations (i.e., appear due to physically different mechanisms of intermolecular interactions). The sum of all contributions with $K > 2$ constitutes the (pairwise) nonadditive component of the interaction energy of an N-mer. It should be noted that the many-body expansion of the interaction energy can be defined only when the quantum states of all subsystems are unambiguously defined. For strongly interacting systems such as liquid metals or chemically bound molecules, this condition is not fulfilled and the applicability of the many-body expansion can be questioned. In most applications of (1.2), the number of molecules N will be constant and therefore the index N in $[K, N]$ will be omitted in the remainder of this Chapter.

The interaction potential $E_{\text{int}}(\boldsymbol{Q}_1, ..., \boldsymbol{Q}_N)$ depends on internal geometries of all monomers. For many purposes the dependence of $E_{\text{int}}(\boldsymbol{Q}_1, ..., \boldsymbol{Q}_N)$ on intramonomer coordinates can be neglected and the resulting intermolecular potentials are referred to as rigid-monomer potentials. Such potentials depend on much smaller number of coordinates and therefore are significantly easier to calculate and use. Already for the water dimer, the complete potential is 12-dimensional whereas the rigid-monomer potential is only 6-dimensional. Further discussions will be limited to rigid-monomer potentials. For issues connected with calculations of flexible-monomer potentials see Refs. [1, 2].

Investigations of the interaction potential of the water dimer date back to 1970s when rotational spectroscopy measurements determined the minimum structure of this system. A similar structure had been predicted even earlier by *ab initio* calculations. The experimental equilibrium angular orientation of the two monomers measured by Odutola and Dyke in 1980 [3] agrees very well with results of modern *ab initio* calculations. Numerous measurements of the infrared spectra of water dimer have been published. These spectra, in particular in the far-infrared region (see Refs. [4, 5] for recent examples), provide a wealth of data characterizing the system. These data allow stringent evaluations of the quality of water pair potentials. However, such evaluations could be done only very recently since only in the late 1990s six-dimensional calculations of the water dimer dynamics became possible [6–9]. These calculations have shown that none of the water dimer potentials published until the late 1990s was able to predict the correct spectra even qualitatively. Only the *ab initio* potential developed in Refs. [10, 11] using symmetry-adapted perturbation theory (SAPT) achieved this goal, and in fact predicted a majority of transitions among the ground-state manifold to within about 0.01 cm^{-1} [8,9,12]. The spectroscopic data were also used to fit an empirical water dimer potential [13]. This fit was recently revised in Ref. [14]. Also a 12-dimensional surface based on the MCY [15] functional form has been fitted to spectral data using an adiabatic (6+6)-dimensional dynamics approach [16].

The SAPT potential for the water dimer, named SAPT-5s due to five symmetry-distinct sites used in the fit, was based on about 2500 *ab initio* grid points, a much larger number of points than in any previous work on this dimer. A medium-size, interaction-optimized basis set with *spdf* angular components and with bond functions was applied in SAPT calculations. Based on the very accurate value of the energy of the global minimum computed by Klopper *et al.* [17], the SAPT interaction energies can be estimated to be accurate to about 0.3 kcal/mol or 5% with respect to basis set and theory level extensions. Very recent high-accuracy supermolecular calculations of the energies of the stationary points on the water dimer potential energy surface [18] provide an additional confirmation of the accuracy of the SAPT-5s potential. The analytic fit reproduces the computed points with root-mean square error of 0.38 kcal/mol (0.10 kcal/mol for points with negative interaction energies). Thus, it appears that the water pair potential is finally known reasonably well. The SAPT-5s potential allowed resolution of some controversies related to the minimum structure of the water dimer. Average values of $1/R^2$ – obtained from the two lowest-energy rovibrational wave functions computed using the SAPT-5s potential – have been combined with the measured value to obtain a new empirical estimate of the equilibrium O-O separation equal to 5.50 ± 0.01 bohr [11], significantly shorter than the previously accepted value, but in a very good agreement with high-level *ab initio* calculations of Klopper *et al.* [17]. The extrapolated *ab initio* calculations [17] together with the zero-point energy computed from SAPT-5s lead to a new prediction of the dissociation energy equal to 1165 ± 54 cm^{-1} [11], close to but significantly more accurate than the best empirical value [19].

Both the SAPT-5s water dimer potential of Ref. [11] and the earlier SAPT-pp potential from Ref. [10] were used to compute the second virial coefficient for water and reproduced the experimental data very accurately. In fact, the discrepancies between theory and experiment are of the same order of magnitude as estimated errors of measurements. Among the *ab initio* potentials only the ASP potentials of Millot *et al.* [20] perform as well as SAPT. Compared to the popular empirical effective pair potentials [21–23] or to the well-known MCY *ab initio* potential [15], the SAPT values are nearly an order of magnitude more accurate.

The comparison of virial coefficients points out the well-known deficiency of effective pair potentials. It is not possible to simultaneously fit such potentials to the virials and to typical bulk properties dependent on the pair-nonadditive effects. Thus, these potentials reproduce dimer properties, including virial coefficients, rather poorly. In contrast, polarizable empirical potentials have to a lesser extent the "effective" character as these potentials, as used in Chapter 2, model explicitly the nonadditive induction energy. Thus, the pairwise-additive component is less biased by efforts to mimic nonadditive forces. The polarizable empirical potentials can be made to reproduce dimer properties and sometimes virial coefficients are used in the fits. When the virial coefficients are not used, as in the case of the polarizable point-charge (PPC) potential of Svishchev *et al.* [24], the computed virial coefficients are much better than obtained with the empirical pair potentials, but still significantly less accurate than predicted by SAPT potentials [10, 11].

The SAPT-5s potential was used to compute spectra of the water dimer [8, 9, 12]. The water dimer potential exhibits eight equivalent minima differing only by a permutation of hydrogen atoms. The monomers can tunnel between these minima through the separating barriers (this tunneling should not be confused with the electron tunneling leading to the exchange forces). The tunneling results in characteristic splittings of the vibrational energy levels. The so-called acceptor tunneling splittings are of the order of 10 cm^{-1}. The donor-acceptor interchange splittings are of the order of 1 cm^{-1}. The donor (bifurcation) tunneling does not introduce more splitting but leads to shifts of levels. The characteristic splitting patterns significantly help in assigning the spectra. Therefore the name vibration-rotation-tunneling (VRT) spectra is sometimes used.

The agreement between the ground vibrational state tunneling splittings measured [4, 5, 25, 26] and calculated from the SAPT-5s potential [8, 9] is very good for most levels. In particular the small interchange splittings were recovered to about 0.01 cm^{-1}. Also the $B + C$ rotational constant agreed with experiment to about 0.5%. Only the sum of large splittings (individual components have not been measured) was less accurate and differed from experiment by about 40%. As already mentioned, SAPT-5s was the first *ab initio* potential that achieved quantitative agreement with experimental spectra. Previous potentials have not reached even a qualitative agreement, giving results typically a few times too small or too large. The sum of acceptor splittings (a single experimental number) was used together with SAPT computed energies to tune the SAPT-5s potential. The tuned potential, denoted by SAPT-5st, reproduced this value to 0.06 cm^{-1} and at the same time did not lose the high accuracy of SAPT-5s for other transitions. In a more recent publication, Smit *et al.* [12] used the SAPT-5st potential to predict the vibrational excitations of the water dimer. In general, a very good agreement with experiment [4, 5] was found. In a few cases a reassignment of experimental lines was proposed.

The spectral results computed from the SAPT-5s potential could be compared to those produced by the VRT(ASP-W) potential of Fellers *et al.* [13], which was fitted to the dimer spectra. In general, SAPT-5s, SAPT-5st, and VRT(ASP-W) agree with experiment equally well, except for the sum of acceptor splittings in the case of SAPT-5s. However, when the VRT(ASP-W) potential was used together with a three-body nonadditive SAPT potential to calculate the spectra of the water trimer (see below), the predictions from VRT(ASP-W) were significantly worse than those given by the SAPT-5s plus three-body potential. Thus, SAPT-5s/st provide probably the best current characterization of the water dimer interaction.

The next term in the many-body expansion of the water potential is the three-body interaction. Similarly as for the water dimer, spectral data are available for the water trimer [27, 28]. As it will be discussed below, the three-body term makes a significant contribution to the water potential. An accurate three-body potential for water has recently been developed [29]. One might expect that for clusters larger than trimer and for condensed phase four- and higher-body interactions may become important. However, this does not appear to be the case. The nonadditive effects in water clusters will be the subject of the remaining sections of this Chapter.

It has been long recognized that many-body forces are critical for describing bulk materials; see Refs.[30–32] for reviews of literature in this field. Availability of accurate *ab initio* intermolecular potentials including pairwise additive and three-body nonadditive forces opens the possibility of reliable first-principle simulations of condensed phases. Molecular simulations based on empirical potentials have been an active field of research for a long time. The best known work using *ab initio* potentials is that of Clementi, Corongiu, and collaborators [33] on water. Recently the number of such simulations has been increasing rapidly. Examples include work by Huber and collaborators [34, 35] as well as on-the-fly simulations by Parrinello *et al.* [36]. Water remains the most investigated system [36–42]. Several first-principle simulations have recently been performed based on SAPT potentials [43–46], including simulations for water. The latter predict most of the observed values very well; however, some discrepancies remain.

1.2 Perturbation theory of intermolecular interactions

Since interactions between water molecules are much weaker than the chemical bonds inside a water molecule, these interactions can naturally be described in terms of a perturbation theory treating isolated monomers as the zeroth-order approximation. If the standard Rayleigh-Schrödinger perturbation theory were applied, the resulting wave functions of the N-mer would not fulfill the Pauli's exclusion principle. To overcome this problem, explicit antisymmetrization operators have to be applied, leading to symmetry-adapted perturbation theory. For pair interactions SAPT is a well-developed method, see Refs. [47–50] for reviews. More recently, symmetry-adapted perturbation theory of three-body interactions has been developed [51–54]. In the SAPT expansion of the interaction energy for an N-mer, the total Hamiltonian is partitioned as

$$H = F + V + W \tag{1.5}$$

where F is the sum of the Fock operators for all monomers, $V = V_{AB} + V_{BC} + V_{AC} + ...$ is the sum of all binary intermolecular interaction operators V_{XY}, and $W = W_A + W_B + W_C + ...$ is the sum of the intramolecular correlation operators (Møller-Plesset fluctuation potentials) for all monomers. This partitioning of the total Hamiltonian allows the interaction energy to be written as a double perturbation expansion in V and W. The latter expansion is analogous to that used in the standard many-body perturbation theory and coupled-cluster methods [55, 56]. The K-body interaction energy can be written as

$$E_{\mathrm{int}}[K] = \sum_{i=1;j=0} \left(E^{(ij)}_{\mathrm{pol}}[K] + E^{(ij)}_{\mathrm{exch}}[K] \right) \tag{1.6}$$

where subscripts "pol(exch)" denote the polarization (exchange) energy components. The former components can be found by an application of the Rayleigh-Schrödinger perturbation theory, the corresponding exchange energy components are defined as the difference between the SAPT and polarization energies. The superscripts i and j refer to the order with respect to the intermolecular interaction V and the intramolecular correlation operator W, respectively. It is also convenient to consider the interaction energy components which fully include intramonomer correlation effects

$$E^{(i)}_{\mathrm{pol(exch)}}[K] = \sum_{j=0}^{\infty} E^{(ij)}_{\mathrm{pol(exch)}}[K]. \tag{1.7}$$

1.3 Physical origins of pair contributions

The two-body interaction energies are well described by an expansion terminating at the second-order in V and include the well-known electrostatic, induction, dispersion, and exchange components:

$$E_{\text{int}}[2] = E^{(1)}_{\text{elst}}[2] + E^{(1)}_{\text{exch}}[2] + E^{(2)}_{\text{ind}}[2] + E^{(2)}_{\text{disp}}[2] + E^{(2)}_{\text{exch-ind}}[2] + E^{(2)}_{\text{exch-disp}}[2] + \cdots . \quad (1.8)$$

For the water dimer all of the components contribute significantly to the interaction energy. These components provide a clear physical picture of the two-body interaction in water. The physical interpretation of the water dimer energy has been the subject of great interest as it advances our understanding of the hydrogen bond [37,57–59].

Table 1.1. Components of the SAPT interaction energy of the water dimer for three intermolecular separations R. The angular coordinates are close to those of the global minimum configuration. All energies in kcal/mol.

Component	$R = 2.5$ Å	$R = 3.0$ Å	$R = 3.6$ Å
$E^{(10)}_{\text{elst}}[2]$	-19.536	-8.493	-2.938
$E^{(12)}_{\text{elst,resp}}[2]$	-0.562	-0.114	0.091
$E^{(13)}_{\text{elst,resp}}[2]$	0.430	0.205	0.034
$E^{(1)}_{\text{elst}}[2]$	-19.669	-8.402	-2.813
$E^{(20)}_{\text{ind,resp}}[2]$	-13.102	-3.034	-0.337
${}^{t}E^{(22)}_{\text{ind}}[2]$	-2.317	-0.589	-0.059
$E^{(2)}_{\text{ind}}[2]$	-15.419	-3.623	-0.396
$E^{(20)}_{\text{disp}}[2]$	-6.356	-2.548	-0.572
$E^{(21)}_{\text{disp}}[2]$	0.324	0.028	-0.007
$E^{(22)}_{\text{disp}}[2]$	-1.239	-0.551	-0.144
$E^{(2)}_{\text{disp}}[2]$	-7.272	-3.071	-0.723
$E^{(10)}_{\text{exch}}[2]$	29.939	7.205	0.583
$E^{(1)}_{\text{exch}}(\text{CCSD})[2]–E^{(10)}_{\text{exch}}[2]$	4.427	1.622	0.208
$E^{(1)}_{\text{exch}}[2]$	34.365	8.827	0.791
$E^{(20)}_{\text{exch-ind,resp}}[2]$	8.446	1.677	0.101
${}^{t}E^{(22)}_{\text{exch-ind}}[2]$	1.493	0.326	0.018
$E^{(20)}_{\text{exch-disp}}[2]$	1.513	0.461	0.049
$\delta^{\text{HF}}_{\text{int,resp}}[2]$	-3.744	-0.898	-0.088
$E^{(2)}_{\text{exch}}[2] + \delta^{\text{HF}}_{\text{int,resp}}[2]$	7.709	1.566	0.080
$E_{\text{int}}[2]$	-0.285	-4.703	-3.062

Table 1.1 shows interaction energy components of the water dimer on a cut of the potential passing near the global minimum. The subscripts "resp" indicate that a given term was computed with orbital relaxation effects [60, 61].

The correction $E^{(1)}_{\rm exch}({\rm CCSD})$ was computed with the monomer wave functions correlated at the coupled-clusters with single and double excitations level [62]. The term $\delta E^{\rm HF}_{\rm int,resp}$ collects contributions beyond second order in V to the supermolecular Hartree-Fock interaction energy $E^{\rm HF}_{\rm int}$ [63, 64]:

$$\delta E^{\rm HF}_{\rm int,resp}[2] = E^{\rm HF}_{\rm int}[2] - E^{(10)}_{\rm elst}[2] - E^{(10)}_{\rm exch}[2] - E^{(20)}_{\rm ind,resp}[2]$$
$$-E^{(20)}_{\rm exch-ind,resp}[2]. \qquad (1.9)$$

Asymptotically, for large intermonomer separations R, the interaction energy in the water dimer is completely dominated by the electrostatic interactions due to the slow, R^{-3} decay of the dipole-dipole potential. Already at R = 3.6 Å the electrostatic component constitutes 92% of the interaction energy, cf. Table 1.1. As R approaches the minimum configuration, the induction and dispersion energies of nearly equal magnitude begin to provide larger and larger contributions – at the minimum their sum is approximately equal to the electrostatic contribution. However, as R decreases, the exponentially decaying exchange component very strongly quenches the energy lowering provided by the polarization components. At the minimum distance, the exchange component is about twice as large in magnitude as the total interaction energy and therefore it cancels about two thirds of the attractive effect. For still shorter R, the exchange energy becomes even more important. At R = 2.5 Å it is about as large as the sum of attractive components.

The interaction energy components can also be split into the Hartree-Fock and correlation effects. The total electron correlation contribution at the minimum is −1.3 kcal/mol or 25% of interaction energy [65]. The value quoted from Ref. [65] was obtained in a larger basis set than used to calculate the results shown in Table 1.1 and also the dimer geometries used in the two calculations were somewhat different. The same effect is -1.2 kcal/mol for the near-minimum configuration listed in Table 1.1 (the results from this table will be given below in parentheses). The size of the correlation contribution shows that even for this strongly polar dimer the Hartree-Fock level of theory would not be satisfactory for most purposes. The largest correlation effect comes from the dispersion energy equal to -2.77 (-3.07) kcal/mol. This attractive component is partly quenched by the intramolecular correlation correction to the first-order exchange energy equal to 1.05 (1.6) kcal/mol. The intramonomer correlation correction to the electrostatic energy is also positive but much smaller, as it amounts to 0.23 (0.09) kcal/mol. Another correlation component, the exchange-dispersion energy, contributes 0.41 (0.46) kcal/mol. The overall correlation correction to induction is -0.19 (-0.26) kcal/mol.

1.4 Physical origins of nonadditive contributions

The components of three-body interaction energy are quite different from those in the two-body case and to get a reasonable description of this interaction one has to include the effects of the third order in V:

$$E_{\rm int}[3] = E^{(1)}_{\rm exch}[3] + E^{(2)}_{\rm ind}[3] + E^{(2)}_{\rm exch-ind}[3] + E^{(2)}_{\rm exch-disp}[3] + E^{(3)}_{\rm disp}[3]$$

$$+E^{(3)}_{\rm ind}[3] + \cdots . \tag{1.10}$$

The nonadditive electrostatic interaction and the dispersion interaction in the second order are exactly equal to zero. It is often convenient to compute the Hartree-Fock three-body contribution, $E^{\rm HF}_{\rm int}[3]$. This quantity can be decomposed as:

$$E^{\rm HF}_{\rm int}[3] = E^{(10)}_{\rm exch}[3] + E^{(20)}_{\rm ind}[3] + E^{(20)}_{\rm exch-ind}[3] + \delta E^{\rm HF}_{\rm int}[3], \tag{1.11}$$

where $\delta E^{\rm HF}_{\rm int}[3]$ contains mostly $E^{(3)}_{\rm ind}[3]$ and its exchange counterpart.

The knowledge of relative importance of nonadditive components of the interaction energy is critical for our ability to describe this energy via expansion as in (1.2). In practice, the only way to learn how large these contributions are is by *ab initio* calculations, although for some systems empirical fits to spectroscopic data can also provide useful information [66, 67]. However, generally not much information is available at the present time. Some of this information is shown in Fig. 1.1, where the major nonadditive components are displayed for four different near-equilibrium trimers. As one can see, the ratio of the three-body component to the sum of two-body interactions changes dramatically from system to system: from 0.4% for He_3 to 16% for the water trimer. The total nonadditive contribution can also be of either sign, even for similar systems like rare gas trimers He_3 and Ar_3. Notice that the signs of the major components like $E^{(1)}_{\rm exch}[3]$ or $E^{(3)}_{\rm disp}[3]$ are the same for both trimers and the opposite signs of percentage contributions result from the total nonadditive energies being of opposite sign (due to a different balance of positive and negative components). The increased role of nonadditive terms is clearly correlated with the polarity of a system. The more polar a system is, the larger are the relative contributions of nonadditive terms.

The relative importance of various components of nonadditive interaction energies appearing in (1.4) also varies dramatically between different trimers. In fact, since only the largest components are shown, the set of displayed components is not always the same. For rare gas trimers the nonadditive contribution is dominated by the first-order exchange and third-order dispersion energies, although for Ar_3 the exchange-dispersion contribution is quite large. This contribution is critical for predicting the correct crystal structure of argon [70]. For Ar_2–HF all these components are still important but a very significant role is played also by the induction and exchange-induction nonadditive contributions. The situation changes completely in the water

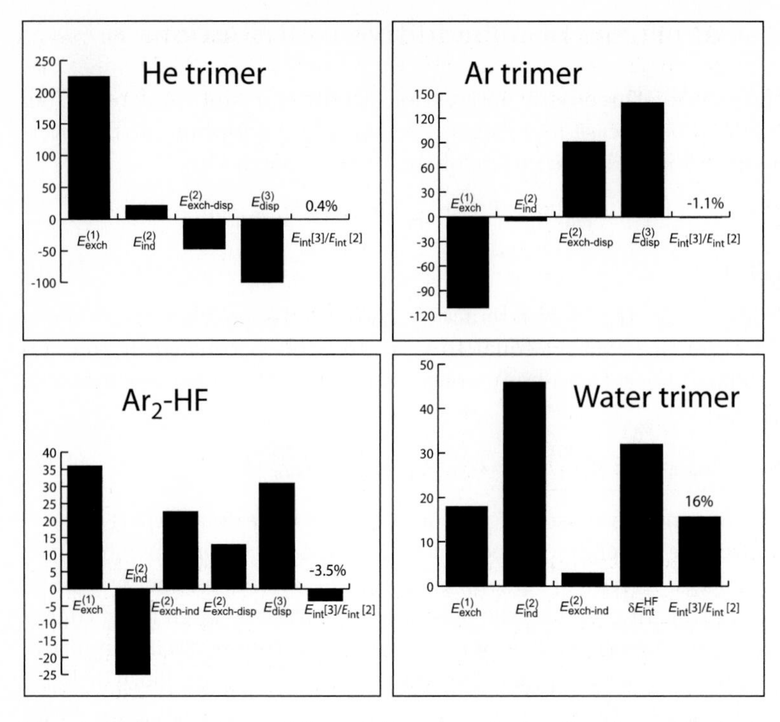

Fig. 1.1. Comparison of the relative magnitude of various 3-body interactions. The bars show the percentage contribution of a given 3-body component to the total 3-body interaction except for the last bar which shows percentage ratio of the 3-body to 2-body contribution. The data are from Ref. [53] for He_3, Ref. [68] for Ar_3, Ref. [69] for Ar_2–HF, and Ref. [29] for the water trimer.

trimer. The dispersion contributions are dwarfed by the very significant induction effects. Even the third-order induction nonadditivity, the major part of $\delta E_{\rm int}^{\rm HF}[3]$, is very important. Figure 1.1 shows also that the first-order exchange nonadditivity is important for all systems.

The individual components of the 3-body contribution to the interaction energy can be given a physical interpretation. This physical interpretation not only allows one to better understand the mechanisms of nonadditive forces and their relation to monomer properties, but also to propose analytic forms of fitting functions which are the most appropriate for a given type of interactions. Figure 1.2 illustrates such an interpretation for the components important for the water trimer and Table 1.2 presents some numerical results. All the displayed terms are of zeroth order with respect to the intramonomer correlation operator. The first-order nonadditive exchange energy can be decomposed into terms resulting from different types of electron exchanges: the contributions from single and double electron exchanges, as well as from the cyclic permutations involving three electrons at a time. The resulting contri-

Table 1.2. SAPT decomposition of interaction energies (in kcal/mol) for various stationary points on the water trimer potential energy surface. SAPT results are from *ab initio* calculations at stationary points determined on a fitted surface [29].

	uud	*bif*	*ada*	*dad*
2-body (SAPT-5s)	-13.197	-11.595	-9.101	-8.909
$E^{(10)}_{\mathrm{exch}}[3](S^2)$	-0.047	-0.006	0.034	-0.078
$E^{(10)}_{\mathrm{exch}}[3](S^3)$	-0.317	-0.225	0.005	0.006
$E^{(10)}_{\mathrm{exch}}[3](S^4)$	-0.007	0.000	-0.011	-0.010
$E^{(10)}_{\mathrm{exch}}[3]$	-0.371	-0.231	0.027	-0.082
$E^{(20)}_{\mathrm{ind}}[3]$	-0.960	-0.752	0.186	0.143
$E^{(20)}_{\mathrm{exch-ind}}[3]$	-0.063	-0.027	0.011	-0.064
$\delta E^{\mathrm{HF}}_{\mathrm{int}}[3]$	-0.676	-0.444	0.197	0.189
Total 3-body ($E^{\mathrm{HF}}_{\mathrm{int}}[3]$)	-2.070	-1.454	0.421	0.186
Total (SAPT-5s + $E^{\mathrm{HF}}_{\mathrm{int}}[3]$)	-15.267	-13.048	-8.681	-8.723

butions differ by the powers of a typical intermolecular orbital overlap integral S. The components can be quadratic (single exchanges), cubic (three-electron cycles), quartic (double exchanges), etc. in S. For the water trimer at cyclic configurations the quadratic and cubic terms dominate. Since the quadratic terms result from only a single exchange, such exchange can involve only two water molecules. As shown in Fig. 1.2, the exchange deforms the wave function of the dimer. This effect can be described in the first approximation as leading to creation of a quadrupole moment. This moment interacts with multipole moments of a third monomer, giving a nonadditive contribution proportional to S^2. This contribution has an interesting property that it decays as an inverse power of the two of the intermonomer separations, a rather atypical feature for an exchange component. The next term, the S^3 contribution, results from a pure exchange interaction: a triple of electrons is cyclically permuted among the monomers A, B, and C, see Fig. 1.2. This process is realized if, e.g., two electrons are exchanged between monomers A and B and then the electron originally from monomer A is exchanged with an electron from monomer C. The first-order nonadditive exchange energy is an important component of the nonadditive energy in the equilibrium water trimer since it contributes 15% to this energy. At this configuration, the exchange is dominated by the S^3 contribution. For large trimers, the S^2 contribution increases in importance due to its slower decay.

The most important component of the water trimer nonadditive energy is the induction interaction of the second order in V. Its simple mechanism is shown in Fig. 1.2: a permanent multipole moment on monomer A induces multipole moments on monomer B which in turn interact with the permanent multipole moments of monomer C. Higher orders involve interactions between induced moments. The nonadditive induction energy is the most important nonadditive component for hydrogen-bonded systems. It is the only term used

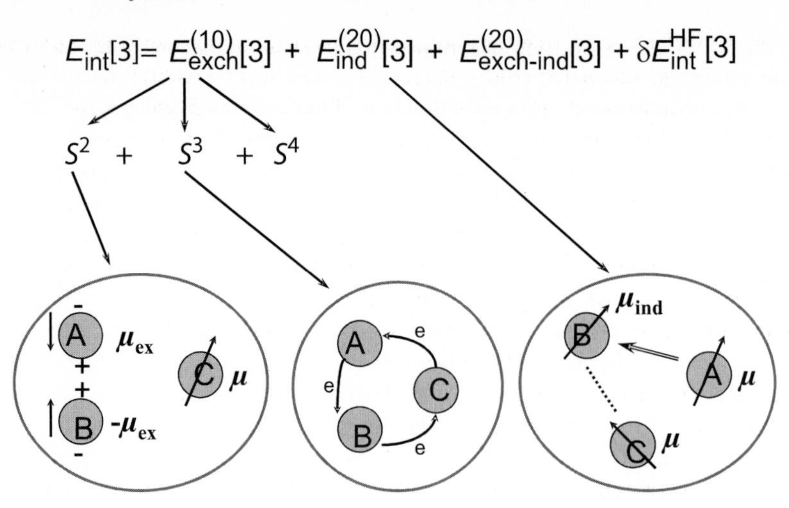

Fig. 1.2. Physical interpretation of major components of 3-body interaction energy for water.

– and only in the asymptotic approximation, i.e., neglecting charge-overlap effects – in the so-called polarizable intermolecular potentials.

Table 1.2 gives the decomposition of the water trimer nonadditive interaction energy for various stationary points on the trimer potential surface. The configurations are as follows: *uud*: global minimum (a cyclic hydrogen-bonded structure), *bif*: the bifurcated structure, *ada* and *dad*: two quasi-linear structures where *a* stands for acceptor and *d* for donor. The components listed in Table 1.2 show several interesting characteristics. As mentioned above, for all cyclic configurations the negative second-order induction energy provides the largest contribution. The third-order induction included in $\delta E_{\text{int}}^{\text{HF}}[3]$ is of comparable size. The nonadditivity of the first-order exchange interaction is also negative and is dominated by the S^3 term. For the quasi-linear trimers the situation is quite different. First, the induction contribution is positive and the third-order effect is even larger than the second-order one. The nonadditivity of exchange interactions is almost negligible for these trimers.

1.5 Three-body potential for water

The nonadditive part of the interaction energy in the water trimer was thoroughly investigated in Ref. [29]. The nonadditive energies were computed at 7533 grid points using the three-body SAPT developed in Ref. [51]. A 12-dimensional, physically motivated functional form was fitted to these points. This form is significantly more sophisticated than any published potential of this type. It involves the exchange-quadrupole model for the nonadditive exchange components proportional to the second power of the overlap integrals

S, a Legendre polynomial times an exponential model for the S^3 and other short-range effects, and a damped polarization model. The overall root mean square error of the fit was only 0.07 kcal/mol. This potential was applied, together with the SAPT-5s pair potential, to evaluate the effects of nonadditive forces on the structure and energetics of the water trimer. Good agreement was found with literature calculations for the characteristic structures of the trimer [71–73]. The primary effect of the nonadditive forces is to change intermolecular distances, whereas the orientations of the monomers are less affected. The three-body effects have significant influence on the heights of the barriers on the trimer potential surface. For example, on the nonadditive surface, the *uud-upd* hydrogen-flipping barrier is 50% higher than the corresponding barrier on the pairwise-additive surface (u, d, and p refer to up, down, and planar free-hydrogen positions, respectively).

Reference [8] describes an earlier SAPT-based work on the nonadditive effects in the water trimer. These contributions to the potential were computed in the same way as in Ref. [29], but only on a three-dimensional grid with 568 symmetry-distinct points. These three-body energies – together with the SAPT-5s pair potential – were employed in a three-dimensional dynamical calculation in which the oxygen atoms and the bonding hydrogen atoms were forming a rigid frame while the free hydrogens were undergoing flipping motion. The water trimer spectrum resulting from this calculation turned out to be in a very good agreement with experiment.

1.6 Simulations of liquid water

Monte Carlo (MC) simulations of liquid water, incorporating the new three-body potential described above, were carried out [43] with 512 molecules in the canonical ensemble at ambient conditions. In addition, the four-body and higher nonadditive terms were included via the polarization model. Analysis of radial distribution functions from these simulations illustrates the profound effect the nonadditive forces have on the hydrogen-bonded structure of liquid water. Simulations using only the two-body potential give one hydrogen bond per molecule less than observed experimentally, radial distribution functions far from the measured ones, and the internal energy underestimated relative to the experimental value. When three-body effects were introduced, all these quantities have become significantly closer to experimental ones. In particular, it was found that (in contrast to an isolated trimer) three-body effects result in a reorientation of water molecules leading to significantly increased number of hydrogen bonds. The simulations of Ref. [43] indicate that three-body effects contribute 14.5% to the internal energy of water, whereas four- and higher-body effects contribute 1.4%. Thus, for the liquid water the relative energetic contribution of 3-body interactions is approximately at the same level as in the equilibrium trimer. Clearly, the 3-body effects are very important for water. It is fortunate that the remaining nonadditive contribution appears to be small, although it should be stated that this estimate was obtained by representing the 4-body and higher nonadditive potentials by the polarization model only. However, calculations on small clusters also indicate that these contributions will be small. *Ab initio* calculations of Refs. [74, 75] have found that the 4-body contribution to interaction energies varied between -1.2% and 2.3% for tetramers and between 1.2% and 3.6% for pentamers. The 5-body contribution to the interaction energy of pentamers was completely negligible as it ranged between -0.11% and 0.25%. These percentage contributions are likely to be even smaller if basis sets of sizes larger than double-zeta are used since small basis sets underestimate pair contributions while giving more saturated values of nonadditive terms [29].

The MC simulations predicted the internal energy of water to be -10.76 kcal/mol compared to the experimental value of -9.92 kcal/mole. The difference is very close to the estimated size of the quantum effect, not included by the classical MC method. The number of hydrogen bonds was predicted to be 3.34 per molecule compared to the empirical estimate of 3.58. The H-H and O-H pair distribution functions were found in excellent agreement with the experimental functions of Soper *et al.* [76]. Also the O-O function agreed well in the region of first peak. However, the regions of the first minimum and the second maximum on this curve agreed with experiment poorly. The reasons of this disagreement are unclear but the most likely one is the rigidity of the SAPT potential.

When the three-body SAPT potential was replaced by the classical three-body induction interactions in the form of the polarization model applied to trimers and limited to the third order (or two iterations), the results of simulations changed very little [43]. Thus, the polarization model performs very well in representing the three-body interactions in liquid water, despite only a moderately good representation of near-equilibrium trimer energies, cf. Fig. 1.1. The reason is that in liquid water the number of such trimers is small compared to the number of larger trimers, for which the induction interaction dominates the total nonadditive effect. Thus, the three-body terms which are relevant for simulations can be well reproduced by the polarization model. Therefore, calculations of Ref. [43] confirm to some extent the validity of polarization models of water and explain the considerable success of such models. In contrast, with the nonadditive effects being so large, one cannot expect that the "effective" pair potentials provide a reasonable physical model of water, despite their popularity. These potentials, fitted to thermodynamic data for liquids and solids, approximate the nonadditive effects by an unphysical distortion of the two-body potential relative to the exact two-body potential. There does not exist a mathematical transformation that would allow any sum of 2-body potentials to represent the nonadditive potentials. As a consequence, the effective two-body potentials perform poorly in predicting pure dimer properties such as dimer spectra or second virial coefficients. In fact, the effective two-body potentials perform poorly also in predicting trimer properties (although the three-body component dominates the nonadditive effects). Polarizable water potentials are able to predict the dimer properties much better than the effective two-body potentials.

The nonadditive interactions in water have a dramatic impact on the water structure and energetics. In fact, out of not so many systems for which reliable information is available, water is certainly the system most impacted. Rare gases are at the other edge of the range, with fairly small nonadditive effects. Even in this case, however, nonadditive interactions can lead to qualitative structural effects [70].

References

1. G. Murdachaew, K. Szalewicz, Faraday Discuss. **118**, 121 (2001).
2. G. Murdachaew, K. Szalewicz, R. Bukowski, Phys. Rev. Lett. **88**, eid. 123202 (2002).
3. J. A. Odutola, T. R. Dyke, J. Chem. Phys. **72**, 5062 (1980).
4. L. B. Braly, J. D. Cruzan, K. Liu, R. S. Fellers, R. J. Saykally, J. Chem. Phys. **112**, 10293 (2000).
5. L. B. Braly, K. Liu, M. G. Brown, F. N. Keutsch, R. S. Fellers, R. J. Saykally, J. Chem. Phys. **112**, 10314 (2000).
6. C. Leforestier, L. B. Braly, K. Liu, M. J. Elrod, R. J. Saykally, J. Chem. Phys. **106**, 8527 (1997).
7. H. Chen, S. Liu, J. C. Light, J. Chem. Phys. **110**, 168 (1999).
8. G. C. Groenenboom, E. M. Mas, R. Bukowski, K. Szalewicz, P. E. S. Wormer, A. van der Avoird, Phys. Rev. Lett. **84**, 4072 (2000).
9. G. C. Groenenboom, P. E. S. Wormer, A. van der Avoird, E. M. Mas, R. Bukowski, K. Szalewicz, J. Chem. Phys. **113**, 6702 (2000).
10. E. M. Mas, K. Szalewicz, R. Bukowski, B. Jeziorski, J. Chem. Phys. **107**, 4207 (1997).
11. E. M. Mas, R. Bukowski, K. Szalewicz, G. C. Groenenboom, P. E. S. Wormer, A. van der Avoird, J. Chem. Phys. **113**, 6687 (2000).
12. M. J. Smit, G. C. Groenenboom, P. E. S. Wormer, A. van der Avoird, R. Bukowski, K. Szalewicz, J. Phys. Chem. A **105**, 6212 (2001).
13. R. S. Fellers, C. Leforestier, L. B. Braly, M. G. Brown, R. J. Saykally, Science **284**, 945 (1999).
14. N. Goldman, R. S. Fellers, M. G. Brown, L. B. Braly, C. J. Keoshian, C. Leforestier, R. J. Saykally, J. Chem. Phys. **116**, 10148 (2002).
15. O. Matsuoka, E. Clementi, M. Yoshimine, J. Chem. Phys. **64**, 1351 (1976).
16. C. Leforestier, R. S. Fellers, R. J. Saykally, to be published.
17. W. Klopper, J.G.C.M. van Duijneveldt-van de Rijdt, and F.B. van Duijneveldt, Phys. Chem. Chem. Phys. **2**, 2227 (2000).
18. J.G.C.M. van Duijneveldt-van de Rijdt, W.T.M. Mooij, and F.B. van Duijneveldt, Phys. Chem. Chem. Phys., in press.
19. L. A. Curtiss, D. J. Frurip, M. Blander, J. Chem. Phys. **71**, 2703 (1979).
20. C. Millot, J. C. Soetens, M. T. C. M. Costa, M. P. Hodges, A. J. Stone, J. Phys. Chem. **102**, 754 (1998).

21. W. L. Jorgensen, J. Am. Chem. Soc. **103**, 335 (1981).
22. W. L. Jorgensen, J. Chandrasekhar, J. D. Madura, R. W. Impey, M. L. Klein, J. Chem. Phys. **79**, 926 (1983).
23. H. J. C. Berendsen, J.R. Grigera, T.P. Straatsma, J. Phys. Chem. **91**, 6269 (1987).
24. I. M. Svishchev, P. G. Kusalik, J. Wang, R.J. Boyd, J. Chem. Phys. **105**, 4742 (1996).
25. E. Zwart, J. J. ter Meulen, W. L. Meerts, L. H. Coudert, J. Mol. Spectrosc. **147**, 27 (1991).
26. G. T. Fraser, Intern. Rev. Phys. Chem. **10**, 189 (1991).
27. M. R. Viant, M. G. Brown, J. D. Cruzan, R. J. Saykally, M. Geleijns, A. van der Avoird, J. Chem. Phys. **110**, 4369 (1999).
28. M. G. Brown, M. R. Viant, R. P. McLaughlin, C. J. Keoshian, E. Michael, J. D. Cruzan, R. J. Saykally, A. van der Avoird, J. Chem. Phys. **111**, 7789 (1999).
29. E. M. Mas, R. Bukowski, K. Szalewicz, J. Chem. Phys., to be published (paper I).
30. M. J. Elrod, R. J. Saykally, Chem. Rev. **94**, 1975 (1994).
31. W. J. Meath, R. A. Aziz, Mol. Phys. **52**, 225 (1984); W.J. Meath, M. Koulis, J. Mol. Struct. (Theochem) **226**, 1 (1991).
32. M. M. Szczesniak, G. Chalasinski in *Molecular Interactions – From van der Waals to Strongly Bound Complexes*, (Ed. S. Scheiner) Wiley, New York, 1996, p. 45.
33. U. Niesar, G. Corongiu, M.-J. Huang, M. Dupuis, E. Clementi, Int. J. Quantum Chem. Symp. **23**, 421 (1989); G. Corongiu and E. Clementi, J. Chem. Phys. **97**, 2030, 8818(E) (1992).
34. R. Eggenberger, H. Huber, M. Welker, Chem. Phys. **187**, 317 (1994); E. Ermakova, J. Solca, H. Huber, D. Marx, Chem. Phys. Lett. **246**, 204 (1995).
35. G. Steinebrunner, A.J. Dyson, B. Kirchner, H. Huber, J. Chem. Phys. **109**, 3153 (1998).
36. K. Laasonen, M. Spirk, M. Parrinello, R. Car, J. Chem. Phys. **99**, 9080 (1993); M. Spirk, J. Hutter, M. Parrinello, J. Chem. Phys. **105**, 1142 (1996); P.L. Silvestrelli, M. Parrinello, Phys. Rev. Lett. **82**, 3308 (1999).
37. S. Scheiner, in *Theoretical Models of Chemical Bonding*, edited by Z.B. Maksic, Springer-Verlag 1991, p. 171; S. Scheiner, Ann. Rev. Phys. Chem. **45**, 23 (1994).
38. S. Scheiner *Hydrogen Bonding. A Theoretical Perspective*, Oxford Univ. Press, Oxford and New York, 1997.
39. J.G.C.M. van Duijneveldt-van de Rijdt, F.B. van Duijneveldt, in *Theoretical treatments of hydrogen bonding*, edited by D. Hadzi, Wiley 1997, p. 13.
40. C. J. Burnharm, J. Li, S.S. Xantheas, M. Leslie, J. Chem. Phys. **110**, 4566 (1999).
41. A. Famulari, R. Specchio, M. Sironi, M. Raimondi, J. Chem. Phys. **108**, 3296 (1998).
42. Y.-P. Liu, K. Kim, B. J. Berne, R. A. Friesner, S. W. Rick, J. Chem. Phys. **108**, 4739 (1998).
43. E. M. Mas, R. Bukowski, K. Szalewicz, J. Chem. Phys., to be published (paper II).
44. R. Bukowski, K. Szalewicz, J. Chem. Phys. **114**, 9518 (2001).
45. A. K. Sum, S. I. Sandler, R. Bukowski, K. Szalewicz, J. Chem. Phys. **116**, 7627 (2002).

46. A. K. Sum, S. I. Sandler, R. Bukowski, K. Szalewicz, J. Chem. Phys. **116**, 7637 (2002).
47. B. Jeziorski, R. Moszynski, K. Szalewicz, Chem. Rev. **94**, 1887 (1994).
48. K. Szalewicz, B. Jeziorski, in *Molecular Interactions - from van der Waals to strongly bound complexes*, edited by S. Scheiner, (Wiley, New York, 1997), p. 3.
49. B. Jeziorski, K. Szalewicz, in *Encyclopedia of Computational Chemistry*, edited by P. von Ragué Schleyer *et al.*, Wiley, New York, 1998, vol. 2, p. 1376.
50. B. Jeziorski, K. Szalewicz, in *Handbook of Molecular Physics and Quantum Chemistry*, edited by S. Wilson, Wiley, 2002, Vol. 3, Part 2, Chap. 8, p. 37.
51. V. F. Lotrich, K. Szalewicz, J. Chem. Phys. **106**, 9668 (1997).
52. R. Moszyński, P. E. S. Wormer, B. Jeziorski, A. van der Avoird, J. Chem. Phys. **103**, 8058 (1995); Erratum: **107**, 672 (1997).
53. V. F. Lotrich, K. Szalewicz, J. Chem. Phys. **112**, 112 (2000).
54. P. E. S. Wormer, R. Moszyński, A. van der Avoird, J. Chem. Phys. **112**, 3159 (2000).
55. J. Paldus, J. Cizek, Adv. Quantum Chem. **9**, 105 (1975).
56. R. J. Bartlett, Ann. Rev. Phys. Chem. **32**, 359 (1981).
57. B. Jeziorski, M. C. van Hemert, Mol. Phys. **31**, 713 (1976).
58. K. Szalewicz, S. J. Cole, W. Kolos, R.J. Bartlett, J. Chem. Phys. **89**, 3662 (1988).
59. K. Szalewicz, in *Encyclopedia of Physical Science and Technology*, edited by R.A. Meyers *et al.*, Academic Press, San Diego, CA, 2002, Third Edition, Vol. 7, p. 505.
60. R. Moszynski, B. Jeziorski, A. Ratkiewicz, S. Rybak, J. Chem. Phys. **99**, 8856 (1993).
61. R. Moszynski, S.M. Cybulski, G. Chalasinski, J. Chem. Phys. **100**, 4998 (1994).
62. R. Moszynski, B. Jeziorski, K. Szalewicz, J. Chem. Phys. **100**, 1312 (1994).
63. M. Jeziorska, B. Jeziorski, J. Cizek, Int. J. Quantum Chem. **32**, 149 (1987).
64. R. Moszynski, T. G. A. Heijmen, B. Jeziorski, Mol. Phys. **88**, 741 (1996).
65. E. M. Mas, K. Szalewicz, J. Chem. Phys. **104**, 7606 (1996).
66. A. Ernesti, J. M. Hutson, Phys. Rev. A **51**, 239 (1995).
67. A. Ernesti, J.M. Hutson, J. Chem. Phys. **106**, 6288 (1997).
68. V. F. Lotrich, K. Szalewicz, J. Chem. Phys. **106**, 9688 (1997).
69. V. F. Lotrich, P. Jankowski, K. Szalewicz, J. Chem. Phys. **108**, 4725 (1998).
70. V.F. Lotrich, K. Szalewicz, Phys. Rev. Lett. **79**, 1301 (1997).
71. T. Bürgi, S. Graf, S. Leutwyler, W. Klopper, J. Chem. Phys. **103**, 1077 (1995).
72. J.G.C.M. van Duijneveldt-van de Rijdt, F.B. van Duijneveldt, Chem. Phys. **175**, 271 (1993); J.G.C.M. van Duijneveldt-van de Rijdt, F.B. van Duijneveldt, Chem. Phys. Lett. **273**, 560 (1995).
73. I. M. B. Nielsen, E. T. Seidl, C. L. Janssen, J. Chem. Phys. **110**, 9435 (1999).
74. M. P. Hodges, A. J. Stone, S. S. Xantheas, J. Phys. Chem. A **101**, 9163 (1997).
75. A. Milet, R. Moszyński, P. E. S. Wormer, A. van der Avoird, J. Phys. Chem. A **103**, 6811 (1999).
76. A. K. Soper, F. Bruni, M. A. Ricci, J. Chem. Phys. **106**, 247 (1997).

[illegible] J. Chem. Phys. 116, [illegible] (2002)

[illegible] Chem. Rev. 94, [illegible] (1994)

[illegible] Molecular Interactions [illegible]

[illegible] edited by S. Wilson [illegible]

[illegible]

2 Thermochemistry and Kinetics of Evaporation and Condensation for Small Water Clusters

Bruce C. Garrett, Shawn M. Kathmann, and Gregory K. Schenter

Summary. The evaluation of thermochemical properties of small water clusters (e.g., consisting of 2 – 10 water molecules) is complicated by their dissociative nature. At temperatures for which dissociation of the cluster into molecular fragments has significant probability, an "operational definition" of a cluster is required to restrict the phase space and evaluate a finite value for the partition function or cluster property of interest. We will review a theoretical approach, Dynamical Nucleation Theory (DNT), for evaluating rate constants for cluster evaporation and condensation, which allows for a unique definition of clusters and their thermochemical properties. We will also discuss the relevance of DNT to homogeneous vapor-to-liquid nucleation of water.

2.1 Introduction

Equilibrium properties of water clusters are important in a variety of areas. Free energies of formation of water clusters from gas-phase water molecules determine equilibrium cluster populations, which are important in the atmosphere [1-4]. Thermochemical properties, such as enthalpies and free energies of cluster formation and virial coefficients [5-11], also provide important benchmarks for intermolecular interactions [12-14] that are difficult to probe directly by experiment. Dynamical properties of water clusters, such as rate constants for evaporation and condensation of water clusters, are also important. The kinetics of cluster condensation and evaporation determine time-dependent cluster populations and their energy content in expansions [15] and vapor-to-liquid nucleation rates are extremely sensitive to the kinetics of cluster condensation and evaporation [16].

Thermochemical properties of water clusters can be determined from statistical mechanical simulations based upon a molecular interaction potential. Although dynamical properties of water clusters require extensions of statistical mechanical simulation techniques beyond simple equilibrium methods, the accurate treatment of partition functions provides the foundation for these extensions. Molecular-level approaches to determining thermochemical and dynamical properties of clusters require (i) accurate descriptions of the intermolecular interactions, (ii) consistent theories connecting the molecular interaction potentials with observable properties, and (iii) effective simulation methods, which allow efficient sampling of those regions of phase space that contribute to the physical property. In this work we are most concerned with the correct theoretical treatment of thermochemical and dynamical properties of weakly bound clusters at temperatures for which dissociation is a probable event; that is, our focus is on step (ii) above. A major focus of the present work is the description of a theoretical framework for treating dissociative states consistently in calculations of thermochemical and dynamical properties. The treatment of this important issue in the literature has been limited primarily to studies of the effects of quasibound or metastable states of dimers and their effects on thermally averaged effective dissociation energies and absorption spectra [17-24].

There have been many calculations of the structure and energetics of water clusters over the last several years. These studies have allowed estimates of thermodynamic properties of water clusters, with a focus on energetic properties such as the dissociation energy D_0 and the enthalpy of formation ΔH, particularly for the water dimer. Consideration of D_0 or ΔH at sufficiently low temperature requires knowledge of the potential energy surface only in a region around the local minimum. In these cases, ab initio electronic structure methods are powerful tools for determining these types of properties [13, 25-30]. These methods use information only about the interaction energies in regions around local minima, particularly with the use of the harmonic approximation for vibrational energy states, and their extension to calculate thermodynamics properties at room temperature, such as enthapies, entropies, and free energies [31-37], is suspect. Average properties of water clusters have also been calculated for empirical models of water clusters with an aim of understanding the solid to liquid phase transition, which occurs at relatively low temperatures (well below 200 K for the 8- and 20-molecule clusters) [38-42]. There have also been studies of the dynamics of rearrangement between different local minima for water clusters [43-49]. The main emphasis of this paper is the calculation of thermodynamic properties of water clusters at temperatures where dissociation is important and the methods used in these previous thermodynamic studies are not adequate. Furthermore, we limit our kinetic studies to condensation and evaporation (addition and loss) of molecules to and from the water clusters.

A central construct of statistical mechanics is the partition function, since many thermodynamical properties can be related to it. In this work we focus on calculations of the partition function for the intermolecular modes of water clusters, i.e., the rotational-vibrational partition function of the molecular cluster. A standard approach to the calculation of partition functions of molecular clusters is to approximate the full potential by a quadratic expansion around the minimum on the potential energy surface and use the rigid-rotor harmonic oscillator approximation to simplify the statistical mechanical evaluation. This approximate approach is inappropriate for weakly bound clusters as demonstrated for the water dimer in a previous paper [50] and for larger water clusters in the present work. The harmonic approximation to vibrational partition functions effectively includes contributions to the partition function for energies from the minimum of the potential (classical) or zero-point energy (quantum mechanically) up to infinity. As shown below, contributions from vibrational energies above the dissociation energy are appreciable for weakly bound systems such as water clusters, and it is not justified to approximate harmonically the contribution of these energies in the highly anharmonic region near or above dissociation.

A consistent treatment of dissociative states in thermodynamic calculations requires determining which regions of phase space in a classical treatment, or which continuum states in a quantum mechanical treatment, will contribute to the thermodynamical property. Stated another way, one needs an "operational definition" of a cluster that limits the phase space. There have been several suggestions of cluster definitions in the literature, which limit the contribution from dissociative states to finite values. The use of centrifugal potentials to define the regions of phase space revelant to dimer formation for structureless monomers goes back to the pioneering work of Hill [51] and Stogryn and Hirshfelder [52] and has been extended to water dimers [22]. The definition of a cluster has long been recognized as an important issue in molecular approaches of vapor-to-liquid homogeneous nucleation, because clusters that contribute to nucleation are inherently unstable [53-63]. These approaches often employ geometric constraints such as in the definition of 'physically consistent' clusters [53, 54, 56], which are defined such that all molecules in the cluster lie within a spherical volume centered on the center of mass of the cluster. Stillinger's definition [64] offers an alternative geometric constraint in which a molecule must be within a predefined distance of another molecule to be considered part of the cluster. The harmonic approximation can also be viewed as a geometric constraint, since it effectively limits the system to lie between the turning points of the quadratic potential. Constraints on the energy can also be used to avoid divergence of the partition function. Constraining the total energy to lie below dissociation may be too restrictive, since resonances embedded in the dissociative continuum may be relatively long lived, such as orbiting resonances or vibrational excitations in modes that are weakly coupled to dissocative modes. Less restrictive energy constraints can be applied to selective modes (such as vibrations but not rotations), however these constraints require approximate separation of degrees of freedom in the system Hamiltonian for the selective modes. A more compelling approach is to use the lifetimes of dissociative states as the criteria for inclusion in the partition functions. The relationship between collision lifetimes and thermodynamic functions was first recognized by Smith [65]. The time for water dimers at energies above dissociation to dissociate was recently used to determine whether the state should be included in the calculation of the water dimerization enthalpy [66]. Although an approach based upon the lifetime matrix is conceptually compelling, it requires full solution of the collision dynamics.

Rather than focusing on the thermodynamics of the water clusters and trying to resolve the issue of the proper definition of a cluster, we first examine the kinetics of cluster evaporation and condensation, which are important in vapor-to-liquid nucleation. Through these studies we learn what molecular theories of reaction kinetics can tell us about how to define clusters that are important in nucleation. Dynamical Nucleation Theory (DNT) [61-63] provides a convenient theoretical approach for studying water nucleation. Although the simplest form of DNT employs several approximations, DNT provides a theoretical framework that can be systematically improved by relaxing the approximations [67]. Work is currently underway to understand the validity of the approximations we use and the conditions under which more rigorous treatments must be used [68]. For the current work, DNT in its simplest form provides the necessary insight into the thermochemistry and kinetics of water clusters, and we will not address issues about limitations of the method in this contribution.

There are a variety of potentials for water-water interactions, and in this study we restrict our discussion to results obtained for the Dang-Chang water potential [69] in order to illustrate the theoretical framework. Nucleation exhibits an extreme sensitivity to physical conditions such as temperature and pressure of the nucleating gas, and it is not surprising that molecular-level theoretical approaches to nucleation also display an extreme sensitivity to the underlying interaction potential [16]. The exploration of this sensitivity is the topic of another study and will not be discussed here. We note though that once a consistent theory and simulation method for nucleation are developed, comparison of computed and experimental nucleation rates can provide a sensitive test of the accuracy of the molecular interaction potential.

Section 2 provides an overview of water cluster partition functions, demonstrating the importance of the treatment of dissociative states in harmonic approximations to the partition function. Section 3 presents a brief review of Dynamical Nucleation Theory (DNT) and shows how DNT provides a unique definition for the clusters that are important in the nucleation process. Section 4 presents the results of calculations of thermodynamic and kinetic properties of small water clusters, and section 5 provides a summary and concluding remarks.

2.2 Thermodynamics of water clusters: contribution of dissociative states to the partition function

One way to examine the contributions of dissociative states to the partition function is to express it as a convolution of the density of states (DOS) with the Boltzmann factor. The partition function $q_i(T)$ for the rotational-vibrational degrees of freedom of an i-molecule water cluster can be written as

$$q_i(T) = \int_0^\infty dE \exp(-E/k_BT)\, \rho_i(E) \tag{2.1}$$

where T is temperature, E is the total energy, k_B is Boltzmann's constant and $\rho_i(E)$ is the density of rotational-vibrational states for the i-molecule cluster. For energies below the cluster dissociation energy D_i, the classical DOS scales like $E^{M+1/2}$ in the rigid-rotor harmonic-oscillator (RRHO) approximation, where M is the number of active vibrations in the cluster. For energies above dissociation, the unconstrained DOS scales linearly with the total volume of the system and diverges as the volume increases. However, some of the phase points for energies above dissociation, may not be considered a cluster because the distance between molecules is too large, or the lifetime is short for molecules to stay within close proximity of each other. This situation calls for the need to constrain the available phase space, which is equivalent to providing an operational definition of the cluster (i.e., what are the conditions for a molecule to be considered part of a cluster rather than an isolated, gas-phase molecule?). This central issue has been discussed in detail for the water dimer [50] and is discussed in more detail in the next section for larger water clusters. In the current section we show that when we use one method of constraining the phase space (the harmonic approximation), contributions to the partition function from energies above dissociation are substantial for water clusters.

Most previous calculations of thermodynamic properties of water clusters, including partition functions, employ the RRHO approximation [31, 36, 70-73]; although the effects of anharmonicity of selected modes of the water dimer were studied [74] and accurate vibrational energy levels have been used in constructing the partition function for the water dimer [75]. The harmonic oscillator approximation for vibrations implicitly constrains the phase-space for energies above dissociation by extending the quadratic approximation to the potential up to arbitrarily large energies that allow the calculation of the partition function to be converged. The use of a quadratic approximation to the potential for energies above dissociation is highly suspect and this approximation will have the chance of being reliable only if the major contributions to the partition function come from energies well below dissociation. In this section we examine how much contribution comes from energies above dissociation in the RRHO approximation for water clusters with $i = 2 - 6$. In the present work we neglect the effects of the high frequency intramolecular vibrations so that the number of active vibrational modes M_i is the number of intermolecular vibrations, which equals $6(i - 1)$.

The lowest classical dissociation energies (difference in energies for equilibrium structures neglecting explicit contributions from zero-point energies)

Table 2.1. Classical dissociation energies and ratios of partition functions at T = 300K.

i	D_i (cm^{-1})^a	$q^{\mathrm{HO}}_{\mathrm{CH},i}(T)/q^{\mathrm{D}}_{\mathrm{CH},i}(\mathrm{T})$	$q^{\mathrm{HO}}_{\mathrm{QH},i}(T)/q^{\mathrm{D}}_{\mathrm{QH},i}(\mathrm{T})$
2	1644	1.7	2.3
3	3008	1.6	1.4
4	3742	2.6	1.3
5	2728	660	4.4
6	3218	4580	10

a 1 cal/mol = 349.8 cm^{-1}

for water clusters containing 2-6 water molecules, as computed for the Dang-Chang model of water, are listed in Table 2.1 (the values of the partition function ratios are discussed below). All of these dissociation energies correspond to the loss of a single water molecule from the cluster. Dissociation energies for loss of water dimers from clusters with $i = 4 - 6$ or trimers from the $i = 6$ cluster are in the range 4300 – 5100 cm^{-1}. Figure 2.1 shows the classical RRHO density of states (normalized to unity at D_i) as a function of E/D_i, where the zero of energy is taken at the equilibrium geometry of the i-molecule cluster. (Expressions for the classical and quantum mechanical RRHO density of states are standard and can be found in textbooks [76].) The product of the Boltzmann factor at 300 K and the DOS (normalized to unity at D_i) that appears in the integrand of eq (2.1), is also shown in Fig. 2.1. Examination of $exp[-(E - D_i)/k_BT]\ \rho_i(E)/\rho_i(D_i)$ shows that there are significant contributions to the partition function from energies above dissociation ($E/D_i > 1$). Although the curves for $i = 2 - 4$ have their maxima below $E/D_i = 1$, they have appreciable values out to high values of E/D_i. In addition, the curves for $i = 5$ and 6 have their maxima well above $E/D_i = 1$.

We have also calculated the quantum mechanical RRHO density of states and report these results in Fig. 2.2 (We employ an expression for the DOS that treats rotations classically and vibrations by an explicit quantum mechanical sum, which was shown to reproduce accurately the fully quantum mechanical DOS for the water dimer [50].) Quantum mechanically the dissociation energy for the i-molecule cluster is shifted by the difference in zero-point energies for the i-molecule and (i-1)-molecule clusters

$$D_{0,i} = D_i - \varepsilon_{0,i} + \varepsilon_{0,i-1} \tag{2.2}$$

where $\varepsilon_{0,i}$ is the zero-point energy of the i-molecule cluster. In Fig. 2.2 we display the quantum mechanical DOS (normalized to unity at $E = D_{0,i}$) as a function of $(E - \varepsilon_{0,i})/D_{0,i}$, where the zero of energy is again taken at the equilibrium geometry of the i-molecule cluster. The qualitative trend in the

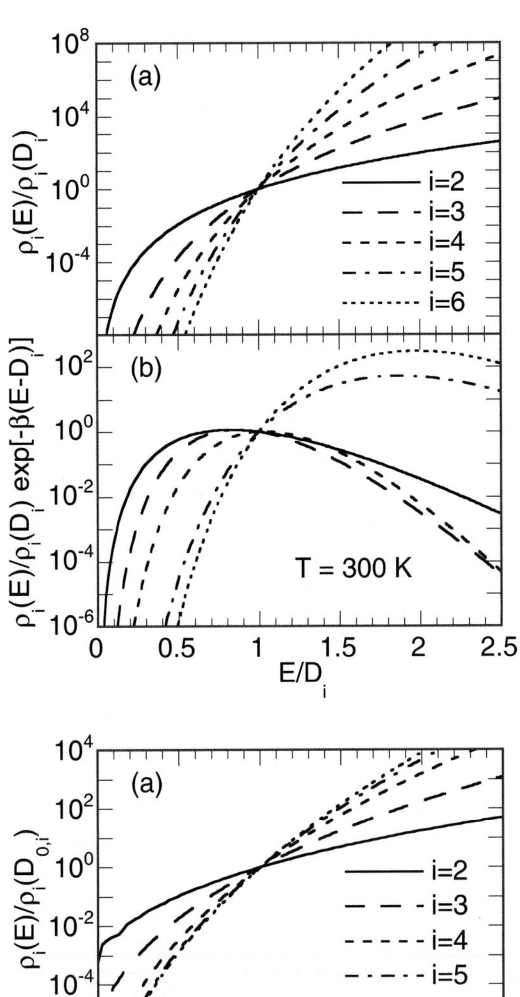

Fig. 2.1. (a) Classical rigid rotor, harmonic oscillator density of states normalized to unity at the dissociation energy D_i as a function of energy E normalized by the dissociation energy D_i for i-molecule water clusters with $i = 2 - 6$. (b) Product of the RRHO DOS and Boltzmann factor at 300 K for i-molecule water clusters with $i = 2 - 6$ as a function of energy E. The product is normalized to be unity at the dissociation energy of the i-molecule cluster. The zero of energy is taken at the equilibrium geometry of the i-molecule cluster.

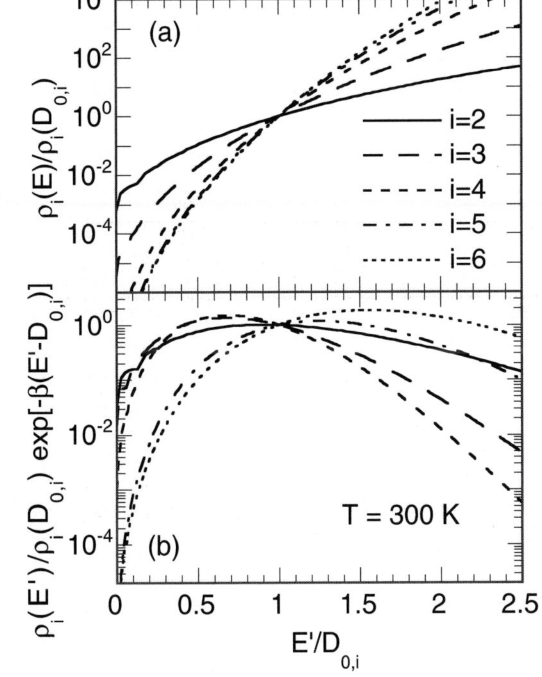

Fig. 2.2. Same as Fig. 2.1, except for the quantum mechanical RRHO DOS as a function of the energy $E' = E - \varepsilon_{0,i}$ relative to the ground state vibrational energy $\varepsilon_{0,i}$.

quantum mechanical DOS is the same as for the classical DOS. Significant contributions to the partition function at 300 K come from energies above dissociation.

The partition function for the i-molecule cluster computed using the RRHO with the harmonic oscillator constraint (i.e., using the quadratic approximation to the potential above dissociation) is denoted $q^{\text{HO}}_{\text{CH},i}(\text{T})$ when computed classically and $q^{\text{HO}}_{\text{QH},i}(\text{T})$ when computed quantum mechanically. (Explicit expressions for these partition functions are standard and can be found in textbooks [76].) We also consider the constraint in which the upper limit in eq (2.1) is replaced by the dissociation energy, i.e., only bound states are included in the partition functions. We denote the classical and quantum mechanical RRHO partition function with this energetic constraint by $q^{\text{D}}_{\text{CH},i}(\text{T})$ and $q^{\text{D}}_{\text{QH},i}(\text{T})$, respectively. (Explicit expressions for these partition functions have been presented elsewhere [50].) Values of the ratios $q^{\text{HO}}_{\text{CH},i}(\text{T})/q^{\text{D}}_{\text{CH},i}(\text{T})$ and $q^{\text{HO}}_{\text{QH},i}(\text{T})/q^{\text{D}}_{\text{QH},i}(\text{T})$, which are presented in Table 2.1 for 300 K, provide a quantitative measure of the contribution to the harmonic partition functions from dissociative states. Figure 2.3 presents the temperature dependence of the percent contribution to the partition functions from energies above dissociation. Both Table 2.1 and Fig. 2.3 clearly show that with the RRHO approximation contributions from energies above dissociation make significant contributions to the partition functions for all but the smallest clusters (2 – 4) at the lowest temperatures. These qualitative calculations clearly show that energies near or above dissociation need to be treated properly to correctly predict thermodynamic properties of these weakly bound systems. It is questionable whether a harmonic treatment is appropriate at these energies where anharmonic effects are expected to be large.

Accurate calculations of thermochemical properties of water clusters will require appropriate treatment of dissociative states. The proper treatment of dissociative states will depend upon the property being calculated. Inclusion of a state or phase point in an average of a property of the cluster should be based on whether the state or phase point contributes to the observed property in a measurement. Rather than examining possible constraints on dissociative states, we will focus on cluster kinetics that are important in vapor-to-liquid nucleation and see what molecular theories of reaction kinetics can tell us about how to define those clusters that play a role in nucleation.

2.3 Water cluster evaporation and condensation kinetics

The kinetics of evaporation and condensation plays an important role in homogeneous vapor-to-liquid nucleation. Most theoretical approaches to nucleation (see review by Laaksonen et al. [77] and references therein) treat

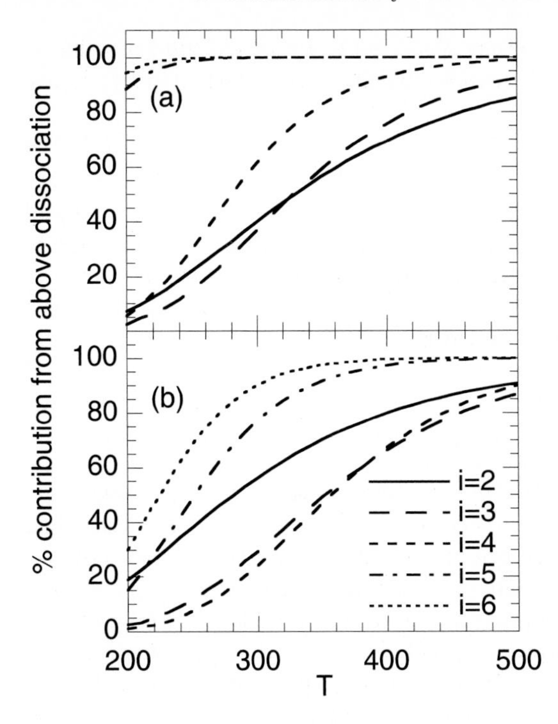

Fig. 2.3. Percent contribution to the rigid rotor, harmonic oscillator partition functions from energies above dissociation. Classical results are presented in part (a) and quantum mechanical results are presented in part (b).

the process as the evolution of cluster populations. In its simplest form the kinetic mechanism involves addition and loss of monomers from clusters

$$A + A_{i-1} \underset{\alpha_i}{\overset{\beta_{i-1}}{\rightleftarrows}} A_i \tag{2.3}$$

where A are monomers, A_i are clusters consisting of i monomers, and α_i and β_{i-1} are the evaporation and condensation rate constants. Note that throughout this paper we assume the concentration of monomers is large and therefore can be treated as constant. We define β_i as the product of the bimolecular condensation rate constant and the monomer concentration so that it has the same units as the unimolecular evaporation rate constant α_i. Most previous theories of nucleation approximate the condensation rate constants by the gas kinetic collision rate of molecules with a liquid droplet, where the surface area of the droplet is determined so that the cluster has it bulk liquid density, i.e.,

$$\beta_i^{\text{coll}} = \frac{\bar{c}}{4}\left(4\pi r_i^2\right)\frac{N_1}{V} \approx \frac{\bar{c}}{4}\left(4\pi r_i^2\right)\frac{p_1}{k_B T} \tag{2.4}$$

where $\bar{c}$ is the average gas phase velocity of the monomers given by $\bar{c} = (8k_{\mathrm{B}}T/\pi\, m)^{1/2}$, m is the monomer mass, N_1 is the number of monomers in the total system of volume V so that N_1/V is the gas-phase concentration of monomers, and the radius of the i-molecule cluster is determined by

$$\frac{4\pi r_i^3}{3} = \sigma \frac{i}{C_{\text{liq}}} \tag{2.5}$$

where C_{liq} is the concentration (i.e., number density) of the bulk liquid and σ is a scaling factor that is generally taken as either 1 or 5. The approximate equality in (2.4) assumes ideal gas behavior where p_1 is the partial pressure of the monomers. The evaporation rate constant is then obtained from the condensation rate constant by detailed balance

$$\frac{\beta_{i-1}}{\alpha_i} = \frac{N_i^{\text{EQ}}}{N_{i-1}^{\text{EQ}}} = K_{i-1,i}^{\text{EQ}}, \tag{2.6}$$

where N_i^{EQ}/V is the concentration of i-clusters in equilibrium with monomers of concentration N_1/V and $K_{i-1,i}^{\text{EQ}}$ is the equilibrium constant. The task then reduces to determining the equilibrium constants or equivalently the equilibrium populations. As discussed above, this task is complicated by the need to define clusters that are important in nucleation. Clearly, clusters with sufficient energy to dissociate need to be included, since they are the ones that will evaporate and the ones that result from a condensation event. Rather than focus on the thermodynamic properties of the clusters, we first examine their kinetics.

In Dynamical Nucleation Theory (DNT), we do not employ the simple approximation of a gas kinetic rate constant for condensation, but instead use variational transition state theory (VTST) [78-83] to evaluate equilibrium rate constants for evaporation and condensation. In this approach we treat evaporation as a gas-phase unimolecular dissociation process and apply VTST to calculate the rate constant. For the dissociation reaction considered here the energy profile from reactants to products is monotonically increasing, so there is no intrinsic barrier maximum for quantum mechanical tunneling to occur. In this case, quantum mechanical effects will only be important for bound motions of the cluster. We treat the molecules as rigid so that the high frequency internal modes of each molecule, which require a quantum mechanical treatment, are not explicitly treated. A classical statistical mechanics description of lower frequency motions between molecules is a reasonable first-order approximation for the systems considered here. This approximation can be systematically improved using Feynman path integral methods [84] to include quantum mechanical effects in equilibrium properties such as free energies.

In transition state theory (TST) [85-87], the net reactive flux is approximated by the one-way flux through a dividing surface separating reactants and products (Wigner's fundamental assumption of TST [87]). The rate constant is the reactive flux divided by the reactant partition function. The fundamental assumption guarantees that the TST approximation to the classical reactive flux is greater than or equal to the exact classical flux and the best estimate of the reactive flux can be obtained by variationally optimizing the dividing surface to minimize the reactive flux [78-81]. The definition of the dividing surface is therefore crucial in determining the rate constant. The VTST estimate of the rate constant can be systematically improved by using more flexible definitions of the dividing surface. Alternatively, the effects of recrossing can be estimated from ensembles of trajectories started at the optimized dividing surface [88].

For evaporation of a monomer from an i-molecule cluster, reactants are the i-molecule cluster and products are a monomer infinitely separated from an (i-1)-molecule cluster. In our previous studies [61-63], we constrained the molecules of a cluster to all lie within a sphere of radius r_{cut} with the center of the sphere at the center of mass of the cluster. The dividing surface is defined by the function

$$f_i^{\mathrm{DS}}\left(r_{cut},\mathbf{r}^{(i)}\right)=0 \tag{2.7}$$

where the function f_i^{DS} is implicitly defined by

$$\theta\left(f_i^{\mathrm{DS}}\right)=\prod_{j=1}^{i}\theta\left(r_{\mathrm{cut}}-\left|\mathbf{r}_j-\mathbf{R}_{\mathrm{CM}}\right|\right) \tag{2.8}$$

$\mathbf{r}^{(i)}$ is the collection of coordinates for each molecule, $(\mathbf{r}_1, \mathbf{r}_2, \ldots, \mathbf{r}_i)$, $\mathbf{r}_j$ are the coordinates for molecule j, $\mathbf{R}_{CM}$ is the coordinate of the center of mass of the i molecules, and the Heaviside step function is defined by $\theta(x) = 0$, $x < 0$ and $\theta(x) = 1$, $x > 1$. The location of the dividing surface is determined by one parameter, r_{cut}. This definition of the cluster is equivalent to the 'physically consistent cluster' of Reiss and coworkers [53, 56], and therefore allows us to make the connection to previous attempts to provide an operational definition of clusters that are important in nucleation. The cluster partition function (or the reactant partition function for evaporation from an i-molecule cluster) for this choice of dividing surface is written

$$q_i^{\mathrm{TOT}}\left(r_{\mathrm{cut}},T\right)= \left(i!h^{3i}\right)^{-1}\int d\mathbf{r}^{(i)}\int d\mathbf{p}^{(i)}\,\exp\left(-H_i^{\mathrm{TOT}}/k_{\mathrm{B}}T\right)\theta\left(f_i^{\mathrm{DS}}\right) \tag{2.9}$$

where h is Planck's constant, $\mathbf{p}^{(i)}$ is the collection of momenta conjugate to $\mathbf{r}^{(i)}$, and H_i^{TOT} is the Hamiltonian for the i-molecule cluster. As noted previously, just examining the partition function in (2.9) does not give us any clue of how to determine the optimum value of the parameter r_{cut}, so we turn to an examination of the rate constant expression.

The TST expression for the evaporation rate constant with a generalized transition state dividing surface defined by f_i^{DS} can be written [89]

$$\alpha_i^{\mathrm{GT}}(r_{\mathrm{cut}},T)=\left[i!h^{3i}q_i^{\mathrm{TOT}}(r_{\mathrm{cut}},T)\right]^{-1}\int d\mathbf{r}^{(i)}.$$

$$\int d\mathbf{p}^{(i)}\ \exp\left(-H_i^{\mathrm{TOT}}/k_{\mathrm{B}}T\right)\frac{1}{2}\left|\left[\left(\mathbf{M}^{(i)}\right)^{-1}\mathbf{p}^{(i)}\right]\frac{\partial\theta\left(f_i^{\mathrm{DS}}\right)}{\partial\mathbf{r}^{(i)}}\right| \tag{2.10}$$

where $\mathbf{M}^{(i)}$ is the diagonal mass matrix. Note that the integral in this expression includes a contribution from the center-of-mass translation of the cluster, which is trivially evaluated as $i^{3/2}\gamma V$, where $\gamma=(2\pi mk_{\mathrm{B}}T)^{3/2}$, since the integrand is invariant with respect to center-of-mass translations. The total partition function in (2.9) also includes a contribution from the center-of-mass translation of the cluster, which is identical and cancels the contribution from the integral in (2.10). Using the dividing surface in (2.8), the generalized TST rate constant takes the compact form [61-63]

$$\alpha_i^{\mathrm{GT}}(r_{\mathrm{cut}},T)=\frac{\bar{c}}{4}\left(4\pi r_{\mathrm{cut}}^2\right)\left(-\frac{1}{k_{\mathrm{B}}T}\frac{\partial A_i}{\partial v}\right)_{i,T} \tag{2.11}$$

where the volume is defined by $v=\frac{4}{3}\pi r_{\mathrm{cut}}^3$ and the subscript i,T indicates that these thermodynamic variables are held constant. The internal Helmholtz free energy for the i-molecule cluster constrained within the sphere of radius r_{cut} is given by

$$A_i(r_{\mathrm{cut}},T)=-k_{\mathrm{B}}Tln\left[q_i(r_{\mathrm{cut}},T)\right]=-k_{\mathrm{B}}Tln\left[\frac{q_i^{\mathrm{TOT}}(r_{\mathrm{cut}},T)}{i^{3/2}\gamma V}\right] \tag{2.12}$$

where q_i denotes the internal (rotational-vibrational) partition function of the i-cluster. Since the derivative of the Helmholtz free energy with respect to volume is the negative of the pressure, (2.11) takes a form that resembles the collision frequency given in (2.4). The expression for the condensation rate is given in terms of the external pressure of monomers, whereas (2.10) is given in terms of the internal pressure of molecules in the constraining volume.

VTST gives a prescription for defining the optimum value of the radius of the sphere, r_{cut}, or equivalently the optimum location of the dividing surface for each dissociation reaction. In canonical variational theory the optimum location for a given i-cluster is determined for each temperature. The optimum value of r_{cut} for the i-cluster, r_i^{CVT}, minimizes the reactive flux, i.e.,

$$q_i^{\mathrm{TOT}}\left(r_i^{\mathrm{CVT}},T\right)\alpha_i^{\mathrm{CVT}}(T)=\min_{r_{\mathrm{cut}}}\left[q_i^{\mathrm{TOT}}(r_{\mathrm{cut}},T)\ \alpha_i^{\mathrm{GT}}(r_{\mathrm{cut}},T)\right] \tag{2.13}$$

Equation (2.11) illustrates the importance of knowing the dependence of the Helmholtz free energy on the radius of the sphere or the constraining volume. Many previous studies assumed that the Helmholtz free energy is relatively insensitive to variations of r_{cut}. In contrast, our approach has shown that it is crucial to know the r_{cut}-dependence of $A_i(r_{cut},T)$ since it determines the rate constant. Furthermore, the variational principle provides a theoretically justified procedure to identify the optimum value of r_{cut} for each cluster. This prescription selects those states that are the most stable relative to evaporation, since the reactive flux is a minimum for the optimum choice of r_{cut}.

Once the evaporation rate constant is determined, the condensation rate constant is obtained consistently using detailed balance, (2.6), which requires evaluation of equilibrium constants or equivalently the equilibrium concentrations of i-clusters, N_i^{EQ}, so that thermochemistry is still needed. By considering the partition function for the entire (supersaturated) vapor system as a product of partition functions for bath molecules, monomers of condensing molecules, and clusters of varying sizes (assuming ideal behavior of each), the equilibrium distribution is the most probable distribution subject to the constraints that the total number of molecules and volume are constant [56,63]. The equilibrium concentration of an i-cluster is given by

$$\frac{N_i^{\mathrm{EQ}}(T)}{V} = \left[\frac{N_1}{V\,q_1(T)}\right]^i i^{3/2}\gamma\ \exp\left\{-\left[A_i\left(r_i^{\mathrm{CVT}},T\right)+pv_i^{\mathrm{CVT}}\right]/k_{\mathrm{B}}T\right\}$$

$$= \left[\frac{N_1}{V\,q_1(T)}\right]^i i^{3/2}\gamma\ q_i\left(r_i^{\mathrm{CVT}},T\right)\ \exp\left(-pv_i^{\mathrm{CVT}}/k_{\mathrm{B}}T\right) \tag{2.14}$$

where N_1/V is the concentration of monomers, q_1 is the monomer partition function including the partition function for the center-of-mass translation, γ, as well as the partition function for internal modes, p is the total external pressure, and the volume of the i-cluster is given by $v_i^{\mathrm{CVT}} = \frac{4}{3}\pi\left(r_i^{\mathrm{CVT}}\right)^3$. The term pv_i^{CVT} represents the work to create a cavity of radius r_i^{CVT} in the vapor and for the clusters important in water nucleation it is sufficiently small that it can be neglected. Consistent with the expression for the evaporation rate constant, the partition function for the i-molecule cluster in (2.14) is evaluated at the optimum value of r_{cut}.

2.4 Thermodynamic and kinetic properties of small water clusters

Both the evaporation rate constant, given by (2.11) and (2.13), and the equilibrium concentration, given by (2.14), are given in terms of the partition function or equivalently the Helmholtz free energy of the i-molecule cluster. Therefore, these free energies are the initial focus of our calculations. The rate constant requires knowledge only about the derivative of q_i or A_i, so we first examine the dependence of the Helmholtz free energy curves on constraining radius, r_{cut}, which have been computed for water clusters with i = 2-10 at 243

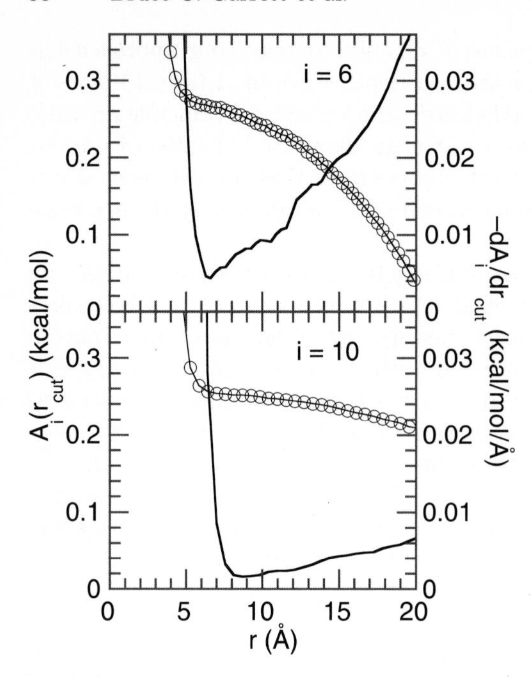

Fig. 2.4. Calculated Helmholtz free energies (open circles connected by dashed curve) and derivatives of the free energies with respect to r_{cut} (solid curves) for the Dang-Chang water model at 243 K as a function of the constraining radius r_{cut} for i-molecule clusters with i = 6 (part a) and i = 10 (part b). The zero of energy was arbitrarily chosen so that the values in parts a and b were comparable.

K using the Dang-Chang water potential. Details of the simulation methods are presented elsewhere [62, 63]. In summary, Monte Carlo simulations performed in the isobaric-isothermal ensemble provide relative probabilities of finding a given cluster volume (or value of r_{cut}) and therefore the free energy as a function of constraining radius. Figure 2.4 presents examples of the dependence of the free energy $A_i(r_{cut},T)$ on the constraining radius for the i = 6 and 10 water clusters at 243 K. These free energy curves are typical of those for other size clusters. The curves show a rapid decrease in free energy at small values of the constraining radius, r_{cut}. This repulsive inner wall arises from the repulsive interactions between water molecules that are sampled more frequently as the cluster is compressed. The curves also show plateau regions where the free energies are relatively slowly varying. In this region the water molecules prefer to form relatively compact structures that would easily be identified as clusters using a geometric constraint such as Stillinger's criteria [64], and the free energy computed from these structures is relatively insensitive to the volume of the phase space. As r_{cut} increases further, the volume of phase space becomes sufficiently large so that configurations with 'isolated' water molecules become increasingly probable. These 'isolated' molecules are outside hydrogen bonding distances with any other water molecules in the system. As the volume of phase space increases even more the probably of having all the molecules isolated from one other becomes appreciable and the free energy approaches the limit of the noninteracting ideal gas, for which the free energy monotonically decreases as the entropy increases with increasing volume. The plateau region extends over wider ranges of r_{cut} as the number of molecules in the cluster increases. Figure 2.4 also shows derivatives of the free energy curves with respect to r_{cut}. The minimum of $-\partial A_i/\partial r_{cut}$, which gives the minimum evaporation rate constant by (2.11), occurs in the plateau region. Since this region gives rise to the smallest rate constants, it is the most stable with respect to evaporation and is identified as the optimum definition of clusters for the nucleation process.

The cluster partition functions and equilibrium concentrations require evaluation of absolute values of the Helmholtz free energies. For small clusters the absolute scale of the Helmholtz free energy curve can be set by requiring the Helmholtz free energy to match the ideal gas value at large values of r_{cut}. The internal Helmholtz free energy for i non-interacting molecules in a volume $v = 4\pi r_{\rm cut}^3/3$ is given by the expression [54]

$$\exp\left[-A_i^{\rm IG}\left(r_{\rm cut}, T\right)\right] = \frac{\left(4\pi q_1 r_{\rm cut}^3\right)^{i-1}}{i!} i^{3/2} \frac{2}{\pi} \int_0^\infty dy\, y^2 \left(\frac{\sin y - y\cos y}{y^3}\right)^i . \quad (2.15)$$

Setting the absolute scale of the Helmholtz free energy by matching it to the ideal gas limit at large r_{cut} requires evaluating the Helmholtz free energy curve over large ranges of r_{cut}. Figure 2.5 demonstrates how this procedure works for the $i = 2 - 8$ clusters. The ideal-gas Helmholtz free energy curves are seen to overlap the $A_i(r_{cut},T)$ curves at sufficiently large values of r_{cut}. The range of overlap moves to larger values of r_{cut} as i increases and for $i =$ 8 this range is off the scale of the plot. The minimum in $-\partial A_i/\partial r_{cut}$ occurs

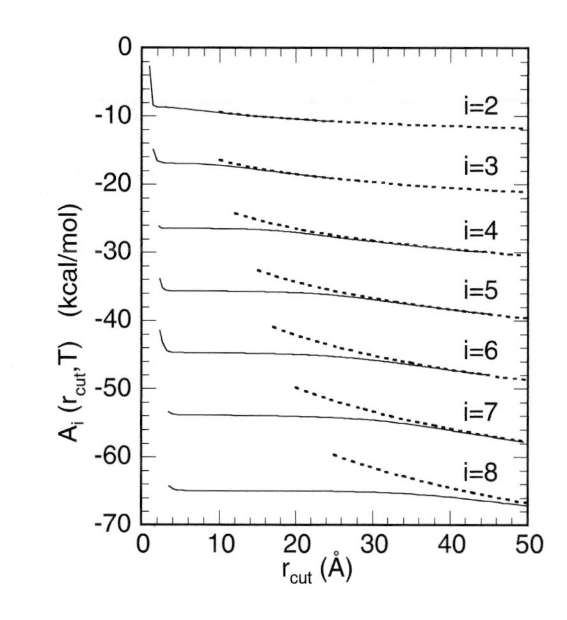

Fig. 2.5. Comparison of absolute internal Helmholtz free energy curves at 243 K computed with the Dang-Chang water model (solid curves) and for the ideal gas, e.g., with the interaction turned off, (dashed curve) as a function of constraining radius r_{cut} for i-molecule clusters with $i = 2 - 8$.

in the region from 5 to 9 Å, while the region of overlap of the ideal gas and interacting gas free energy curves increases from 15-20 Å for $i = 2$ to 45 – 50 Å for $i = 7$. Therefore, it becomes necessary to compute the free energy curves over larger and larger distances as i increases and it is difficult to implement an efficient numerical algorithm to do this for larger clusters [63].

An alternate procedure to set the absolute scale of A_i computes the difference in free energies between adjacent size clusters for a fixed value of r_{cut}. The absolute free energy of the i-molecule cluster at its optimized value of r_{cut}, r_i^{CVT}, is obtained from the absolute free energy of the (i-1)-molecule cluster at r_{i-1}^{CVT} and free energy difference as follows:

$$\begin{aligned} A_i\left(r_i^{\mathrm{CVT}}, T\right) &= A_{i-1}\left(r_{i-1}^{\mathrm{CVT}}, T\right) + \left[A_{i-1}\left(r_{\mathrm{cut}}, T\right) - A_{i-1}\left(r_{i-1}^{\mathrm{CVT}}, T\right)\right] \\ &+ \left[A_i\left(r_{\mathrm{cut}}, T\right) - A_{i-1}\left(r_{\mathrm{cut}}, T\right)\right] + \left[A_i\left(r_i^{\mathrm{CVT}}, T\right) - A_i\left(r_{\mathrm{cut}}, T\right)\right] . \end{aligned} \tag{2.16}$$

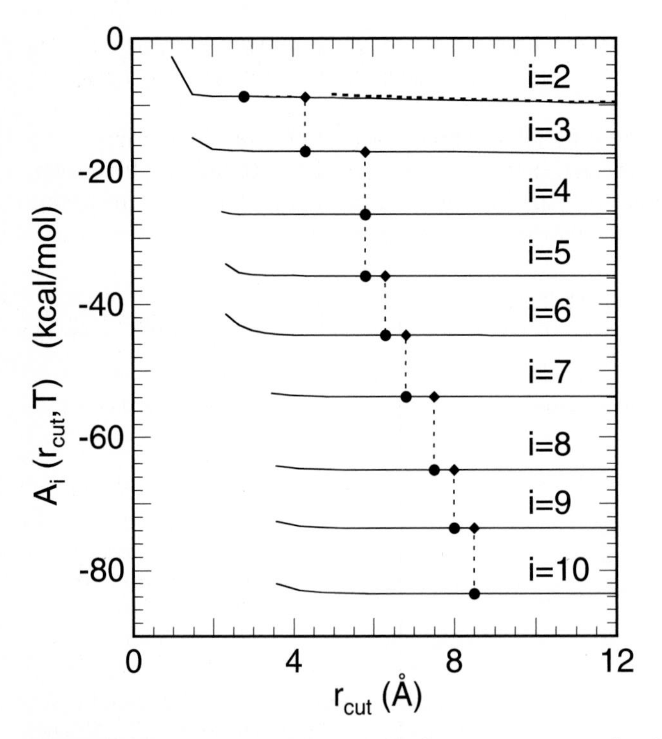

Fig. 2.6. Absolute internal Helmholtz free energy curves for the Dang-Chang water model at 243 K (solid lines). The absolute internal Helmholtz free energy curve for the ideal gas of the dimer ($i = 2$) is shown as the thick dashed line. Filled circles indicate the value of the free energy at the optimum value of r_{cut} for each i-cluster, r_i^{CVT}. The free energy differences between adjacent sized clusters, $A_i\left(r_i^{\mathrm{CVT}}, T\right) - A_{i-1}\left(r_i^{\mathrm{CVT}}, T\right)$, are indicated by the vertical dashed lines between a diamond and a circle.

Energy differences for the same size clusters but at different values of r_{cut}, e.g., $A_i\left(r_i^{\mathrm{CVT}},T\right) - A_i\left(r_{\mathrm{cut}},T\right)$, are obtained using the method described in the previous paragraph to obtain the relative free energy curves as a function of constraining radius. Energy differences for different size clusters but at the same value of r_{cut}, e.g., $A_i\left(r_{\mathrm{cut}},T\right) - A_{i-1}\left(r_{\mathrm{cut}},T\right)$, are obtained using the modified Bennett technique as described elsewhere [63]. Figure 2.6 demonstrates how this procedure works for the $i = 2 - 10$ clusters, with r_{cut} set to r_i^{CVT} in (2.16), so that the last term in brackets vanishes. First, relative values of the free energy curves, $A_i(r_{cut},T)$, are obtained over relatively narrow ranges of r_{cut} (in the range 3 – 9 Å for the $i = 2 - 10$ clusters) and the minima in $-\partial A_i/\partial r_{cut}$, r_i^{CVT} are located. Filled circles in Fig. 2.6 indicate the values of r_i^{CVT}. Free energy differences $A_{i-1}\left(r_i^{\mathrm{CVT}},T\right) - A_{i-1}\left(r_{i-1}^{\mathrm{CVT}},T\right)$ for the same size cluster are given by the difference between the filled circles and diamonds on the $A_{i-1}(r_{cut},T)$ curve. The free energy differences between adjacent sized clusters, $A_i\left(r_i^{\mathrm{CVT}},T\right) - A_{i-1}\left(r_i^{\mathrm{CVT}},T\right)$, are indicated by the vertical dashed lines between a diamond and a circle. The absolute scale of all free energies is set by the absolute free energy of the dimer at its optimum value of r_{cut}, $A_2\left(r_2^{\mathrm{CVT}},T\right)$, which is obtained by fitting to the ideal gas value (shown by the thick dotted curve in Fig. 2.6) at larger r_{cut}, as described in the previous paragraph. The absolute free energies for higher clusters are obtained using a bootstrap procedure by repeated application of (2.16). This procedure suffers from the need to obtain free energies for all clusters up to the ones of interest.

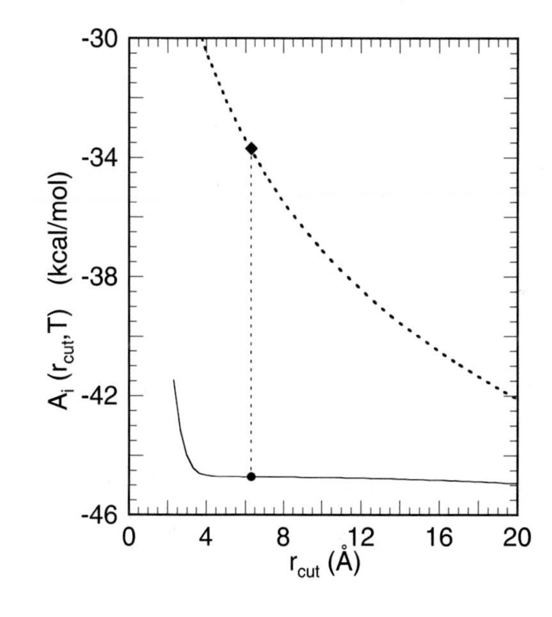

Fig. 2.7. Comparison of absolute internal Helmholtz free energy curves at 243 K computed with the Dang-Chang water model (solid curves) and for the ideal gas, e.g., with the interaction turned off, (dashed curve) as a function of constraining radius r_{cut} for the water hexamer. The filled circle indicates the value of the free energy at the optimum value of r_{cut} for the hexamer, r_6^{CVT}. The difference between the interaction and ideal gas free energies $A_i\left(r_i^{\mathrm{CVT}},T\right) - A_i^{\mathrm{IG}}\left(r_i^{\mathrm{CVT}},T\right)$ is indicated by the vertical dashed line between the diamond and the circle.

Table 2.2. Optimum constraining volume, r_i^{CVT}, volume derivative of the internal Helmholtz free energy at r_i^{CVT}, $-\partial A_i/\partial v$, evaporation and condensed rate constants, α_i and β_i , and equilibrium constant, $K_{i-1,i}^{\mathrm{EQ}}$, for clusters size $i = 2 - 10$ calculated with the Dang-Chang water model at 243 K and a monomer concentration of 1.5 $\times$ 10^{17} cm^{-3}.

i	r_i^{CVT}(Å)	$-\partial \mathrm{A}_i/\partial \mathrm{v}$ (torr)	α_i (ps^{-1})	β_i (ps^{-1})	$K_{i-1,i}^{\mathrm{EQ}}$
2	2.8	32000	160	0.069	0.0004
3	4.3	4000	49	0.17	0.0034
4	5.8	210	4.7	0.18	0.039
5	5.8	450	10.0	0.22	0.023
6	6.3	490	12.8	0.26	0.021
7	6.8	310	9.4	0.29	0.031
8	7.5	190	7.2	0.26	0.037
9	8.0	120	5.2	0.33	0.064
10	8.5	96	4.6	0.28	0.061

The finite-time external-work method of Reinhardt and coworkers [90, 91] is an attractive procedure to directly calculate the absolute free energy of the i-molecule cluster without the need to evaluate free energy difference between different cluster sizes or over large ranges of r_{cut}. Figure 2.7 indicates how this procedures works for the $i = 6$ cluster. The absolute energy scale is set by the ideal gas limit of the free energy for the i-cluster as given in (2.14), as shown by the thick dashed line in Fig. 2.7 The external-work method allows a direct calculation of the difference between the interaction cluster and the noninteracting (ideal gas) cluster at the same location of r_{cut}, which is depicted as the dashed vertical line in Fig. 2.7. The relative free energy curve still needs to be computed over a small range of r_{cut} to locate r_i^{CVT}, then the external work method is used to calculate the difference $A_i\left(r_i^{\mathrm{CVT}}, T\right) - A_i^{\mathrm{IG}}\left(r_i^{\mathrm{CVT}}, T\right)$ to set the absolute scale.

Table 2.2 provides a summary of the kinetic properties of clusters containing $2 - 10$ water molecules. The optimum size of each cluster r_i^{CVT} is seen to increase with increasing i. The derivative of the Helmholtz free energy with respect to the constraining volume, which is a key element in the evaporation rate constant expression shown in (2.11), can be interpreted as the internal pressure of the cluster. This internal pressure is seen to decrease markedly, as does the evaporation rate constant, with increasing i indicating that the clusters become more stable as i increases. Condensation rate constants are computed for a monomer concentration of 1.5×10^{17} cm^{-3}, which corresponds to a supersaturation of about 10. Even at these high monomer concentrations the condensation rate constants are significantly smaller than the evaporation rate constants leading to equilibrium constants that are much less than unity.

Table 2.3. Thermodynamic properties of water clusters for $i = 2 - 6$ calculated classically using the Dang-Chang water model at 243 K. Global potential energy minimum $V_{0,i}$, average potential energy above minimum $\langle \Delta U_i \rangle$, average internal energy E_i, entropy S_i, and Helmholtz free energy A_i (see text for a description of the different quantities given below). The zero of energy for the potential energy surface is taken as the energy with all monomers at infinite separation from each other. Energies are in units of kcal mol^{-1} and entropies are in units of cal mol^{-1} K^{-1}.

i	2	3	4	5	6
$V_{0,i}$	-4.7	-13.3	-24.0	-31.8	-41.0
$\langle \Delta U^{R}_{CA,i} \rangle$	2.0	5.1	6.4	7.8	11.2
$\langle \Delta U^{HO}_{CH,i} \rangle$	1.4	2.9	4.3	5.8	7.2
$E^{IG}_{C,i}$	2.2	3.6	5.1	6.5	8.0
$E^{R}_{CA,i}$	-0.5	-4.6	-12.5	-17.5	-21.8
$E^{HO}_{CH,i}$	-1.1	-6.8	-14.6	-19.5	-25.8
$E^{D}_{CH,i}$	-1.6	-7.2	-15.6	-24.6	-32.4
$S^{IG}_{C,i}$	40.2	72.5	107.7	138.4	171.5
$S^{R}_{CA,i}$	34.0	51.1	57.5	75.0	94.9
$S^{HO}_{CH,i}$	29.0	35.4	42.9	59.7	58.5
$S^{D}_{CH,i}$	26.6	33.3	38.4	30.8	21.0
$A^{IG}_{C,i}$	-7.6	-14.0	-21.1	-27.1	-33.7
$A^{R}_{CA,i}$	-8.8	-17.0	-26.5	-35.7	-44.9
$A^{HO}_{CH,i}$	-8.1	-15.4	-25.0	-34.0	-40.0
$A^{D}_{CH,i}$	-8.0	-15.3	-24.9	-32.1	-37.5

Table 2.3 and Fig. 2.8 provide a summary of the average energy, entropy, and Helmholtz free energy for the internal (vibrational and rotational) modes of clusters containing 2 – 10 water molecules calculated classically using the Dang-Chang water potential. The table presents these quantities for $i = 2 - 6$ because we have only evaluated the RRHO approximations to these quantities through the $i = 6$ cluster. Figure 2.8 presents incremental energies and entropies for adjacent size clusters up to $i = 10$. These thermodynamic properties are related by the standard thermodynamic relationship

$$A_i = E_i - T\,S_i = -k_{\mathrm{B}}T \ln q_i \tag{2.17}$$

where S_i is the internal entropy and the average internal energy (kinetic plus potential) of an i-cluster, E_i, is defined by

$$E_i = k_{\mathrm{B}}T^2 \left(\frac{\partial \ln q_i}{\partial T} \right)_{i,v} \tag{2.18}$$

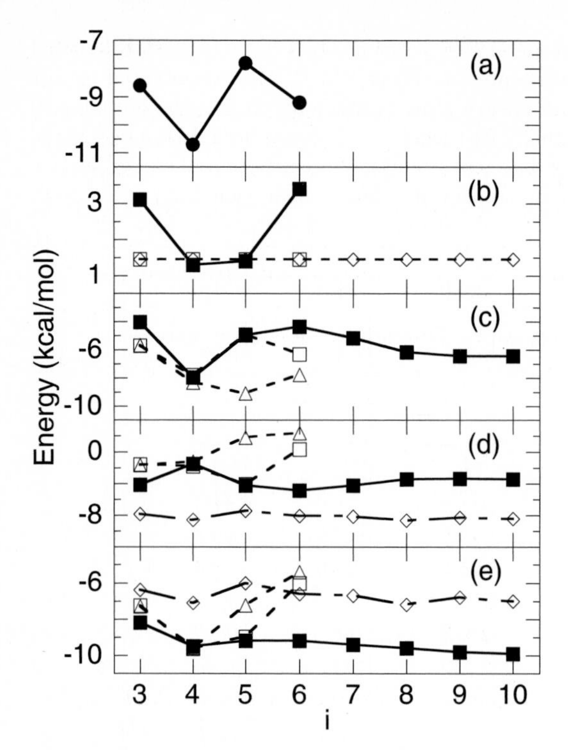

Fig. 2.8. Incremental energies for adjacent sized clusters. (a) Global potential energy minimum $\Delta V_{0,i,i-1}$. (b) Average potential energy above minimum $\langle \Delta\Delta U_{i,i-1} \rangle$ (squares) and the incremental average kinetic energy (open diamonds), which is the same as the incremental average internal energy for the ideal gas case. (c) Average internal energy, $\Delta E_{i,i-1}$. (d) Entropy contribution, $\mathrm{T}\Delta S_{i,i-1}$. (e) Incremental Helmholtz free energy, $\Delta A_{i,i-1}$. Solid squares, open squares, open triangles, and open diamonds are results for the anharmonic potential with geometric constraint, the harmonic potential with the harmonic oscillator constraint, the harmonic potential with the energy constraint, and the ideal gas case, respectively.

We report the average internal energy, entropy, and Helmholtz free energy for the anharmonic potential, the ideal gas in which the potential is set to zero, and in the rigid-rotor, harmonic-oscillator approximation. For the anharmonic potential the quantities are denoted $E^{\mathrm{R}}_{\mathrm{CA},i}$, $S^{\mathrm{R}}_{\mathrm{CA},i}$, and $A^{\mathrm{R}}_{\mathrm{CA},i}$ to indicate that they are calculated classically with the full anharmonic potential and the geometric constraint defined in (2.13). Note that $A^{\mathrm{R}}_{\mathrm{CA},i}$ is the same as $A_i\left(r_i^{\mathrm{CVT}}, T\right)$ defined in (2.12). We use the same geometric constraint for the ideal gas case and denote the quantities $E^{\mathrm{IG}}_{\mathrm{C},i}$, $S^{\mathrm{IG}}_{\mathrm{C},i}$, and $A^{\mathrm{IG}}_{\mathrm{C},i}$. Note that $A^{\mathrm{IG}}_{\mathrm{C},i}$ is equivalent to $A^{\mathrm{IG}}_i\left(r_i^{\mathrm{CVT}}, T\right)$ as defined in (2.15). The two constraints discussed in section 2 are used with the RRHO approximation. For the RRHO approximation with the coordinates constrained to lie within their harmonic turning points, the quantities are denoted $E^{\mathrm{HO}}_{\mathrm{CH},i}$, $S^{\mathrm{HO}}_{\mathrm{CH},i}$, and

$A^{\mathrm{HO}}_{\mathrm{CH},i}$ and with the total energy constrained to lie below dissociation they are denoted $E^{\mathrm{D}}_{\mathrm{CH},i}$, $S^{\mathrm{D}}_{\mathrm{CH},i}$, and $A^{\mathrm{D}}_{\mathrm{CH},i}$.

The internal Helmholtz free energies for the anharmonic potential are obtained numerically by Monte Carlo simulation procedures discussed in detail in previous publications [62, 63] and those for the ideal gas case are obtained by evaluating the one-dimensional integral in (2.15). For the RRHO approximation, the partition functions are obtained from explicit expressions [50, 76] and the internal Helmholtz free energies are obtained from the partition functions using (2.17).

It is convenient for the discussion that follows to define the average internal energy in terms of contributions from the potential and kinetic energy terms

$$E_i = \langle H_i \rangle = \langle U_i \rangle + \langle T_i \rangle = V_{0,i} + \langle \Delta U_i \rangle + \langle T_i \rangle \tag{2.19}$$

where $\langle \cdots \rangle$ denotes an ensemble average, H_i is the internal Hamiltonian, U_i is the potential, T_i is the internal kinetic energy of the i-cluster, $V_{0,i}$ is the energy at the global minimum of the potential energy surface of the i-cluster, and $U_i = V_{0,i} + \Delta U_i$. Note that H_i differs from H_i^{TOT} in (2.9) by the removal of the kinetic energy term for the center-of-mass motion of the i-cluster. Averages of ΔU_i provide a measure of the contributions of thermal excitations above the global minimum to the average potential. The average internal energy goes to the limit $V_{0,i}$ as temperature goes to zero since both and $\langle \Delta U_i \rangle$ and $\langle T_i \rangle$ goes to zero in this limit. For the ideal gas case ($U_i = V_{0,i} = \Delta U_i = 0$) the average internal energy is just the average internal kinetic energy given by

$$E^{\mathrm{IG}}_{\mathrm{C},i} = \left\langle T^{\mathrm{IG}}_{\mathrm{C},i} \right\rangle = (6i-3)\,\frac{k_{\mathrm{B}}T}{2}. \tag{2.20}$$

The averages of the potential and kinetic energy can be performed independently and for the methods that employ geometric constraints on the phase space the average internal kinetic energy is the same as for the ideal gas as given in (2.20), i.e.,

$$E^{\mathrm{R}}_{\mathrm{CA},i} = \left\langle U^{\mathrm{R}}_{\mathrm{CA},i} \right\rangle + (6i-3)\,\frac{k_{\mathrm{B}}T}{2} \tag{2.21}$$

$$E^{\mathrm{HO}}_{\mathrm{CH},i} = \left\langle U^{\mathrm{HO}}_{\mathrm{CH},i} \right\rangle + (6i-3)\,\frac{k_{\mathrm{B}}T}{2}. \tag{2.22}$$

The average potential energies for the anharmonic potential are obtained numerically by Monte Carlo simulation procedures to evaluate the ensemble averages. In the classical rigid-rotor, harmonic oscillator approximation with the harmonic oscillator, $\left\langle \Delta U^{\mathrm{HO}}_{\mathrm{CH},i} \right\rangle$ is given by

$$\left\langle \Delta U^{\mathrm{HO}}_{\mathrm{CH},i} \right\rangle = 3ik_{\mathrm{B}}T. \tag{2.23}$$

For the RRHO approximation with the total energy constraint the evaluation of the average internal kinetic energy is more complicated and we report only the average internal energy for that case

$$E^{\rm D}_{{\rm CH},i} = E^{\rm HO}_{{\rm CH},i} - \frac{q^{\rm HO}_{{\rm CH},i}2D}{q^{\rm D}_{{\rm CH},i}}\sqrt{\frac{D}{k_BT}}\exp\left(-\frac{D}{k_BT}\right) \times\left[1+\sum_{m=1}^{M}\left(\frac{D}{k_{\rm B}T}\right)^{m-1}\left(\frac{D}{k_{\rm B}T}-m\right)\prod_{l=0}^{m-1}\left(l+\tfrac{3}{2}\right)^{-1}\right] \quad (2.24)$$

where

$$q^{\rm D}_{{\rm CH},i} = q^{\rm HO}_{{\rm CH},i}\times \left[erf\left(\sqrt{\frac{D}{k_{\rm B}T}}\right)-2\sqrt{\frac{D}{\pi k_{\rm B}T}}\left(1+\sum_{m=1}^{M}\left(\frac{D}{k_{\rm B}T}\right)^{M}\prod_{l=0}^{m-1}\left(l+\tfrac{3}{2}\right)^{-1}\right)\right] \quad (2.25)$$

and $erf(x)$ is the error function. Both the average potential relative to the minimum, $\langle \Delta U_i \rangle$, and the average kinetic energy for the ideal gas, $E^{\rm IG}_{{\rm C},i}$, are monotonically increasing functions of i, but the global minimum energy decreases sufficiently fast with cluster size so that the average internal energy E_i is a monotonically decreasing function of energy for the anharmonic and harmonic potentials. The harmonic oscillator gives smaller values for $\langle \Delta U_i \rangle$ than the anharmonic potential, so that the average internal energy is more negative for the harmonic oscillator approximations, e.g., $E^{\rm HO}_{{\rm CH},i} < E^{\rm R}_{{\rm CA},i}$. Because the total energy constraint eliminates any contributions from energies above dissociation, whereas the harmonic oscillator constraint includes contributions for $E > D_i$, $E^{\rm D}_{{\rm CH},i} < E^{\rm HO}_{{\rm CH},i}$. The average internal energy is seen to obey the relationship $E^{\rm IG}_{{\rm C},i} > E^{\rm R}_{{\rm CA},i} > E^{\rm HO}_{{\rm CH},i} > E^{\rm D}_{{\rm CH},i}$ for $i = 2-6$.

The entropy for the ideal gas and anharmonic potential is seen to increase monotonically as the cluster size increases, which reflects the increase in phase space with increasing number of degrees of freedom in the cluster. The internal entropy for the different methods obeys the relationship $S^{\rm IG}_{{\rm C},i} > S^{\rm R}_{{\rm CA},i} > S^{\rm HO}_{{\rm CH},i} > S^{\rm D}_{{\rm CH},i}$. With the potential turned off and the geometric constraint employed, the cluster is able to sample the maximum amount of phase space. Using the same geometric constraint and the anharmonic potential, the entropy is decreased significantly, by 15 – 50% for $i = 2-10$. The harmonic potential with the harmonic oscillator constraint restrict the phase space even more, decreasing the entropy by 28 – 66%, relative to the ideal gas entropy, for $i = 2-6$. By eliminating all contributions from phase space points with energy above dissociation, the harmonic oscillator entropy for $S^{\rm D}_{{\rm CH},i}$ is reduced by 8 – 64%, relative to $S^{\rm HO}_{{\rm CH},i}$, for $i = 2-6$.

The contributions to the internal Helmholtz free energy, E_i and $-TS_i$, decrease (become more negative) with increasing cluster size for the ideal gas and anharmonic potential cases , so that the internal Helmholtz free energy calculated using these methods also becomes more negative for larger clusters. Also, the relationships of these two components for the different methods have opposite trends: $E^{\rm IG}_{{\rm C},i} > E^{\rm R}_{{\rm CA},i} > E^{\rm HO}_{{\rm CH},i} > E^{\rm D}_{{\rm CH},i}$ but $-TS^{\rm IG}_{{\rm C},i} < -TS^{\rm R}_{{\rm CA},i} < -TS^{\rm HO}_{{\rm CH},i} < -TS^{\rm D}_{{\rm CH},i}$. It is the anharmonic internal Helmholtz free energy, $A^{\rm R}_{{\rm CA},i}$, which is intermediate in the trends for the average internal energy and internal entropy, that gives the most negative free energies. The internal Helmholtz free energies are seen to obey the relationship $A^{\rm R}_{{\rm CA},i} < A^{\rm HO}_{{\rm CH},i} < A^{\rm D}_{{\rm CH},i} < A^{\rm IG}_{{\rm C},i}$, so that except for the ideal gas case where the potential is turned off, the energetic contributions to the free energy are larger than the entropic contributions, for the constraints employed in this study.

Figure 2.8 shows incremental energies or energy differences between adjacent sized clusters. These quantities are the relative energies for adding a monomer to an existing cluster, $A + A_{i-1} \rightarrow A_i$. The minimum in $\Delta V_{0,i,i-1}$, for $i = 4$ indicates the relative stability of this cluster size. Because the average potential energy above the minimum $\langle \Delta\Delta U_{i,i-1} \rangle$ for the anharmonic potential is about equal to the value for the harmonic potential for $i = 4$ and 5, it may be tempting to assume that the harmonic approximation works well for these cluster sizes. However, as shown in Table 2.3, the values of $\left\langle \Delta U^{\rm HO}_{{\rm CH},i} \right\rangle$ and $\left\langle \Delta U^{\rm R}_{{\rm CA},i} \right\rangle$ differ significantly for cluster sizes from 2 – 6, so that any agreement in the values for $\langle \Delta\Delta U_{i,i-1} \rangle$ for the different methods is a result of a cancellation of errors. Note that $\left\langle \Delta\Delta U^{\rm HO}_{{\rm CH},i,i-1} \right\rangle = \Delta E^{\rm IG}_{{\rm C},i} = 3k_B T$, so that these curves overlap in Fig. 2.8(b). Cancellation of errors also leads to apparent agreement between $\Delta E^{\rm R}_{{\rm CA},i,i-1}$ and $\Delta E^{\rm HO}_{{\rm CH},i,i-1}$ for $i = 4$ and 5. The harmonic methods tend to show less systematic behavior for $\Delta E_{i,i-1}$, $-T\Delta S_{i,i-1}$, and $\Delta A_{i,i-1}$ than do the anharmonic and ideal gas results. The incremental anharmonic free energies are lower than the ideal gas ones by 1.8 to 3.2 kcal/mol for $i = 3 - 10$, and only 2.4 to 3.0 kcal/mol for the larger clusters ($i = 6 - 10$).

2.5 Conclusions

In this paper we demonstrate by use of simple harmonic approximations to the intermolecular potential for water clusters that energy states above the dissociation limit can contribute significantly to the partition function of the cluster. Since the harmonic approximation is suspect for energies above dissociation, another procedure is required for evaluating the thermodynamic properties of the clusters. We require the method to treat dissociative states in a consistent manner, using a theoretically-justified basis, to provide an operational definition of a cluster that limits the regions of phase space contributing to the thermodynamic properties. Towards this end, we review Dynamical Nucleation Theory (DNT) [61-63], which is a consistent procedure for calculating the condensation and evaporation rate constants for water clusters. DNT uses variational transition state theory to determine the evaporation rate constant, and the variational procedure, which is used to obtain the lowest rate constant, provides a unique definition of those clusters that are important in the nucleation process. The unique definition of the cluster allows calculation of the partition function and other equilibrium thermodynamic properties.

DNT is used to calculate condensation and evaporation rate constants at 243 K for water clusters containing 2 to 10 water molecules, using the Dang-Chang [69] interaction potential for water. Using the cluster definition obtained from the DNT calculations, we compute the average internal energy, internal entropy, and internal Helmholtz free energy for the vibrational and rotational modes (i.e., neglecting the center-of-mass motion of the cluster) for these water clusters using the full anharmonic potential. For comparison, thermodynamic properties are also computed with the potential turned off (the ideal-gas case) using the same cluster definition, and using a harmonic approximation to the potential, using two different cluster definitions. For the harmonic approximation, the phase space is constrained either by restricting the total energy to be less than dissociation or implicitly by the harmonic turning points. The harmonic methods tend to show less systematic behavior for these thermodynamic properties than do the anharmonic and ideal gas results.

Acknowledgments

This work was supported by the Division of Chemical Sciences, Office of Basic Energy Sciences, of the U.S. Department of Energy. This research was performed in part using the Molecular Science Computing Facility (MSCF) in the William R. Wiley Environmental Molecular Sciences Laboratory, a national scientific user facility sponsored by the Department of Energy's Office of Biological and Environmental Research and located at Pacific Northwest National Laboratory. Battelle operates the Pacific Northwest National Laboratory for the Department of Energy.

References

1. A. D. Devir, M. Neumann, S. G. Lipson, U. P. Oppenheim : Opt. Eng. **33**, 746 (1994).
2. P. Pilewskie, F. P. J. Valero: Science **267**, 1626 (1995).
3. A. Arking: Science **273**, 779 (1996).
4. J. S. Daniel, S. Solomon, R. W. Sanders, R. W. Portmann, D. C. Miller, W. Madsen: J. Geophys. Res-Atmos. **104**, 16785 (1999).
5. F. G. Keyes: J. Chem. Phys. **15**, 602 (1947).
6. G. S. Kell, G. E. McLaurin, E. Whalley: J. Chem. Phys. **48**, 3805 (1968)
7. J. H. Dymond, E. B. Smith: The Virial Coefficients of Gases. Clarendon, Oxford (1969).

8. H. A. Gebbie, W. J. Burroughs, J. Chamberlain, J. E. Harries, R. G. Jones: Nature (London) **221,** 143 (1969).
9. L. A. Curtiss, D. J. Frurip, M. Blander: J. Chem. Phys. **71,** 2703 (1979).
10. V. I. Dianov-Klokov, V. M. Ivanov, V. N. Arefev, N. I. Sizov: J Quant. Spectrosc. Radiat. Transfer **25,** 83 (1981).
11. G. V. Bondarenko, Y. E. Gorbaty: Mol. Phys. **74,** 639 (1991).
12. R. S. Fellers, C. Leforestier, L. B. Braly, M. G. Brown, R. J. Saykally: Science **284,** 945 (1999).
13. E. M. Mas, R. Bukowski, K. Szalewicz, G. C. Groenenboom, P. E. S. Wormer, A. van der Avoird: J. Chem. Phys. **113,** 6687 (2000).
14. C. J. Burnham, S. S. Xantheas: J. Chem. Phys. **116,** 1500 (2002).
15. O. M. Cabarcos, C. J. Weinheimer, J. M. Lisy: J. Phys. Chem. A **103,** 8777 (1999).
16. S. M. Kathmann, G. K. Schenter, B. C. Garrett: J. Chem. Phys. in press (2002).
17. A. A. Vigasin Chem. Phys. Lett. **117,** 85 (1985).
18. A. A. Vigasin: Infrared Phys. **32,** 461 (1991).
19. S. Y. Epifanov, A. A. Vigasin: Chem. Phys. Lett. **225,** 537 (1994).
20. A. A. Vigasin: Chem. Phys. Lett. **242,** 33 (1995).
21. A. A. Vigasin: J. Quant. Spectros. & Rad. Transfer **56,** 4097 (1996).
22. S. Y. Epifanov, A. A. Vigasin: Mol. Phys. **90,** 101 (1997).
23. A. A. Vigasin: Chem. Phys. Lett. **290,** 495 (1998).
24. A. A. Vigasin: J. Mol. Spectros. **205**, 9 (2001). .
25. D. Feller: J. Chem. Phys. **96**, 6104 (1992).
26. S. S. Xantheas, T. H. Dunning: J. Chem. Phys. **99**, 8774 (1993).
27. S. S. Xantheas: J. Chem. Phys. **100,** 7523 (1994).
28. S. S. Xantheas: J. Chem. Phys. **102,** 4505 (1995).
29. M. W. Feyereisen, D. Feller, D. A. Dixon: J. Phys. Chem. **100,** 2993 (1996).
30. W. Klopper, J. van Duijneveldt-van de Rijdt, F. B. van Duijneveldt: Phys. Chem. Chem. Phys. **2,** 2227 (2000).
31. Z. Salnina: J. Atmos. Chem. **6**, 185 (1988).
32. J. O. Jensen, A. C. Samuels, P. N. Krishnan, L. A. Burke: Chem. Phys. Lett. **276,** 145 (1997).
33. P. Chylek, D. J. W. Geldart: Geophys. Res. Lett. **24,** 2015 (1997).
34. W. Tso, D. J. W. Geldart, P. Chylek: J. Chem. Phys. **108,** 5319 (1998).
35. P. Chylek, Q. Fu, W. Tso, D.J.W. Geldart: Tellus **51A,** 304 (1999).
36. V. Vaida, J. E. Headrick: J. Phys. Chem. A **104,** 5401 (2000).
37. L. A. Montero, J. Molina, J. Fabian: Int. J. Quant. Chem. **79,** 8 (2000).
38. C. J. Tsai, K. D. Jordan: J. Chem. Phys. **95,** 3850 (1991).4
39. C. J. Tsai, K. D. Jordan: J. Chem. Phys. **99,** 6957 (1993).
40. J. M. Pedulla, K. D. Jordan: Chem. Phys. **239,** 593 (1998).
41. D. J. Wales, I. Ohmine: J. Chem. Phys. **98,** 7245 (1993).
42. M. Svanberg, J. B. C. Pettersson: J. Phys. Chem. A **102,** 1865 (1998).
43. D. J. Wales: J. Am. Chem.Soc. **115,** 11180 (1993).
44. D. J. Wales, I. Ohmine: J. Chem. Phys. **98,** 7257 (1993).
45. D. J. Wales, T. R. Walsh: J. Chem. Phys. **105,** 6957 (1996).
46. T. R. Walsh, D. J. Wales: J. Chem. Soc.- Faraday Trans. **92,** 2505 (1996).
47. D. J. Wales, T. R. Walsh: J. Chem. Phys. **106,** 7193 (1997).
48. S. J. Saito, I. Ohmine: J. Chem. Phys. **101,** 6063 (1994).
49. A. Baba, Y. Hirata, S. Saito, I. Ohmine, D. J. Wales: J. Chem. Phys. **106,** 3329 (1997).

50. G. K. Schenter, S. M. Kathmann, B. C. Garrett: J. Phys. Chem. A in press (2002).
51. T. L. Hill: Statistical Mechanics. McGraw-Hill, New York (1956).
52. D. E. Stogryn, J. O. Hirschfelder: J. Chem. Phys. **31,** 1531 (1959).
53. H. Reiss, J. L. Katz, E. T. Cohen: J. Chem. Phys. **48,** 5553 (1968).
54. J. K. Lee, J. A. Barker, F. F. Abraham: J. Chem. Phys. **58,** 3166 (1973).
55. M. Rao, B. J. Berne, M. H. Kalos: J. Chem. Phys. **68,** 1325 (1978).
56. H. Reiss, A. Tabazadeh, J. Talbot: J. Chem. Phys. **92,** 1266 (1990).
57. H. M. Ellerby, C. L. Weakliem, H. Reiss: J. Chem. Phys. **95,** 9209 (1991).
58. H. M. Ellerby, C. L. Weakliem, H. Reiss: J. Chem. Phys. **97,** 5766 (1992).
59. C. L. Weakliem, H. Reiss: J. Chem. Phys. **99,** 5374 (1993).
60. C. L. Weakliem, H. Reiss: J. Chem. Phys. **101,** 2398 (1994).
61. G. K. Schenter, S. M. Kathmann, B. C. Garrett: Phys. Rev. Lett. **82,** 3484 (1999).
62. G. K. Schenter, S. M. Kathmann, B. C. Garrett: J. Chem. Phys. **110,** 7951 (1999).
63. S. M. Kathmann, G. K. Schenter, B. C. Garrett: J. Chem. Phys. **111,** 4688 (1999).
64. F. H. Stillinger: J. Chem. Phys. **38,** 1486 (1963).
65. F. T. Smith: J. Chem. Phys. **38,** 1034 (1963).
66. G. K. Schenter: J. Chem. Phys. **108,** 6222 (1998).
67. B. C. Garrett, S. M. Kathmann, G. K. Schenter: In Proceedings of the 15th International Conference on Nucleation and Atmospheric Aerosols B. N. Hale, M. Kulmala (eds), American Institute of Physics, Melville, New York, (2000) p 201.
68. G. K. Schenter, S. M. Kathmann, B. C. Garrett: J. Chem. Phys. in press (2002).
69. L. X. Dang, T. M. Chang: J. Chem. Phys. **106,** 8149 (1997).
70. Z. Salnina, J. F. Crifo: Int. J. Thermophys. **13,** 465 (1992).
71. K. S. Kim, B. J. Mhin, U-S. Choi, K. Lee: J. Chem. Phys. **97,** 6649 (1992).
72. B. J. Mhin, S. J. Lee, K. S. Kim: Phys. Rev. A **48,** 3764 (1993).
73. P. Hobza, O. Bludsky, S. Suhai: Phys. Chem. Chem. Phys. **1,** 3073 (1999).
74. C. Munoz-Caro, A. Nino: J. Phys. Chem. A **101,** 4128 (1997).
75. N. Goldman, R. S. Fellers, C. Leforestier, R. J. Saykally: J. Phys. Chem. A **105,** 515 (2001).
76. T. Baer, W. L. Hase: Unimolecular Reaction Dynamics. Oxford University Press, New York (1996).
77. A. Laaksonen, V. Talanquer, D. W. Oxtoby: Annu. Rev. Phys. Chem. **46,** 489 (1995).
78. E. Wigner: J. Chem. Phys. **5,** 720 (1937).
79. J. Horiuti: Bull. Chem. Soc. Japan **13,** 210 (1938).
80. J. C. Keck: J. Chem. Phys. **32,** 1035 (1960).
81. J. C. Keck: Adv. Chem. Phys. **13,** 85 (1967).
82. D. G. Truhlar, A. D. Isaacson, B. C. Garrett: In Theory of Chemical Reaction Dynamics. M Baer (ed), CRC Press, Boca Raton, FL, (1985) p 65.
83. D. G. Truhlar, B. C. Garrett, S. J. Klippenstein: J. Phys. Chem. **100,** 12771 (1996).
84. R. P. Feynman, A. R. Hibbs: Quantum Mechanics and Path Integrals. McGraw-Hill, New York (1965).
85. H. Eyring: J. Chem. Phys. **3,** 107 (1935).

86. H. Eyring: Trans. Faraday Soc. **34,** 41 (1938).
87. E. Wigner: Trans. Faraday Soc. **34,** 29 (1938).
88. J. B. Anderson: J. Chem. Phys. **58,** 4684 (1973).
89. W. H. Miller: J. Chem. Phys. **61,** 1823 (1974).
90. W. P. Reinhardt, J. E. Hunter: J. Chem. Phys. **97,** 1599 (1992).
91. W. P. Reinhardt, M. A. Miller, L. M. Amon: Acc. Chem. Res. **34,** 607 (2001).

3 Vibrational Spectroscopy and Reactions of Water Clusters

Udo Buck and Christof Steinbach

3.1 Introduction

The condensed phase of water is probably the most investigated substance in Physical Chemistry. On the one hand, there are numerous anomalous properties of the liquid like the heat capacity and the density as well as the many configurations of ice that make water with its tendency to form a network of hydrogen bonds so interesting [1]. On the other hand, water plays a key role as ubiquitous solvent on earth and as promotor of reactions in atmospheric and extraterrestrial chemistry. In spite of all the efforts and the progress made in the last 20 years, a unified description of all the phenomena starting from the basic molecular interaction is still missing. We do not have a consistent description of liquid water nor do we understand completely the spectral and surface properties of the different conformations of ice. One of the reasons is certainly the lack of good, flexible interaction potentials that correctly account for all the many-body effects which play a crucial role in the hydrogen-bonded network of the water interaction [2]. In many of the simulations many-body effects are included by effective two-body interactions and are thus mostly valid in that range of applications to which their parameters are fitted. The other problem is the lack of detailed experimental results that would allow us to derive explicit conclusions about the underlying molecular models. In liquids, for instance, the information is often restricted by averaging processes and only a global picture results.

In both respects clusters provide a much better tool to study the microscopic behavior of a number of macroscopic phenomena. The advantage of clusters is the possibility to simply vary the size and to investigate step by step the development of properties of the condensed phase that are otherwise difficult to disentangle. In addition, the finite number of particles in a well-defined environment helps appreciably to find theoretical concepts which are easier to apply than the direct treatment of the many-body problem of bulk liquids or solids.

In this chapter we will treat mainly two examples of water cluster research and we will try to make contributions to a better understanding of spectra, structure and their theoretical prediction as well as to clarify concepts and to elucidate the underlying mechanism of dynamical processes. The first one is the spectroscopy of the OH-stretch vibration. The measurement of this vibration by infrared spectroscopy has proved to be a very sensitive indicator of the strength and the coordination of the hydrogen bond [3]. Therefore it gives direct information on the underlying force field and thus the interaction potential.

One has, however, to make sure that the experiments are size selective. This is not a simple task for neutral clusters, since the techniques for generating free cluster beams, the supersonic adiabatic expansion or the aggregation in cold gas flows, produce in almost all cases a distribution of cluster sizes [4]. In principle, the problem could be solved by a size-specific detection method. The most commonly used detection method, the ionization and the subsequent mass selection in a mass spectrometer is hampered by fragmentation during the ionization process. It is caused by the energy released into the system as the clusters change from their neutral to their ionic equilibrium structure. This excess energy then leads to the evaporation of neutral subunits and thus fragmentation occurs [5, 6]. For molecular species like water, the resulting fragmentation pattern is often modified by fast chemical reactions of one of the ionized molecular units with neutral partner molecules within the cluster [6, 7] that leads to protonated ions [8]. In any case, a simple mass spectrum does not characterize at all the neutral cluster distribution. Therefore we will concentrate here on the measurement of neutral clusters with known sizes.

Small clusters in the size range from $n = 7 - 10$, that exhibit three-dimensional structures, are size selected by momentum transfer from a helium beam [6, 9, 10]. The vibrational spectroscopy is carried out by tunable laser excitation and depletion spectroscopy based on vibrational predissociation [11, 12]. With accompanying calculations, all the measured features could be assigned and traced back to the type of bonding and the environment [13]. The clusters in this size range are dominated by 3-dimensional cage structures with the octamer cube as the central building block.

Similar approaches in the size range from $n = 2 - 5$ have been followed by several groups using mass spectrometric [12, 14, 15] and optothermal [16, 17] detection methods with and without size selection. A different concept has been persued in the vibration-rotation-tunneling spectroscopy in the terahertz regime [18, 19] where the high resolution spectra themselves are used as fingerprints for the sizes $n = 2 - 6$. Another possibility to overcome the size problem is to attach an aromatic molecule to the water clusters. These molecules can be ionized by resonant two-photon ionization in a size and even isomer specific way [20, 21]. Applications are available for benzene-water [22, 23], phenol-water [24–27], and other substituted benzene-water [28] systems.

A further variable is the cluster temperature. By changing the source conditions, we were able to increase the cluster temperature of the water nonamer from about 60 K to about 180 K as was checked by model calculations of the adiabatic expansion. The OH stretch spectra change dramatically and calculations by V. Buch reveal that with increasing temperature more and more isomers contribute to the spectrum which finally resembles that of the surface of bulk liquid water [29].

For larger clusters, the experiments are carried out for a distribution of sizes with the average values in the range of $\langle n \rangle = 20 - 2000$. Here the sizes are determined, free of fragmentation, by doping them with one sodium atom. For convenience, we will present a correlation that relates the water cluster sizes to the source conditions of the adiabatic expansion. The OH stretch spectra were taken by observing the emitted products. To our surprise, we detected in a special size range mainly hexamer products that we attibute to originate from amorphous structures. There are several approaches in the literature using the same experimental concepts [12, 16, 30], but all with unknown cluster sizes. In this way we try to bridge the gap between the well known results for small clusters and those obtained by Devlin and co-workers for ice nanoparticles [31].

The second group of experiments is related to the well known reaction of (solid) sodium with (liquid) water. The products of this textbook reaction are solvated sodium hydroxide and molecular hydrogen gas. Surprisingly, the mechanism of this reaction is not really known. Therefore we started a series of experiments with sodium and water clusters under single and multiple collision conditions [32]. Multiple collisions proved to be necessary for the production of solvated sodium hydroxide and at least three sodium atoms and six water molecules were necessary to reach this step [33]. These experimental findings were nicely confirmed by extensive calculations of the Parrinello group using ab initio molecular dynamics simulations [34]. We also investigated what happens when we offer more Na atoms and when we replace water by ammonia.

We start the presentation of the results obtained for the vibrational spectroscopy and the reaction dynamics with a short description of the experimental arrangements and close with the discussion of the goals that we have achieved. A smaller part of the spectroscopic portion has recently been reviewed as part of a larger enterprise [35, 36]. Here the emphasis will be on the most recent developments.

3.2 Vibrational spectroscopy of the OH stretch mode

3.2.1 Overview

The OH stretch frequency has proved to be a very sensitive measure of the strength of the hydrogen bond, since it directly probes the bonding. The water monomer has the two fundamental frequencies, the symmetric stretch at 3657.0 cm^{-1} and the asymmetric stretch at 3756.0 cm^{-1}. Within a hydrogen bond, usually a red shift is measured that originates from the deeper potential well in the vibrationally excited state. The origin of such behavior is traced back to the elongation of the O–H distance in the molecule and the shortening of the O$\cdots$H hydrogen bond distance. This leads to an increase of the attractive forces caused by the electrostatic, the induction and the dispersion forces. The structure of the dimer, a linear hydrogen bond, has been predicted in many *ab initio* calculations and was confirmed in the experiments [14, 16, 17]. Based on this structure, one would expect four frequencies, the free and hydogen bonded one of the donor and the asymmetric and symmetric stretch of the acceptor. While the free and the asymmetric OH stretch were correctly assigned in these measurements, the frequency of the bonded OH, 3601 cm^{-1}, was first obtained in an experiment where the clusters were analyzed according to their size by atomic beam deflection [15]. The water trimer, tetramer, and pentamer are cyclic with only two frequencies for each cluster [15]. The bonded OH stretch frequencies decrease from 3533 cm^{-1} to 3360 cm^{-1} thus indicating the increasing strength of the hydrogen bond based on the cooperativity of the underlying forces. There is good agreement with experiments based on infrared cavity ring down absorption spectroscopy [37] that are, however, not size selective. Also many calculations confirm the structures and the trend in the line shifts [38–40]. The same structure has also been observed by a series of experiments on the vibration-rotation-tunneling in the terahertz frequency regime [19, 41, 42]. From these experiments the only information on hexamers, that form three-dimensional cage structures, is obtained [43]. We note that in calculations also the prism and the cyclic form are very close in energy to that of the cage [44]. The latter structure has recently been observed in helium droplets [45]. Clusters in the size range between $n = 7 - 10$ have been measured by applying the technique of size selection by momentum transfer [46, 47]. The resulting structures can be derived from the octamer cube by adding or subtracting molecules [13]. Similar results have been obtained by adding an aromatic chromophore to the water clusters. Results are available for benzene-water [22, 23], phenol-water [24, 26, 27], and other substituted benzene-water [28] systems.

For larger clusters, only data are available where the sizes are not known. Here, aside from the isolated peak of the free OH stretch around 3720 cm^{-1}, mainly broad distributions were measured [12, 14, 30, 37, 48]. For even larger clusters FTIR measurements of ice nanoparticles in the size range of several nanometers were carried out [31]. We note that the IR spectrum of the OH stretch mode of liquid water shows an unstructured distribution peaked at 3450 cm^{-1}. The corresponding spectrum of hexagonal ice I_h exhibits a further red-shifted peak at 3220 cm^{-1} with two shoulders at 3100 cm^{-1} and 3380 cm^{-1}. The general shift is an indication of the stronger hydrogen bonds in the tetrahedral arrangement of the solid, while the relatively broad distribution can be traced back to the proton disorder [49].

3.2.2 Clusters $n = 7 - 10$

Experiment: The experiments in the size range from $n = 7$ to $n = 10$ have been carried out for fully size-selected clusters [46, 47]. The method of size selection by momentum transfer in a scattering experiment with atoms under single collision conditions has been described in detail in the literature [6, 9, 10, 50, 51] so that we give here only a short account of the principle. The method is based on the fact that the heavier clusters are scattered into smaller angular ranges with different final velocities compared to the lighter clusters. It is immediately obvious from this picture that larger clusters can easily be excluded by choosing the correct *angle* for detection. For the necessary additional discrimination against smaller clusters we use the *mass* [6, 7, 10]. This procedure only works, if at least a small fraction of the cluster M_n is detected at the nominal mass of the ion M_n^+ or at a fragment mass M_k^+ which is larger than that of the next smaller cluster M_{n-1}^+. Water clusters are usually detected at the mass of the protonated ion $(H_2O)_{n-1}H^+$ of the next smaller cluster and thus fulfill this condition.

The procedure requires, in general, a high-resolution molecular beam apparatus and intense cluster beams with good expansion conditions to minimize the angular ($\Delta\Theta < 0.1°$) and velocity spread ($\Delta v/v < 0.05$) of the colliding beams. The first part is realized by skimmed beams introducing additional apertures. The latter is usually achieved by expanding a mixture of 2 to 10% of gas in helium or neon. In this way we were able to separate water clusters up to $n = 10$. For larger water clusters we had problems generating enough intensity at the selected size.

The experimental arrangement is shown in Fig. 3.1. The clusters are selected by the scattering process before they are probed by the laser. By choosing a deflection angle appropriate for selecting cluster size n, the laser beam will interact only with clusters of size $k \leq n$. If the quadrupole mass spectrometer with electron impact ionization is tuned to a mass larger than that of size $n-1$, the depletion is monitored only for the clusters of size n. Since the laser is pulsed, the interaction with the scattered cluster beam results in a transient depletion signal which is measured by time-of-flight analysis as is schematically shown in the inset of Fig. 3.1. The depletion is measured for each laser frequency, and in this way a size-selected cluster spectrum is obtained.

The infrared radiation used to excite and dissociate the clusters is obtained from a Nd:YAG-laser-pumped optical parametric oscillator (OPO) [52]. It consists of a master oscillator containing a $LiNbO_3$ crystal, which is pumped by the fundamental of a Nd:YAG laser. The master oscillator is seeded by the narrow-bandwidth infrared radiation obtained by difference frequency mixing the output of a pulsed dye laser and the 532 nm radiation of the same Nd:YAG laser in a $LiIO_3$ crystal. The typical output energy for the low frequency component (idler) is ≥ 4 mJ per pulse in the entire spectral range covered in this study. The bandwidth of the infrared radiation, which is determined by the bandwidth of the dye laser, is chosen to be 0.5 cm^{-1}.

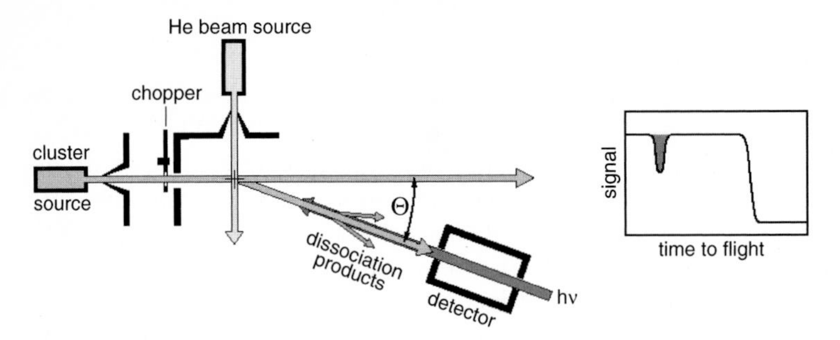

Fig. 3.1. Schematic view of the experimental set-up for depletion spectroscopy with size selected clusters by momentum transfer.

Results: The results for $n = 7 - 10$ are presented in Fig. 3.2. At first glance the spectra exhibit more structure than the two groups, the free and the hydrogen-bonded OH bands, that are observed in the spectra of the cyclic clusters from $n = 3$ to $n = 5$. This is clearly a reflection of their more complicated 3-dimensional structure which starts with $n = 6$ [43]. For the heptamer, a sort of line spectrum was measured in which each of the 14 OH groups is reflected in a different line position ranging from 3710 to 2935 cm^{-1}. The latter is the largest line shift ever measured for water clusters, the bulk condensed phase included. For the larger clusters the spectra are again simplified and consist essentially of three groups, now an indication of collective vibrations of the OH groups in similar environments. We note the remarkable fact that the spectra of the nonamer and decamer are, aside from some intensity around 3130 cm^{-1}, not very different from that of the octamer.

In a very detailed effort, the experimental spectra could be completely explained by combined high level *ab initio* calculations and simulations based on a polarizable model potential, using permanent charges and induced dipoles [13]. In the latter case the frequencies were calculated in a Morse oscillator basis with the help of a parametrized function of the electric field component parallel to the OH bond. The zero-point motion was explicitly taken into account. First the contribution to the binding energies was calculated using the rigid body diffusion Monte Carlo method. Later also the force constant was averaged over the intermolecular motion.

The structures that reproduce the measured spectra and correspond mainly to the minimum energy configurations are shown in Fig. 3.3 and Fig. 3.4. They can be considered as members of a series, derived from the octamer cube by either insertion or removal of water molecules. The structural units of the octamer are threefold-coordinated molecules acting as double donor and single acceptor (DDA) (grey), or as single donor and double acceptor (DAA) (blue). All DDA molecules are connected only to DAA molecules, and vice versa. As already noted in the past, the cube is not perfect and the

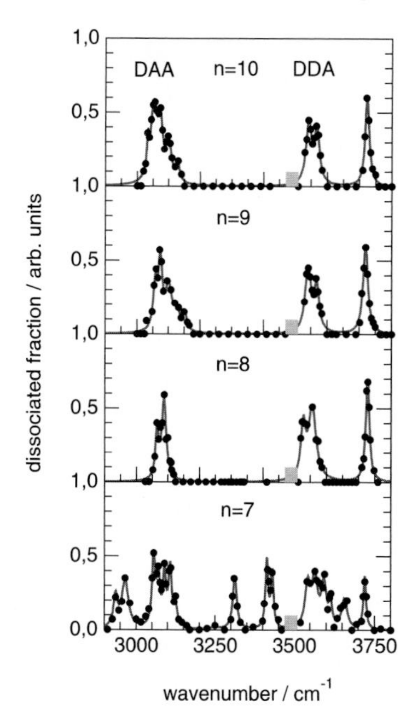

Fig. 3.2. Measured depletion spectra of the OH stretch vibration for size selected water clusters. The block on the wavenumber axis indicates the gap of the OPO.

O$\cdots$O distances emanating from bonds between DDA and DAA molecules tend to be longer (2.8 Å) than bonds emanating from DAA to DDA (2.6 Å) and thus the OH frequencies tend to be larger, since the hydrogen bond is weaker [22, 38, 53–55]. The reason is that the DAA bonds manage to optimize the single, linear hydrogen bond geometry much better than the DDA molecules which have to accommodate two bonds which are bent. As a result the OH stretch spectrum contains two well separated bands one at 3065 cm^{-1} which can be traced back to the DAA configuration and one at 3560 cm^{-1} which corresponds to DDA molecules. The third one close to 3720 cm^{-1} is that of the unshifted free OH stretch of the DAA molecules. The splitting of the DDA band is caused by the interaction of the symmetric and asymmetric stretch of the two adjacent OH bonds. It is, at 30 cm^{-1}, much smaller than the 100 cm^{-1} of the free molecules and even smaller than the 45 cm^{-1} of the solid. The reason is the change of the sign in one of the potential coupling constants which reduces the size. The splitting of the DAA band originates from two different isomers with nearly the same minimum energy but a different orientation of the hydrogen bonds in the upper and lower four-membered rings of the octamer (see Fig. 3.3). The one with D_{2d} symmetry has the opposite and the one with S_4 symmetry the same orientation of the bonds.

The simple spectroscopic pattern of the three clearly separated bands is also observed for the two larger clusters $n = 9$ and $n = 10$. Here, the rings are extended by adding one or two twofold-coordinated DA molecules (green, see Fig. 3.4 (no. 1) and Fig. 3.3 (bfl)). This leads to a new line around 3130

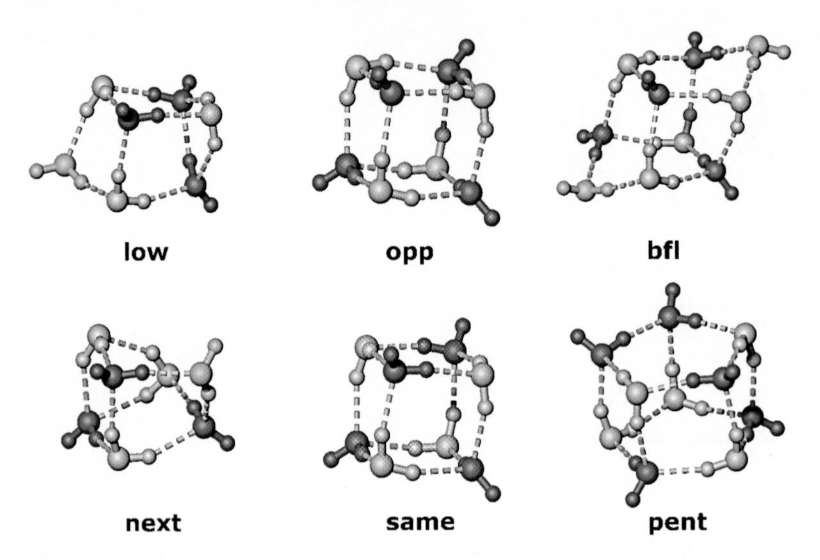

Fig. 3.3. Calculated minimum energy structures of water clusters for $n = 7$, 8, and 10. Color code of the coordination numbers: DAA, blue; DDA, grey; DA, green.

cm^{-1}. All the other molecules are more or less in the same environment as they are in the octamer so that the same spectra result. We note that for the octamer and nonamer other isomers have distinctly larger minimum energies and are thus clearly separated from the global minimum configurations. This is not the case for the decamer. Here, the configuration shown in Fig. 3.3, the "butterfly (bfl)" structure with two DA molecules, is 2.6 kcal/mol higher in energy than the configuration with two five-membered rings (pent). This difference is reduced to 0.08 kcal/mol when the zero point energy is taken into account. Since in the latter configuration two DDA and two DAA molecules are found in neighbored positions, the calculation gives additional spectral features between 3200 and 3500 cm^{-1}, in clear contrast to the experimental result. Obviously, this configuration can be ruled out, and the potential model is still not optimal.

The spectrum of the heptamer looks completely different. There are two energetically close-lying minima corresponding to a fused three- and four-membered ring which is obtained from the S_4 (same) octamer by removing a DDA or a DAA molecule from the back right corner, respectively. The two lines with the largest shifts around 2950 cm^{-1} result from DAA molecules which connect the two rings in each isomer. Aside from bands in the usual DAA (3065 cm^{-1}) and DDA (3560 cm^{-1}) frequency range, there appear additional lines in the gap around 3300 and 3400 cm^{-1} which result from AD molecules and DAA or DDA molecules with adjacent molecules of the same type. It is noted that the excellent agreement of the theoretical predictions

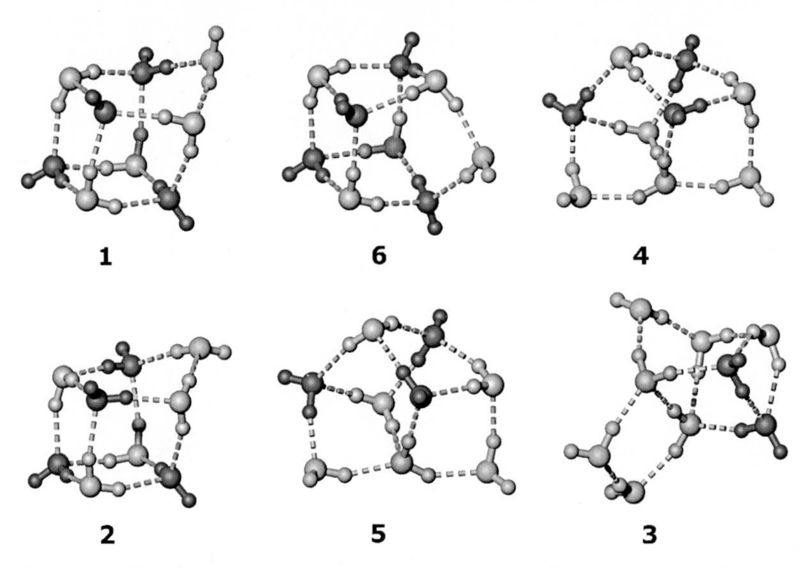

Fig. 3.4. Calculated minimum energy structures of water clusters for $n = 9$. Color code of the coordination numbers: DAA, blue; DDA, grey; DA, green; DDAA red. The minimum energies in kcal/mol are 1: 80.54; 2: 80.24; 6: 79.24; 5: 78.37; 4: 77.89; 3: 75.31. The numbers are those of Ref. [29] for better identification.

with the experimental results is only achieved by averaging over the zero-point motion [13].

The frequency of the free OH stretch mode is always found in the range of 3720 cm^{-1}. This is in contrast to the 3695 cm^{-1} observed for the free OH mode for ice or liquid water surfaces [54, 56] (see also Fig. 3.5). The reason is twofold. For clusters, the reduced directionality of the hydrogen bond and the lower coordination of the neighbor molecules leads to the blueshift. In bulk water the next neighbors of the surface molecules are fourfold-coordinated, while in the small clusters we find many threefold coordinated molecules which are absent in crystalline and liquid water [57].

Temperature effects: When we ask the question what will happen, if the temperature of the clusters is increased, it is obvoius that an increasing number of isomers will be available with different spectra than those of the minimum configurations which will start to fill the gap between the signatures of the DAA and DDA molecules [46]. Melting-like transitions have been predicted for clusters in this size range with temperatures between 105 and 170 K [58–62]. Very recently we were able to detect such a transition for $(H_2O)_9$. By varying the source conditions for the cluster production (slightly higher concentration of water vapor in the helium mixture and a factor of two lower total pressure), the spectrum shown in the lower part of the right panel of Fig. 3.5 resulted [29]. The pronounced peak structure of the low temperature

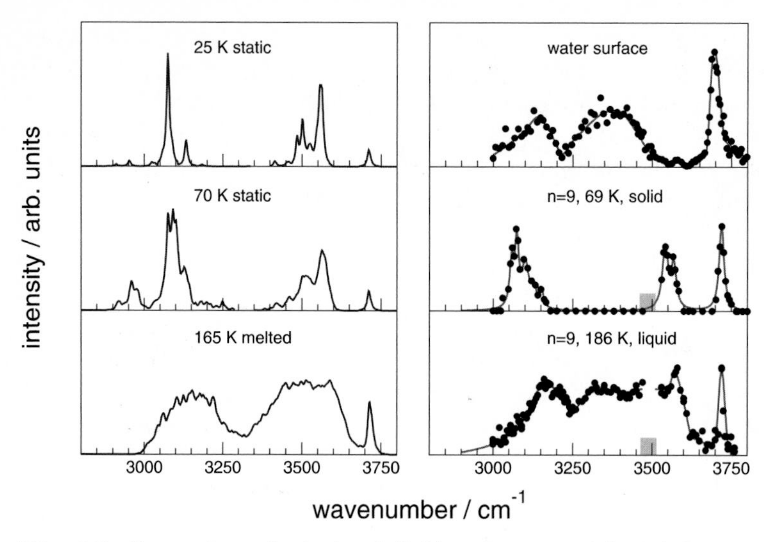

Fig. 3.5. Comparison of calculated (left) and measured (right) OH stretch spectra of $(H_2O)_9$ for different temperatures. In addition, the measured surface spectrum of liquid water from Ref. [56] is presented.

nonamer, exhibited in the middle part of the right panel of Fig. 3.5, nearly disappears and is replaced by a broad distribution.

To get a qualitative understanding of the measured spectrum, we have carried out calculations that included classical trajectory simulations of the cluster structure and dynamics as a function of temperature, computation of the minimum energy structures, and quantum calculations of temperature dependent spectra of selected minima, employing the model that was already devised for the analysis of $n = 7 - 10$ clusters. Six selected configurations found in the simulations are presented in Fig. 3.4. The two lowest energy structures (1 and 2) can be viewed as products of the insertion of a DA molecule to either the D_{2d} or the S_4 octamer with their opposite and same orientation of the upper and lower rings. At 70 K these two isomers contribute to the signal. This leads to the broadening of the DAA band compared to the octamer spectrum, in agreement with the experimental results. In contrast to the conclusion in Refs. [46] and [13], apparently both low lying isomers contribute to the spectrum and explain the unresolved structure. Similar conclusions have been obtained in a study of $(H_2O)_9$ complexes with benzene [63].

As the simulation temperature was raised to 140 K, the trajectories were still localized predominantly near the two lowest energy configurations. Near 165 K a qualitative change occured: the cluster left the vicinity of the initial low energy structure. In the course of each of the trajectories the cluster became localized in the vicinity of one dominant high energy structure, while visiting neighboring minima as well. Four of these structures are also shown in

Fig. 3.4 (3 - 6). One is a high energy version of the cube, the remaining three correspond to more open "amorphous" structures including 2-4 coordinated water molecules, and 3-5 membered rings. The qualitative effect of the structure distribution is assessed by overlapping the four corresponding spectra that, in turn, are each thermally broadened (since at 165 K a large number of vibrational intermolecular states are populated). The result is shown in Fig. 3.5. The agreement with the measured high temperature spectrum is good, except that the calculated dip in the spectrum is too large.

The preliminary analysis of the temperature based on the adiabatic expansion gives about 69 K for the measurement of the "solid-like" and 186 K for the "liquid-like" nonamer, in good agreement with the temperatures used in the simulations. Thus the melting point is considerably reduced in small clusters compared to bulk water but still larger than in some of the theoretical predictions.

For comparison we have also plotted a recent measurement of the spectrum of the surface of water in the liquid state [56]. The similarities are quite remarkable although the method applied, the sum frequency generation as described in Chapter 11, has different selection rules than the IR spectroscopy and the same spectrum obtained by a different group is less structured [64]. Nevertheless, the two broad peaks at 3150 and 3400 cm^{-1} and the single peak close to 3700 cm^{-1} are very similar. They are attributed to different types of hydrogen bonding near the liquid surface and the free OH stretch, respectively [65]. The main difference is the DDA peak at 3550 cm^{-1} which is not present in the spectrum of bulk liquid.

3.2.3 Clusters $\langle n \rangle = 20 - 650$

Experiment: In this size range we do not work with size selected clusters. The tunable IR laser radiation is the same as was used in the size selected experiments. It interacts with the cluster beam at a certain angle. To increase the sensitivity for very large clusters, we do not detect the depletion of the signal by the dissociation process, but we measure directly the outcoming fragments. In this way we combine low background signals with a high probability to observe the dissociation event also for very large clusters. The experimental arrangement is depicted in Fig. 3.6 [66]. The detector is set on a small scattering angle, usually 3° to 4°, and the mass spectrometer signal gave the largest intensity on mass $(H_2O)_4H^+$ that we attribute, because of fragmentation [8], to the hexamer. In contrast to the results that were obtained previously [12], the fragment intensities of larger complexes from $n = 4$ to 7 were an order of magnitude higher than that of the dimer. Although, based on pure energetical considerations, two photons are necessary to dissociate the clusters and to form pentamers or hexamers, it could very well be that we are particularly sensitive to preformed cyclic structures of this kind, as an indication of amorphous behavior.

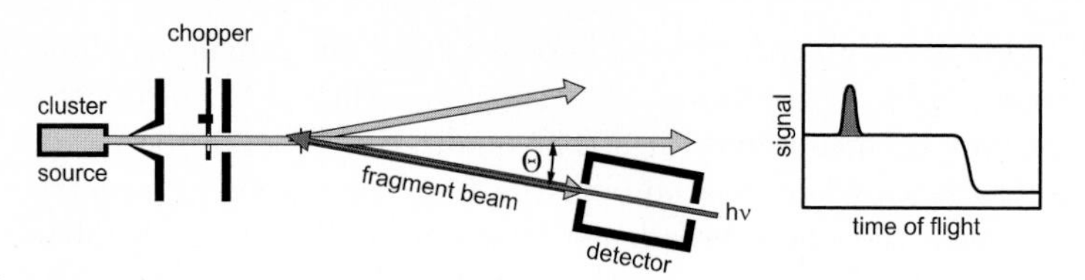

Fig. 3.6. Schematic view of the experimental set-up for fragment spectroscopy of large clusters.

Size distributions: In contrast to all previous experiments, we know the size distribution of our clusters. The usual mass spectrometer detection will not work, since water clusters are known to fragment upon electron impact ionization [8]. Therefore we used a trick for a fragmentation-free detection. The water clusters were doped by one sodium atom and the resulting cluster is ionized by a one photon process close to the threshold using a dye-laser [67]. Typical examples for measured water cluster distributions are shown in Fig. 3.7. Here we used a conical nozzle with a diameter of 50 μm, an opening angle of $2\beta = 41°$, and a length of 2 mm. The left part exhibits the pressure dependence and the right part the temperature dependence. In general, the expected behavior is observed. At small cluster sizes, the exponential behavior dominates and, at large cluster sizes, the measured curves can be fitted nearly perfectly by log-normal distributions. In the intermediate range, we observe, interestingly, a bimodal distribution. The two peaks can both be fitted by individual log-normal distributions. They occur when the pressure is increased at constant temperature (left panel) or when the temperature is lowered at constant pressure (right panel). The origin of this effect is obviously a different production process, since fragmentation can be ruled out as origin of this behavior. We think that the peak at lower size can be traced back to the growing of cold solid clusters by addition of small complexes, say dimers or trimers, to the cluster. In contrast, the peak at larger sizes is definitely based on the coagulation of larger pieces. A plausible explanation might be that because of the addition of larger parts with larger energy release the cluster is first liquid and then cools off by evaporation. The first mechanism is usually assumed to be the main process in the cluster production of small systems. The latter one has been invoked by the extensive investigation of the condensation of large water clusters by electron diffraction [68] (see also chapter 16). We note that in some cases the intensity gap between the two peaks is filled up and that, in any case, the average size fits very well into the general scaling laws for cluster sizes.

To relate, in general, the measured size distributions to the source conditions of the cluster production, we apply the scaling law originally derived by Hagena for rare gas clusters [69, 70]. The key parameter that correlates the flow which produces the same cluster size is given by

$$\Gamma = n_0 d_{eq}^q T_0^\alpha (0 < q \leq 1; \alpha = q - 3). \tag{3.1}$$

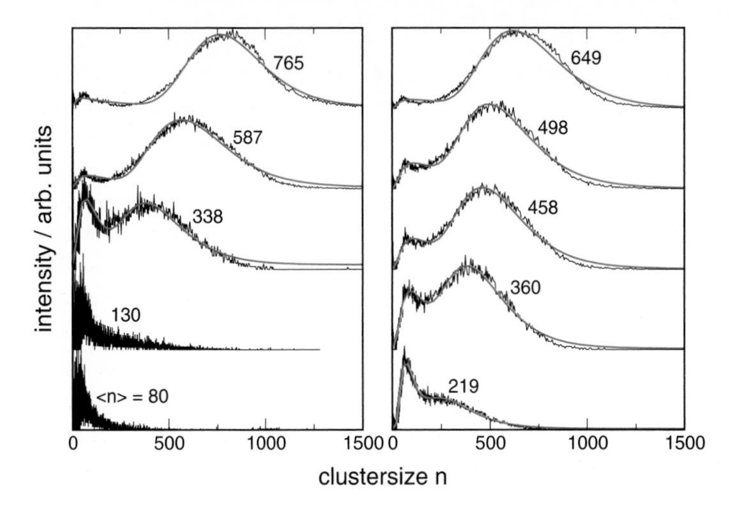

Fig. 3.7. Measured mass spectra of water clusters doped by Na atoms and detected by single photoionization. Left panel: variation of pressure from 5.1 to 13.7 bar at 495 K; right panel: variation of temperature from 544 to 694 K at 17.6 bar. The average sizes indicated are determined from fitted log-normal distributions specified by the solid lines.

Here n_0 and T_0 are the source density and nozzle temperature, and $d_{eq} = 0.933d/\tan\beta$ is the equivalent nozzle diameter for a conical nozzle with a cone angle of 2β and a throat diameter d. These relations are valid for axial symmetric flows and $f = 6$ energetically active degrees of freedom [67]. The variables q and α are fitted to the data with the results $q = 0.643 \pm 0.063$ and $\alpha = -2.655 \pm 0.309$. For α the rule given in (3.1) is indeed fulfilled within the given error limits. Finally the mean cluster size is given by

$$\langle n \rangle = D(\frac{\Gamma^*}{1000})^a, \tag{3.2}$$

where $\Gamma^* = \Gamma/K_{ch}$ and $K_{ch} = r_{ch}^{q-3}T_{ch}^{\alpha}$ with $r_{ch} = 3.19$ Å and $T_{ch} = 5684$ K. The two variables are again fitted to the data with the result $D = 2.63$ and $a = 1.872$ [67]. This new scaling law for water clusters is also in nice agreement with the results of two electron diffraction studies [71, 72] and two mass spectrometer investigations [73, 74].

Results: With cluster size distributions known from sodium labeling results for the source conditions, the OH stretch spectra were obtained by fragment spectroscopy of the average sizes $\langle n \rangle = 40, 111$, and 648 as depicted in Fig. 3.8 [66]. The three spectra are dominated by peaks at 3720 cm^{-1}, 3550 cm^{-1}, and a variable one moving from 3420 to 3350. In addition, they all exhibit a sort of shoulder below 3200 cm^{-1}. While the first two features can definitely be attributed to the free OH (dangling H) and the double donor molecules (DDA or dangling O) at the surface, the interesting part is the maximum around 3400 cm^{-1} that shifts to smaller frequencies with increasing cluster size. There is clearly more intensity in this frequency range than around 3200 cm^{-1} where we expect the signature of 4-coordinated cristalline ice. Clusters in this size range are usually considered to be of amorphous structure [71], although the peak of bulk amorphous water is close to 3260 cm^{-1}. The reason for this discrepancy has been presented by V. Buch and P. Devlin in Sec. 17.2.6 of this book. In the calculation of OH stretch water spectra this blue shift is attributed to the decrease of interior 4-coordinated molecules with decreasing size. The calculated numbers are near 3350 cm^{-1}, in the frequency range measured here.

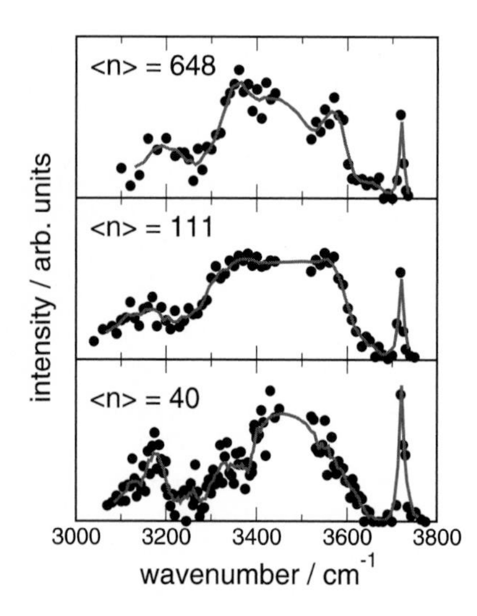

Fig. 3.8. Measured OH strech spectra of large $(H_2O)_n$ clusters detected at the fragment mass $(H_2O)_4H^+$ for the average sizes indicated.

Our experimental results are in qualitative agreement with those obtained by the same predissociation technique [12, 16, 30, 48], although the size is not known in all these contributions. We note that the amplitudes in these measurements do not necessarily reflect the IR absorption cross sections, since what is finally measured is the dissociated fraction after the excitation [75]. Nevertheless, the experiments based on Fourier transform IR [31, 76] and cavity ring down [37] spectroscopy exhibit similar features, although the occurrence of the largest intensity at 3200 cm^{-1} was never observed in dissociation experiments. The only other attempt to derive size selective information in this size range is that of Devlin and coworkers described in Ref. [31] and in Chapter 17 of this book. The results, peak intensities at 3300 cm^{-1} for $\langle n \rangle = 150$ and 3400 cm^{-1} for $\langle n \rangle = 48$, are in reasonable agreement with our results.

With our ability to provide information about the size dependence, we carried out an experiment to determine the range of completely amorphous clusters. The results are depicted in Fig. 3.9. We detected again the fragments

$(H_2O)_4H^+$ for two different wavenumbers in the range of the free (3720 cm^{-1}) and hydrogen bonded (3200 cm^{-1}) OH stretch frequencies. The distributions starts at about $\langle n \rangle = 25$, exhibits a pronounced peak around $\langle n \rangle = 65$ and falls off at about $\langle n \rangle = 300$ with a long tail. Apparently, our experimental method is quite sensitive to the amorphous structure. With increasing size the amount of this structure gets smaller and smaller.

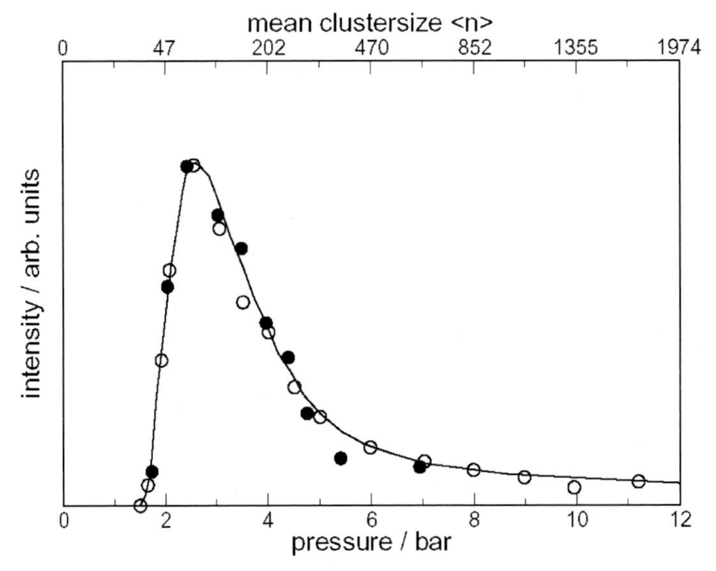

Fig. 3.9. Cluster size dependence of the OH stretch intensity taken at 3720 cm^{-1} (full circles) and 3200 cm^{-1} (open circles) at the mass $(H_2O)_4H^+$.

3.3 Reactions of sodium with water clusters

3.3.1 Overview

The reaction of solid sodium with liquid water is one of the best known exothermic reactions in chemistry that leads to NaOH and H_2 gas. The products are solvated Na^+ and OH^- ions that convert to NaOH only after all the solvent molecules have been removed. Surprisingly, the detailed mechanism of this reaction is not really known. Macroscopic experiments have been conducted in sodium hydroxide solutions. Here, after the production of solvated electrons by photolysis of OH^-_{aq}, neutral metal atoms are found as intermediates which rapidly decay to the known reaction products [77, 78]. On the other hand, it was observed that the scattering of single Na atoms with isolated H_2O molecules in a crossed molecular beam experiment did not lead to any reaction products [79]. Therefore clusters were thought to be the key in solving the problem microscopically. But these experiments also failed to produce any remarkable concentration of species containing NaOH. The first results were obtained in a pick-up experiment in which a pulsed water cluster beam seeded in Ar was crossed with a supersonic sodium cluster beam under multiple collision conditions. Solvated sodium in water $Na(H_2O)_n$ and with much lesser intensity also solvated sodium dimers $Na_2(H_2O)_n$ were the only products observed [80, 81]. Recently, we have investigated the scattering of sodium clusters Na_n, $n \leq 21$, with water clusters $(H_2O)_m$, $m \leq 40$, in a very detailed crossed molecular beam experiment under single collision conditions [32]. To detect the products, they have to be formed in the collision region which corresponds to a reaction time scale of picoseconds. The only products observed were again solvated sodium atoms in water clusters $Na(H_2O)_n$. By measuring their angular and velocity distribution after the collision, the reaction mechanism could be clarified. We observed narrow distributions with forward-backward symmetry near the center-of-mass. This is a clear indication of a complex forming reaction of the type

$$Na + (H_2O)_{n+x} \rightarrow Na(H_2O)^*_{n+x} \rightarrow Na(H_2O)_n + x\ H_2O, \tag{3.3}$$

with high internal energies being stabilized by the isotropic evaporation of $x = 2$ or 3 water molecules. Products containing Na clusters or NaOH did not appear in the mass spectra, although clusters up to $n = 21$ were offered as reactants.

An experimental limitation in the search for products in these experiments should be mentioned. The measured ionization potential of $Na(H_2O)_n$, $n \geq 4$, is 3.2 eV [80], which is confirmed by calculations [82–86]. This low value makes them accessible for photoionization by a single laser photon. Perhaps the ionization potential of the $(NaOH)_n(H_2O)_m$ clusters is higher than the applied photon energies of 3.5 to 4.7 eV. On the other hand, these products are accessible for detection, if additional Na atoms are attached. Thus some very spurious amounts of masses that might be assigned to $Na_n(NaOH)(H_2O)_m$, $n, m = 1, 2$ [80] and products of the type $Na_n(OH)_m$ arising from surface reactions with water impurities [87] were reported.

The generally negative experimental results with respect to the detection of a large amount of NaOH containing products are confirmed by a recent theoretical study of the direct reaction [88]

$$Na_2 + 2H_2O \rightarrow (NaOH)_2 + H_2. \tag{3.4}$$

Although this reaction is exothermic by 1.81 to 2.0 eV, it is hindered by a barrier of 1.28 to 1.56 eV, depending on the method of calculation. This is caused by the fact that the minimum energy configuration of Na_2 and the chain-like $(H_2O)_2$ dimer is far away from the transition state of the reaction.

We note that there are two further differences of the beam experiment compared to that of the bulk. Sodium is offered as clusters and not solvated, and water is not liquid. To overcome the problems that arise in the single collision experiment, we were looking for an experimental arrangement in which some of the solvation properties of the macroscopic experiments are present: preparation of the reactants as solution, multiple collisions and longer interaction times. Such an arrangement is described in the next section. Indeed, in this experiment we were able to detect solvated NaOH products which we propose as being produced in two sequential reaction steps.

3.3.2 Experiment and results

The heart of the new experiment is a sodium pick-up source with a nozzle diameter of 2 mm and capacities for heating up to 1000 K as is depicted in Fig. 3.10 [33]. Increasing the Na vapor pressure results in the formation of dimers leaving the source. Thus the continuous water cluster beam crosses the Na vapor for about 1 cm length before passing the skimmer. It contains Na atoms or a mixture of Na atoms and Na_2 dimers depending on the pressure. In this way the interaction time is expanded to several microseconds. The complete experiment is carried out in a molecular beam machine with a water cluster source equipped with a conical nozzle and a reflectron time-of-flight mass spectrometer (RETOF) for detection [67]. The products are ionized by 360 nm single photon ionization.

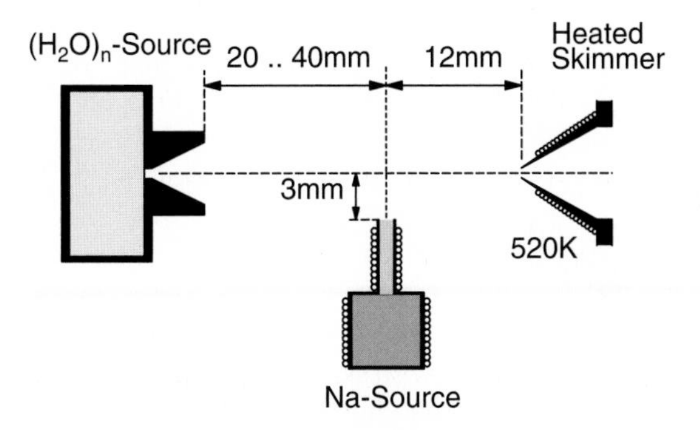

Fig. 3.10. Experimental arrangement for conducting the sodium water reaction.

At a Na pressure of 3 mbar only Na atoms leave the pick-up source. The result of the product measurement is, as we would have expected from the previous experiments [32, 80], pure $Na(H_2O)_n$ complexes. Despite the much longer interaction time, we still get the same products as in the single collision experiment. The result is a spectrum of Na doped water clusters which represents the nearly undisturbed size distribution of the pure water cluster beam. In fact, we used this method to determine the size of the water cluster distributions in this way in Sec. 3.2.3. The mean cluster size of the water cluster beam at the given conditions turns out to be $\langle n \rangle = 63$ with the extension to clusters $n = 140$. The smallest complex detected is $Na(H_2O)_3$.

With an ionization potential (IP) of 3.48 eV the 360 nm photons just reach the energy required, whereas $Na(H_2O)$ (IP= 4.38 eV) and $Na(H_2O)_2$ (IP= 3.80 eV) cannot be ionized. Since the corresponding ions could not be detected, their formation by ionization induced fragmentation of larger complexes can also be excluded.

At a temperature of 919 K corresponding to a sodium vapor pressure of 66 mbar Na atoms and dimers are generated in the pick-up source. This was confirmed by a measurement in which the pick-up source was mounted in front of the skimmer. The product spectrum consists now, aside from the sodium doped water clusters, of a series of new masses that are assigned to the products $Na(NaOH)_2(H_2O)_m$ and $Na(NaOH)_4(H_2O)_p$. These are shown in Fig. 3.11. Their intensities are even higher than those of the primary products $Na(H_2O)_n$. The onset for both reactive species is at $m = 0$ (no solvating water molecules). The high intensities of the reactive products indicate very large reaction cross sections. It is striking that only products with an even number of NaOH are found, thus pointing to H_2 molecules as further products. This is also inferred from the macroscopic reaction and (3.4).

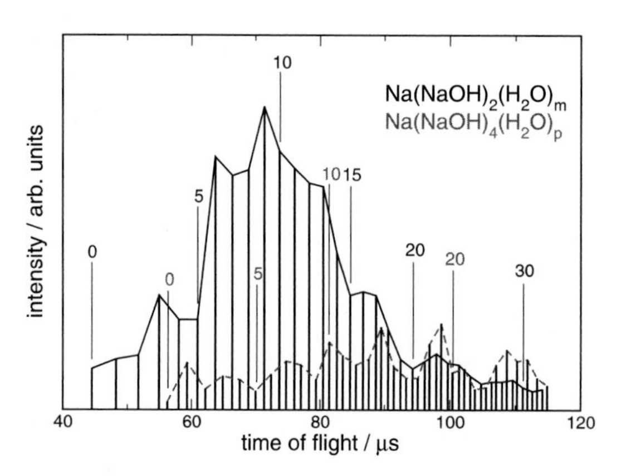

Fig. 3.11. Mass spectrum taken under condition with Na and Na_2 present. The products $Na(NaOH)_2(H_2O)_m$ are connected by a solid line, $Na(NaOH)_4(H_2O)_p$ by a dashed line.

Raising the sodium vapor pressure from 3 to 66 mbar leads to the following development of the mass spectra: As soon as Na_2 is present in the interaction region, the products $Na(NaOH)_2(H_2O)_m$ appear. The range of masses covered corresponds roughly to that of the $Na(H_2O)_n$ clusters. With increasing intensity of Na_2 the reactive products become more dominant with peak intensities between $m = 6$ and 14, and, in addition, $Na(NaOH)_4(H_2O)_p$ is observed with smaller intensities. Thus we write the two step reaction equations as follows: In the first step $Na(H_2O)_n$ is formed according to

$$Na + (H_2O)_{n+x} \rightarrow Na(H_2O)_n + xH_2O, x = 2, 3 \quad (3.5)$$

which then reacts with Na_2 to give the final products

$$Na_2 + Na(H_2O)_n \rightarrow Na(NaOH)_2(H_2O)_m + H_2 + (n - m - 2)H_2O. \quad (3.6)$$

If we increase the amount of Na, the reaction proceeds further as follows

$$2Na_2 + Na(H_2O)_n \rightarrow Na(NaOH)_4(H_2O)_p + 2H_2 + (n - p - 4)H_2O. \quad (3.7)$$

Under the present multiple collision conditions reaction products containing NaOH are formed according to the multi-step reaction (3.5) and (3.6) or (3.7). In the single collision experiment no such products have been observed. Apparently for their production previous solvation of a Na atom in the water clusters is necessary. The large intensities of the reactive products indicate very large reaction cross sections for the second step, whereas the first step, the solvation of the Na atoms, takes place with relatively small cross sections [32]. We note that, because of the exothermicity of the reaction, a couple of water molecules evaporate so that even the smallest product $m = 0$ stems from a precursor with $n \geq 6$. The product sizes of maximal intensities $m = 6$ to 14 are therefore traced back to the correspondingly larger initial clusters.

Based on the results of Ref. [32] processes of the kind $Na_3 + (H_2O)_{m+x+2} \rightarrow Na(NaOH)_2(H_2O)_m + H_2 + xH_2O$ are also excluded. Thus we propose the reaction mechanism of two sequential steps and consider the direct reaction of (3.4) as an unlikely process. Apparently, the structure of the water clusters doped with one sodium atom brings them in a position closer to the transition state to overcome possibly existing barriers more easily and to generate in the reaction with Na_2 the products of (3.6) and (3.7).

To investigate the influence of an increasing amount of sodium on the reaction products, we have carried out experiments for $(H_2O)_n$ clusters with the average size of $\langle n \rangle = 65$ and maximum values of $n = 140$ using a scattering cell for the pick-up process. For a pressure of $0.14 \cdot 10^{-5}$ mbar, we get the expected result of sodium doped water clusters. The mass spectra obtained for $17.0 \cdot 10^{-5}$ mbar are depicted in the left panel of Fig. 3.12. We only observe products where NaOH is solvated in the form $Na(NaOH)_{2n}(H_2O)_m$ as even number and H_2 as the other product. This is a clear indication that the same mechanism is operating as was found for pure Na_2 molecules as reactants and that the additional Na atoms are used to produce larger aggregates of the products. At very high sodium pressures all water molecules have reacted and are finally evaporated so that products of the form $Na(NaOH)_{2m}$ result.

As a counter-example we have carried out the same measurements for ammonia clusters with $\langle n \rangle = 95$ and a maximal $n = 200$. Here it is known from the bulk measurement that no chemical reaction occurs and sodium is easily solvated in ammonia [89]. At small sodium pressures we observed the expected result, ammonia clusters doped with one sodium atom. The mass spectrum taken at higher pressures of $1.51 \cdot 10^{-5}$ mbar is shown in the right panel of Fig. 3.12. We observe completely mixed clusters of the form $Na_m(NH_3)_n$ with no restrictions for m and n. If we increase the pressure to

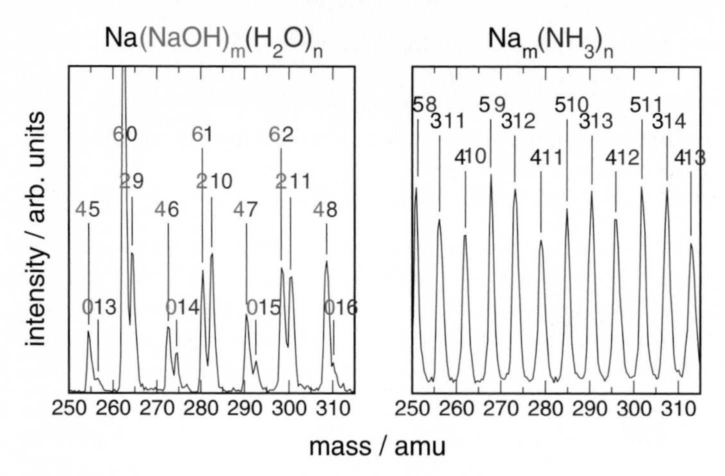

Fig. 3.12. Mass spectra taken with multiply Na doped water (left) and ammonia (right) clusters. The products are $Na(NaOH)_m(H_2O)_n$ clusters with m even and $Na_m(NH_3)_n$ clusters with no restriction for n and m.

$1.5 \cdot 10^{-5}$ mbar, only pure sodium clusters survive. They have to be fragments of the mixed clusters with ammonia, since the ionization wavelenghth of 500 nm used in our experiments is not sufficient to ionize pure sodium clusters. We conclude that the cluster systems behave in the same way as the bulk systems concerning their solubility and their chemical reactivity.

3.3.3 Calculation of the reaction mechanism

Given the main experimental results, the presence of a solvated sodium atom and its interaction with a sodium dimer, the questions to be answered by a theoretical treatment are as follows. (1) What is the reaction path and what are the reaction intermediates? (2) Why is such a long reaction time necessary? (3) What is the role of the additional Na atom observed in the products. To solve these problems a detailed calculation was performed using the Car-Parrinello molecular dynamics (CPMD) method [34]. The starting point was an equilibrated $Na(H_2O)_6$ and Na_2. The CPMD dynamics were carried out at 100 K with plane waves and pseudopotentials aside for the outer electron of sodium. For the description of the electron distributions localized orbitals are used based on the Boys criterion [90] that minimizes the spread in the position operator [91]. The structure of $Na(H_2O)_6$ is given by a tetrahedral coordination of the sodium atom with delocalized electrons over most of the water molecules in agreement with previous investigations [83]. The approaching sodium dimer starts to dissociate into a dipolar anion Na^- and a solvated Na^+. Independent calculations show that the charge-separated pair is a local minimum in the energy with six solvating water molecules [92]. It is, however, not formed spontaneously in the beam without the aid of the solvated sodium that acts as a catalyst. The next step is the attack of the Na^- on a water proton to form NaH. This is demonstrated pictorially in Fig. 3.13. To cope with the long time scales of the experiment the coordination number constraint is employed [93]. It gives a barrier of 14 kcal/mol that corresponds at the finite temperature of 100 K to a reaction time in the order of $1 \mu s$, in agreement with experiment.

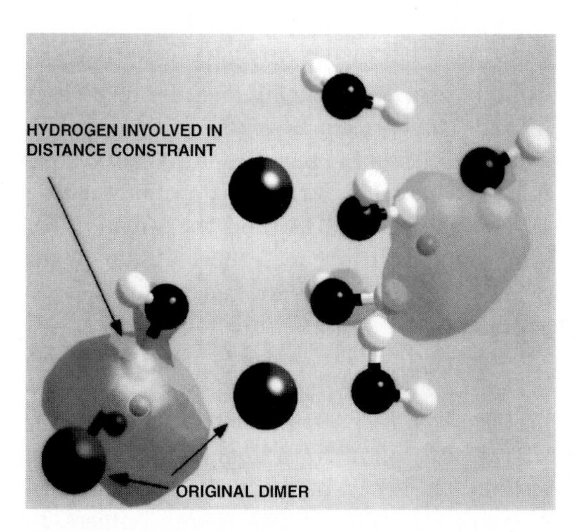

Fig. 3.13. Snapshot of the system $Na_3(H_2O)_6$ during the formation of NaH. The electrons close to the Na in the lower left corner are spin paired thus forming the reactive intermediate Na^-.

Once NaH is formed, the reaction proceeds spontaneously to the well known products. In particular, the following steps are executed:

(1) Cleavage and solvation of the sodium dimer

$$Na_2 + Na(H_2O)_6 \rightarrow Na_3(H_2O)_6. \tag{3.8}$$

(2) Formation of sodium hydride and sodium hydroxide:

$$Na_3(H_2O)_6 \rightarrow (NaH)(NaOH)(H_2O)_5 + Na. \tag{3.9}$$

(3) Formation of the second sodium hydroxide and molecular hydrogen:

$$(NaH)(H_2O)_5 \rightarrow (NaOH)(H_2O)_4 + H_2. \tag{3.10}$$

The reaction intermediate is the dipolar anion Na^- and the necessary cleaving of the Na_2 bond is catalyzed by $Na(H_2O)_6$. The analogy of the catalyst is further emphasized by the fact that Na is consumed by the reaction, regenerated after the final step, and is thus available for additional formation of $(NaOH)_2$. This behavior is definitely confirmed in the experiment with a larger amount of sodium. It would also be helpful to investigate these intermediate products spectroscopically in order to get some information about their structure and their charge distribution.

3.4 Concluding remarks

The intramolecular OH stretching vibrations of small water clusters $(H_2O)_n$ from $n = 7$ to 10 have been investigated in molecular beams using infrared depletion spectroscopy. Size-specific information was obtained by dispersing the water clusters by a helium beam. The clusters in this size range have 3-dimensional cage structures that appear in two different versions. The clusters $n = 8$ to 10 show well ordered spectra consisting of three groups, originating from the free, the 3-coordinated double donor DDA, and the single donor DAA bonds which reflect their environment. In contrast, the spectrum of the heptamer exhibits seven bands which are spread over the range from 2935 to 3720 cm^{-1}. The correponding minimum energy structures for $n = 7, 9$, and 10 can be derived from the octamer cube by the removal or addition of one or two water molecules.

The new measurement of the OH stretch spectrum of $(H_2O)_9$ at higher temperature exhibits the ingredients of a solid-like to liquid-like transition. The simulations indicate that the spectral broadening is due to a superposition of thermally broadened spectra of a number of high energy isomers. In contrast to the two low energy structures which are derived from the compact octamer cubes, the high energy structures are more open and include fused 3-5 membered rings. The estimated temperature of the cluster in the liquid-like spectra and thus the upper limit for the melting transition is in the range between 165 K (simulation) and 186 K (experiment). It is lower than that calculated for the more symmetric octamers and much lower than that of the bulk.

In all cases investigated, the structures determined for the low temperature solids are not observed for the small clusters. Therefore we started to conduct measurements in the size range from $n = 40$ to 650. We combined the reliable size determination of doping the water clusters with one Na atom and detecting them fragmentation-free by single photionization with IR-fragment spectrocopy. Thus we measured the OH stretch spectra for known size distributions. The results exhibit, aside from the free and DDA bands, mainly peaks in the range from 3420 to 3350 cm^{-1} that are attributed to amorphous behavior with increasing size and decreasing influence of 3-coordinated molecules. Apparently the fragments that correspond to neutral hexamers are in particular a sensitive indicator of amorphous structures. The size distribution peaks at $n = 65$ and goes to smaller values near $n = 25$ and $n = 300$.

The well known reaction of sodium with water was investigated in molecular beams. While single collision experiments at the time scale of several ps with sodium and water clusters up to $n = 40$ did not give the desired NaOH products, an experimental arrangement that allowed us to include multiple collisions and to extend the interaction time to μs gave products of the type $Na(NaOH)_{2,4}(H_2O)_m$ with H_2 as second product. The minimum starting configuration turned out to be a sodium dimer Na_2 and a solvated sodium atom $Na(H_2O)_6$. Accompanying calculations using Car-Parrinello molecular dynamics simulations revealed that the reactive intermediate is the sodium dipolar anion Na^- that reacts further with one of the H atoms of water to form NaH, and, subsequently, the products noted above result. The solvated Na atom acts as catalyst for producing the dipolar anion and the solvated Na^+ and is regenerated after the final products are formed.

Acknowledgement: The authors acknowledge with gratitude the many contributions of their former coworkers on the experimental part of the work presented here, Dr. I. Ettischer, Dr. J. Brudermann, M. Melzer, Dr. S. Schütte, and Dr. C. Bobbert. They are thankful for the input from very fruitful cooperations with Prof. V. Buch and Prof. J. Sadlej on the interpretation of the data, and the helpful support from calculations of the group of Prof. M. Parrinello. This work has been supported by the Deutsche Forschungsgemeinschaft.

References

1. in *Water: A comprehensive Treatise*, ed. by F. Franks (Plenum, New York, 1972).
2. E. M. Mas, K. Szalewicz: J. Chem. Phys. **104**, 7606 (1996).
3. S. S. Xantheas, T. H. Dunning: in *Advances in Molecular Vibrations and Collision Dynamics*, ed. by J.M. Bowman, Z. Bačič (JAI Press, Stamford, 1998), p. 281.
4. M. Kappes, S. Leutwyler: in *Atomic and Molecular Beam Methods*, ed. by G. Scoles (Oxford University Press, New York, 1988), p. 380.
5. H. Haberland: Surf. Sci. **156**, 305 (1985).
6. U. Buck: J. Phys. Chem. **92**, 1023 (1988).
7. U. Buck: in *The Chemical Physics of Atomic and Molecular Clusters*, ed. by G. Scoles (North-Holland, Amsterdam, 1990), p. 543.
8. U. Buck, M. Winter: Z. Phys. D **31**, 291 (1994).
9. U. Buck, H. Meyer: Phys. Rev. Lett. **52**, 109 (1984).
10. U. Buck, H. Meyer: J. Chem. Phys. **84**, 4854 (1986).
11. T. E. Gough, R. E. Miller, G. Scoles: J. Chem. Phys. **69**, 1588 (1978).
12. M. F. Vernon, D. J. Krajnovich, H. S. Kwok, J. M. Lisy, J. R. Shen, Y. T. Lee: J. Chem. Phys. **77**, 47 (1982).
13. J. Sadlej, V. Buch, J. Kazimirski, U. Buck: J. Phys. Chem. A **103**, 4933 (1999).
14. R. H. Page, J. G. Frey, Y. R. Shen, Y. T. Lee: Chem. Phys. Lett. **106**, 373 (1984).
15. F. Huisken, M. Kaloudis, A. Kulcke: J. Chem. Phys. **104**, 17–25 (1996).
16. D. F. Coker, R. E. Miller, R. O. Watts: J. Chem. Phys. **82**, 3554 (1985).
17. Z. S. Huang, R. E. Miller: J. Chem. Phys. **91**, 6613 (1989).

18. F. Keutsch, R. Saykally: Proc. Nat. Acad. Sci.(USA) **98**, 10 533 (2001).
19. N. Pugliano, R. J. Saykally: J. Chem. Phys. **96**, 1832 (1992).
20. K. O. Börnsen, L. H. Lin, H. L. Selzle, E. W. Schlag: J. Chem. Phys. **90**, 1299 (1989).
21. S. Leutwyler: J. Chem. Phys. **90**, 489 (1990).
22. R. N. Pribble, T. S. Zwier: Science **265**, 75 (1994).
23. C. J. Gruenloh, J. R. Carney, C. A. Arrington, T. S. Zwier, S. Y. Fredericks, K. D. Jordan: Science **276**, 1678 (1997).
24. T. Watanabe, T. Ebata, S. Tanabe, N. Mikami: J. Chem. Phys. **105**, 408 (1996).
25. C. Jacoby, W. Roth, M. Schmitt, C. Janzen, K. Kleinermanns: J. Phys. Chem. **102**, 4471 (1998).
26. W. Roth, M. Schmitt, C. Jacoby, D. Spangenberg, C. Janzen, K. Kleinermanns: Chem. Phys. **239**, 1 (1998).
27. C. Janzen, D. Spangenberg, W. Roth, K. Kleinermanns: J. Chem. Phys. **110**, 9898 (1999).
28. H. D. Barth, K. Buchhold, S. Djafari, B. Reimann, U. Lommatzsch, B. Brutschy: Chem. Phys. **239**, 49 (1998).
29. J. Brudermann, U. Buck, V. Buch: J. Phys. Chem. **106**, 453 (2002).
30. R. Page, M. F. Vernon, Y. R. Shen, Y. T. Lee: Chem. Phys. Lett. **141**, 1 (1987).
31. J. Devlin, C. Joyce, V. Buch: J. Phys. Chem. A **104**, 1974 (2000).
32. L. Bewig, U. Buck, S. Rakowsky, M. Reymann, C. Steinbach: J. Phys. Chem. **102**, 1124 (1998).
33. U. Buck, C. Steinbach: J. Phys. Chem. A **102**, 7333 (1998).
34. C. J. Mundy, J. Hutter, M. Parrinello: J. Am. Chem. Soc. **122**, 4837 (2000).
35. U. Buck, F. Huisken: Chem. Rev. **100**, 3863 (2000).
36. U. Buck, F. Huisken: Chem. Rev. **101**, 205 (2001)
37. J. B. Paul, C. P. Collier, R. J. Saykally, J. J. Scherer, A. O. O'Keefe: J. Phys. Chem. **101**, 5211 (1997).
38. R. Knochenmuss, S. Leutwyler: J. Chem. Phys. **96**, 5233 (1992).
39. S. S. Xantheas, T. H. Dunning, Jr.: J. Chem. Phys. **99**, 8774 (1993).
40. H. M. Lee, S. B. Suh, J. Y. Lee, P. Trarakeshwar, K. S. Kim: J. Chem. Phys. **112**, 9759 (2000).
41. J. D. Cruzan, M. G. Brown, K. Liu, L. B. Braly, R. J. Saykally: J. Chem. Phys. **105**, 6634 (1996).
42. K. Liu, J. D. Cruzan, R. J. Saykally: Science **271**, 929 (1996).
43. K. Liu, M. G. Brown, C. Carter, R. J. Saykally, J. K. Gregory, D. C. Clary: Nature **381**, 501 (1996).
44. J. K. Gregory, D. C. Clary: J. Chem. Phys. **105**, 6626 (1996).
45. K. Nauta, R. E. Miller: Science **287**, 293 (2000).
46. U. Buck, I. Ettischer, M. Melzer, V. Buch, J. Sadlej: Phys. Rev. Lett. **80**, 2578 (1998).
47. J. Brudermann, M. Melzer, U. Buck, J. Kazimirski, J. Sadlej, V. Buch: J. Chem. Phys. **110**, 10 649 (1999).
48. F. Huisken, S. Mohammed-Pooran, O. Werhahn: Chem. Phys. **239**, 11 (1998).
49. V. Buch, J. P. Devlin: J. Chem. Phys. **110**, 3437 (1999).
50. U. Buck: Ber. Bunsenges. Phys. Chem. **96**, 1275 (1992).
51. F. Huisken: Adv. Chem. Phys. **81**, 63 (1992).
52. F. Huisken, A. Kulcke, D. Voelkel, C. Laush, J. M. Lisy: Appl. Phys. Lett. **62**, 805–807 (1993).

53. J. O. Jensen, P. N. Krishnan, L. A. Burke: Chem. Phys. Lett. **246**, 13 (1995).
54. B. Rowland, S. Kadagathur, J. P. Devlin, V. Buch, T. Feldmann, M. Wojcik: J. Chem. Phys. **102**, 8328 (1995).
55. K. Kim, K. D. Jordan, T. S. Zwier: J. Am. Chem. Soc. **116**, 11 568 (1994).
56. S. Baldelli, C. Schnitzer, D. J. Campbell, M. J. Shultz: J. Phys. Chem. B **103**, 2789 (1999).
57. J. C. Jiang, J. C. Chang, B. C. Wang, S. H. Lin, Y. T. Lee, H. C. Chang: Chem. Phys. Lett. **289**, 373 (1998).
58. S. C. Farantos, S. Kapetanikis, A. Vegiri: J. Phys. Chem. **97**, 1993 (1993).
59. D. J. Wales, I. Ohmine: J. Phys. Chem. **98**, 7245 (1993).
60. C. J. Tsai, K. D. Jordan: J. Chem. Phys. **99**, 6957 (1993) .
61. J. M. Pedulla, K. Jordan: Chem. Phys. **239**, 593 (1999).
62. A. V. Egorov, E. N. Brodskaya, A. Laaksonen: Mol. Phys. **100**, 941 (2002).
63. C. J. Gruenloh, J. R. Carney, F. C. Hagemeister, T. Zwier: J. Chem. Phys. **113**, 2290 (2000).
64. Q. Du, R. Superfine, E. Freyez, Y. R. Shen: Phys. Rev. Lett. **70**, 2313 (1993).
65. A. Motita, J. T. Hynes: J. Phys. Chem. B **106**, 673 (2002).
66. P. Andersson, C. Steinbach, U. Buck: unpublished results (2002).
67. C. Bobbert, S. Schütte, C. Steinbach, U. Buck: Eur. Phys. J. D **19**, 183 (2002).
68. J. Huang, L. S. Bartell: J. Phys. Chem. **99**, 3924 (1995).
69. O. F. Hagena: Surf. Sci. **106**, 101 (1981).
70. O. F. Hagena: Z. Phys. D **4**, 291 (1987).
71. G. Torchet, P. Schwartz, J. Farge, M. F. de Feraudy, B. Raoult: J. Chem. Phys. **79**, 6196 (1983).
72. G. D. Stein, J. Amstrong: J. Chem. Phys. **58**, 1999 (1973).
73. M. Ahmed, C. J. Apps, C. Hughes, J. C. Whitehead: J. Phys. Chem. **98**
74. A. A. Vostrikov, D. Y. Dubov: Z. Phys. D **20**, 429 (1991).
75. U. Buck: Adv. At. Mol. Opt. Phys. **35**, 121 (1995).
76. L. Goss, S. W. Sharpe, T. A. Blake, V. Vaida, J. W. Brault: J. Phys. Chem. A **103**, 103 (1999).
77. T. Telser, U. Schindewolf: J. Phys. Chem. **90**, 5378 (1986).
78. C. Gopinathan, E. J. Hart, K. H. Schmidt: J. Phys. Chem. **74**, 4169 (1970).
79. R. Düren, U. Lachschewitz, S. Milošević, H. J. Waldapfel: Chem. Phys. **140**, 199 (1990).
80. C. P. Schulz, R. Haugstätter, H. U. Tittes, I. V. Hertel: Phys. Rev. Lett. **57**, 1703 (1986).
81. C. P. Schulz, R. Haugstätter, H. U. Tittes, I. V. Hertel: Z. Phys. D **10**, 279 (1988).
82. R. Dhar, N.R. Kestner: Radiat. Phys. Chem. **32**, 355 (1988).
83. R. N. Barnett, U. Landman: Phys. Rev. Lett. **70**, 1775 (1993).
84. R. N. Barnett, C. Yannouleas, U. Landman: Z. Phys. D **26**, 119 (1993).
85. K. Hashimoto, K. Morokuma: J. Am. Chem. Soc. **116**, 11 436 (1994).
86. K. Hashimoto, K. Morokuma: Chem. Phys. Lett. **223**, 423 (1994).
87. E. C. Honea, M. L. Homer, R. L. Whetten: Phys. Rev. B **47**, 7480 (1993).
88. R. N. Barnett, U. Landman: J. Phys. Chem. **100**, 13 950 (1996).
89. Z. Deng, G. J. Martyna, M. L. Klein: J. Chem. Phys. **100**, 7590 (1994).
90. S. F. Boys: Rev. Mod. Phys. **32**, 296 (1960).
91. P. L. Silvistrelli, M. Mazarini, D. Vanderbilt, M. Parrinello: Solid State Commun. **7**, 107 (1998).
92. F. Mercuri, C. J. Mundy, M. Parrinello: J. Phys. Chem. A **105**, 8423 (2001).
93. M. Sprik: Faraday Discuss. **110**, 437 (1998).

4 Solvent Effects of Individual Water Molecules

Fu-Ming Tao

4.1 Introduction

Water as a solvent plays a broad range of roles in virtually every aspect of physical, chemical, and biological processes. The recognition of the solvent effect, however, has traditionally been limited to systems of condensed phase, usually aqueous solutions, where water molecules are in a vast majority and are collectively represented by the dielectric constant of the liquid water. The interactions of water with solute species are typically modeled as the latter is placed in a void of the continuum medium formed by the bulk water and characterized by the dielectric constant[1, 2]. Accordingly, the solvent effect of water has generally been understood as resulting from the macroscopic properties of a bulk system. For the past decade or so, however, the understanding of the solvent effect of water has rapidly advanced to the molecular level, thanks to the development of new theoretical and experimental methodologies [3–6], particularly to the widespread applications of quantum mechanical electronic structure methods [7-30].

What are the solvent effects of water at the molecular level? What can be assessed as the solvent effects at the molecular level and in a consistent and quantitative manner? Is a solvent effect mainly the effect of local interactions or must it be associated with a bulk system? If a solvent effect is a local phenomenon, then can the bulk properties be effectively represented by the properties of a limited and well-defined molecular cluster that contains the solute and a small number of solvent molecules? How small can such a molecular cluster be that still possesses the main characteristics of the bulk system? What roles do the individual solvent molecules assume in the small clusters and what effects do the solvent molecules have on the solute species in the clusters? How great are the solvent effects of a single water molecule? What are the corresponding relationships between the solvent effects of the bulk system and those of a single solvent molecule? Is it meaningful and practically useful to pursue the solvent effects of individual molecules? Are there any practical applications that may take advantage of this understanding of solvent effects at the molecular level?

With the fast development of computer resources, the direct calculation of molecular clusters consisting of solvent and solute molecules has become a practical effective approach for the study of solvent effects. Different from the continuum model, the molecular cluster approach treats the solvent molecules around the solute on the same footing as the solute. It allows the full characterization of molecular properties of a molecular cluster of arbitrary size and, therefore, is capable of answering all of the above questions. To take further advantage of this approach, one can consider a series of molecular clusters that contain a given solute species and an increasing number of solvent molecules. The solvent effects of individual solvent molecules can be examined and recognized from the incremental changes in molecular properties of the cluster series. This approach has been applied successfully to several prototypical systems that accommodate various chemical reactions, such as the proton transfers in acid-base systems [12–14, 16, 18, 20, 27] and the hydrolysis of oxides [15, 19, 21, 22]. The exciting results may provide insights into our fundamental understanding of the solvent effects of water.

This chapter introduces the quantum chemical calculation of molecular clusters as an effective and reliable approach for the study of the solvent effects of water. The approach will be shown particularly useful in determining the effects of individual water molecules and providing a fundamental understanding of the solvent effect at the molecular level. The chapter is organized as follows. The next section (Section 2) outlines the appropriate construction of molecular clusters and the quantum mechanical methods available for the reliable calculation of the clusters. Section 3 discusses the calculated molecular properties of the clusters relevant to the solvent effects of water molecules. Section 4 highlights several recent applications.

4.2 Computational Methods

4.2.1 Construction of molecular clusters

Molecular clusters consist of the chemically active species (the solute) and the inactive water molecules (the solvent). They are constructed to simulate molecular clusters in the gas phase or the local environment around a solute species in the aqueous system intended for study. The molecules present in a cluster interact with one another, and the interactions among the molecules should well represent the short-range interactions in the aqueous state. Considerations for long-range interactions are limited with the clusters because the size of the allowed cluster is limited by computer resources. A large cluster not only demands excessive computer resources, but also presents too many competing geometries of nearly equal stabilities. Nevertheless, long-range effects can still be evaluated by use of a series of clusters with a successively increasing number of water molecules. The convergence behavior of calculated results from the cluster series may indicate how significant the long range effects are and whether the largest cluster used is adequate. It has been shown that the convergence is rather fast and short-range effects are dominant.

The use of small clusters inadvertently has many unique and important advantages. First of all, small clusters may represent actual or likely species in the gas phase as a result of interactions of the active gas molecules and water vapor. The clusters bridge the gap between the gas phase and a condensed phase. Secondly, small clusters can easily be calculated at a high level of quantum mechanical theory, yielding accurate and reliable results for the clusters. The specific roles of individual water molecules in the clusters can be fully examined from the calculated results, providing fundamental insights into the solvent effects. It is particularly true with the clusters that are in a systematic series with a successive number of water molecules, n = 0, 1, 2, 3, ... The first few clusters in such a series, with n = 0, 1, 2, 3, respectively, usually show pronounced effects of each water molecule and reveal the specific roles each water plays as solvent. This is because the small clusters have simple characteristics and the water molecules are in closest possible contact with the solute.

In many cases, a series of small clusters are able to show the critical transition for a solute species from the isolated state to completely solvated state. For example, the acid-base unit of NH_3-HCl in the cluster series NH_3-HCl-$(H_2O)_n$, n = 0 – 3, exists in one of two distinct states depending on the number of water molecules present [13,14,18]. For n = 0 and 1, the NH_3-HCl unit exists as a usual hydrogen-bonded system, similar to the structure of isolated NH_3-HCl determined from gas-phase experiments [31-34]. For n = 2 and 3, on the other hand, the NH_3-HCl unit becomes an ionic pair, $NH_4^+ \ldots Cl^-$, resulting from a proton transfer from HCl to NH_3, consistent with the acid-base reaction between HCl and NH_3 in aqueous solution.

The equilibrium geometry for a given cluster is that which maximizes the total attractions among the molecules within the cluster. In other words, only stable geometries are considered for the clusters. Water is a highly polar molecule and the attractions between any polar species and water are dominated by electrostatic interactions. Hydrogen bonding is the most important type of interactions expected in the clusters and, therefore, is a leading consideration in determining the stable structures for the clusters. Several stable structures may often be expected for a given cluster. The most stable one usually contains a network with the overall strongest hydrogen bonding. However, exceptions are common, especially in cases where multiple hydrogen bonds can be formed. Accordingly, all candidate structures should be considered, although only some of the stable structures may be reported (usually within 5 kcal mol^{-1} in total energy).

4.2.2 Quantum mechanical methods

A molecular cluster contains two or more closed-shell stable molecules. The structure and the stability of a cluster are determined by four types of intermolecular interactions: exchange repulsion, electrostatics, induction, and dispersion. A reliable theoretical method should accurately recover all the four types of interaction energies. The exchange repulsion and electrostatic energies can be calculated at a relatively lower level of theory, but the induction and dispersion energies require evaluation at a high level of theoretical treatment. Wide choices of quantum mechanical methods are available for the calculation of molecular clusters. The following briefly discusses some of the most commonly used methods.

Ab initio molecular orbital theory [35] and density-functional theory (DFT) [36] are two major branches of modern electronic structure theory based on quantum mechanics. Ab initio theory explicitly uses the electronic wave function, which is expanded by one or more antisymmetrized products of one-electron wave functions (molecular orbitals) with electron spin. Density functional theory, on the other hand, calculates the molecular electron probability density. Both of the theories attempt to treat all electrons in a given molecular system and are appropriate for the calculation of molecular clusters. Other theories without explicit treatment of all electrons, such as molecular mechanics and semiempirical quantum mechanics, are not reliable for the calculation of molecular clusters.

Either ab initio theory or DFT gives the molecular electronic energy and electron distribution that ultimately lead to other molecular properties. For example, the equilibrium geometry of a molecule can be determined by minimizing the electronic energy (including nuclear repulsions) with respect to the nuclear coordinates. The harmonic frequencies of nuclear vibrations at the equilibrium geometry, on the other hand, can be found from the second derivatives of the electronic energy with respect to the normal mode nuclear coordinates.

The Hartree-Fock method [37] is the simplest *ab initio* method that uses the single determinant (Hartree-Fock) wave function to approximate the many-electron wave function of a molecule. The method adequately treats the core and valence electrons of a closed-shell molecule at the ground state and fully recovers the exchange repulsion energy between any two molecules. It also recovers the bulk of electrostatic interaction energy based on the charge distributions calculated using the Hartree-Fock wave function. The Hartree-Fock method may thus produce reasonable results on molecular clusters at the stable configurations. However, the method treats electron-electron interactions as a mean field and neglects the instantaneous interactions or electron correlation. Electron correlation is responsible for the induction energy and the dispersion energy. Accordingly, the treatment of electron correlation beyond the Hartree-Fock level is required for the accurate and reliable calculation of molecular clusters.

The second-order Moller-Plesset perturbation approximation (MP2) [38, 39] is a simple and popular *ab initio* method beyond the Hartree-Fock level. The method gives the bulk of the correlation energy and is reliable for closed-shell molecules. The MP2 method is appropriate for molecular clusters of stable closed-shell species and is reasonably accurate in recovering the induction energy and the dispersion energy. Electron correlation effects from core electrons are not significant and, as a result, the MP2 method is usually used with frozen core approximation.

High level *ab initio* methods beyond MP2 are demanding in computer resources and not practical for calculating large clusters of multiple solvent molecules. However, higher-level methods may be necessary for radicals and transition state species. High-level methods may also be used for small clusters of stable molecules to verify the reliability of MP2 results. High-level methods that are widely available include third and fourth-order Moller-Plesset perturbation methods (MP3 and MP4, respectively), quadratic configuration interaction with single and double excitations (QCISD), and coupled cluster with single and double excitations (CCSD), along with the QCISD and CCSD variations including the perturbative triple excitations ((QCISD(T) and CCSD(T), respectively)[40–42].

Methods based on density functional theory include some electron correlation effects. DFT methods are highly efficient and reliable and have become increasingly popular. The methods are particularly appropriate for the calculation of molecular clusters containing multiple solvent molecules. Among the many variations, the DFT method using Becke's three-parameter functional [43–45]with the nonlocal correlation provided by Lee, Yang and Parr (B3LYP) [46] is the best established and most widely used. The B3LYP method requires moderate computer resources (less than for Hartree-Fock method) and yet offers to produce highly accurate results (equivalent to or better than MP2 results).

4.2.3 Basis sets

Basis sets are used to expand molecular orbitals in *ab initio* theory or electron probability densities in density function theory. A wide variety of basis sets are available for quantum mechanical calculations. The accuracy of a calculation at a given level of theory depends strongly on the size and quality of the basis set used. Larger basis sets usually give more accurate results, but are more demanding in computer resources. As a result, it is important to choose the efficient basis set for accurate, yet tractable calculations.

A basis set for the calculation of a molecular cluster should provide an adequate description of electron distributions within the individual molecules and in the intermolecular regions. Molecular clusters of mostly water molecules are dominated by electrostatic interactions, determined primarily by the electron distributions of the individual molecules in the clusters. Basis sets of the double-zeta quality extended with a small set of polarization and diffuse functions are expected to give reasonably accurate results. A typical example of such basis sets is 6-311++G(d,p), a basis set of Pople and coworkers [47–49]. Another example is the aug-cc-PVDZ basis set, an augmented correlation-consistent double-zeta basis set of Dunning and coworkers [50–52].

Although a median basis set, such as 6-311++G(d,p) or aug-cc-pVDZ, serves well for the reasonable calculation of molecular clusters, it is important to ensure that the results from a given basis set is adequately converged. Several strategies are available to examined the convergence of a result. If the clusters are small enough, additional calculations can be carried out on the same clusters with larger basis sets, such as aug-cc-PVTZ. On the other hand, if the clusters are too large, calculations using smaller basis sets, such as 6-31+G(d) [47–49], may be valuable. If the 6-31+G(d) and 6-311++G(d,p) basis sets provide essentially the same results, the results from either basis set are likely well converged. Otherwise, the results may not be well converged and a more extended basis set should be used. For small clusters, larger basis basis sets, such as aug-cc-pVTZ and aug-cc-pVQZ, may also be used to ensure the convergence of the results.

4.3 Molecular Properties Relevant to Solvent Effects

4.3.1 Potential energy surface

What are the specific molecular properties of the clusters that are closely associated with the solvent effect of water molecules at the molecular level? In principle, all the molecular properties of a cluster reflect or are related to the solvent effect of the water molecules present in the clusters. The potential energy surface, however, is the most important molecular property of the cluster that is fundamentally related to the solvent effects of the water molecules in the cluster. As will be shown, the potential energy surface responds accurately and systematically to the presence of any water molecules. A series of other molecular properties, such as molecular geometries, binding energies, and harmonic frequencies, originate from the potential energy surface and can also be used to study the effects of the solvent molecules.

Because of the large dimensions of the surface, it is not practical to calculate the full potential energy surface even for a small cluster that contains only one or two water molecules along with the solvent. Instead, the potential energy surface is calculated in significantly reduced dimensions, usually explicit in only one or two geometric parameters. Typically, the active geometric parameters are chosen to be the reaction coordinate or other key geometric parameters of the solute species in the cluster. The remaining geometrical parameters either are optimized to produce a minimum energy surface/curve or are frozen at certain preset values.

Freezing geometric parameters at preset values can be useful in examining a potential energy surface in its reduced dimensions. One situation is to constrain the structure of the solvent molecules in a cluster to model the solvent environment in a protein or other polymer. The solvent molecules can be locked in the framework of a large polymer and may not respond freely to the change of an adjacent solute. A similar situation applies to the solvent molecules in a solid state. Another situation is to deliberately place solvent molecules in a certain arrangement. One can use the different arrangements to study how the change in solvent structure affects the solute in the cluster. One can also use this approach to model a given environment of a bulk solvent.

4.3.2 Equilibrium geometry

An equilibrium geometry of a cluster represents a minimum point on the potential energy surface of the cluster. It shows the structural relationships among the species in the cluster. It not only provides the structural information about the solute in the cluster but also reveals how the solute and solvent molecules interact with each other.

The effects of the solvent molecules can reliably be shown on the bond lengths and angles of the solute. The lengthening of a bond indicates the weakening of the bond by the solvent molecules. Such an effect may eventually lead to the dissociation of the bond induced by the solvent. The progress of the event may be demonstrated by the equilibrium geometries for a cluster series with an increasing number of solvent molecules.

4.3.3 Binding energy

The stability of a cluster can be measured in its relative energy such as the binding energy. The binding energy of a cluster, D_0, can be defined relative to the energies of the molecules in the monomeric form at infinite separations. In this definition, the binding energy is directly related to the total interactions of the cluster, ΔE, given as the energy difference between the cluster and all component molecules at infinite separations, i.e.

$$\Delta E = -D_0 = E_{cluster} - \Sigma E_{molecule}. \tag{4.1}$$

It should be pointed out that the molecular energies must include the zero-point energies (ZPE), which are usually calculated from the harmonic frequencies for the equilibrium geometries.

The binding energy of a cluster reliably indicates the stability of the cluster and the strength of the overall interactions among the molecules. Favorable hydrogen bonding and other strong interactions within the cluster are supported by a large value of D_0. Conversely, unfavorable interactions are reflected by a relatively small value of D_0.

Another useful quantity associated with the binding energy is the incremental change in D_0 resulting from the addition of a single water molecule to a cluster. Such a quantity is directly related to the specific effect of a given water molecule.

4.3.4 Harmonic vibrational frequencies and IR intensities

Harmonic frequencies may be highly valuable in revealing the bonding characteristics of a molecular system. In particular, the strength of a specific bond is strongly correlated to the harmonic frequencies of the normal modes associated with the bond. The effect of solvent molecules on the bond strength is reliably reflected by the frequency shifts of the associated normal modes. A red shift in frequency indicates a weakening effect of the bond from the solvent while a blue shift means a strengthening effect of the bond from the solvent.

While the frequencies may shift successively with the number of solvent molecules introduced in the cluster, some vibrational modes may disappear along with the emergence of new vibrational modes as the cluster increases to a certain size. Such an occasion is indicative of the formation of new species in the cluster.

The IR intensities are associated with the changes in bond polarity in the normal mode vibrations, and are useful in recognizing the bonding characteristics of a molecular system. A large IR intensity is usually associated with the presence of a strongly polarized bond. An enhancement in the IR intensity due to the introduction of a solvent molecule indicates a shift of electric charge distribution, producing further polarization of the bond.

Clearly, the frequency calculation provides a major connection between theory and experiment. Harmonic vibrational frequencies and IR intensities represent a large set of characteristic properties of a cluster directly observable in laboratory experiments using spectroscopic and other techniques. The calculation of these quantities may serve as an important guide for experimental studies.

4.3.5 Other properties

Atomic charge distributions of the solute in a cluster are sensitive to the presence of solvent molecules. Accordingly, Mulliken population analysis can be used to show the solvent effects. It should be pointed out that in any scheme atomic charges are arbitrarily assigned and should be used in a relative sense.

The molecular dipole moment of a cluster is a more reliable measure of the overall polarity of the cluster. Such a quantity may reveal a significant change in the polarity of the solute. For a cluster containing polar solvent molecules, however, the overall dipole of the cluster is strongly affected by the orientations of the solvent molecules. As a result, the determination of the solute polarity in a cluster is often complicated by changes in the orientations of the solvent molecules.

4.4 Applications

4.4.1 Acid-base reactions

Acid-base reactions are a class of the most general chemical reactions and play a critical role in a wide range of chemical and biological processes. An acid-base reaction can simply be described as the transfer of a proton (H^+) from the acid molecule (HA) to the base (B). It involves the break of an old bond (H-A) and the formation of a new bond (H-B), along with the shift of the charge carried by the proton between the two reacting molecules. A simple prototype of such a reaction is the reaction between hydrogen chloride (HCl) and ammonia (NH_3) to form ammonium chloride ($NH_4^+ \ldots Cl^-$).

It is widely known that the reaction of HCl and NH_3 is instantaneous in aqueous solution, forming the solvated NH_4^+ and Cl^- ions. However, microwave spectra of mixtures of the pure HCl and NH_3 gases strongly indicate that the union of HCl and NH_3 simply results in the traditional hydrogen-bonded system of NH_3-HCl, with HCl as the hydrogen bond donor and NH_3 as the acceptor. Clearly, water plays a critical role in the reaction of HCl and NH_3 and the formation of stabilized NH_4^+ and Cl^- ions, as shown from the reaction in aqueous solutions. However, it is unclear about the exact roles of individual water molecules and whether the reaction would take place in the gas phase in the presence of water vapor. More specifically, what is the minimum number of water molecules required to covert NH_3-HCl into stabilized $NH_4^+ \ldots Cl^-$?

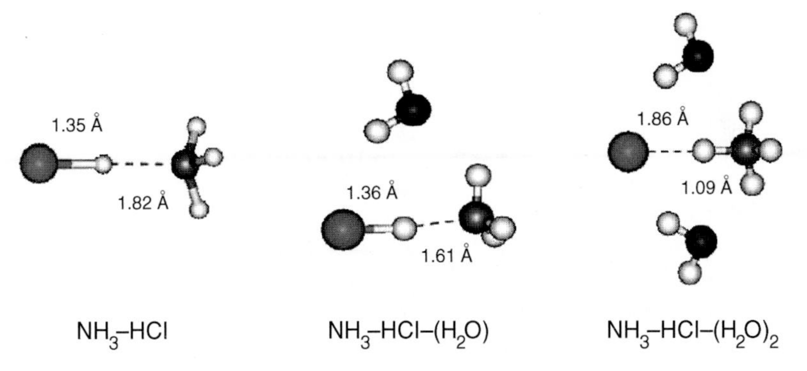

Fig. 4.1. The equilibrium geometries of NH_3-HCl-$(H_2O)_n$ (n = 0, 1, 2). Bond distances are in Angstrom (Å).

The answers to these and many other questions were recently pursued using MP2/6-311++G(d,p) calculations of the clusters NH_3-HCl-$(H_2O)_n$, n = 0, 1, 2, 3 [13,14,18]. The equilibrium geometries of the first three clusters, shown in Fig. 4.1, indicate a major geometrical transition for the NH_3-HCl unit. For the n = 0 and 1 clusters, the NH_3-HCl unit is hydrogen-bonded, consistent with the gas-phase microwave spectroscopic experiments, but for the n = 2 and 3 clusters, the NH_3-HCl unit is converted into $NH_4^+ \ldots Cl^-$, similar to the solvated NH_4^+ and Cl^- ions. It is surprising that only two water molecules are required to promote the proton transfer and stabilize the NH_4^+ and Cl^- ions, producing much of the solvent effect of bulk water. The four clusters dramatically show the transition of NH_3-HCl from the gas phase to the aqueous state.

The structures of the NH_3-HCl-$(H_2O)_n$ clusters also reveal in detail the effects of H_2O as a solvent molecule. Apparently, each H_2O is positioned and oriented to maximize the (attractive) electrostatic interactions within the clusters. The H_2O molecules bridge between HCl and NH_3 and create an electric field to promote the transfer of a proton from HCl to NH_3. The NH_4^+ and Cl^- ions resulting from the proton transfer, on the other hand, further enhance the electrostatic interactions with the H_2O molecules. Clearly, only a few H_2O molecules are needed to produce the key event of the acid-base reaction.

One might wonder about the abrupt, dramatic effect from the second H_2O molecule as introduced in the n = 2 cluster, when contrasted with the effect from the first or third H_2O. The second H_2O seems to play a unique and more important role compared to the others. It turns out to be not true if other properties of the clusters are examined. Each H_2O molecule in the clusters contributes the same differential effect in stabilization as the neutral NH_3-HCl is dissociated into the corresponding ion pair. The potential energy curves of the clusters along the proton-transfer pathway of NH_3-HCl

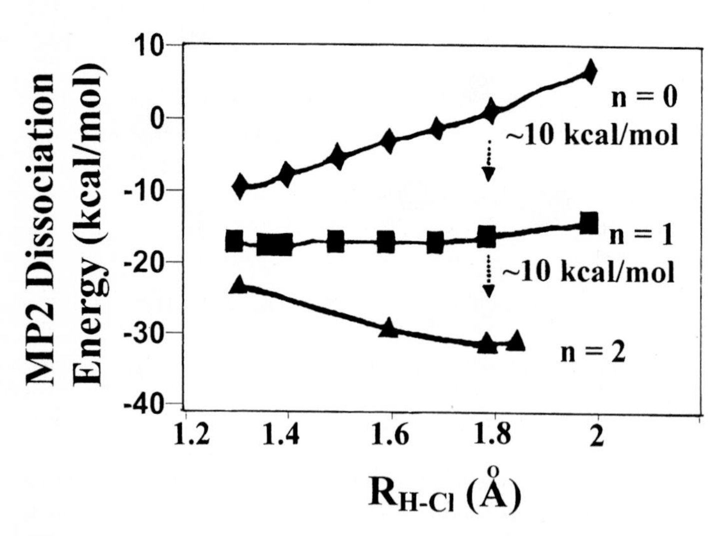

Fig. 4.2. Potential energy curves of NH_3-HCl-$(H_2O)_n$ (n = 0, 1, 2) along the R(H-Cl) coordinate.

are shown in Fig. 4.2. It can be seen that the curve for the NH_3-HCl system ($n = 0$) has only one minimum corresponding to the usual hydrogen-bonded geometry. The curve has a positive slope and is rather steep, indicating that the energy of NH_3-HCl increases steadily and monotonically as the proton from HCl is forced to transfer to NH_3. In contrast, the curve for the $n = 1$ cluster is virtually flat, indicating that the proton is no longer strongly bonded and can move freely. The curve for the $n = 2$ cluster, on the other hand, appears to follow the same trend of change in the slope, pushing the slope from positive to negative in sign. As a result, the proton from HCl is spontaneously transferred to NH_3, forming $NH_4^+ \ldots Cl^-$ as the stable product. The curve for the $n = 3$ cluster continues the trend of change in the slope, further stabilizing the ion pair product. It is therefore clear that each H_2O molecule provides additional stabilization energy in favor of the $NH_4^+ \ldots Cl^-$ ion pair. It is interesting to note that such an effect is nearly equal for each of the H_2O molecules. The first H_2O produces the same energetic effect on NH_3-HCl as the second H_2O, despite the fact that the former has no dramatic geometric effect.

The large, constant effect of each of the H_2O molecules on the NH_3-HCl unit is further demonstrated by the harmonic frequencies and IR intensities. The most relevant mode of vibration is the stretching of HCl along the proton-transfer pathway. The calculated harmonic frequency for the HCl monomer is 3091 cm^{-1} with an IR intensity of 35 km mol^{-1}. The calculated harmonic frequencies for the $n = 0$ and 1 clusters are 2528 cm^{-1} and 1901 cm^{-1}, respectively, with the corresponding IR intensities of 1572 km mol^{-1} and 2888 km mol^{-1}. It is seen that the introduction of each H_2O results in a large redshift in the HCl stretching frequency, along with an enhancement in the IR intensity. The large redshift indicates a significant weakening in the HCl covalent bond while the enhancement in the IR intensity means an enhancement in the polarity of the HCl bond. It is interesting to note again that the magnitude of redshift is nearly equal for each of the first and second H_2O molecules.

The HCl stretching mode vanishes in the $n = 2$ and 3 clusters as the covalent bond is completely dissociated within the clusters. New vibrational modes characteristic of the NH_4^+ ion emerge in these clusters.

Similar studies [14, 18] have been carried out for the other hydrogen halides, HF and HBr, in place of HCl. Hydrogen bromide is a stronger acid than HCl while hydrogen fluoride is a weaker acid than HCl. It is reasonable to expect that the stronger is an acid, the easier would be the transfer of a proton from the acid. This means that NH_3-HBr requires fewer water molecules to convert itself into $NH_4^+ \ldots Br^-$ while NH_3-HF requires more water molecules to convert itself into $NH_4^+ \ldots F^-$. Such an expectation turns out to be exactly true from the calculations of the water clusters containing these species.

Just as for NH_3-HCl, the equilibrium geometries for the pure NH_3-HF and NH_3-HBr systems are both hydrogen-bonded, despite the high acidity of HBr. However, NH_3-HBr is converted into $NH_4^+ \ldots Br^-$ in the presence of only one H_2O molecule. On the other hand, NH_3-HF remains hydrogen-bonded until three H_2O molecules are present. Even in the n = 3 cluster, the proton from HF is not completely transferred to NH_3.

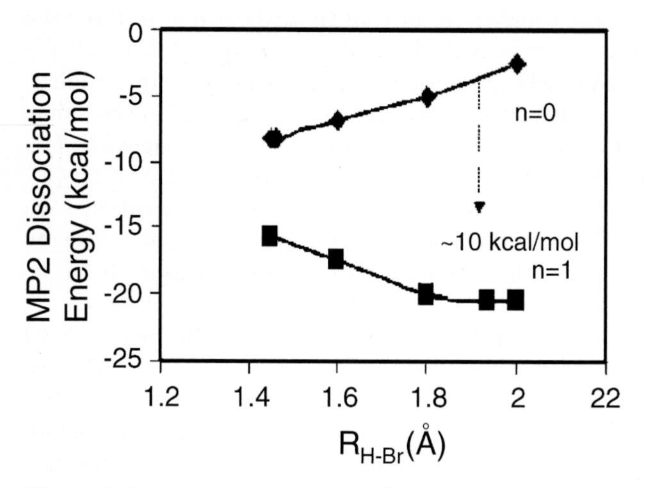

Fig. 4.3. Potential energy curves of NH_3-HBr-$(H_2O)_n$ (n = 0, 1) along the R(H-Br) coordinate.

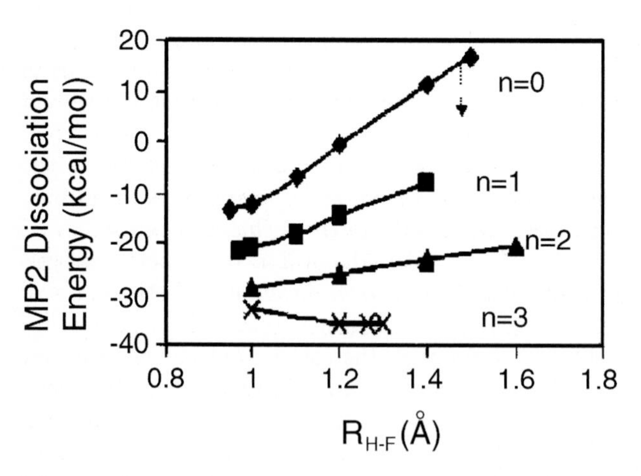

Fig. 4.4. Potential energy curves of NH_3-HF-$(H_2O)_n$ (n=0,1,2,3) along the R(H-F) coordinate.

The potential energy curves for the NH_3-HBr and NH_3-HF clusters with water, shown in Fig. 4.3 and Fig. 4.4 respectively, are also revealing of the differences among the three different halides. They also affirm the roles of the H_2O molecules present in the clusters. The curve for the pure NH_3-HBr (n = 0) system is much flatter than that for the pure NH_3-HF (n = 0) system. This means that fewer H_2O molecules are required to reverse the sign of the curve slope (to become negative) for the HBr systems than for the HF systems. As it turns out, one H_2O molecule is enough in the case of HBr but at least three H_2O molecules are required in the case of HF.

From the potential energy curves for the clusters of the three halides, one might notice that the magnitude of change in the curve slope resulting from each H_2O molecule is approximately constant among all of the clusters considered. The slopes of the curves may be quantitized in the following approach. By treating each curve to be linear, the slope of a given curve is determined as the incremental change in energy (vertical axis) divided by the corresponding span in the horizontal coordinate (the distance bond H-X). For example, the slopes for the n = 0 clusters are found to be 59, 21, and 9 kcal mol^{-1} $Å^{-1}$, for NH_3-HF, NH_3-HCl, and NH_3-HBr, respectively. The change in the slope resulting from each H_2O molecule is found to be approximately 20 kcal mol^{-1} $Å^{-1}$ for all of the clusters considered. This suggests that the change in the slope may be regarded as the fundamental effect from an individual H_2O molecule. Using such a quantity, one may also predict the number of H_2O required to promote the proton transfer in a similar system based on the slope determined for the pure acid-base complex.

4.4.2 Hydrolysis reactions

Sulfur trioxide (SO_3) may react with water to form sulfuric acid (H_2SO_4). Such a reaction is not only a prototype of hydrolysis reactions but also has a practical significance in atmospheric chemistry. Sulfuric acid is a major component of some atmospheric aerosols and plays important roles in atmospheric chemistry. The reaction takes place spontaneously with liquid water. It is unclear whether or not this reaction takes place in water vapor. If it does, what would be the reaction mechanism and how many water molecules would be involved? The one-step 1:1 addition of H_2O to SO_3 was considered but was ruled out. A mechanism involving SO_3 and two H_2O molecules was proposed and supported by a reduced energy barrier from ab initio calculations [53].

The reaction $H_2O + SO_3 \rightarrow H_2SO_4$ requires the formation of a new covalent S–O bond between H_2O and SO_3 and transfer of a proton (H^+) from H_2O to SO_3 to form a second O–H group of H_2SO_4. A high energy barrier is expected for the elementary bimolecular mechanism because there is a long pathway for the proton in the course of reaction. Additional H_2O molecules would be effective in stabilizing the transient proton and reducing the reaction energy barrier. These additional H_2O molecules can naturally be regarded as playing the role of solvent. It is reasonable to expect that the barrier would be reduced progressively as the number of H_2O molecules involved in the reaction increases. One may even expect the barrier to disappear completely if enough H_2O molecules are involved. All these expectations are confirmed by B3LYP and MP2 calculations of the cluster series $SO_3(H_2O)_n$, n = 1-4 [22].

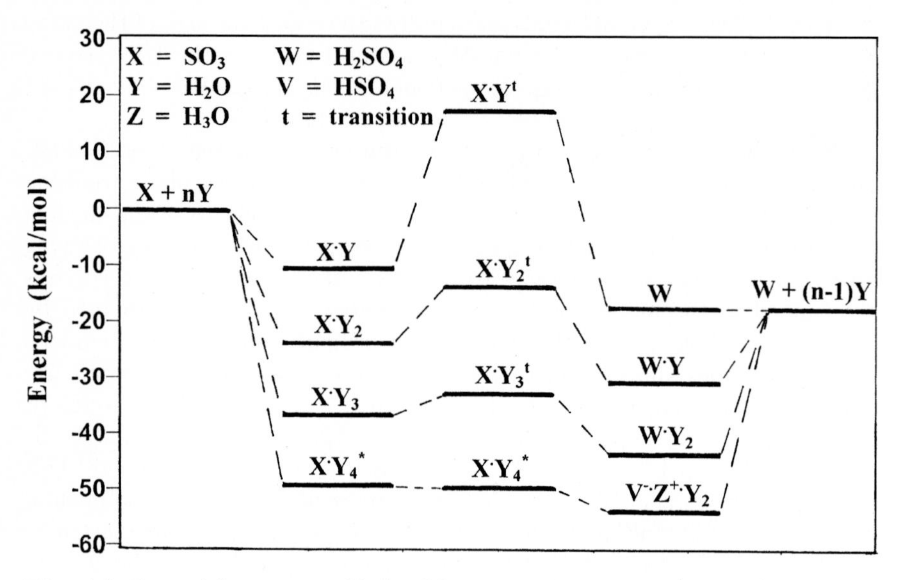

Fig. 4.5. Potential energy profile for SO_3 + n H_2O reactions (n = 1-4).

Figure 4.5 shows the relative energies of $SO_3(H_2O)_n$ at different stationary structures in the course of reactions, including the transition states and separate reactants and products. The energy barriers calculated at the MP2/6-311++G(d,p) level are 31.4, 14.0, and 6.1 kcal mol^{-1} for the number of H_2O molecules n = 1, 2, and 3, respectively, and the barrier disappears for n = 4. Very similar results are obtained from B3LYP calculations with the basis sets 6-31+G(d) and 6-311++G(d,p). It is clear that the energy barrier decreases progressively with the number of H_2O molecules present in the cluster. It is interesting to note that the barrier vanishes completely with only four H_2O molecules. No equilibrium structure exists for the reactant cluster $SO_3(H_2O)_4$ which, by geometry optimization, leads directly to the product cluster, consisting of $H_3O^+ \ldots HSO_4^-$ solvated by two remaining H_2O molecules.

The hydrolysis of dinitrogen pentoxide (N_2O_5) is a reaction similar to the hydrolysis of SO_3. The reaction leads to nitric acid, HNO_3. Like sulfur oxides and sulfuric acid, nitrogen oxides and nitric acid are major players in urban atmospheric chemistry, and similar questions may be asked, such as whether or not this reaction takes place in the water vapor and how many water molecules must be involved. The study was also similarly carried out, using the cluster series $N_2O_5(H_2O)_n$, n = 1-4 [15,19].

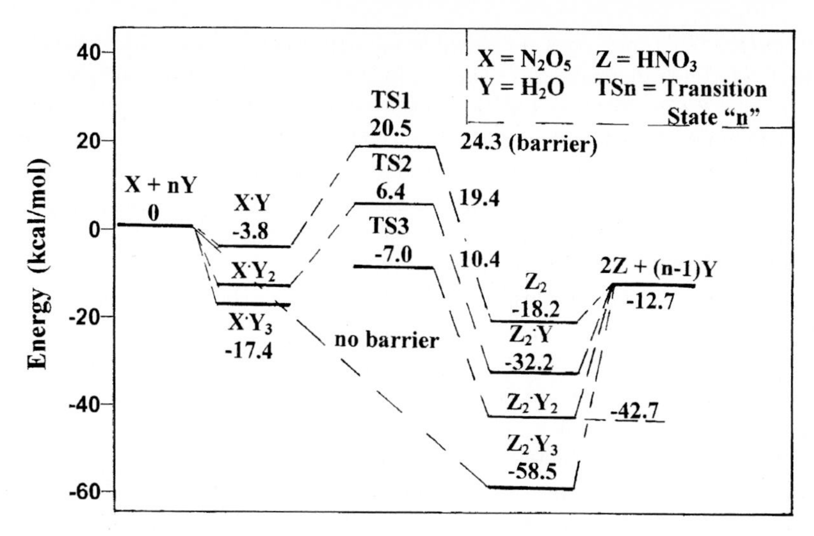

Fig. 4.6. Potential energy profile for N_2O_5 + n H_2O reactions (n = 1-4). See Ref. [19] for zero-point corrections.

Figure 4.6 shows the relative energies of $N_2O_5(H_2O)_n$ at different stationary structures in the course of reactions, including the transition states and separate reactants and products. The energy barriers calculated at the B3LYP/6-311++G(d,p) level are 24.3, 20.1, and 11.6 kcal mol^{-1} for the number of H_2O molecules n = 1, 2, and 3, respectively, and the barrier disappears for n = 4. It is unclear that the results are very similar to those for the $SO_3(H_2O)_n$ clusters.

One common feature shared in the course of the reactions for the two systems, $SO_3(H_2O)_n$ and $N_2O_5(H_2O)_n$, is that these reactions involve an unstable ion pair as the transition state. For the SO_3 + H_2O reaction, the oxygen of H_2O approaches the S atom in the direction perpendicular to the SO_3 molecular plane. A proton is forced to leave from H_2O and is transferred to an O atom of SO_3, creating a transition state similar to $H^+ \ldots HSO_4^-$. For the N_2O_5 + H_2O reaction, the N_2O_5 begins to dissociate in the direction of forming $NO_2^+ \ldots NO_3^-$ while the oxygen of H_2O approaches the N atom of NO_2^+. This leads to a transition state similar to a protonated nitric acid plus nitrate, $HNO_3H^+ \ldots NO_3^-$. The proton then couples with NO_3^- to form a second HNO_3 molecule. It should be pointed out that the evidence of the ionic species at the transition state was provided by Mulliken population analysis. The ion pairs are unstable in vacuum but can be stabilized by a polar solvent. As a result, the ion pair transition state for the SO_3 + H_2O or N_2O_5 + H_2O reaction can be stabilized by the additional H_2O molecules, and the energy barrier decreases with the number of H_2O molecules involved.

It appears that reactions are strongly affected by polar solvents if the reacting species change polarities during the course of the reaction. If a reaction of neutral reactants does not involve an ionic pair, or an enhancement in polarity, at the transition state, the reaction may not be influenced significantly by water as a solvent. This is supported by a theoretical exploration of the possible reaction of nitrogen dioxide (NO_2) with water to form nitrous acid (HONO). Because of its critical implications in atmospheric chemistry, the reaction has long attracted the attention of atmospheric chemists. Nevertheless, it is still unclear whether NO_2 reacts with water (liquid or vapor) to produce HONO. The reaction was studied by B3LYP calculations on two series of clusters [21]. One series, NO_2 $(H_2O)_n$, follows the mechanism that NO_2 reacts first with a given number of H_2O molecules to form an intermediate system containing HONO and an OH radical, followed by the reaction of the OH radical with a second NO_2 to form HNO_3. The other series, N_2O_4 $(H_2O)_n$, considers an alternative mechanism that N_2O_4 forms first by dimerization of NO_2 and then reacts with H_2O to form HONO + HNO_3. These reactions involve the production of a neutral OH radical, rather than an ionic pair or a significant enhancement of bond polarity, during the course of reactions. As expected, the calculations confirm that neither reaction is affected by the presence of additional H_2O molecules. As shown in Fig. 4.7, the energy barriers remain very high ($\sim$30 kcal mol^{-1}) and are nearly constant for the different number of H_2O, n = 1 to 3. This result rules out the possibility of the production of HONO from the reaction of NO_2 with pure water (liquid or vapor) and is consistent with recent experiments indicating that the reaction is likely catalyzed by aerosols, particularly carbon-containing soot particles.

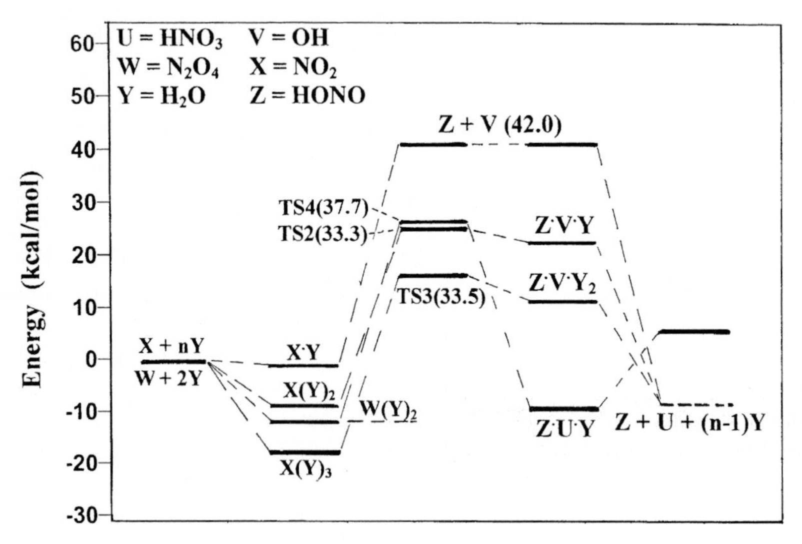

Fig. 4.7. Potential energy profile for NO_2 + n H_2O (n = 1-3) and N_2O_5 + 2 H_2O reactions.

4.4.3 Gas-to-solid transitions in $(NH_3\text{-}HX)_n$ ($X = F$, Cl, Br; $n = 1, 2, 4$)

As demonstrated in the previous cases, water as a solvent molecule essentially provides an electric field that influences a nearby solute species. This electric field may alternatively be provided by other polar molecules including a solute of the same kind. In other words, solute molecules may mutually act on each other as solvent molecules. This argument is demonstrated in a study of the gas to solid transition using the cluster series $(NH_3\text{-}HX)_n$ ($X = F$, Cl, Br; $n = 1, 2, 4$) [17,25].

Water as a solvent is required for the acid-base reaction between NH_3 and hydrogen halide to form the ion-pair product $NH_4^+ \ldots X^-$. As shown previously, a very small number of water molecules are enough to convert a given NH_3-HX unit into $NH_4^+ \ldots X^-$. It is reasonable to expect that another NH_3-HX unit might take the role of the water molecules and help convert the given NH_3-HX into $NH_4^+ \ldots X^-$. The effect is mutual and the NH_3-HX units would help one another convert into $NH_4^+ \ldots X^-$. Since $NH_4^+ \ldots X^-$ is

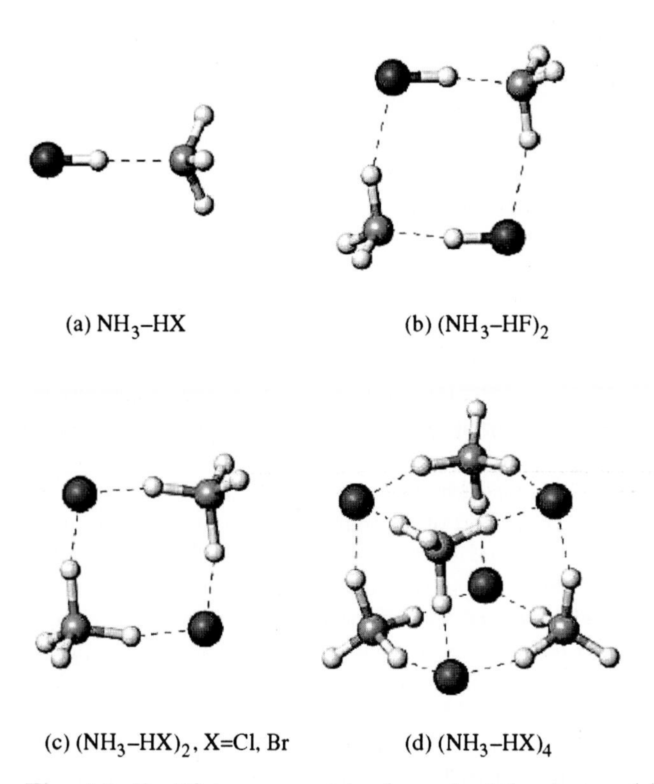

Fig. 4.8. Equilibrium geometries for each of the clusters (a) NH_3-HX, (b) $(NH_3\text{-}HF)_2$, (c) $(NH_3\text{-}HX)_2$ ($X = $ Cl, Br), and (d) $(NH_3\text{-}HX)_4$. Halogen atoms are in dark gray, nitrogen atoms are in light gray, and hydrogen atoms are in white.

much more polar than NH_3-HX, there would be a cooperative or domino effect to turn all NH_3-HX units into the $NH_4^+ \ldots X^-$ ion pairs. Such an event is precisely demonstrated by the cluster series $(NH_3\text{-}HX)_n$ (X = F, Cl, Br; n = 1, 2, 4).

The equilibrium geometries for the $(NH_3\text{-}HX)_n$ clusters are shown in Fig. 4.8. Quite surprisingly, the clusters of two NH_3-HX units, except for $(NH_3\text{-}HF)_2$, are already converted into the ion pairs. The clusters of four NH_3-HX units, including $(NH_3\text{-}HF)_4$, are composed completely of ion pairs. The potential energy curves of the ion pair clusters along the proton transfer pathway show that the neutral hydrogen bonded structure is not at a local minimum, but corresponds to the transition state between two equivalent forms of the stable ion pair structures. The presence of distinct NH_4^+ and X^- is supported by the calculated geometric parameters, binding energies, and harmonic frequencies of the clusters as well as the known experimental data for the crystals.

The system NH_3-HX may serve as an important model for homogenous nucleation processes of gas-phase species. The results for the $(NH_3\text{-}HX)_n$ clusters may provide insights to the mechanisms for gas-to-solid transitions and help with understanding a series of natural phenomena, such as the formation of polar stratospheric clouds and the generation of atmospheric aerosols.

4.5 Conclusion and Remarks

Quantum mechanical calculation of molecular clusters is introduced as a reliable methodology for studying the solvent effects of water molecules. In contrast to the various methods based on the continuum model, the method of molecular clusters focuses on the specific roles of individual water molecules present in the clusters. Molecular clusters are constructed to simulate the local environment around a solute in an aqueous solution. The stable clusters may also represent the likely cluster species present in water vapor mixtures. A series of molecular clusters are used to examine the expanding environment in the aqueous state or the transition of a homogenous gas phase system to the aqueous solution. The potential energy surface, particularly associated with the solute species, is identified as the fundamental molecular property that can be directly related to the solvent effects of water molecules. A series of other molecular properties, such as molecular geometry, binding energy, harmonic frequency, IR intensity, and charge distribution, can also be used to reveal the solvent effects. Some of these properties respond dramatically to the small number of water molecules in the cluster. For example, the onset of the proton transfer in NH_3-HCl with two water molecules shows dramatically the roles that the water molecules play in the acid-base neutralization reaction.

A number of chemical reactions are highlighted to show that the solvent effect of water is evident in small clusters containing only a few water molecules, and converges quickly to that of bulk water. Each of the first few water molecules, permitted close contact with the solute species, produces a great and nearly equal solvent effect. The solvent effects contributed from individual water molecules are additive and the overall effect is cumulative as shown by the potential energy surface and other calculated molecular properties.

The relatively large solvent effect of the single water or the first few water molecules has broad practical significance. It provides us unique evidence and a strong basis for considering solvent effects in molecular systems beyond traditional aqueous solutions. Homogeneous gas-phase reactions, such as the reaction of ammonia and nitric acid in the natural atmosphere, can be dramatically influenced by the presence of only one or two water molecules. On the other hand, a single water molecule, such as one isolated in a hydrophobic environment of a protein or DNA strand, may considerably alter the local characteristics of the biochemical species. As a result, the solvent effect derived for individual water molecules is fundamentally useful in understanding the physical and chemical properties of systems beyond the aqueous state.

Acknowledgements

The author is a Henry Dreyfus Teacher-Scholar (2001-2005) and gratefully acknowledges support from the Camille and Henry Dreyfus Foundation. He would also like to acknowledge support from the Research Corporation, and the Petroleum Research Fund, administrated by the American Chemical Society.

References

1. L. Onsager: J. Am. Chem. Soc. **58**, 1486 (1938).
2. J. G. Kirkwood: J. Chem. Phys. **2**, 351 (1934).
3. T. S. Zwier: Annu. Rev. Phys. Chem. **47**, 205-241 (1996).
4. D. L. Freeman, J. D. Doll: Annu. Rev. Phys. Chem. **47**, 43-80 (1996).
5. T. Schindler, C. Berg, G. N. Schatteburg, V. E. Bondybey: Chem. Phys. Lett. **229**, 57-64 (1994).
6. F. R. Tortonda, J. L. Pascual-Ahuir, E. Silla and I. Tuñón: Chem. Phys. Lett. **260**, 21-26 (1996).
7. R. Knochenmuss, S. Leutwyler: J. Chem. Phys. **91**, 1268 (1989).
8. D. Wei, D. R. Salahub: J. Chem. Phys. **101**, 7633 (1994).
9. W. Siebrand, M. Z. Zgierski, Z. K. Smedarchina, M. Vener, J. Kaneti:Chem. Phys. Lett. **266**, 47-52 (1997).
10. M. P. Hodges, A. J. Stone, and S. S. Xantheas: J. Phys. Chem. A **101**, 9163 (1997).

11. M. A. Vincent, I. J. Palmer, E. Akhmatskaya: J. Am. Chem. Soc., **120**, 3431 (1998).
12. F.-M. Tao: J. Chem. Phys. **108**, 193-202 (1998).
13. R. A. Cazar, A. J. Jamka, F. M. Tao: Chem. Phys. Lett. **287**, 549-552 (1998).
14. R. A. Cazar, A. J. Jamka, F. -M. Tao: J. Phys. Chem. A **102**, 5117-5123 (1998).
15. D. Hanway, F. -M. Tao: Chem. Phys. Lett. **285**, 459-466 (1998).
16. C. Conley, F. -M. Tao: Chem. Phys. Lett. **301**, 29-36 (1999).
17. F. -M. Tao: J. Chem. Phys. (Communication) **110**, 11121-11124 (1999).
18. J. A. Snyder, R. A. Cazar, A. J. Jamka, F. -M. Tao: J. Phys. Chem. *A* **103**, 7719-7724 (1999).
19. J. A. Snyder, D. Hanway, J. Mendez, A. J. Jamka, F. -M. Tao: J. Phys. Chem. A **103**, 9355-9358 (1999).
20. L. J. Larson, A. Largent, F. -M. Tao: J. Phys. Chem. A **103**, 6786-6792 (1999).
21. A. Chou, Z. Li, and F.-M. Tao: J. Phys. Chem. A **103**, 7848-7855 (1999).
22. L. J. Larson, M. Kuno, F.-M. Tao: J. Chem. Phys. **112**, 8830-8838 (2000).
23. J. E. M. Cabaleiro-Lago, J. M. Hermida-Ramon, A. Pena-Gallego, E. Martinez-Nunez, A. Fernandez-Ramos: J. Mol. Struct. (Theochem) **498**, 21-28 (2000).
24. E. Kassab, J. Langlet, E. Evleth, Y. Akacem: J. Mol. Struct. **531**, 267-282 (2000).
25. B. Cherng, F.-M. Tao: J. Chem. Phys. **114**, 1720-1726 (2001).
26. Y. Kurosaki: J. Phys. Chem. A **105**, 11080-11087 (2001).
27. K. H. Weber, F.-M. Tao: J. Phys. Chem. A **105**, 1208-1213 (2001).
28. L. J. Larson, F. -M. Tao: J. Phys. Chem. A **105**, 4344-4350 (2001).
29. K. B. Wiberg, D. J. Rush: J. Am. Chem. Soc. **123**, 2038-2046 (2001).
30. O. Lehtonen, J. Hartikainen, K. Rissanen, O. Ikkala, L. Pietila: J. Chem. Phys. **116**, 2417-2424 (2002).
31. A. C. Legon, C. A. Rego: J. Chem. Phys. **47**, 3837 (1967).
32. A. C. Legon, A. L. Wallwork, C. A. Rego: J. Chem Phys. **92**, 6397 (1990).
33. A. C. Legon, C. A. Rego: J. Chem. Phys. **99**, 1463 (1993).
34. A. C. Legon: Chem. Soc. Rev. **22**, 153 (1993).
35. W. J. Hehre, L. Radom, P. v. R. Schleyer, J. A. Pople, Ab initio molecular orbital theory, Wiley, (1986).
36. R. G. Parr, W. Yang, Density-functional theory of atoms and molecules, Oxford University Press,(1989).
37. C. C. J. Roothaan: Rev. Mod. Phys. **23**, 69 (1951).
38. C. Moller, M. S. Plesset: Phys. Rev. **46**, 6189 (1934).
39. R. Krishnan, M. J. Frisch, J. A. Pople: J. Chem. Phys. **72**, 4244 (1980).
40. J. Cizek: Adv. Chem. Phys. **14**, 35 (1969).
41. G. D. Purvis, R. J. Bartlett: J. Chem. Phys. **76**, 1910 (1982).
42. J. A. Pople, M. Head-Gordon, K. Raghavachari: J. Chem. Phys. **87**, 5968 (1987).
43. A. D. Becke : J. Chem. Phys. **96**, 2155 (1992).
44. A. D. Becke : J. Chem. Phys. **97**, 9193 (1992).
45. A. D. Becke : J. Chem. Phys. **98**, 5648 (1993).
46. C. Lee, W. Yang, R. G. Parr: Phys. Rev. B, **37**, 785 (1988).
47. R. Krishnan, J. S. Brinkley, R. Seeger, J. A. Pople: J. Chem. Phys. **72**, 650 (1980).
48. M. J. Frisch, J. A. Pople, J. S. Brinkley: J. Chem. Phys. **80**, 3265 (1984).

49. T. Clark, J. Chandrasekhar, G. W. Spitznagel, P. v. R. Schleyer: J. Comp. Chem. **4**, 294 (1983).
50. T.H. Dunning, Jr: J. Chem. Phys. **90**, 1007 (1989).
51. R.A. Kendall, T.H. Dunning, Jr., and R.J. Harrison: J. Chem. Phys. **96**, 6796 (1992).
52. D.E. Woon, T.H. Dunning, Jr: J. Chem. Phys. **98**, 1358 (1993).
53. K. Morokuma, C. Muguruma: J. Am. Chem. Soc. **116**, 10316 (1994).

5 Solvation Effects on the Properties and Reactivities of Ionic and Neutral Water Clusters

J.J. Gilligan and A.W. Castleman, Jr.

5.1 Introduction

Elucidating factors that influence differences in the behavior of matter in the gaseous compared to the condensed state is a subject of fundamental as well as practical importance with implications to most areas of chemistry ranging from biological to environmental science to numerous industrial chemical processes. Indeed, determining the influence which solvation has on the properties and reactivity of both neutral and ionic systems is one of the challenging problems in the field of chemical physics. Studies of isolated clusters in the gas phase are providing a wealth of new information that is serving to bridge an understanding of the contributions which solvation has on the course of reactions ranging from simple effects due to caging to complex ones which may involve alterations in the potential energy surfaces and/or the thermochemistry of the reaction processes. New insights at the molecular level are made possible through a large arsenal of techniques now available for studying clusters of wide ranging sizes, and in this chapter we endeavor to demonstrate the value of applying these techniques to explore the various roles which solvation plays.

With water undisputedly being the most ubiquitous and important solvent, it is natural that much research focuses on water's role in affecting the properties and the course of wide-ranging classes of reactions. Our group at Penn State has had interests in many aspects of the field ranging from basic studies of thermochemistry to reaction dynamics, and in applications to atmospheric science as well as to insights into fundamental mechanisms of biochemical significance. This chapter gives an overview of some of our findings that offer particular insight into aspects of the field of water cluster research.

5.2 Thermochemical properties and structure of water cluster ions

The study of cluster ion thermochemistry is a topic, which has been actively pursued for several decades and provides invaluable quantitative information on the energetics of interactions between ions and molecules. Employing techniques such as high pressure mass spectrometry, through a judicious selection of experimental conditions, it is possible to produce ions of most desired types, and to hydrate them under equilibrium conditions. Studies of the intensities of the acquired equilibrium distributions made over a range of temperatures, allows the requisite enthalpies and entropies of individual hydration steps to be determined. Such data have found wide applications in comparison with theoretical calculations, thereby gaining insight into the structure and bonding of hydration complexes, especially those existing in electrolyte solutions. Other applications include an accounting of mass spectral distributions measured in the atmosphere in problems related to communication, species detection, heterogeneous chemistry, and nucleation phenomena to name a few. Most importantly, the use of theoretical concepts of hydration, in combination with thermochemical data acquired for progressively larger clusters, enables a connection between the gas and condensed phase as described below.

5.2.1 Thermochemical experiments / measurements

2.1.1 Thermodynamics of Cluster Reactions. Cluster formation can be represented by a series of stepwise association reactions of the form represented by the general reaction:

$$IL_{n-1} + L \rightleftharpoons IL_n. \tag{5.1}$$

Collisions of energetically activated intermediates in the clustering sequence with a third-body are necessary for stabilization of the complexes. Taking the standard state to be 1 atm., and making the usual assumptions [1] concerning ideal gas behavior and the proportionality of the chemical activity of an ion cluster to its measured intensity, the equilibrium constant $\mathrm{K}_{n-1,n}$ for the nth clustering step is given by:

$$lnK_{n-1,n} = ln\frac{[\mathrm{IL}_n]}{[\mathrm{IL}_{n-1}][\mathrm{P}_L]} = -\frac{\Delta \mathrm{G}^o_{n-1,n}}{\mathrm{RT}} = -\frac{\Delta \mathrm{H}^o_{n-1,n}}{\mathrm{RT}} + \frac{\Delta \mathrm{S}^o_{n-1,n}}{\mathrm{R}}. \tag{5.2}$$

Here, $[\mathrm{IL}_{n-1,n}]$ and $[\mathrm{IL}_n]$ represent the respective measured ion intensities, P_L the pressure (atm.) of the clustering species L, $\Delta \mathrm{G}^o_{n-1,n}$, $\Delta \mathrm{H}^o_{n-1,n}$, and $\Delta \mathrm{S}^o_{n-1,n}$the standard Gibbs free energy, enthalpy, and entropy changes, respectively, R the gas-law constant, and T absolute temperature. By measuring the equilibrium constant $\mathrm{K}_{n-1,n}$ as a function of temperature, the enthalpy and entropy change for each sequential association reaction can be obtained from the slope and intercept of the van't Hoff plot ($ln\mathrm{K}_{n-1,n}$ versus 1/T, see Fig. 5.1).

Alternatively, ligand exchange reactions can also be used to obtain thermodynamic information [2, 3]. Experimental techniques that employ van't Hoff plots, which are often represented as straight lines over moderate temperature ranges, lead to enthalpy changes derived from slopes. In actual fact,

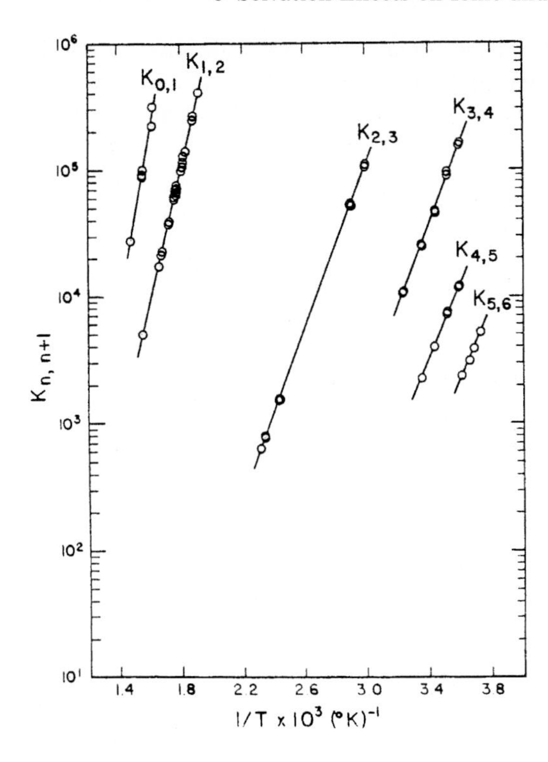

Fig. 5.1. van't Hoff plot for the reaction $Ag^+(H_2O)_n$ + H_2O.

the enthalpy change is a weak function of temperature due to the difference in heat capacity, ΔC_p between products and reactants [4]:

$$\Delta H_{T_2} = \Delta H_{T_1} + \int_{T1}^{T2} \Delta C_P(T)dT. \tag{5.3}$$

Correction to a zero temperature is necessary in order to obtain bond energies, but numerically these values are generally only slightly different from the measured enthalpies. Various experimental techniques measure and report related values: the enthalpy change ΔH_T^o of association, the bond dissociation energy D_0 (= -ΔH_0^o) or the potential well depth D_e (= $D_0 + \Sigma_i h\nu_i/2$), where ν_i denotes the frequencies of the vibrational modes related to the association bond.

High-pressure mass spectrometry (HPMS) [1,3,5] has been a valuable tool in quantitatively determining the thermochemical properties of ion clusters. In this technique, ions effuse from a high-pressure source, typically at a few Torr, through a small aperture into a mass filter where the equilibrium distribution of ion clusters is determined. The pressure of the ion source is maintained sufficiently high such that ions reside in a region of well-defined temperature for a time adequate to ensure the attainment of equilibria among the various ion cluster species of interest. The pressure and concentration of the clustering component must be low enough to avoid additional clustering by adiabatic expansion as the gas exits the sampling orifice.

Comparing the relative bond strengths for a variety of ligands about a given positive ion is very instructive in elucidating the role of the ligand. For this discussion, the enthalpy change $\Delta H^o_{0,1}$, is assumed to approximate the ion-molecule bond dissociation energy. This assumption is reasonable since ΔH^o is expected to be weakly dependent on the formation of the ion-molecule complex. See Table 5.1 for dissociation energies of some ionic water complexes.

Table 5.1. Dissociation Energies (kcal/mol) of Ion-Neutral Complexes[a]

Ion	H_2O
OH^-	25
O^-	<30
F-	23.3
O_2-	18.4
$O_2^-(H_2O)$	17.2
NO_2^-	15.2
Cl^-	14.9
NO_3^-	14.6
CO_3^-	14.1
SO_4^-	$\sim$ 12.5
HSO_4^-	11.9
SO_3^-	$\sim$ 13
I^-	11.1

a. Keesee, R. G.; Castleman, A. W., Jr.[2].

Entropy values find use in elucidating the structure of cluster ions through comparison with calculations. There are three major contributions to the entropy change for a clustering reaction. These are translational, rotational, and vibrational. In the case of ion-neutral association reactions, the translation contribution is dominant so the combining of two particles into one results in the overall negative sign in the entropy change. The magnitude of the translational contribution to the entropy change is generally – 35 to – 40 cal/K·mol. The entropy change for various reactions differs primarily due to the rotational and vibrational contributions. As clustering proceeds, the motions of the ligands are constricted due to crowding by neighboring ligands. As a result, internal rotational and bending frequencies become higher; but this is countered by a decrease in the stretching frequencies due to weaker bonding. Experimentally, the $\Delta S^o_{n-1,n}$ values are often observed to at first become more negative upon successive clustering. This is an indication that crowding is overcompensating for the effect of weaker bonding. Such trends are instructive regarding cluster structure. In other cases, a shift to smaller absolute entropy as well as enthalpy change values happens upon reaching some degree of clustering; this is attributed to the onset of formation of a second solvation layer around the ion where the newly bound molecule is attached to the solvent shell and not the ion itself. The first such molecule to occupy the periphery of the first solvation shell will have no restricted motions due to neighboring molecules.

2.1.2 ΔH^o correlation: evolution toward bulk solvation. The experimental data for gas-phase clustering onto ions do not extend to sufficiently large clusters where droplet formation can be supposed. One approach for relating the gas-phase data to solvation is to consider the differences between ions and compare these to the differences expected for solution [6, 7]. By considering the differences, the condensation contribution to the gas-phase data is cancelled, but the solvation term is still incomplete. For very large clusters, the gas-phase and solution data are expected to converge. Approximately 60% of the differences between single-ion heats of solvation for the halides are found to be accounted for by the first four gas-phase hydration steps [8]. Comparison of the difference between halide and alkali ions provided a consistency check on the results obtained with various methods from the solvation of salts [9].

A useful correlation is obtained by comparing ΔH^o_h (enthalpies of ion hydration) and $\Delta H^o_{0,n}$ (sum over the individual stepwise gas phase enthalpies of hydration) with size, n, for various ligands [10]. For a hydrate of very large size n corresponding to bulk liquid, the overall enthalpy change for a process starting from water monomers and an ion in the gas phase is estimated as $\Delta H^o_{0,n} \sim \Delta H^o_h + n\Delta H^o_v$ (enthalpy of condensation of water). Therefore, $\Delta H^o_h \,/\, \Delta H^o_{0,n}$ will asymptotically approach zero as n goes to infinity. This is due to the fact that the contribution from the heat of condensation of water molecules eventually overtakes that from the heat of hydration of the ion. The experimentally investigated size n is not large enough to reveal this convergence. At small cluster size, the hydration effect of an ion does reflect its value for both $\Delta H^o_{0,n}$ and $\Delta H^o_h \,/\, \Delta H^o_{0,n}$. $\Delta H^o_h \,/\, \Delta H^o_{0,n}$ converges rapidly despite the very different features for water binding to various ions. The correlation between ΔH^o_h and $\Delta H^o_{0,n}$ can be used to predict the heat of hydration once experimental $\Delta H^o_{0,n}$ is obtained by gas phase ion equilibria studies.

Another approach has been to consider the ratio $\Delta H^o_{solv} \,/\, \Delta H^o_{0,n}$ as a function of cluster size [11]. The ratio is calculated by the Born (liquid phase) and Thomson (cluster) equations, compared to experimental data for the hydration of positive ions. The ratio is seen to converge for a range of ionic radii for clusters of about five solvent molecules. See Fig. 5.2. Similar results are

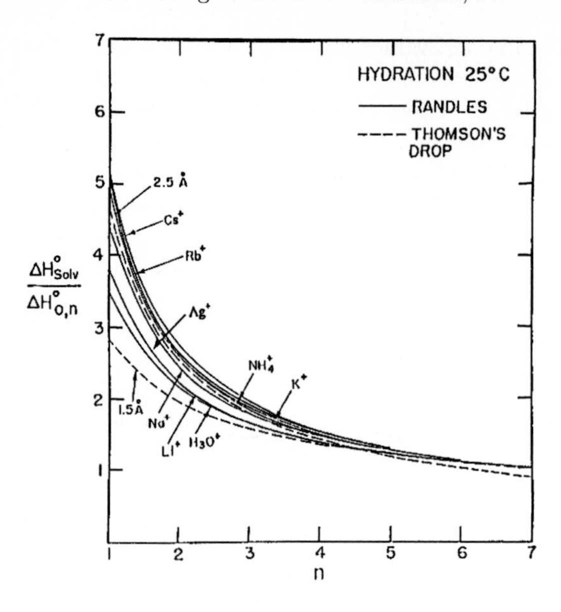

Fig. 5.2. Ratio of Randle's total enthalpy of solvation to the partial gas-phase enthalpy of hydration for positive ionic cluster size, n [17].

also obtained for the ammoniation of positive ions and hydration of negative ions. The findings suggest that valuable solvation data can be deduced from the gas-phase experiments.

2.1.3 Mixed solvents: non-ideal behavior. Another interesting aspect of cluster ions deals with binary solvents. A question arises as to the composition of the solvent in the neighborhood of the solute, and the extent to which it differs from the bulk solution. Stace and co-workers [12] have shown that cluster ions formed by ionizing mixed neutral clusters of alcohols and water dissociate with preferred loss channels of either alcohol or water, depending on the degree of aggregation. Using experimental techniques such as collision induced dissociation, metastable decay, and computational methods [13], Garvey and co-workers have observed stable mixed water-alcohol clusters whose stability shows a dependence on the degree of hydration.

Detailed insight often can be gained into the nature of such complexes by considering the total enthalpy of clustering ($-\Delta H^o_{fc}$) as a function of the composition of mixed cluster ions. As an example, Fig. 5.3 gives the heat for the mixed clustering of H_2O and SO_2 with Cl^-. The values for the mixed systems are higher than the compositionally weighted average value of the pure systems which suggests an enhancement in the overall clustering of the mixed system [14] due to the interactions between the two different ligands. By contrast, the exchange reactions involving mixed clusters of benzene and water bound to K^+ show a slight decrease or no deviation from the average value [15]. Plots of this type are valuable in assessing compositional effects on the bonding of mixed cluster systems, and on trends expected for complexes of liquid phase interest.

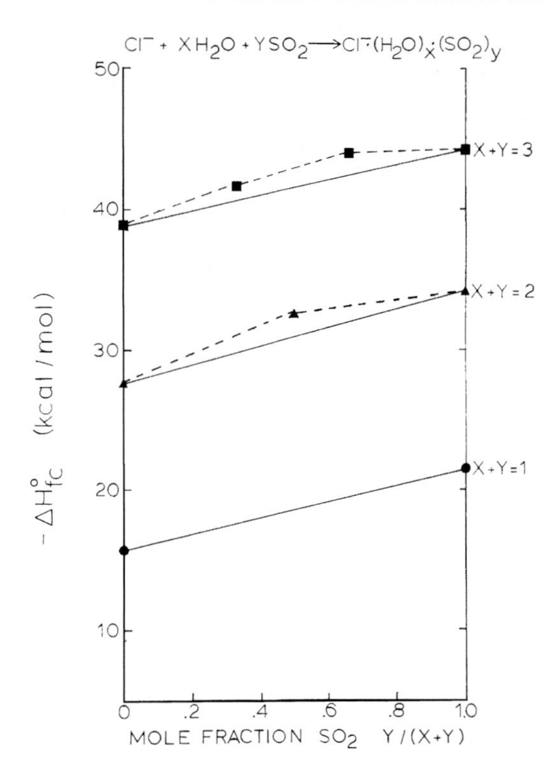

Fig. 5.3. The heats of clustering, -ΔH^o_{fc} for $Cl^-(H_2O)_x(SO_2)_y$. Note the enhancement, over a weighted-average, in the enthalpy of clustering for this system comprised of two different ligands. The enhancement is due to ligand-ligand attractive interaction [14].

2.1.4 Solvation of multiple-charged ions. The field of cluster ion thermochemistry has progressively developed through the contributions of a number of researchers, and the findings are too extensive to enable a detailed survey of the many contributions to be given here [3,5,16-20]. Until recently, all studies of cluster ions have been confined to systems of cations (or anions) having a single charge. This is primarily due to the fact that the early techniques involved the successive addition of the clustering ligand to the ion core. Since the second ionization potential of most species exceeds the first ionization potential of the clustering ligand, experimental limitations of charge transfer prevented the attainment of multiply charged clusters without some major change in technique whereby they could be formed directly in a multisolvated condition. Likewise, except for large systems, doubly charged anions are generally unstable with respect to dissociation to individual cluster anions. A major advancement in methods to study solvated ions came with the development of the technique of electrospray mass spectrometry. This provided a solution to the difficulty of solvating doubly-charged ions in the past few years [21].

2.1.5 Solvation thermochemistry of propanediol $H^+(H_2O)_n$, $n = 0$–3. Knowledge of the thermochemical properties of non-covalent bonds, particularly those involving water and biologically significant molecules, is fundamental to understanding the molecular interactions and changes in conformations. The study of thermochemical properties of hydrates of 1,2-propane and 1,3-propane diols are conducted with a quadrupole mass spectrometer using a variable-temperature high pressure equilibrium ion source [22]. These diols have the same atomic composition, differing only by the position of one of their hydroxyls. Leikin and co-workers have reported the differential effects of 1,2- and 1,3-propane diol on collagen self-assembly [23]. 1,2-propane diol was found to weakly inhibit fiber assembly while, at the same concentration, 1,3-propane diol strongly suppressed fibrillogenesis. Hydrogen-bonded water clusters bridging opposing helices were proposed to be disrupted with the addition of 1,3-propane diol.

Table 5.2. Thermochemical Data for Reaction $H^+(\text{propane diol})(H_2O)_{n-1} + H_2O \rightleftharpoons H^+(\text{propane diol})(H_2O)_n$

Ions	$(n-1, n)$	$-\Delta G^o_{298K}$ kcal/mol	$-\Delta H^o$ kcal/mol	$-\Delta S^o$ cal/(deg. mol)
(1,2-propanediol)H^+	$(0, 1)$	9.2*	17.1	26.5**
	$(1, 2)$	7.2±0.8	14.0±0.4	23.0±1.3
	$(2, 3)$	5.4±1.0	11.6±0.5	21.1±1.6
(1,3-propanediol)H^+	$(0, 1)$	9.3±2.0	15.1±1.1	19.4±3.1
	$(1, 2)$	5.8±1.4	13.4±0.6	25.6±2.2
	$(2, 3)$	4.4±1.2	14.6±0.6	34.1±2.0

* Free energy was calculated at 118°C

** Entropy change was estimated from the average of previous $(0, 1)$ alkyl ammonium hydration results.

The stepwise addition of water to the (1,2-propane diol)H^+ results in a decreasing trend in the binding enthalpy. See Table 5.2. For the $(0, 1)$ reaction of (1,2-propane diol)$H^+(H_2O)_n$, reliable ion intensity ratios at temperatures greater than 118°C were unattainable because of the weak ion intensity of (1,2-propane diol)H^+. Favorable internal hydrogen bonds may be formed leading to a stable monohydrated protonated 1,2-propane diol. The difference in the enthalpy and entropy changes of the second and third water addition to (1,3-propane diol)H^+ is indicative of (1,3-propane diol)$H^+(H_2O)_3$ having a more stable structure than (1,3-propane diol)$H^+(H_2O)_2$. Results suggest that 1,3-propane diol is capable of forming favorable structures with an increasing degree of hydration. This would support the aforementioned proposal of 1,3-propane diol incorporating into and disrupting hydrogen bonded water clusters bridging collagen helices.

5.2.2 Evidence for the structure of magic-numbered water clusters

2.2.1 ***ΔH^o, ΔS^o, ΔG^o, studies.*** A major limitation of the high-pressure mass spectrometry technique is imposed by the inability to work very close to a point of condensation around the cluster ions. The practical consequence of this limitation is the inability to work at sufficiently low temperatures and high ligand partial pressures such that clusters beyond about 10 or so ligands can be produced and studied under equilibrium conditions. An important development in overcoming this limitation came from a marriage of the reflectron time-of-flight technique [24-30] and the theory of the evaporative canonical ensemble [31-33] which opened up the possibility of investigating the molecular details of cluster dissociation dynamics and using the results to deduce thermochemical data. It has now been demonstrated that measurement of the relative dissociation fractions in a metastable time window, preferably in combination with kinetic energy release measurements, [27, 30] enables the evaluation of bond energies of clusters having relatively large degrees of solvation [34]. See Fig. 5.4(b).

To further understand the origin of magic numbers, the thermochemical properties of water cluster ions was investigated [34]. A TOF reflectron technique, whereby the decay fractions of dissociating clusters could be measured and the binding energies derived from the evaporative ensemble model was used [35-38]. Then, through an investigation of the steady-state distribution of clusters in a flow tube reactor under well-defined thermal conditions [Fig. 5.4(a)], we ascertained intensity ratios and concentrations which enabled us to deduce values of their free energy of formation. By combining these measurements, we were able to ascertain bond energies and entropies as a function of cluster size for species ranging from 6 to 28 water molecules.

An important development from our laboratory has been the development of a titration technique [39] for use in elucidating the structures of clusters from which hydrogen atoms or protons extend. This work provided the first definitive evidence concerning the structure of the magic number protonated water cluster $(H_2O)_{21}H^+$[Fig. 5.4(c)]. Many speculations existed in the literature concerning the stability of this species and the reasons for its observation in studies ranging from those of expansions of neutral water molecules and subsequent electron impact ionization, to expansion of protons in molecular beams, and even following the sputtering of ions from ice-like surfaces. The correlation between the observed magic numbers of cluster ion intensity distributions and the stable hydrogen-bonding structures is valid for the mixed cluster ions $n = 1 - 34$. For cluster ions, $(H_2O)_{21}(TMA)_mH^+$, TMA is trimethylamine, the ion intensity distribution shows a maximum at $(H_2O)_{21}(TMA)_{10}H^+$ and then an abrupt drop at $(H_2O)_{21}(TMA)_{11}H^+$. The findings establish that $(H_2O)_{21}H^+$ ion has a total number of 10 hydrogen-bonding sites available for TMA to form the fully solvated hydrogen-bonding structure. The titration method provides convincing evidence that the 21-mer

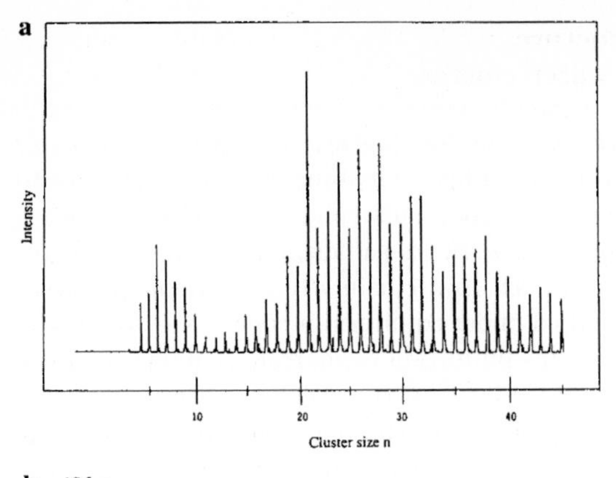

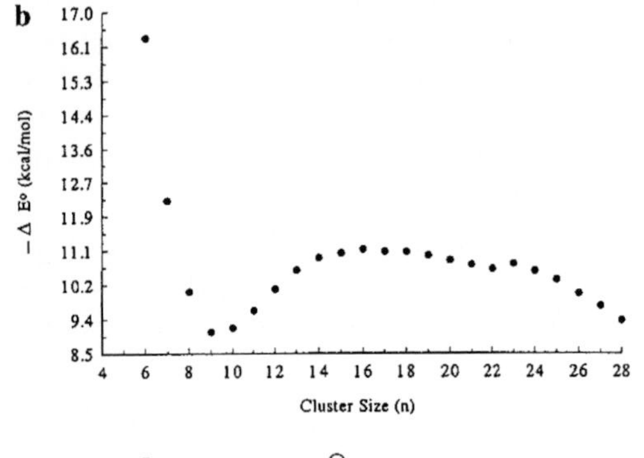

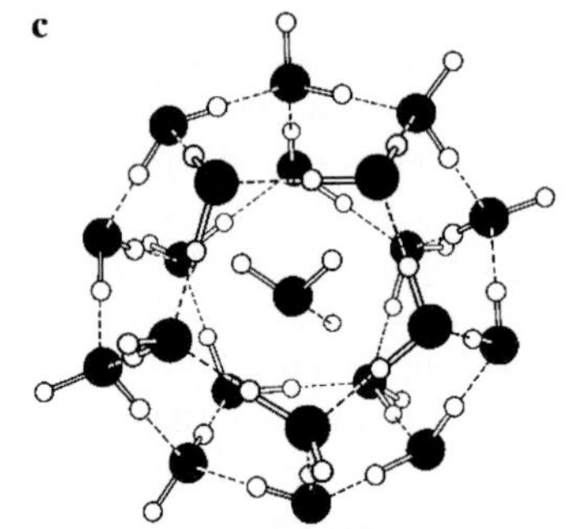

Fig. 5.4. (a) Mass spectrum of protonated water clusters $H^+(H_2O)_n$ (n= 4 – 45) at 119K and 0.3 Torr of He in a flow reaction. Note the prominence of $H_3O^+(H_2O)_{20}$ even under quasi-equilibrium conditions [52]. (b) Relative binding energies of H_2O clusters [34]. (c) Encagement of H_3O^+ within the pentagonal dodecahedral hydrogen-bonded structure of $(H_2O)_{20}$.

is a clathrate-like cage with an encaged ion. For the clathrate structure of $(H_2O)_{20}$, there are 30 self-hydrogen-bonded (forming 12 five-member rings) and 10 nonhydrogen-bonded hydrogen atoms.

2.2.2 Mixed water-alcohol studies. Mixed water-methanol clusters were investigated utilizing a laser based reflectron time-of-flight mass spectrometer [40]. The intensity distributions of $(H_2O)_n(CH_3OH)_mH^+$ shows magic numbers at $n + m = 21$, $0 \leq m \leq 8$ attributed to the enhanced stabilities of the dodecahedral cage structures in the mixed clusters. The magic peaks of 21-mer of protonated mixed water-methanol clusters $(H_2O)_{21-m}(CH_3OH)_mH^+$

($0 \leq m \leq 8$) are caused by stable dodecahedral cage structures. Studies of the metastable dissociations of $(H_2O)_n(CH_3OH)_mH^+$ ($n+m \leq 40$) provided evidence that the dissociation channels are governed by competition of effects dominated by dipole-dipole and dipole-polarizability interactions compared to those arising due to the hydrogen-bonded nature of the cluster, and the cluster structure. The metastable studies suggest the water molecules have a tendency to form an outer shell in the large mixed clusters, and that polarizability is important in governing the bonding strength in water-methanol mixed clusters. The metastable dissociation channels reveal the effects of the competition between pure electrostatic bonding and hydrogen bonding in the water-rich mixed clusters.

Gas phase ion-molecule reactions of protonated water clusters with methanol were studied in a flow reactor under thermal conditions [41]. See Fig. 5.5. A distinct feature at $H^+(H_2O)_{21}$ corresponding to a stable clathrate was ob-

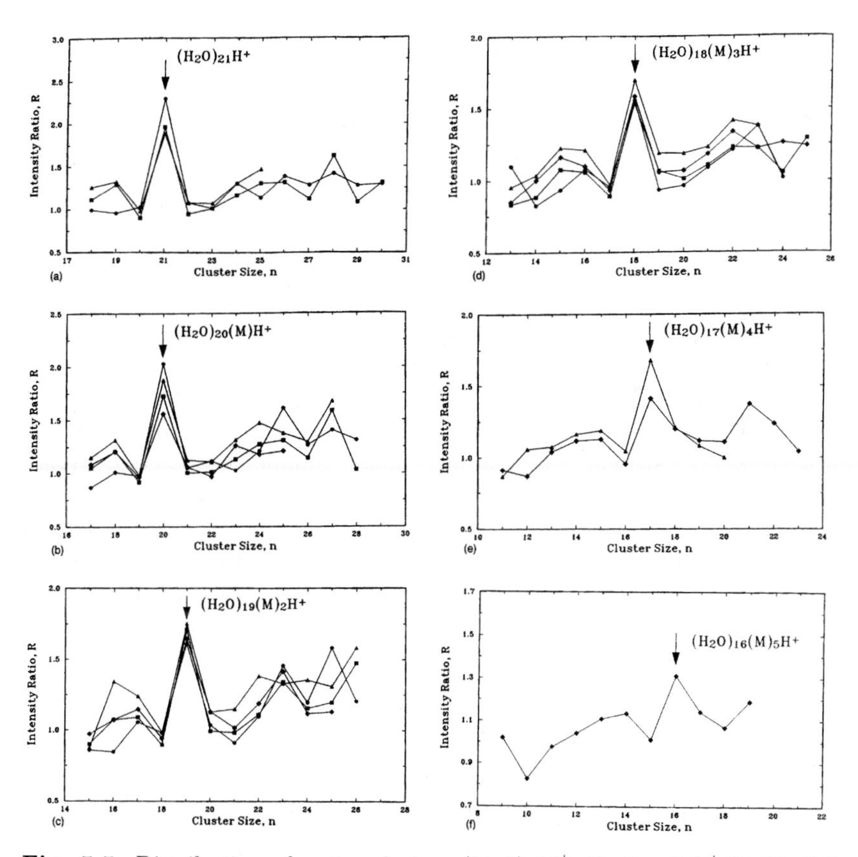

Fig. 5.5. Distribution of water clusters $(H_2O)_nH^+$. Ratio of $H^+(H_2O)_n(M)_x$ / $H^+(H_2O)_{n+1}(M)_{x+1}$ [Zhang, X.; Castleman, A.W., Jr. *J. Chem. Phys.* **1994**, 101, 1157.].

served. Upon addition of methanol vapor into the fast flow reactor by a reactant gas inlet, products of the general form $H^+(H_2O)_n(CH_3OH)_m$ ($m = 1–5$) are detected. Distinct features at $n + m = 21$ are observed, which is indicative of the formation of clathrate-like structures for mixed water-methanol cluster ions. One final question is whether the intensity distributions resulting from the switching reactions merely reflect those of the original water cluster reactant, or whether the enhanced peaks for $n + m = 21$ are indicative of the enhanced stability of the mixed cluster. To address this issue, we also conducted experiments in which both water and methanol were added into the ion source, and the resulting clusters were introduced into the flow tube and subsequently detected. These experiments revealed magic numbers at $n + m = 21$ for $m = 1 - 3$ (and indications up to around seven, which were the largest mixed clusters that could be formed by this method). The finding that methanol molecules can replace water molecules in the clathrate-like structure and form "mixed" clathrate species is fascinating. The results from the present study demonstrate that the stable clathrate-like structures can exist not only for water clusters, but also for other mixed composition hydrogen-bonded complexes. The occurrence of the substitution of methanol for water in the clathrate structures under thermal conditions indicates that this transformation takes place at thermal energy with very little barrier to substitution and reorganization. The finding that methanol can replace water molecules in the clathrate structure implies that the structure of "mixed" clathrates of $H^+(H_2O)_{21-m}(CH_3OH)_m$ are also very stable.

2.2.3 Alkali metal ion–water complexes. The first evidence for the gas-phase encagement of atomic ions in clathrate-like clusters has been reported previously by our research group [42–44] on the basis of observation of especially stable magic-number clusters of the form $M^+(H_2O)_n$ for M = Cs, K, Li, Rb, and Na, where n is up to 40, using a variable-temperature flow reactor experimental technique. See Fig. 5.6. A series of magic numbers could be observed, especially pronounced for $n = 20$ for Cs, K, Rb, and to a lesser degree Li. All of the alkali metal cations except for sodium display a tendency to form distinct and reproducible magic number species at $n = 20$. Results of molecular dynamics calculations suggest that a pentagonal dodecahedral network of water molecules surrounding a central ion is a possible stable structure for $M^+(H_2O)_{20}$.

2.2.4 Results of an ion core interacting with water clathrates. Beauchamp and co-workers [45, 46] investigated the solvent evaporation from extensively hydrated peptides generated by electrospray ionization and detected using an external ion source Fourier transform ion cyclotron resonance mass spectrometer. Using a technique called freeze-drying, water evaporation from the clusters formed at room temperature by appropriate operation of an electrospray ion source is initially rapid and results in an evaporative cooling of the cluster. The cluster temperature is determined by the balance between

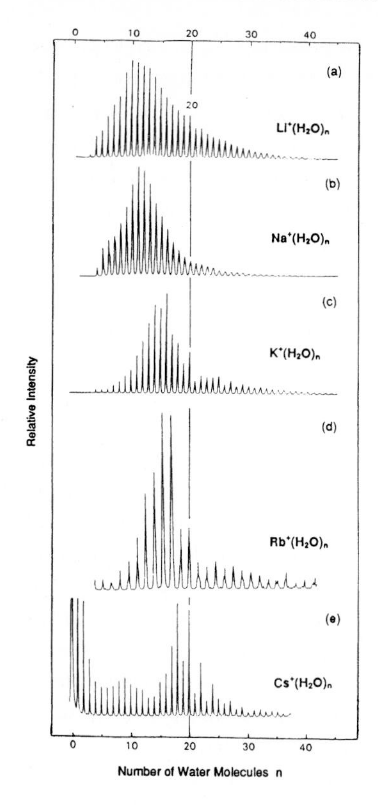

Fig. 5.6. Ion abundance in arbitrary units as a function of mass for the cluster of water molecules to alkali metal cations [4].

evaporative cooling and energy input due to background blackbody radiation. During this process, clusters with special stability are identified as prominent features in the overall cluster distribution. The signature of a stable cluster is an enhanced abundance accompanied by an unusually low abundance for the cluster with one additional solvent molecule. In agreement with earlier studies of the hydrated proton, [39, 47–49] pure water clusters exhibit the special characteristics of clathrate formation. Similar clustering of water occurs around protonated primary alkylamines where the protonated amine replaces one of the water molecules in the clathrate structures, which encapsulate one or more neutral water molecules. Also, doubly protonated cyclic decapeptide gramicidin S with 40 water molecules attached has both protonated ornithine residues solvated by pentagonal dodecahedron clathrate structures. The observations suggest the structure is comprised of neutral clathrates attached to a "spectator" ion.

In studies of organic ions, these hydrated clusters become significant for biology especially in understanding processes occurring at the interfaces of lipid bilayers. The relationship between alkylammonium ions $NR_mH^+_{4-m}$, m = 1-3, and the charged portion of some lipid molecules, i.e., phosphatidylcholines and phosphatidylethanolamines are increasingly important

[50]. Magic number clusters also have been found for $H_3O^+(H_2O)_n$ and $NH_4^+(H_2O)_n$ at $n = 20$, and 27 under nonanalytical electrospray ionization conditions and a single quadrupole mass spectrometer used in the electrospray ionization mode. Hydrates of $NH_3(C_mH_{2m+1})^+(H_2O)_n$ displayed magic-numbered clusters at $n = 20$ and 27 in their mass spectra. Interestingly, both the dimer and trimer forms of the alkyl ammonium ions exhibit magic-numbered clusters at $n = 20$ and 27.

5.3 Reaction kinetics and mechanisms

5.3.1 Ligand-switching reactions

The production of large water clusters [51-53] under thermal reaction conditions enables the investigation of reactions at sites of selected degrees of hydration. This provides some insight into reactions that may occur in the condensed state. During the course of these studies, large water clusters bound to H^+ or OH^-, with sizes ranging up to 50 molecules, were produced and various [51, 54, 55] reactions were carried out at well-defined temperatures using the fast-flow reactor technique. One test case for which extensive studies were made concerned their reaction with CH_3CN. While, for protonated water clusters, CH_3CN undergoes a proton transfer reaction at small water-cluster sizes $[H^+(H_2O)_{n=2-4}]$ and a ligand-switching reaction occurs at intermediate sizes, for large water clusters an association mechanism was found to dominate. Further reactions of protonated water clusters with CH_3COCH_3 and CH_3COOCH_3 were investigated. The reaction mechanisms were found to change as a function of increasing cluster size from proton-transfer reaction to ligand switching and ultimately to an association process, which would be equivalent to adsorption in the case of bulk systems.

With regard to anionic water clusters $[X^-(H_2O)_n]$, all cluster ions($X =$ OH, O, O_2, and O_3) were found to undergo a proton transfer reaction. Small hydrates were found to react with CH_3CN at near collision rate via proton-transfer and ligand-switching mechanisms; further hydration of the anions greatly reduces their reactivity due to the thermodynamic instability of the products compared to the reactants.

Dinitrogen pentoxide N_2O_5 has received considerable attention because of its importance in atmospheric chemistry. Heterogeneous reactions are believed to occur on the surface of polar stratospheric cloud particles, and are thought to be important in polar regions in winter where PSCs are formed. *In situ* measurements confirm the importance of PSC surfaces in atmospheric reactions, which lead to conversion of active nitrogen species (NO, NO_2, N_2O_5, $ClONO_2$) to reservoir forms, namely HNO_3 [56]. Laboratory studies [57-59] suggest that N_2O_5 also may be catalytically converted to HNO_3 in the stratosphere by a mechanism involving protonated water clusters, which are known to exist in the stratosphere.

Reactions of N_2O_5 with protonated water clusters $X^+(X_2O)_{n=3-30}$, X = H or D, were studied at temperatures ranging from 128 to 300 K and at pressures between 0.23 and 0.66 Torr using a fast flow reactor [60]. For cluster ions with $n \geq 5$, the reaction $X^+(X_2O)_n + N_2O_5 \rightarrow X^+(X_2O)_{n-1}XNO_3 + XNO_3$ was observed to occur at temperatures below 150 K. At larger values of n which were acquired at temperatures of about 130 K, the product ions $X^+(X_2O)_{n-2}(XNO_3)_2$ and $X^+(X_2O)_{n-3}(XNO_3)_3$ were also observed. The rate constants of the thermal-energy reactions of N_2O_5 with $X^+(X_2O)_{n=5-21}$ were found to display both a size and pressure dependence. This study provided the first experimental evidence for the reaction of protonated water clusters ($n \geq 5$) and N_2O_5 under laboratory conditions. The results provide further information on the chemistry of the upper atmosphere.

5.3.2 Ion core transformation reactions in water clusters

Thermal reaction techniques enable a quantification of the influence of solvation on reactivities [51,54,55,61-67]. One particular reaction, which is a good example of how solvation can affect the nature of an ion core reaction site, comes from a study [62] of the interaction of OH^- with CO_2. The gas-phase reaction between the individual species is very exothermic (ΔH_r^o = - 88 kcal/mol) and can only take place by a three-body association mechanism. The reaction proceeds very slowly in the liquid phase and has been calculated [68] to have a barrier of about 13 kcal/mol. In biological systems, the reaction rate is enhanced by about four orders of magnitude through the enzyme carbonic anhydrase. Recent studies carried out in our laboratory provide detailed information on the influence of hydration on the reaction kinetics, and are supportive of the suggested role played by the enzyme in facilitating this reaction which is important in respiration.

Hydration can have a pronounced effect on reaction thermodynamics, but in the case of $OH^-(H_2O)_n$ the clustering of a number of water molecules is still insufficient to cause the reaction to become endothermic. See Fig. 5.7(left). Owing to the formation of stable products, namely $HCO_3^-(H_2O)_m$ the reaction enthalpy is still very exothermic even if as many as three water ligands were to become replaced by one CO_2 molecule. Thus an explanation for the discrepancy between the experimentally measured rate constants and the theoretically predicted values must be attributed to reaction kinetics. A general formulation of the reaction can be written [62] in terms of a Lindemann-type mechanism:

$$CO_2 + OH^-(H_2O)_n \rightleftharpoons \{OH^-(H_2O)_n(CO_2)\}^* \tag{5.4}$$

$$\{OH^-(H_2O)_n(CO_2)\}^* + He \xrightarrow{k_S} OH^-(H_2O)_n(CO_2) + He \tag{5.5}$$

$$\{OH^-(H_2O)_n(CO_2)\}^* \xrightarrow{k_r} HCO_3^-(H_2O)_m + (n-m)(H_2O) \tag{5.6}$$

where the intermediate reaction complexes can undergo unimolecular dissociation (k_{-1}) back to the original reactants, collisional stabilization (k_s) via a third-body, and intermolecular reaction (k_r) to form stable products $HCO_3^-(H_2O)_m$ with the concomitant displacement of water molecules. By using a steady state approximation on the concentration of $\{OH^-(H_2O)_n(CO_2)\}^*$, the experimentally measured rate constant, k_{exp}, can be related to the rate constants of the elementary steps by the following equation:

$$k_{exp} = k_1\{k_s[\mathrm{He}] + k_r\}/\{k_{-1} + k_s[\mathrm{He}] + k_r\} \tag{5.7}$$

according to four possible situations; but for the experimental conditions used, the following applies: $k_{-1} > k_r$ and $k_s[\mathrm{He}]$, and $k_r > k_s[\mathrm{He}]$, then $k_{exp} = k_1 k_r / k_{-1}$.

The experimental findings are of value in unraveling the details of the enzymatic hydrolysis dynamics of CO_2. In basic solution, CO_2 can react directly with OH^- to form HCO_3^-:

$$\mathrm{CO_2(aq) + OH^-(aq)} \xrightarrow{k_{OH^-}} \mathrm{HCO_3^-(aq)} \tag{5.8}$$

The second-order rate constant k_{OH^-} has been measured to be 8.5×10^3 $M^{-1}s^{-1}$[69], while the rate constant for the enzymatic hydrolysis of CO_2 has been measured to be 7.5×10^7 $M^{-1}s^{-1}$ [70], or 1.2×10^{-13} cm^3s^{-1} in units employed by gas-phase chemists, which is about four orders of magnitude larger than k_{OH^-}. By contrast, the gas-phase collision limit for CO_2 with a large anion is about 10^{-10} cm^3s^{-1} and the experimentally observed values decrease with increased hydration. It has been found [71] that the key aspect of the enzyme catalysis is the presence of a Zn^{2+} center which reduces the hydration of the OH^-, enabling rapid reaction with CO_2 to form HCO_3^-.

It is interesting to compare the reactions of CO_2 and the water cluster of OH^- with the results of the analogous reactions involving SO_2 [67]. Both reactions are highly exothermic. Trends in hydration are not much different from that for the CO_2 case, at least for small clusters. See Fig. 5.7(right). Nevertheless, SO_2 displays a very large reaction rate independent of cluster size over the full range investigated. The trends are in good accord with computations made employing results of trajectory studies.

Measurements made in our laboratory, as well as those by others, reveal that SO_2 interacts with bare OH^- via an association reaction. However, the reaction mechanism changes from one of association to switching upon hydration:

$$\mathrm{OH^-(H_2}O)_n + \mathrm{SO_2} \rightarrow \mathrm{HSO_3^-(H_2O)}_m + (n-m)(\mathrm{H_2O}), n \geq 1 \tag{5.9}$$

With additional water ligands bound to the OH^-, the reaction enthalpy becomes more positive owing to the net stabilization of OH^- by water. For example, in the case where all the water ligands are replaced by an SO_2 molecule during the reaction

$$\mathrm{OH^-(H_2O)}_n + \mathrm{SO_2} \rightarrow \mathrm{HSO_3^-} + n(\mathrm{H_2O}) \tag{5.10}$$

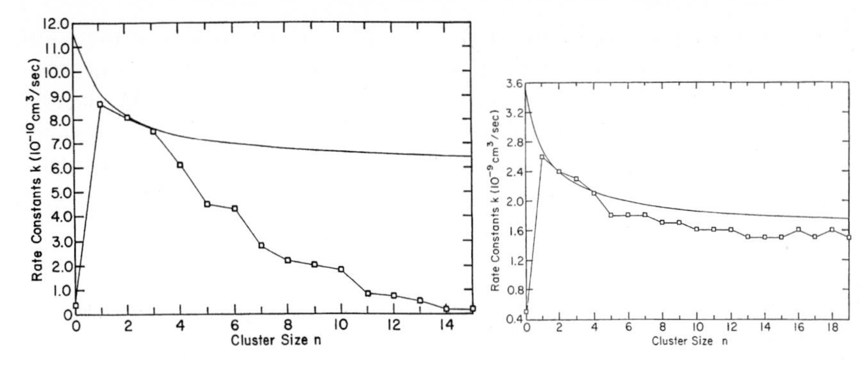

Fig. 5.7. Left - size dependence of the rate constants for the reactions of CO_2 with large hydrated hydroxyl anions at T = 130 K: ○, experimental values for $OH^-(H_2O)_n$; -- , calculated values for $OH^-(H_2O)_n$. Right - dependence of rate constants on cluster size for the reactions of $OH^-(H_2O)_n$ with SO_2 at T = 135 K [62,67].

the reaction enthalpy switches from being exothermic to being endothermic between $n = 3$ and $n = 4$. In hydrated clusters, only reactions leading to partial replacement of the water molecules maintain thermodynamic exoergicity:

$$OH^-(H_2O)_4 + SO_2 \rightarrow HSO_3^-(H_2O) + 3(H_2O) + 46.6\text{kJmol}^{-1} \quad (5.11)$$

Observations are in accord with this fact. When $n = 59$, the anionic clusters $OH^-(H_2O)_n$ are still found to react every fast toward SO_2, approaching the gas phase collision limit of about 10^{-9}cm^{3}s^{-1}. Since SO_2 is a strong reductant in basic solution [72], no kinetic data are directly available for the reaction:

$$SO_2(aq) + OH^-(aq) \rightarrow HSO_3^-(aq) \quad (5.12)$$

However, the rate constant for the following reaction has been measured [73]:

$$SO_2(aq) + H_2O \rightarrow HSO_3^-(aq) + H^+(aq) \quad (5.13)$$

It is 3.4×10^6 M^{-1}s^{-1} (5.6×10^{-15}cm^{3}s^{-1}), which is about ten orders of magnitude faster than the comparable reaction with CO_2.

The rate constant for reaction (5.13) is expected to be larger than 3.4 $\times$ 10^6 M^{-1}s^{-1}, not only because reactions involving ions often have much smaller activation energies than those involving only neutrals, but also because the very stable products $HSO_3^-(H_2O)_n$ are formed. The observation that association becomes the dominant reaction channel at large cluster sizes in the present experiments indicates the possibility of formation of the products $OH^-(H_2O)_n(SO_2)$. It is interesting to note that in solutions the $SO_2(OH^-)$ form exists and the $[SO_2(OH^-)]/[HSO_3^-]$ ratio is about five at 20°C [74].

Usually, the hydration of negative ions will make the reactions more endothermic owing to the stabilization of the reactant ions. However, the products of the reaction between SO_2 and $X^-(H_2O)_n$ are very stable, but can be observed as the reaction accommodates their exothermocity by "boiling off" a certain number of water molecules from the products. A factor, which may contribute to the interaction between SO_2 and the hydrated anion cluster, is attraction of the large quadrupole moment to the negative ion.

5.3.3 The size specific uptakes and dissolution of HNO_3, HCl, and HBr in water clusters; ionic clusters used as probes

In an earlier study by the Bondybey research group using a Fourier transform cyclotron resonance, apparatus protonated water clusters with HCl were investigated and the dissolution of one and two HCl was observed [75]. A detailed description of the further uptake of HCl was left unanswered. The reaction kinetics and mechanisms of protonated water clusters and HCl have been investigated in our laboratory using a variable-temperature fast-flow reactor [76, 77]. Two distinctly dominant mechanisms of HCl uptake are operative: the bimolecular uptake of HCl in a 1:6 ratio with water and a subsequent association or adsorption mechanism of HCl binding to water in a 1:3 ratio. The critical number of water molecules required for the dissolution of one, two, three, and four HCl molecules by protonated water clusters was determined by the fast-flow-reactor experiments as displayed in column 4 of Table 5.3. In Table 5.3, m is the number of water molecules required for the dissolution of an acid species, n is the number of acid molecules that dissolve. The

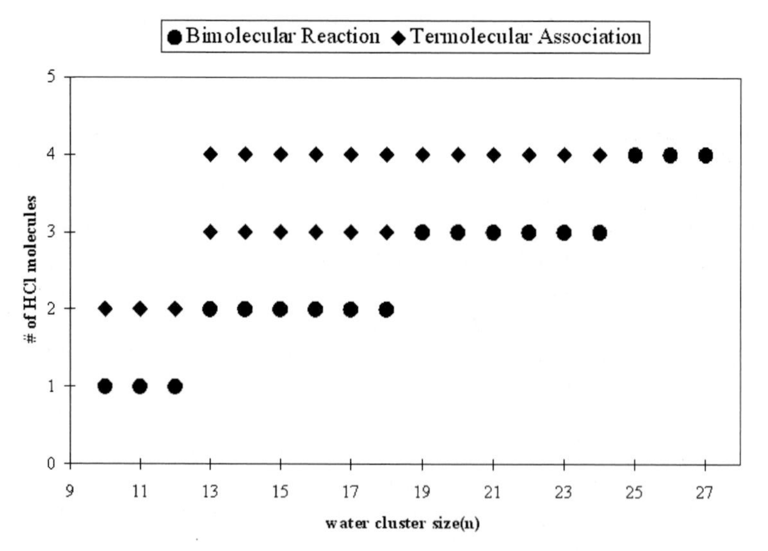

Fig. 5.8. Uptake of HCl cluster size dependence of $D^+(HCl)_y(D_2O)_n$.

Table 5.3. Size Dependence for Acid-Molecule Dissolution in Cationic Water Clusters

$A_3O^+(A_2O)_{m+k} + X \rightarrow A_3O^+(A_2O)_m(\mathbf{X})_n + kA_2O$

(A = H or D)

Uptake Values (m)

X

n	DNO_3^a	HBr^b	HCl^c
1	4	7	9
2	8	11	12
3	11	15	18
4	15	19	24

Uptake Values (m)

$A^+(H_2O)_{k+m} + X \rightarrow A^+(H_2O)_m(\mathbf{X}) + kH_2O$

X

A	HNO_3	HCl
H_3O	4	9
K	4	9
Na	4	9

a Zhang, X.; Mereand, E.; Castleman, A.W., Jr. *J. Phys. Chem.* **1994**, 98, 3554; Gilligan J.J.; Castleman, A.W., Jr. *J. Phy. Chem. A* **2001**, 105, 5601.
b Gilligan J.J.; Castleman, A.W., Jr. *J. Phy. Chem. A* **2001**, 105, 5601.
c Ref.[76] and Ref.[77]

association of HCl to protonated water clusters appears in a ratio of 1:1 with dissolved HCl. Molecular HCl was found not to associate onto protonated water clusters under any conditions of pressure, temperature, or concentration accessible in our experiments, which prompted the conclusion that dissolved HCl plays a role in the association of subsequent HCl uptake beyond the first. See Fig. 5.8. In a similar manner, reactions of other acid species (HNO_3 and HBr) have been investigated in our laboratory to elucidate the nature of the uptake of HCl by protonated water clusters at these distinct cluster sizes.

Monitoring the formation of the product cluster ions as a function of pressure, concentration of reactant gas, and ion residence time enabled the determination of the points of dissolution of the various acid molecules [77]. The points of dissolution of these acid species do not display any dependence on the temperature over the range studied. However, the rate of initial uptake of these acid species displays a temperature dependence [76] as expected. The points of dissolution are presented in Table 5.3. In the hydrogen bromide system, a minimum of 7 water molecules bound to a hydronium ion is required for the dissolution of one HBr, 11 for two, 15 for three, and 19 for four HBr. See Fig. 5.9. In another similar study, reactions of nitric acid and protonated water clusters were investigated to determine the rate coefficients and to evaluate the points of dissolution of nitric acid within the water clusters. In the nitric acid system, a minimum of 4 water molecules bound to a hydronium ion was required for the dissolution of one DNO_3, 8 for two, 11 for three, and 15 for four DNO_3. The initial points of dissolution (the minimum water cluster size) were found to be pressure independent, but dependent on the solvation size of the water cluster ions. The partial pressure of the acid species in the flow tube was found to influence the distribution and amount of acid uptake by water clusters. Rate coefficients are determined in the usual manner by measuring the decrease in reactant ion intensity as a function of the concentration of the reactant gas and by measuring the ion residence time [78]. Reactant ions are not in equilibrium with product ions in this study. Reaction conditions normally required for equilibrium studies were not accessible. The bimolecular rate coefficients, at T = 156 K, between protonated water cluster and nitric acid were found to increase as cluster size increased.

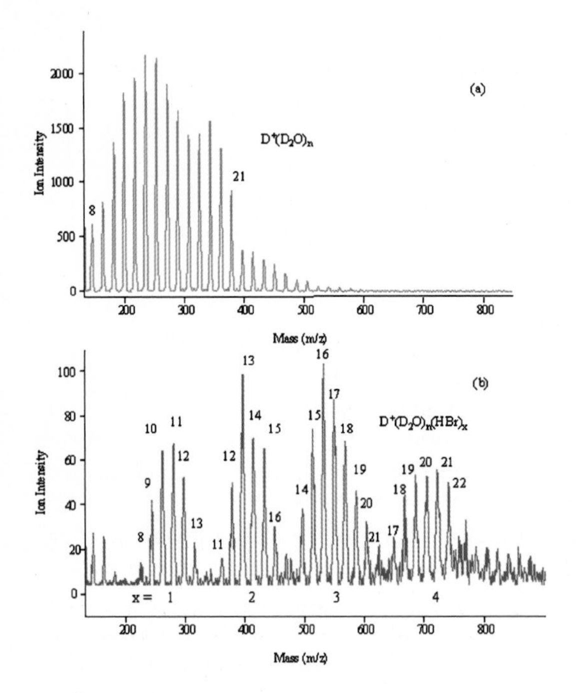

Fig. 5.9. Mass spectra of $D^+(D_2O)_n$ obtained before (a) and after (b) the addition of 5.74×10^{12} cm^{-3} HBr. Pressure: 0.3 Torr; Temperature: 146 K.

The critical size necessary for the first uptake of HCl and HNO_3 by hydrated alkali metal ions was also investigated. For hydrated potassium and sodium ions, uptake with these acid species was found to occur at the same cluster sizes for protonated water cluster (9 waters for HCl, and 4 waters for HNO_3). See Table 5.3. The addition of HNO_3 to the flow-tube was observed to deplete sodium ion water clusters starting at $n = 5$ which led to the formation of $Na^+(H_2O)_{n\geq4}(HNO_3)$ as is evident from the findings.

As a first step, experiments were conducted to investigate whether reactions occur between $D_3O^+(D_2O)_n$ (n = 7- 16) and $ClONO_2$ under thermal conditions; no chemical reactions were observed to occur. See Fig. 5.10. Figure 5.10(a) shows a typical distribution of protonated water cluster before the addition of 1.23×10^{12} cm^{-3} $ClONO_2$. After the addition of $ClONO_2$ [Figure 5.10(b)], the ion intensity of protonated water clusters did not decrease and the formation of DNO_3 reaction product ions did not occur. These results are consistent with previous work from our research group [79] which served to explain several conflicting observations of the reactivity of water clusters and $ClONO_2$ [80-82]. An acid catalyzed mechanism is not supported by these results and is not expected to be involved in the conversion of $ClONO_2$ to HNO_3 on polar stratospheric cloud surfaces as had previously been suggested. Moreover, reactions between the waters of hydrations and $ClONO_2$ were too slow to be observed under chosen experimental conditions.

In the next series of experiments, DCl was used to dope deuterated protonated water clusters. These subsequent studies were conducted at a similar temperature, pressure, and helium flow as the experiments completed in Fig. 5.11(b) and revealed that reactions occur between the reactant ions con-

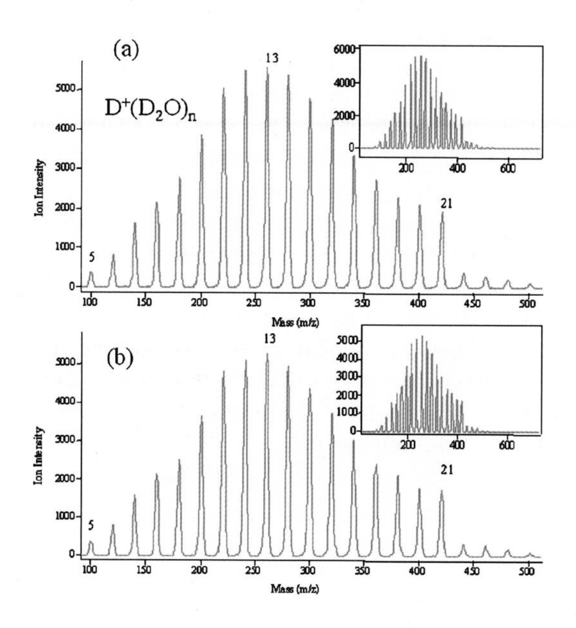

Fig. 5.10. Mass spectra of $D^+(D_2O)_n$ obtained before (a) and after (b) the addition of 1.4×10^{13} cm^{-3} $ClONO_2$. Pressure: 0.3 Torr; Temperature: -127°C

taining DCl and $ClONO_2$. Water cluster ions containing DCl (B_n) appeared 3 amu less than the protonated water clusters. Product ions of the form $D^+(D_2O)_n(DNO_3)$ (D_n) appeared at $n = 8$. Product ions (D_n) from reactions of $ClONO_2$ with dissolved deuterium chloride are clearly visible and contain DNO_3. A reaction mechanism has been proposed to occur between the hydrated chloride anions and chlorine nitrate [83]:

$$\begin{aligned} &Cl^-(D_2O)_n + ClONO_2 \quad \rightarrow [(D_2O)_nCl^- \cdots ClONO_2]^* \rightarrow \\ &NO_3^-(D_2O)_m + Cl_2 + (n-m)D_2O \end{aligned} \tag{5.14}$$

Since deuterium chloride dissolves within the water cluster, the product ions observed could be formed by the following reaction mechanisms:

$$\begin{aligned} &D^+(D_2O)_n(DCl)\ [\leftrightarrow D^+(D_2O)_xD^+(D_2O)_yCl^-(D_2O)_z] + ClONO_2 \quad \rightarrow \\ &D^+(D_2O)_p(DNO_3) \leftrightarrow D^+(D_2O)_xD^+[(D_2O)_yNO_3^-(D_2O)_m] \\ &+Cl_2 + (z-m)D_2O \end{aligned} \tag{5.15}$$

where $n = x + y + z$; $p = x + y + m$.

As determined by thermodynamic information [2, 84-86], reaction (5.14) is exothermic for $(n - m) \leq 2$. Therefore, reaction (5.15) is also expected to be exothermic and water molecules are expected to evaporate from the product cluster ion. From these experiments, we believe DNO_3 product ions [$D_{n\geq 8}$ from Fig. 5.11(b)] are from reactions with DCl doped water clusters [$B_{n\geq 11}$ from Fig. 5.11(b)].

To accurately determine product ions generated when $ClONO_2$ reacted with DCl-doped water clusters, a dilution of HNO_3 was prepared in a second separate mixing chamber and added to the flow-tube to a distribution of protonated water clusters under similar conditions as the dissolved DCl and $ClONO_2$ experiments. The product ions formed from reactions with HNO_3 were compared to the product ions of $ClONO_2$, and were found to be different by 1 amu for the first uptake of nitric acid. Reaction product ions from $ClONO_2$ are from reactions with DCl doped water clusters as determined from available thermodynamic information.

To further address the issue of reactivity of dissolved deuterium chloride, experiments were performed with $ClONO_2$ in which the first and second uptake of DCl by protonated water clusters was present. These findings provide evidence for the observed reaction product ions containing DNO_3 by way of the formation of $D^+(D_2O)_n(DCl)(DNO_3)$ cluster ions.

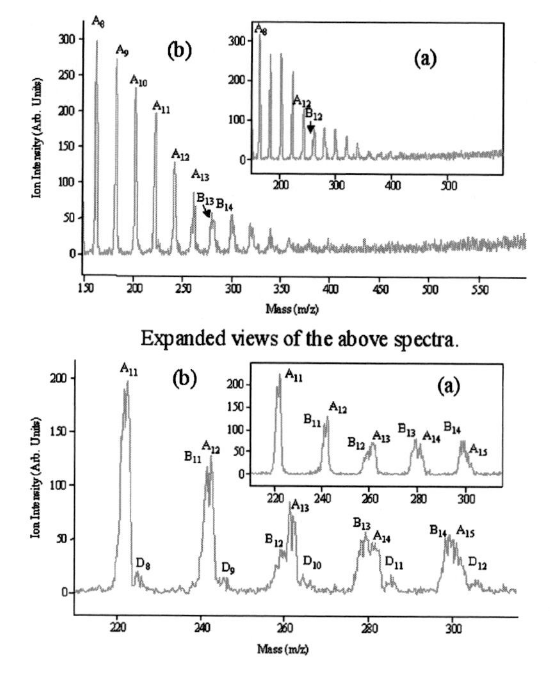

Fig. 5.11. Mass spectra of $D^+(D_2O)_n(DCl)$ obtained before (a) and after (b) the addition of 1.23×10^{12} cm^{-3} $ClONO_2$. Pressure: 0.3 Torr; Temperature: 154 K. $A_n = D^+(D_2O)_n$; $B_n = D^+(D_2O)_n(DCl)$; $D_n = D^+(D_2O)_n(DNO_3)$. The product ions, D_n, from reactions of $ClONO_2$ with dissolved deuterium chloride are clearly discernable.

5.3.4 Solvation effects on base-pair proton transfer and tautomerization reactions from femtosecond pump-probe spectroscopy studies

An interesting system is the 7-azaindole dimer, which has become a model for understanding the nature of proton transfer in base pairs such as DNA. In our laboratory, we have been able to study the excited-state double proton transfer (ESDPT) of the 7-azaindole dimer under conditions ranging from an isolated dimer up to a state of solvation where the hydrogen-bonded dimer is hydrated with as many as nine water molecules. One significant question concerned the possible existence of isomers and their influence on transfer rates. The issue first arose as a result of studies by Kaya and co-workers who found different behavior for species produced under varying stagnation pressure conditions employed in the supersonic expansion [87]. One species was found to be reactive and the other nonreactive. The reactive dimer appeared to undergo a normal double proton transfer, whereas the nonreactive dimer did not. Studies by several research groups led to varying explanations for the observed differences, most pointing to the likelihood that geometrically different isomers were responsible for the reported observations. Some suggestions [88] were made that the attachment of a weakly bound water molecule might influence the photoexcitation process, thereby accounting for the nonreactive species. However, we investigated the dimer over the same range of conditions and established that the presence of a water molecule is not responsible for the formation of the nonreactive dimer. Species are considered reactive if they exhibit a fast decay in their pump-probe trace, which can be fit best using a biexponential function, while species are considered nonreactive if their pump-probe transients show a long-lived lifetime with respect to the reactive dimer. If a water molecule was responsible for the formation of the nonreactive dimer, then the species (7-Aza)$_2$(H_2O) should show a long-lived transient, which is not observed. See Table 5.4. Our studies, in fact, established that the presence of a water molecule clustered with the nonreactive dimer actually facilitates proton transfer [89].

Table 5.4. Proton Transfer Times for the Solvated 7-Azaindole Dimers[a] [89].

Number of Waters	First transfer (fs)	Second transfer (fs)	Transfer time (fs)
1	550 ± 100	2800	
2	565 ± 100	2000	
3	560 ± 100	2000	
4			
5			1800
6			1600
7			1300
8			1100
9			1000

[a] Note that no definite values of lifetimes could be deduced from the data obtained for n = 4, a fact attributed to a change in cluster structure at this degree of hydration.

Subsequently, we investigated species composed of (7-Aza)$_2$(H_2O)$_n$, where $n = 2$ - 9, with the femtosecond pump-probe technique. Table 5.4 displays data for the case where water molecules were attached to a species corresponding to a reactive dimer. As the number of water molecules on the dimer increases, we begin to see evidence that the dimer molecule is behaving more as it would in a fully solvated condensed-phase environment. Transfer times correspond to 550 fs for the first transfer and about 2.8 ps for the second. As seen from data in Table 5.4, subsequent hydration has little effect on the transfer times until a change occurs at four waters, and a single transfer step is seen for additional degrees of hydration. Especially interesting is the observation that the rate changes from a stepwise to a single step (concerted) mechanism after hydration by 5 waters. This suggests that the double proton transfer is undergoing a transition from a two-step process to a one-step concerted mechanism. It is thought that a change in behavior is being caused by progressive clustering that leads to a card-stacked arrangement with waters solvating each of the 7–Azaindole constituents. This consideration would be in accord with suggestions [90] that a change from a hydrogen-bonded dimer to a stacked dimer occurs once enough water molecules become associated with the complex.

This suggestion gains some support from pump-probe studies of the monomer of 7-Azaindole. Referring to the data shown in Fig. 5.12, it is seen that upon sufficient hydration, namely with four water molecules, the monomer of the 7–Azaindole displays a single-exponent decay. This is indicative of an internal proton transfer via a proton translocation mechanism [89, 91].

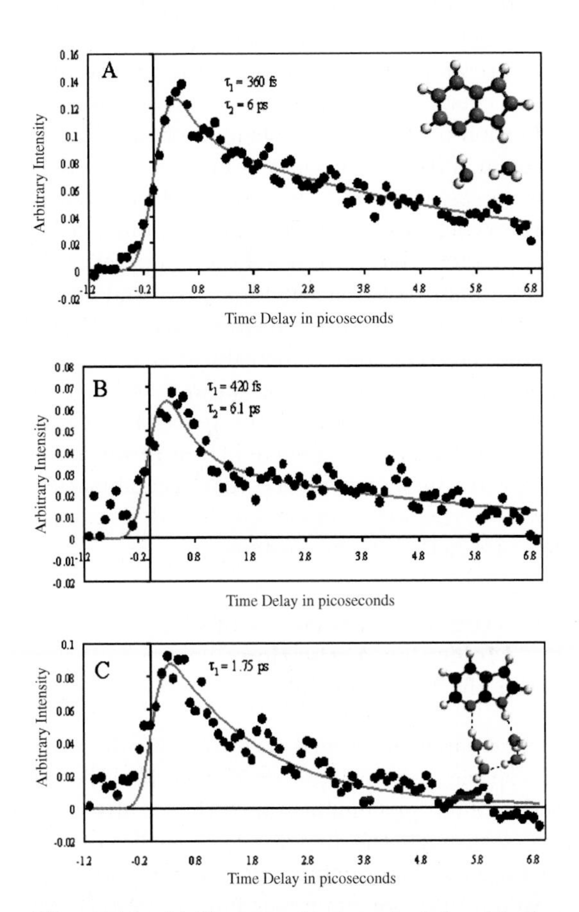

Fig. 5.12. A) Pump-probe transient for dihydrated 7-Azaindole monomer. The data was fit with a biexponential function resulting in first and second time constants of 360 femtoseconds and 6 picoseconds respectively. B) Pump-probe transient for 7-azaindole monomer with three waters clustered to it. The data was fit with a biexponential function yielding time constants of 420 femtoseconds and 6.1 picoseconds. C) Pump-probe transient for 7-azaindole solvated with four water molecules.

5.4 Implications to atmospheric chemistry

5.4.1 Formation and growth of protonated water clusters: Information on molecular mechanisms of noctilucent cloud formation

The role of large protonated water clusters in the appearance of noctilucent clouds (NLC) has been studied in a fast flow tube reactor [92]. Bimodal cluster distributions obtained in laboratory experiments were found to be remarkably similar to those found *in situ* and were also in accordance with predictions based on ion-induced nucleation theories. The experiments show a strong dependence of temperature and water partial pressure on the production of large protonated water clusters of sizes similar to those observed during NLC displays. A consideration of the thermodynamics of the clustering reactions and their unimolecular dissociation kinetics provides a further understanding of the conditions necessary for the observation of protonated water clusters both in the laboratory and *in situ*. Clustering and unimolecular dissociation kinetics analysis predicts that only when the temperature is below 150 K can clusters with more than six water ligands survive and be detected. A theoretical model predicts that cold water clusters prefer clathrate structures and nucleate to ice-like clusters from supersaturated water vapor. This study provides direct evidence of the formation and stability of clathrate-like-structured clusters. The comparison of the reactivity of the clusters toward neutral reactants and the *in situ* observed clean protonated-water-cluster mass spectrum provides further evidence that recombination with anion species and formation of NLC are dominant factors determining the lifetime of the prenucleation protonated water clusters.

5.4.2 Molecular details of heterogenous processes of electrolytes with aqueous surfaces: Molecular Activation by Surface Coordination (MASC) model

One of the more active areas of research in surface chemistry involves the interaction /accommodation of a species (gas or liquid) with a surface (liquid or solid). The common reference points for comparison of the results from various experimental endeavors is through the use of values such as mass accommodation and uptake coefficients. The bimolecular reaction rate constants we have determined provide a basis for numerous interpretations. As an initial approach, it is useful to evaluate the efficiency of these reactions. The means by which this is accomplished is to compare the measured reaction rate to a calculated collision rate, which takes into account the average number of collisions an ion and molecule will experience in the reaction region; this sets an upper bound for a measured reaction rate. The collision rate is the evaluated rate constant one would measure if every ion-molecule encounter resulted in a chemical reaction. These results can also be considered in the larger context of the numerous studies that have been directed toward gaining a more complex picture of the uptake and subsequent reactivity of HCl on PSC surfaces. PSC-mimic work provides information on the observed uptake of HCl by ice (water-ice and NAT) surfaces. The results of the work presented herein can be interpreted as mass accommodation coefficients, which are important values necessary to treat heterogeneous processes involving HCl.

The water cluster/HCl kinetic data can assist in the attempt to build the foundations necessary for the development of new treatments for heterogeneous processes involving the uptake of HCl. Perhaps some simple modifications to the gas-liquid interface models can facilitate these advancements. More results over a greater range of temperatures and cluster sizes will aid in the analysis of this field of inquiry from a water cluster aspect. Furthermore, treatment of the surface data utilizing the results from cluster studies does appear to yield some insight into the mechanisms of HCl uptake by PSCs.

A Molecular Activation by Surface Coordination (MASC) model for HCl surface reactions on PSCs [77] has been proposed. This model accounts for the trend in HCl uptake observed as an initial dissolving of HCl into a PSC followed by subsequent coordination of HCl onto the PSC surface. It is apparent from the experimental data that two distinct mechanisms of HCl uptake are operative; the bimolecular dissolution of HCl and subsequent association of HCl in a one-to-one ratio with dissolved HCl. The fact that HCl does not associate onto bare water clusters prompts the conclusion that dissolved HCl plays a role in the coordination (adsorption) of associated HCl. One could easily take this conclusion one step further to declare that dissolved Cl^- is involved, since no association product is observed on the protonated water clusters in the absence of dissolved HCl.

There has been a multi-pronged research effort aimed at gaining a greater understanding of the nature and reactivity of PSCs. These results are in accord with the ice film work previously conducted, and it would appear that the uptake trends extend from the ion cluster regime to the bulk surface realm. Including consideration of the molecular dynamics simulation results, the data indicates gas-phase (ion-cluster), liquid-phase (concentrated acid solution), and solid-phase (ice-film) consistency in regard to the nature of the HCl-water system.

5.4.3 Reaction mechanisms relevant to the atmosphere: Uptake of acid species and subsequent reaction with atmospheric species

Through a study of water clusters bound to ions of differing nature, we are able to ascertain the general influence that the ion-water bonding may have on the trends observed. Since alkali metal ions are analogous to a rare gas atom with a central charge, covalent bonding with the waters of hydration is not present. On the other hand, the proton of H_3O^+ is relatively strongly bonded to one water molecule, but further interactions with water are comparable to that of K^+/Na^+ with water. See Table 5.3. Our results indicate the dissolution of acid molecules is dependent on the number of water molecules present in the water cluster, but not on the nature of the ion core. Hence, the results from the present study are believed to accurately reflect the critical first uptake of HCl by $Na^+(H_2O)_n$ clusters. Our data conclusively shows that a switch from hydronium ion to sodium ion does not change the acid dissolution phenomena displayed in these cluster systems. The ion core is interpreted to act as a spectator ion, having a small effect on reactions but not playing a major role in the dissolution of acid species.

The lower degree of solvation required for the uptake of HBr compared to HCl by water clusters is similar to the findings obtained for the uptake of the acid species on ice films [93]. Tolbert and co-workers, using a Knudsen cell reactor coupled to Fourier transform infrared-reflection absorption spectroscopy, found HBr exposure to ice films at 110 K yielded surface coverage greater than observed for HCl by approximately a factor of 2. They suggested that fewer water molecules are required to solvate impinging HBr species than the HCl species. HCl and HBr were observed to react efficiently with crystalline and amorphous microporous ice at 110 K to form hydronium ion reaction products. Tolbert and co-workers' results strongly suggest solvation and ionization in interactions of HCl and HBr with ice films representative of atmospheric aerosols.

Using a flow-tube kinetic technique, Chu and Chu investigated the uptake rate of HI on ice surfaces and compared their results with several other uptake studies [94]. The rate of uptake of HX (X = Br, I, Cl, F) on ice was observed to follow the trend: HI $\cong$ HBr $>$ HCl $>$ HF. The authors suggested the uptake trend of acid halides was a result of the acidity and solvation of HX. The dissolution of acid species by water cluster ions is a convolution of the acid species' intensive properties, and the points of dissolution observed in the present study follows a trend in the strength of ion-molecule interaction. The order of increasing polarizability of the acid species (HNO_3 $>$ HBr $>$ HCl) is consistent with earlier uptake results for the acid species. The dissolution of an acid species in the water cluster is dependent, to an extent, on the ability of the cluster ion to distort the electron distribution of the acid species, which results ultimately in the formation of an ion-pair. Lisy and co-workers have suggested that the size and polarizabilty of an anion may play an important role in the structure of water molecules about the anion [95]. The bond dissociation energies, enthalpy of ionization, and the -pK_a's of the acid species did not follow the uptake trend of the various acid species into water cluster ions.

Rate coefficients of protonated water clusters and nitric acid were observed to increase as a function of cluster size. See Fig. 5.13. Figure 5.13(bottom) displays the ratio of the experimental rate coefficient to the calculated

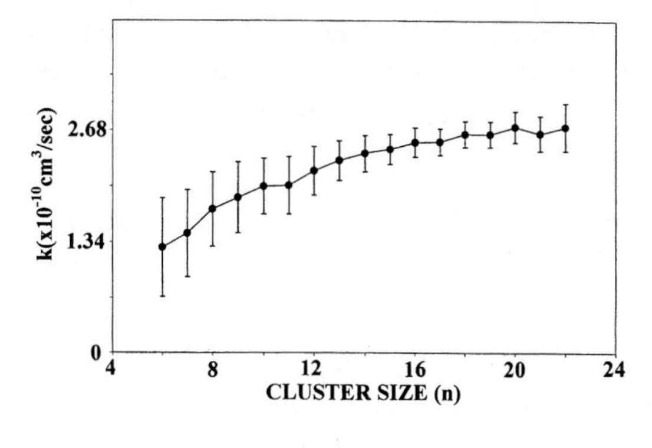

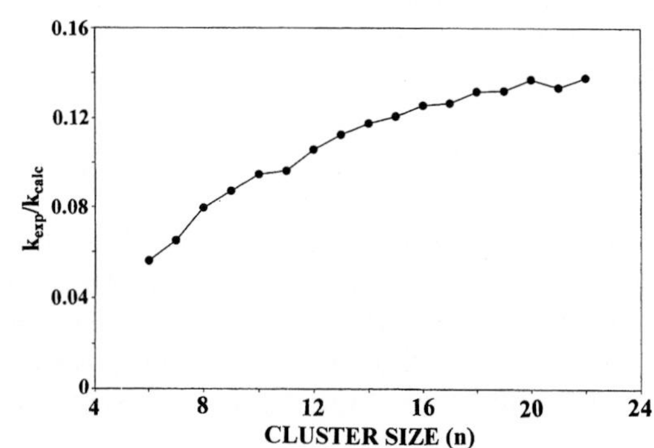

Fig. 5.13. Top - Cluster size dependence of the rate coefficients for the bimolecular reactions of DNO_3 and $D^+(D_2O)_n$ ($n = 6 - 22$) at T = 156 K and P = 0.3 Torr. Bottom - The ratio of experimental rate coefficient to the calculated collisional rate coefficient for the reactions of DNO_3 and $D^+(D_2O)_n$ ($n = 6 - 22$) at T = 156 K and P = 0.3 Torr.

collisional rate coefficient [96, 97] as a function of cluster size. The ratio can be viewed as the reaction efficiency of these cluster ions to nitric acid where the calculated collisional rate coefficient is the rate if every collision led to a reaction. As cluster ions grow, their properties gradually approach mimicry of a bulk system. Based on the rate coefficients, the initial dissolution of nitric acid by type II PSCs should be a rapid reaction. Laboratory studies showed that nitric acid uptake on ice films is very efficient [98].

Two major findings were obtained from the investigations of water clusters with acid species. Firstly, the points of dissolution of nitric acid, hydrogen bromide, and hydrogen chloride in water clusters were elucidated, providing further strong evidence for a dissolution process occurring. Secondly, the kinetics of protonated water clusters with nitric acid provided further support for a rapid dissolution process occurring at large water clusters sizes, which are representative of the condensed phase. We believe that ionic reactions on the surfaces of atmospheric particles could contribute significantly to the observed reactivity of reservoir species as suggested by other theoretical and experimental considerations. The results of the present study offer new insights into the uptake of acid molecules by water clusters and contribute to a more complete understanding of heterogeneous processes of atmospheric significance.

The reactivity of dissolved deuterium chloride water clusters to $ClONO_2$ compared to protonated water clusters found in the present study clearly demonstrates the reactivity of the solvated chloride ions to $ClONO_2$. This observation strongly suggests the reactivation of $ClONO_2$ on PSCs with HCl species will proceed by an ionic solvation mechanism as opposed to a mechanism involving intact molecular HCl. Based on our results of protonated water clusters and $ClONO_2$, an acid catalysis mechanism is not expected to be involved in the hydrolysis of $ClONO_2$. The waters of hydration in these experiments were not sufficient to reduce the barrier for reactions between protonated water clusters and $ClONO_2$. These results would suggest that the role of water is to aid in the dissociation of $ClONO_2$, and limitations of our technique of generating larger water cluster ions prevented us from observing such reactions.

5.4.4 Acid-water clusters: A probe of the gas phase acid loading of the atmosphere

Gas phase molecular aggregates that contain acid molecules have been produced with free jet expansion techniques and detected by using electron impact ionization mass spectrometry. Water clusters of aqueous nitric acid parallel many properties of the condensed phase. Multiple nitric acid molecules were found in the clusters that were sufficiently dilute. Experiments demonstrated the reactivity of sulfur trioxide with water clusters. The natural occurrence of acid cluster negative ions offers a means to probe the gas phase acid loading of the atmosphere through laboratory and field studies of the ion chemistry.

The laboratory investigation of a phenomenon such as acid rain may proceed along several avenues. An approach is devoted to the understanding of basic properties, such as a specific reaction rate, upon which the larger picture may be built. Molecular clusters can be considered to be the smallest size range of an aerosol particle size distribution. Nucleation from the gas phase to particles or droplets involves, in the initial stages, the formation of clusters. The propensity of sulfuric acid molecules to form small hydrated clusters is important to the nucleating ability of sulfuric acid [99].

In aqueous nitric acid studies [100], deuterated species were used to avoid ambiguity in stoichiometry. The basic experiment involves expansion of vapor through a nozzle, collimation of the jet with a skimmer to form a well-directed molecular beam, and detection of clusters via electron impact ionization and quadrupole mass spectrometry. A variation includes the implementation of an electro-static quadrupolar field to examine the polarity of the neutral clusters. The electric deflection technique is described by Klemperer and co-workers [101]. Clusters were produced by the expansion of the vapor from heated concentrated aqueous nitric acid. Electron impact ionization of the clusters produced ions of the form $D^+(DNO_3)_x(D_2O)_y$. In the case of the clusters containing one nitric acid molecule, a distinct local minimum of the signal intensity in the size distribution occurs at the cluster size $D^+(DNO_3)(D_2O)_4$. Since no explanation based on ion stability seems feasible, an attractive explanation is that the position of the minimum is indicative of some rather abrupt transformation of the precursor neutrals in this size range. Such a situation might occur if a complex became sufficiently hydrated to enable the formation of solvated ion pairs. A large change in the charge distribution within the cluster should occur upon solvation to form an ion pair $NO_3^-(H_2O)_nH_3O^+$. A more polar species should have a larger collision rate and hence a faster growth rate. This effect should manifest itself in the observed intensity distribution as an increase in intensity of the larger sized cluster just beyond the cluster that underwent the ion-pair formation. For clusters with more than one nitric acid molecule, the species $(DNO_3)_x(D_2O)_y$ $(1 < x < 6)$ each have a minimum degree of hydration above which clusters are observed. Both this minimum y and the most probable degree of hydration for a cluster of x DNO_3 molecules increase with x. The homomolecular species $D^+(DNO_3)_x$ $(x>1)$ are not detected even though the HNO_3 dimer has been observed [102] in expansion involving anhydrous HNO_3. The exothermermicity of the clustering and electron impact ionization supplies the energy which initiates the decompostion in the clusters containing a large number nitric acid molecules compared to the number of solvent water molecules.

Examination of the clustering of neutral molecules onto ions is another approach to the study of acid clusters. Cluster ions observed in the atmosphere reflect the role of acids. Strong acids are preferentially clustered to negative ions, which act as bases. Under normal atmospheric conditions, ambient ions of the type $HSO_4^-(H_2SO_4)_x(HNO_3)_y$ or $NO_3^-(HNO_3)_n$ (the latter ion is probably hydrated in the lower troposphere) have been observed throughout the lower atmosphere [103, 104]. Laboratory studies have been made of the thermodynamic stability of $NO_3^-(HNO_3)_n$ clusters [10, 105] and the reactivity of nitric acid [106] and sulfuric acid [107] with negative ions. Through these studies, an understanding of the pathways to the productions of the acid cluster and their relationship to the gas-phase acid loading of the atmosphere has been developed.

5.5 Conclusion and Outlook

From the overview of work on water cluster thermochemistry and reactions given in this chapter, it is clear that much has been learned over the years about the role of solvation on processes of wide ranging interest. Yet from a consideration of the individual findings, one becomes intrigued by the many other systems that are yet to be explored in similar ways. New insights are certain to emerge and the prospects stimulate us, and we hope the reader, to join in this fruitful and exciting field of research.

The rapidly developing area of femtosecond laser spectroscopy is yielding insights of unprecedented detail into the influence of solvation on reaction dynamics, but only the "bare surface" of this promising approach has been touched upon thus far. Additionally, it is clear that there is much to be learned from the more well developed high pressure mass spectrometry and flow reactor techniques as can be seen from the findings which have become available that relate to molecules of biological significance and others of atmospheric importance, for example.

The subject is rich and the prospects great. Indeed, the field of water cluster research is alive and healthy, and the prognosis for the future of the field is that it will have a long and prosperous life.

Acknowledgements

Financial support by the U.S. National Science Foundation, Atmospheric Sciences Division, Grant No. ATM-97-11970, is gratefully acknowledged.

References

1. A. W. Castleman Jr, P. M. Holland, D. M. Lindsay, K. I. Peterson: J. Am. Chem. Soc. **100**, 6039 (1978).
2. R. G. Keesee, A. W. Castleman Jr: J. Phys. Chem. Ref. Data **15**, 1011 (1986).
3. A. W. Castleman Jr, R. G. Keesee: Chem. Rev. **86**, 589 (1986).
4. A. W. Castleman Jr, P. M. Holland, R. G. Keesee: Radiat. Phys. Chem. **20**, 57 (1982).
5. P. Kebarle: Annu. Rev. Phys. Chem. **28**, 445 (1977).
6. I. Dzidic, P. Kebarle: J. Phys. Chem. **74**, 1466 (1970).
7. C. E. Klots: J. Phys. Chem. **85**, 3585 (1981).
8. R. G. Keesee, N. Lee, A. W. Castleman Jr: J. Am. Chem. Soc. **101**, 2599 (1979).
9. M. Arshadi, R. Yamdagni, P. Kebarle: J. Phys. Chem. **74**, 1475 (1970).
10. N. Lee, R. G. Keesee, A. W. Castleman Jr: J. Chem. Phys. **72**, 1089 (1980).
11. N. Lee, R. G. Keesee, A. W. Castleman Jr: J. Colloid Interface Sci. **75**, 555 (1980).
12. A. J. Stace, A. K. Shukla: J. Am. Chem. Soc. **104**, 5314 (1982).
13. W. R. J. Herron, T. Coolbaugh, G.Vaidyanathan, W.R. Peifer, J.F.Garvey: J. Am. Chem. Soc. **114**, 3684 (1992).
14. B. L. Upshulte, F. J. Schelling, A. W. Castleman Jr: Chem. Phys. Lett. **111**, 389 (1984).
15. J. Sunner, K. Nishizawa, P. Kebarle: J. Phys. Chem. **85**, 1814 (1985).
16. A. W. Castleman Jr., T. D. Märk: Adv. At. Mol. Phys. **20**, 65 (1985).
17. A. W. Castleman Jr., R. G. Keesee: Acc. Chem. Res. **19**, 413 (1986).
18. R. G. Keesee, A. W. Castleman Jr: In Ion and Cluster Spectroscopy and Structure. J.D.Maier, Ed. Elsevier Science: Amsterdam, 1989, 275.
19. A. W. Castleman Jr: In Kinetics of Ion Molecule Reactions, P. Ausloos, Ed., Plenum: New York, 1979; 295.
20. X. B. Wang, X. Yang, J. B. Nicholas, L. S. Wang: Science **294**,1322 (2001).

21. A. T. Blades, P. Jayaweera, M. G. Ikononmou, P. Kebarle: J. Chem. Phys. **92**, 5900 (1990).
22. J. J. Gilligan, N. E. Vieira, A. L.Yergey: manuscript in preparation.
23. N. Kuznetsova, S. L. Chi, S. Leikin: Biochem. **37**, 11888 (1998).
24. H. Kuhlewind, U. Boesl, R. Weinkauf, H. J. Neusser, E. W. Schlag: Laser Chem. **3**, 3 (1983).
25. A. W. Castleman Jr, O. Echt, S. Morgan, P. D. Dao: Ber Bunsen-Ges. Phys. Chem. **89**, 281 (1985).
26. O. Echt, P. D. Dao, S. Morgan, A. W. Castleman Jr: J. Chem. Phys. **82**, 4076 (1985).
27. S. Wei, W. B. Tzeng, A. W. Castleman Jr: J. Chem. Phys. **92**, 332 (1990).
28. S. Wei, W. B. Tzeng, A. W. Castleman Jr: J. Chem. Phys. **93**, 2506 (1990).
29. S. Wei, K. Kilgore, W. B. Tzeng, A. W. Castleman Jr: J. Phys. Chem. **95**, 8306 (1991).
30. S. Wei, A. W. Castleman Jr: Int. J. Mass Spectrom. Ion Processes **131**, 233 (1994).
31. P. C. Engelking: J. Chem. Phys. **87**, 936 (1987).
32. (a) C. E. Klots: J. Chem. Phys. **83**, 5854 (1985). (b) A. J. Stace: Z. Phys. D **5**, 83 (1983). (c) A. J. Stace: J. Chem. Phys. **85**, 5774 (1986).
33. (a) C. E. Klots: Z. Phys. D **21**, 335 (1991); **20**, 105 (1991), (b) C. E. Klots: J. Chem. Phys. **83**, 5854 (1985), (c) C. E. Klots: Nature **327**, 222 (1987), (d) C. E. Klots: Z. Phys. D **5**, 83 (1987).
34. Z. Shi, J. V. Ford, S. Wei, A. W. Castleman Jr: J. Chem. Phys. **99**, 8009 (1993).
35. C. E. Klots: Z. D. Phys. **21**, 335 (1991); ibid **20**, 105 (1991).
36. C. E. Klots: J. Chem. Phys. **83**, 5854 (1985).
37. C. E. Klots: Nature **327**, 222 (1987).
38. (a) C. E. Klots: Z. Phys. D **5**, 83 (1987) (b) C .E. Klots: Kinetic methods for quantifying magic. East Coast Symposium on the Chemistry and Physics of Clusters and Cluster Ions; Baltimore, MD, 1991.
39. S. Wei, Z. Shi, A. W. Castleman Jr: J. Chem. Phys. **94**, 3268 (1991).
40. Z. Shi, S. Wei, J. V. Ford, A. W. Castleman Jr: Chem. Phys. Lett. **200**, 142 (1992).
41. X. Zhang, A. W. Castleman Jr: J. Chem. Phys. **101**, 1157 (1994).
42. A. Selinger, A. W. Castleman Jr: J Phys. Chem. **95**, 8442 (1991).
43. A. Selinger, A. W. Castleman Jr: Proc., Int. Symp. On the Physics and Chemistry of Finite Systems: From Clusters to Crystals. **1165** (1992).
44. E. A. Steel, K. M. Merz Jr, A. Selinger, A. W. Castleman Jr: J. Phys. Chem. **99**, 7829 (1995).
45. S.-W. Lee, P. Freivogel, T. Schindler, J. L. Beauchamp: J. Am. Chem. Soc. **120**, 11758 (1998).
46. S.-W. Lee, H. Cox, W. A. Goddard, J. L. Beauchamp: J. Am. Chem. Soc. **122**, 9201 (2000).
47. S.-S. Lin: Rev. Sci. Instrum. **44**, 516 (1973).
48. J. Q. Dearcy, J. B. Fenn: J. Chem. Phys. **61**, 5282 (1974).
49. H. Shinohara, U. Nagashima, H. Tanaka, N. Nishi: J. Chem. Phys. **83**, 4184 (1985).
50. V. Q. Nguyen, X. G. Chen, A. L. Yergey: J. Am. Soc. Mass Spectrom. **8**, 1175 (1997).

51. X. Yang, X. Zhang, A. W. Castleman Jr: Int. J. Mass Spectro. Ion Processes **109**, 339 (1991).
52. X. Yang, A. W. Castleman Jr: J. Am. Chem. Soc. **111**, 6845 (1989).
53. X. Yang, A. W. Castleman Jr: J Phys. Chem. **94**, 8500 (1990), "Erratum," J. Phys. Chem. **94**, 8974 (1990).
54. X. Yang, A. W. Castleman Jr: J. Chem. Phys. **95**, 130 (1994).
55. X. Yang, X. Zhang, A. W. Castleman Jr: J. Phys. Chem. **95**, 8520 (1991).
56. D. W. Fahey, S. R. Kawa, E. L. Woodbridge, P. Tin, J. D. Wilson, H. H. Jonsson, J. E. Dye, D. Baumgardner, S. Borrmann, D. W. Toohey, L. M. Avallone, M. H. Proffitt, J. Margitan, M. Loewenstein, J. R. Podoske, R. J. Salawich, S. C. Wolfsy, M. K. W. Ko, D. E. Anderson, M. R. Schoeberl, K. R. Chan: Nature **363**, 509 (1993).
57. D. R. Hanson, A. R. Ravishankara: J. Geophys. Res. **96**, 5081 (1991).
58. H. Bohringer, D. W. Fahey, F. C. Fehsenfeld, E. E. Ferguson: Planet. Space Sci. **31**, 185 (1983).
59. P. J. Crutzen, F. Arnold: Nature **324**, 651 (1986).
60. H. Wincel, E. Mereand, A. W. Castleman Jr: J. Phys. Chem. **98**, 8606 (1994).
61. S. T. Graul, R. R. Squires: Int. J. Mass Spectrom. Ion Processes **81**, 183 (1987).
62. X. Yang, A. W. Castleman Jr: J. Am. Chem. Soc. **113**, 6966 (1991).
63. X. Yang, A. W. Castleman Jr: J. Chem. Phys. **93**, 2405 (1995).
64. X. Yang, A. W. Castleman Jr: Chem. Phys. Lett. **179**, 361 (1991).
65. A. A. Viggiano, F. Dale, J. F. Paulson: J. Chem. Phys. **88**, 2469 (1988).
66. R. A. Morris, A. A. Viggiano, J. F. Paulson: J. Am. Chem. Soc. **113**, 5932 (1991).
67. X. Yang, A. W. Castleman Jr: J. Phys. Chem. **95**, 6182 (1997).
68. B. R. W. Pinsent, L. Pearson, F. J. W. Roughton: Trans. Faraday Soc. **52**, 1512 (1956).
69. R. B. Martin: J. Inorg. Nucl. Chem. **38**, 511 (1976).
70. R. G. Khalifah: Proc. Natl. Acad. Sci., U.S.A. **70**, 1986 (1973).
71. P. Woolley: Nature **258**, 677 (1975).
72. F. A. Cotton: Advanced Inorganic Chemistry, Wiley, New York, 5th edn., 1988.
73. M. Eigen, K. Kustin, G. Maass: Z. Phys. Chem. **30**, 130 (1961).
74. D. A. Horner, R. E. Connick: Inorg. Chem. **25**, 2414 (1986).
75. T. Schindler, C. B. Berg, G. Niedner-Schatteburg, V. E. Bondybey: Chem. Phys. Lett. **229**, 57 (1994).
76. R. S. MacTaylor, J. J. Gilligan, D. J. Moody, A. W. Castleman Jr: J. Phys. Chem. A **103**, 2655 (1993).
77. R. S. MacTaylor, J. J. Gilligan, D. J. Moody, A. W. Castleman Jr: J. Phys. Chem. A **103**, 4196 (1999).
78. E. E. Ferguson, F. C. Fehsenfeld, A. L. Schmeltekopf: Adv. At. Mol. Phys. **5**, 1 (1969).
79. J. J. Gilligan, D. J. Moody, A. W. Castleman Jr: Z Phys. Chem. **214**, 1383 (2000).
80. T. Schindler, C. Berg, G. Niedner-Schatteburg, V. E. Bondybey: J. Chem. Phys. **104**, 3998 (1996).
81. A. A. Viggiano, R. A. Morris, J. M. Van Doren: J. Geophys. Res. **99**, 8221 (1994).

82. C. M. Nelson, M. Okumura: J. Phy. Chem. **96**, 6112 (1992).
83. H. Wincel, E. Mereand, A. W. Castleman Jr: J. Phy. Chem. A **101**, 8248 (1997).
84. L. C. Anderson, D. W. Fahey: J. Phys. Chem. **94**, 644 (1990).
85. S. G. Lias, J. E. Bartness, J. F. Liebman, J. L. Holmes, R. D. Levin, W. G. Mallard: J. Phys. Chem. Ref. Data, Suppl. 1 **17** (1988).
86. K. Hiraoka, S. Mizuse, S. Yamabe: J. Phys. Chem. **92**, 3943 (1988).
87. K. Fuke, K. Kaya: J. Phys. Chem. **93**, 614 (1989).
88. A. Nakajima, M. Hirano, R. Hasumi, K. Kaya, H. Watanabe, C. C. Carter, J. M. Williamson, T. Miller: J. Phys. Chem. **101**, 392 (1997).
89. D. E. Folmer, E. S. Wisniewski, S. M. Hurley, A. W. Castleman Jr: Proc. Natl. Acad. Sci. U.S.A. **96**, 12980 (1999).
90. P. Hobza: personal communication.
91. D. E. Folmer, E. S. Wisniewski, A. W. Castleman Jr: J. Phys. Chem. A **104**, 10545 (2000).
92. X. Yang, A. W. Castleman Jr: J. Geophys. Res. **96**, 22573 (1991).
93. S. B. Barone, M. A. Zondlo, M. A. Tolbert: J. Phys. Chem. A **103**, 9717 (1999).
94. L. T. Chu, L. Chu: J. Phys. Chem. B **101**, 6271 (1997).
95. O. M. Cabarcos, C. J. Weinheimer, T. J. Martínez, J. M. Lisy: J. Chem. Phys. **110**, 9516 (1999).
96. T. Su, W. J. Chesnavich: J. Chem. Phys. **76**, 5183 (1982).
97. T. Su, W. J. Chesnavich: J. Chem. Phys. **89**, 5355 (1988).
98. D. R. Hanson: Geophys. Res. Lett. **19**, 2063 (1992).
99. R. Heist, H. Reiss: J. Chem. Phys. **61**, 573 (1974).
100. B. D. Kay, V. Herman, A. W. Castleman Jr: Chem. Phys. Lett. **80**, 469 (1981).
101. W. E. Falconer, A. Buchler, J. L. Stauffer, W. Klemperer: J. Chem. Phys. **48**, 312 (1968).
102. W. K. Lee, E. W. Prohofsky: Chem. Phys. Lett. **85**, 98 (1982).
103. H. Heitmann, F. Arnold: Nature **306**, 747 (1983).
104. M. D. Perkins, F. L. Eisele: J. Geophys. Res. **89**, 9649 (1984).
105. J. A. Davidson, F. C. Fehsenfeld, C. J. Howard: Int. J. Chem. Kinet. **9**, 17 (1977).
106. F. C. Fehsenfeld, C. J. Howard, A. L. Schmeltekopf: J. Chem. Phys. **63**, 2835 (1975).
107. A. A. Viggiano, R. A. Perry, D. L. Albritton, E. E. Ferguson, F. C. Fehsenfeld: J. Geophys. Res. **87**, 7340 (1982).

Part II

Water at Interfaces

6 Properties of Water Clusters on a Graphite Sheet

K. Karapetian and K. D. Jordan

Summary. A many-body polarizable potential model has been used to investigate the interaction of the H_2O monomer and the $(H_2O)_{n=2-6}$ clusters with a graphite sheet. The water clusters on the graphite sheet are predicted to have geometries similar to those of the isolated gas-phase clusters. In the case of the water hexamer, the most stable isomer on the polarizable graphite surface is predicted to be the "open-book" species, followed by the ring, "prism" and "cage" isomers in terms of decreasing stability. In contrast, in the gas-phase, the most stable isomers of the water hexamer have prism and cage structures, followed by the book and ring isomers. The changes in the relative stability of water clusters when bound to the graphite surface arise primarily from dispersion interactions between the water molecules and the surface.

6.1 Introduction

Water is ubiquitous in chemistry, biology, and geology, and, as a result, has been the subject of numerous experimental and theoretical studies [1-19]. In recent years, water clusters have received considerable attention [3-10,14-19]. Much is now known about the structures and other properties of low-energy isomers of the $(H_2O)_n$ clusters with n as large as 10. The information on the clusters has provided new insight into the role of cooperative effects in hydrogen bonding and is proving exceedingly valuable in developing improved intermolecular potential models for water [9-13].

One of the most important questions concerning water is the extent to which its properties are modified by interactions with surfaces or by confined environments. We are particularly intrigued by the fact that surprisingly little is known about the water/graphite system. There is evidence of formation of nanometer-sized water clusters on graphite surfaces, and it has been suggested that small (*i.e.*, few molecule) clusters play a role in the growth of these nanodroplets [20]. Moreover, there is recent experimental work which suggests that at low coverage and low temperature, water forms a two-dimensional structure on the surface which converts to a three-dimensional structure upon warming [21]. However, the arrangement of the water monomers in the two-dimensional structure is unknown. In fact, to the best of our knowledge, there are no measurements of the binding energies or characterization of the structures of the water monomer or of small water clusters on the graphite surface.

In this work, we examine the interactions of small $(H_2O)_{n=1-6}$ clusters with a single-layer model of the graphite surface using a many-body polarizable potential model. In the case of the water hexamer, four different low-energy isomers with ring, cage, prism, and open-book structures are examined. These species have been the subject of several theoretical studies in the case of the gas-phase clusters [8-10 ,15-19].

6.2 Potential Models

The first challenge in modeling water on graphite is the choice of a procedure for reliably describing the water-water and water-graphite interactions. In the present study, this is accomplished by the use of model potentials. The alternative approach of using density functional methods was dismissed because of the inability of current functionals to describe long-range dispersion interactions [22] which are expected to be especially important for the water-graphite system.

The Dang-Chang (DC) model potential [10] is used to describe the water-water interactions. The DC model assumes a rigid water monomer, with the experimental gas-phase geometry (OH bond lengths of 0.9572 Å and an HOH angle of 104.5°). It employs three point charges: positive charges 8.3x10^{-20}C associated with each H atom and a negative charge of -16.6x10^{-20}C located 0.215 Å from the O atom, on the bisector of the HOH angle, displaced toward the H atoms. There is a single isotropic polarizable site located at the same position as the negative point charge, with the polarizability ($\alpha = 1.444$ Å^3) being chosen to reproduce experiment [10]. Finally, it includes 12-6 Lennard-Jones (LJ) interactions between the O atoms of different water molecules ($\sigma_{OO} = 3.234$ Å, $\varepsilon_{OO} = 0.1825$ kcal/mol). The DC model has been found to be quite reliable for predicting the structures and relative energies of small water clusters [10, 18].

The graphite lattice is assumed to be rigid. The water-graphite potential allows for Lennard-Jones interactions between the carbon and oxygen atoms, with the Lorentz-Berthelot mixing rules [23] being used to generate the C-O Lennard-Jones interaction parameters. (The carbon atom LJ parameters σ_{CC} = 3.40 Å and $\varepsilon_{CC} = 0.0556$ kcal/mol are taken from Ref. 24). In addition, an isotropic polarizable center ($\alpha = 1.52$ Å^3) is included on each C atom [25]. In a subset of the calculations, quadrupole moments of -3.03xe^{-40} Cm^2 (taken from Whitehouse and Buckingham [26]) are included on the carbon atoms and are allowed to interact with both the point charges and the polarizable centers on the water molecules.

Before describing the geometrical models, some additional comments about the treatment of the polarization of the graphite sheet are in order. On a per C atom basis, graphite is about 3.5 times more polarizable in-plane than perpendicular to the sheet [27]. Thus, in models, such as that employed here, in which interactions between the polarizable sites at the C atom positions are not explicitly treated, the in-plane value of the C-atom polarizability should be about 3.5 times larger than out-of-plane value. However, test calculations on the water cluster/graphite systems using an anisotropic polarizability (α_{zz} = 0.57 Å^3, $\alpha_{xx} = \alpha_{yy}$ = 1.995 Å^3) [25] on the C atoms gave similar structures and binding energies to those obtained using an isotropic polarizability. For example, for the water monomer/graphite system the calculations using the anisotropic polarizability on the carbon atoms gave a binding energy only 4% smaller than obtained from the calculations using the isotropic polarizability. Consequently, only the results obtained with isotropic polarizable C atoms are presented. In order to better assess the role of graphite polarization in determining the structure and stabilities of adsorbed water clusters, we also carried out calculations where polarization of the graphite was excluded.

All results reported in the paper were obtained using a cluster model of the surface. The cluster model consisted of a large acene containing 91 fused rings, comprised of a central ring surrounded by four "shells" of fused rings, with the water clusters being located near the middle of the acene. To check that this approach is indeed suitable for modeling water-graphite interactions, we also carried out test calculations using a slab model, with a 25.56 Å(X) x 29.52 Å(Y) x 100 Å(Z) supercell replicated by means of periodic boundary conditions. The large lattice constant in the Z direction, which is perpendicular to the surface, makes interactions between the layers negligible. The two approaches give very similar structures and energies for the adsorbed water clusters, and, for this reason, we report here only the results obtained from the cluster-model calculations. The cluster- and slab-model calculations were carried out using the Orient [28] and Tinker [29, 30] programs, respectively.

For the water monomer/graphite system both one- and two-layer models of the graphite surface were employed. In the two-layer model of the surface, the spacing between the layers of the graphite was taken from experiment. The introduction of the second graphite layer proved to be relatively unimportant for the binding of a water monomer to the surface, and, as a result, for dimer and larger water clusters, only the single-layer model of the surface was employed.

6.3 Results

6.3.1 The water monomer/graphite system

(i) Calculations using a single-layer model of graphite. The minimum-energy structure for water monomer/single-layer graphite system obtained from the calculations including polarization of the graphite surface has the water molecule oriented perpendicular to the graphite sheet with one OH group

pointed toward the sheet and the other roughly parallel to the sheet. This structure is reminiscent of that for the water/benzene cluster [31-34]. The binding energy of the water monomer/single-sheet graphite system (neglecting the carbon quadrupole moments) is calculated to be -2.50 kcal/mol, with -1.05 kcal/mol due to induction and -1.45 kcal/mol due to the Lennard-Jones interactions. A structure with the O atom pointed toward the surface was also optimized and found to be about 0.23 kcal/mol less stable than the "OH-down" structure. When polarization effects are omitted, the OH-down and O-down structures are calculated to be of comparable stability (binding energy $\approx$ -1.63 kcal/mol). Thus, the preference for the OH-down structure is due to the polarization of the graphite sheet. Interestingly, on metal surfaces water molecules tend to orient so that the dipole moments are parallel to the surface [35]. Our calculations indicate that on the graphite surface an isolated water molecule oriented parallel to the surface is unstable to rearrangement to the OH-down species.

The inclusion of the carbon quadrupole moments in the model potential leads to a small, but non-negligible ($\approx$ 0.46 kcal/mol) increase in the magnitude of the binding energy of an isolated water molecule to the graphite surface. (A comparable contribution of the quadrupole terms to the binding was reported previously by Steele and co-workers [36]). The slight preference for the OH-down structure remains.

High-level electronic structure calculations [31] predict that water/benzene binding energy is -3.9 kcal/mol, which is about 30% larger in magnitude than the binding energy of the water/graphite system calculated in the present study using the model including the carbon quadrupole moments. Although polarization and dispersion interactions are greater in magnitude for a water molecule interacting with graphite than with a benzene molecule, the electrostatic interactions are appreciably greater for the water/benzene system. Model potential calculations by Feller and Dang [37] indicate that the electrostatic interactions contribute about 70% of the binding (or $\approx$ -2.6 kcal/mol) of a water molecule to benzene, as compared to the -0.5 kcal/mol electrostatic contribution to the binding of a water monomer to the graphite surface calculated here. As a result, whether a water molecule binds more strongly to the graphite surface or to a benzene molecule depends on whether the increases in the polarization and dispersion interactions (in going from benzene to graphite) more than compensate for the decrease in the magnitude of the electrostatic interaction [38].

There has been one attempt (involving one of the present authors) to estimate using *ab initio* MP2 calculations the binding energy of a water molecule to a single layer of graphite [32]. This study arrived at a binding energy of -5.8 kcal/mol, almost twice as large in magnitude as that obtained from our model potential calculations. However, the uncertainty in this estimate of the binding energy is difficult to ascertain as it was obtained by applying a correction for basis set superposition error [37] (BSSE) which was quite large. (In principal, BSSE can be eliminated by adoption of large, flexible basis sets. However, in practice, this approach proves unsuccessful for large extended systems, since it is accompanied by linear dependency problems.) There is also the question of the suitability of the MP2 method for describing extended π–electron systems. Thus the "true" binding energy of a water molecule to a graphite sheet could be smaller in magnitude than the MP2 estimate of Ref. 32. On the other hand, our model potential may underestimate the binding in magnitude, possibly due to an inadequacy of the Lorentz-Berthelot mixing rules used to estimate the C-O Lennard-Jones parameters or to the neglect of interactions between the polarizable sites on the graphite surface. It is likely that the "true" binding energy of a water molecule to single-sheet graphite falls between our model potential value and the estimate based on the MP2 calculations. Clearly there is a need for experimental measurements and definitive theoretical calculations of the water/graphite binding energy.

(ii) Two-sheet model of graphite. The binding energy of the water monomer to the two-sheet graphite system (neglecting C quadrupole moments) is calculated to be -2.72 kcal/mol, with -1.16 kcal/mol due to induction and -1.56 kcal/mol due to the Lennard-Jones interactions. Comparison with the results for the single-sheet model shows that the binding of a water molecule to the surface is enhanced by only 0.2 kcal/mol in going from the one-layer to the two-layer model of graphite. Thus, a single-layer model of graphite should be useful for developing a qualitative (even, semi-quantitative) understanding of the properties of water on the graphite surface.

(iii) Diffusion of the water monomer on the graphite surface. The pathway for motion of a water monomer between adjacent rings on the surface is depicted in Fig. 6.1. The barrier for this diffusion process is calculated to be only 0.28 kcal/mol, indicating that water monomers will readily diffuse on the graphite surface except at very low temperatures.

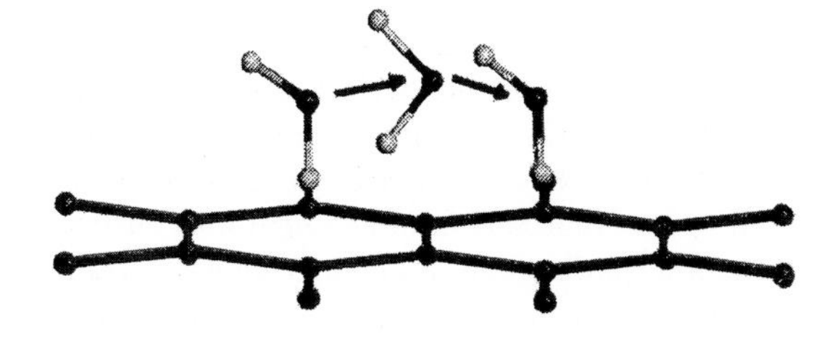

Fig. 6.1. Pathway for diffusion of a water monomer on the graphite surface.

6.3.2 Interaction of $(H_2O)_n$ clusters with a single-layer graphite surface

The global minima of the isolated water trimer, tetramer, and pentamer clusters have cyclic "ring" structures, whereas the hexamer prefers a three-dimensional structure [5, 7, 17, 39]. *Ab initio* MP2 calculations predict that for the hexamer there are cage, prism, and book-like structures very close in energy, with the cage isomer being slightly more stable when vibrational zero-point energy corrections are included [40–42]. The preference for the cage structure is also supported by experimental studies [41, 43]. As seen from Table 6.1, the relative stability of different isomers of $(H_2O)_6$ calculated using the DC model are in accord with the *ab initio* results.

Table 6.1. Interaction energies (kcal/mol) of $(H_2O)_{n=2-6}$ described with the DC model potential.

Species	$E_{induction}$	$E_{(LJ-dispers.)}$	$E_{(LJ-repuls.)}$	$E_{(electrost.)}$	$E_{(total)}$
Dimer	- 0.77	-1.51	3.11	-5.52	-4.69
Trimer	- 3.11	-4.63	9.81	-15.40	-13.33
Tetramer	- 7.57	-7.67	17.97	-26.67	-23.94
Pentamer	-10.84	-9.68	23.07	-34.39	-31.84
Hexamer-cage	-12.16	-14.95	31.11	-44.84	-40.84
Hexamer-ring	-14.20	-11.66	28.42	-41.96	-39.40
Hexamer-book	-13.63	-13.43	30.57	-43.93	-40.42
Hexamer-prism	-12.65	-15.13	31.21	-44.43	-41.00

Our calculations reveal that the structural changes in the clusters brought about by their interaction with the polarizable single-layer model of the graphite are relatively small. The optimized structures of the water cluster/graphite systems are depicted in Fig. 6.2.

Tables 6.2 and 6.3 report the net energies of the clusters interacting with the graphite surface, and Table 6.4 reports the binding energies of the clusters to the graphite surface calculated from

$$\Delta E = E((H_2O)_n/\text{graphite}) - E((H_2O)_n) \tag{6.1}$$

where $E((H_2O)_n/\text{graphite})$ is the energy of a specific $(H_2O)_n$ cluster interacting with the graphite surface, and $E((H_2O)_n)$ is the energy of the corresponding isolated water cluster. The results reported in Tables 6.2 and 6.4 are based on calculations that allow for the polarization of the graphite surface but neglect the influence of the carbon quadrupole moments, whereas Table 6.3 reports, for the graphite species, results obtained for each of the three models considered.

The most stable arrangement of the water dimer on the graphite surface is calculated to have one of the OH groups of the acceptor monomer pointed toward the surface and the other acceptor OH group almost parallel to the surface, reminiscent of the structure of the benzene-$(H_2O)_2$ complex [34]. The dimer is calculated to be bound to the surface by -5.6 kcal/mol, with polar-

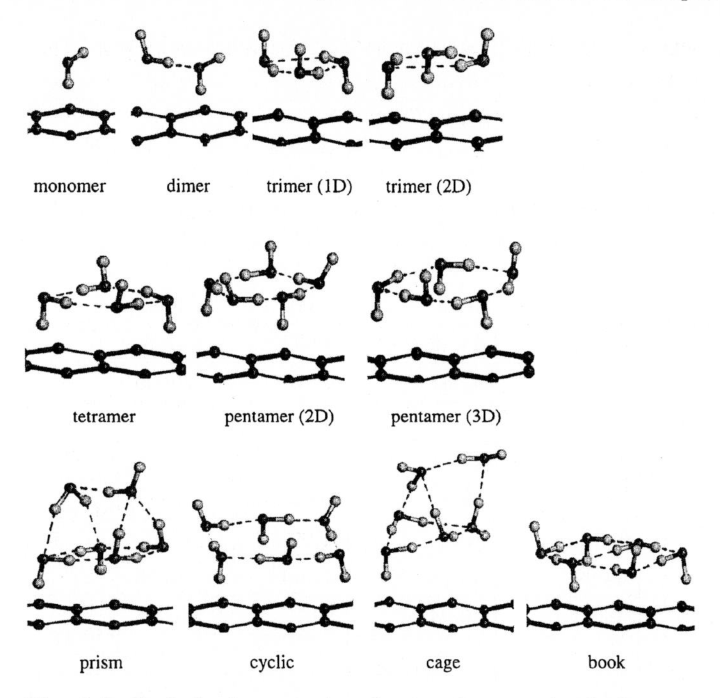

Fig. 6.2. Optimized geometries of water clusters adsorbed on a graphite sheet. Different conformers are distinguished in terms of the number of OH groups (specified in parentheses) pointed towards the surface.

Table 6.2. Interaction energies (kcal/mol) of $(H_2O)_{n=1-5}$/graphite species.[a,b]

Energy Component	monomer $(1Hdown)$	dimer $(1Hdown)$	trimer $(1Hdown)$	trimer $(2Hdown)$	tetramer $(2Hdown)$	pentamer $(2Hdown)$	pentamer $(3Hdown)$
Induction	-1.05	-3.60	-4.88	-5.89	-9.73	-13.01	-13.70
Electrost.	0.00	-5.66	-15.39	-15.22	-26.56	-34.36	-34.12
LJ-dispers.	-3.41	-8.76	-14.18	-14.77	-20.06	-25.04	-25.42
LJ-repuls.	1.96	7.75	14.90	15.85	24.61	30.86	31.59
Total	-2.50	-10.27	-19.55	-20.03	-31.74	-41.50	-41.65

[a] Results obtained using the DC model for water and the polarizable model of the graphite neglecting the quadrupole moments on the C atoms.
[b] The number of H atoms pointed toward the surface is indicated in parentheses.

Table 6.3. Interaction energies (kcal/mol) for the $(H_2O)_6$/graphite system.[a,b]

Model and energy contribution	Cage (1*Hdown*)	Prism (2*Hdown*)	Ring (3*Hdown*)	Book (3*Hdown*)
Model I				
$E_{(induction)}$	-12.23	-12.66	-14.27	-13.89
$E_{(LJ-dispers.)}$	-25.02	-26.65	-27.49	-29.11
$E_{(LJ-repuls.)}$	35.37	35.98	35.08	36.68
$E_{(electrost.)}$	-44.92	-44.48	-42.02	-42.74
$E_{(TOTAL)}$	-46.80	-47.81	-48.70	-49.06
Model II				
$E_{(induction)}$	-14.19	-17.61	-16.58	-17.12
$E_{(LJ-dispers.)}$	-26.09	-29.60	-30.20	-31.57
$E_{(LJ-repuls.)}$	36.94	40.46	38.00	39.87
$E_{(electrost.)}$	-44.99	-44.19	-41.99	-42.56
$E_{(TOTAL)}$	-48.33	-50.94	-50.77	-51.38
Model III				
$E_{(induction)}$	-14.57	19.01	-17.09	-17.48
$E_{(LJ-dispers.)}$	-26.78	-30.88	-30.87	-33.24
$E_{(LJ-repuls.)}$	38.20	43.39	39.36	41.85
$E_{(electrost.)}$	-45.79	-45.26	-42.63	-43.91
$E_{(TOTAL)}$	-48.94	-51.76	-51.23	-52.78

[a] Model I neglects both polarization of graphite and the quadrupole moments on the carbon atoms; model II includes polarization of the graphite but neglects quadrupole moments on the carbon atoms; model III includes both polarization of the graphite and interactions with the quadrupole moments on the carbon atoms.
[b] The number of H atoms pointed toward the surface is indicated in parentheses.

ization contributing -2.8 kcal/mol to the binding. Polarization is important in establishing the orientation of the dimer with respect to the surface.

For the $(H_2O)_3$/graphite system two different conformers were optimized, one with one free OH and the other with two free OH groups pointed toward the surface. The latter species is predicted to be more stable by about 0.5 kcal/mol. For the tetramer, the optimized structure has two free OH groups pointed toward the surface, whereas for the pentamer, conformers with two and three free OH groups pointed toward the surface were optimized, with the conformer with three OH groups pointed toward the surface being slightly ($\approx$ 0.1 kcal/mol) more stable. For each isomer of the water hexamer only a single binding arrangement on the surface was characterized. In the case of the book and ring isomers, the optimized structures have three free OH groups pointed toward the surface, whereas for the prism and cage isomers, the optimized structures have, respectively, two and one free OH groups pointed toward the surface.

Table 6.4. Contributions to the binding energies (kcal/mol) of $(H_2O)_{n=1-6}$ clusters to the graphite surface [a].

cluster	$\Delta E_{(induct.)}$	$\Delta E_{(LJ-dispers.)}$	$\Delta E_{(LJ-repuls.)}$	$\Delta E_{(electrost.)}$	$\Delta E_{(tot.)}$
Monomer	-1.05	-3.41	1.96	0.00	-2.50
Dimer	-2.83	-7.25	4.64	-0.13	-5.57
Trimer	-2.78	-10.14	6.04	0.18	-6.70
Tetramer	-2.16	-12.39	6.64	0.11	-7.80
Pentamer	-2.86	-15.74	8.52	0.27	-9.66
Hexamer-cage	-2.03	-11.14	5.83	-0.15	-7.49
Hexamer-ring	-2.38	-18.54	9.58	-0.03	-11.37
Hexamer-book	-3.49	-18.14	9.30	1.37	-10.96
Hexamer-prism	-4.96	-14.47	9.25	0.24	-9.94

[a] Results obtained using a single-layer model of the graphite surface and neglecting quadrupole moments on the carbons atoms. The energy differences are calculated using (6.1).

As seen from Table 6.4, the strength of the binding of the water cluster to the surface grows monotonically from the monomer (-2.5 kcal/mol) to the cyclic form of the hexamer (-11.4 kcal/mol). The polarization contribution to the binding undergoes a sizable jump between the monomer and dimer, but stays roughly constant from the dimer to the cyclic pentamer, ranging from -2.2 to -2.9 kcal/mol. The growing binding of the cluster to the surface with increasing cluster size is primarily due to the attractive part of the Lennard-Jones interactions (which correspond approximately to the dispersion contribution).

The strength of the binding of $(H_2O)_6$ to the graphite surface is calculated to be -12.4, -11.9 and -10.7 kcal/mol for the book, ring and prism isomers, respectively, but only -8.1 kcal/mol for the cage isomer. (These results are from the calculations that include both polarization of the graphite surface and the interactions with the carbon quadrupole moments.) As a result of the differences in the strengths of their interactions with the surface, the various isomers of the hexamer adopt a different energy ordering on the graphite surface than in the gas phase. Specifically, on the surface the most stable isomer is predicted to be the book, followed by the prism, ring, and cage; 1.0,

1.5, and 3.9 kcal/mol higher in energy, respectively. This is in contrast to the prism, cage, book, and ring ordering predicted for the gas phase cluster.

Dispersion interactions (as estimated by the attractive portion of the Lennard-Jones contributions) are the major factor responsible for the re-ordering of relative stabilities of the hexamer isomers on the surface, being appreciably larger for the book and ring structures (-18.1 and -18.5 kcal/mol, respectively) than for the prism and cage structures (-14.5 and -11.1 kcal/mol, respectively). The greater importance of the dispersion interactions for the book and ring isomers is consistent with these species having more water molecules close to the surface than do the prism and cage isomers. However, induction and, to a lesser extent, the interactions involving the carbon quadrupole moments are also important in establishing the relative stabilities of the different isomers of $(H_2O)_6$ on the graphite surface. The induction contributions to the surface binding range from -5.0 kcal/mol for the prism to -2.0 kcal/mol for the cage isomer, and the carbon quadrupole interactions contribute -1.4 kcal/mol to the binding of the book isomer to the surface but only -0.5 – -0.8 kcal/mol for the other three isomers.

6.4 Conclusions

A many-body polarizable potential model has been used to characterize small water clusters interacting with the graphite surface. The barrier for diffusion of a water molecule on the surface is predicted to be only 0.28 kcal/mol, which indicates that, except at very low temperatures, water molecules should move freely on the graphite surface. The calculations predict that it is energetically favorable for water molecules to assemble into clusters on the graphite surface. Although the dispersion contribution to the binding of the water cluster to the surface grows in magnitude from -7.2 kcal/mol for the dimer to -18.5 kcal/mol for the cyclic form of the hexamer, the induction contribution to the binding stays roughly constant along this series of clusters (ranging from -2.2 to -2.9 kcal/mol). On the other hand, for the different isomers of $(H_2O)_6$, both the induction and dispersion contributions to the surface binding energies differ appreciably from isomer to isomer, indicating that inclusion of polarization is important in establishing the preferred structures of water clusters on the graphite surface. Our model potential calculations indicate that for the optimized structures, the dispersion interactions between $(H_2O)_6$ and the graphite surface are appreciably larger in magnitude (by 3.7 – 7.4 kcal/mol) for the book and ring isomers than for the cage and prism.

Acknowledgements: This research was carried out with the support of a grant from the National Science Foundation. We thank Professor J. Ponder for helpful discussions concerning the Tinker program and Professor K. Johnson for valuable discussions about water/graphite interactions. Some of the calculations were carried out on the IBM 43p computers in the University's Center for Molecular and Materials Simulations. These computers were funded with grants from the NSF and IBM.

References

1. L. D. Barron, L. Hetcht, G. Wilson: Biochemistry, **36**, 13143 (1997).
2. G. W. Robinson, S. B. Zhu, S. Singh, M. W. Evans: Water in *Biology, Chemistry and Physics: Experimental Overview and Computational Methodologies* (World Scientific, Singapore, 1996).
3. "Recent Theoretical and Experimental in Hydrogen-bonded Clusters", NATO ASI Series, Vol 561, ed. S. S. Xantheas (Kluwer, Dordrecht, 2000).
4. F. B. van Duijneveldt, in *Molecular Interactions: From van der Waals to Strongly Bound Complexes*, edited by S. Scheiner, pp.157-179, (Wiley, Susex, 1997).
5. K. Liu, J. D. Cruzan, R. J. Saykally: Science **271**, 929 (1996).
6. J. Sadlej, V. Buch, J. K. Kazimirski, U. Buck: J. Phys. Chem. A **103**, 4933 (1999).
7. T. S. Zwier: Ann. Rev. Phys. Chem. **47**, 205, (1996).
8. J. Kim, K. S. Kim: J. Chem. Phys. **109**, 5886 (1999).
9. C. J. Burnham, J. Li, S. S. Xantheas, M. Leslie: J. Chem. Phys. **110**, 4566 (1999).
10. L. X. Dang, T.-M. Chang: J. Chem. Phys. **106**, 8149 (1997).
11. M. W. Mahony, W. L. Jorgensen: J. Chem. Phys.**112**, 8910 (2000).
12. P. J. van Maaren, D. van der Spoel: J. Phys. Chem. B **105**, 2618 (2001).
13. H. Saint-Martin, J. Hernandez-Cobos, M. I. Bernal-Uruchurtu, I. Ortega-Blake, H. J. C. Berendsen: J. Chem. Phys. **113**, 10899 (2000).
14. K. Koga, R. D. Parra, H. Tanaka, X. C. Zeng: J. Chem. Phys.**113**, 5037 (2000).
15. C.-J. Tsai, K. D. Jordan: Chem. Phys. Lett. **213**, 181 (1993).
16. D. J. Wales, M. P. Hodges: Chem. Phys. Lett. **286**, 65 (1998).
17. S. S. Xantheas, T. H. Dunning: J. Chem. Phys. **99**, 8774 (1993).
18. J. M. Pedulla, K. D. Jordan: J. Chem. Phys. **239**, 593 (1998).
19. K. E. Franken, M. Jalaie, C. E. Dykstra: Chem. Phys. Lett. **198**, 59 (1992).
20. D. Chakarov, B. Kasemo: Phys. Rev. Lett. **81**, 5181 (1998).
21. M. Luna, J. Colchero, A. M. Baro: J. Phys. Chem. **103**, 9576 (1999).
22. S. Kristyan, P. Pulay: Chem. Phys. Lett. **229**, 175 (1994).
23. J. P. Hansen, I. R. McDonald: Theory of Simple Liquids; Academic Press: London, 1986.
24. L. Battezzatti, C. Pisani, F. Ricca: J. Chem. Soc., Faraday Trans. **71**, 1629 (1975).
25. A. D. Crowell: Surface Sci. Letts. **111**, L667 (1981).
26. D. B. Whitehouse, A. D. Buckingham: J. Chem. Soc., Faraday Trans. **89**, 909, (1993).
27. J. C. Phillips, *Covalent Bonding in Crystals, Molecules, and Polymers* (University of Chicago Press, 1969).
28. Orientcode-A.Stone,{http://fandango.ch.cam.ac.uk/}.
29. Tinkercode-J.Ponder,{http://dasher.wustl.edu/tinker/}.

30. J. W. Ponder, F. M. Richards: J. Comput. Chem. **8**, 1016-1024 (1987).
31. D. Feller: J. Phys. Chem. A **103**, 7558 (1999).
32. D. Feller, K. D. Jordan: J. Phys. Chem. A **104**, 9971 (2000).
33. P. Tarakeshwar, H. S. Choi, S. J. Lee, K. S. Kim: J. Chem. Phys. **111**, 5838 (1999).
34. S. Y. Frederick, K. D. Jordan, T. S. Zwier: J. Phys. Chem. **100**, 7810 (1996).
35. T. Lankau, I. L. Cooper:J. Phys. Chem. A **105**, 4084 (2001).
36. A. Vernov, W. A. Steel: Langmuir **8**, 155 (1992).
37. S. F. Boys, F. Bernardi: Mol. Phys.**19**, 553 (1970).
38. L. X. Dang, D. Feller: J. Phys. Chem. B **104**, 4403 (2000).
39. C.-J. Tsai, K. D. Jordan: J. Chem. Phys. **99**, 6957 (1993).
40. J. K. Gregory, D. C. Clary: J. Phys. Chem.**100**, 18014 (1996).
41. 41. K. Kim, K. D. Jordan, T. S. Zwier: J. Am. Chem. Soc.**116**, 11568 (1994).
42. H. M. Lee, S. M. Suh, J. Y. Lee, P. Tarakeshwar, K. S. Kim: J. Chem. Phys.**112**, 9759 (2000).
43. K. Liu, J. D. Cruzan, R. J. Saykally: Science **271**, 929 (1996).

7 Phase Equilibria and Transitions of Confined Systems in Hydrophobic and Aqueous Environments

Hideki Tanaka and Kenichiro Koga

Summary. In the present chapter, we deal with two extreme cases of confinement involving water molecules. In one type, a small hydrophobic molecule is confined in a cage made from hydrogen bonded water molecules, which is called clathrate hydrate. We show a basic statistical mechanical theory to estimate the thermodynamic stability of the clathrate hydrate and how it works and fails for various clathrate hydrates upon conventional assumptions. A new method to estimate the stability of the clathrate hydrates is presented, which turns out to be more powerful to predict dissociation pressures of clathrate hydrates from only intermolecular interactions. In another type, water molecules are confined to cylindrical or slit-like hydrophobic environments and, upon changing temperature or pressure, undergo liquid-solid phase transitions which arise from an interplay between confinement by hydrophobic surfaces and hydrogen bonds among water molecules. Structures of the solid phases are different from those of any bulk ices and are primarily determined by the geometry of confinement. This type of confinement also brings about novel phase behaviors such as liquid-solid continuous phase transformation and liquid-solid polyamorphism.

7.1 Introduction

It has long been recognized that liquid water and ice do not share many physical properties with ordinary substances such as hydrocarbons or noble gases, for examples, the appearance of the density maximum at 4 °C, large heat capacity, the decrease of viscosity of liquid water under pressure , and the negative thermal expansivity at low temperature for ice [1–3]. Much effort has been devoted to accounting for those unique qualities of this ubiquitous substance on the earth, which is either in the form of liquid or crystal. In doing this, use has been invariably made of hydrogen bonds on which various models have been based; most of water molecules in liquid state are connected by hydrogen bonds in ambient temperature and pressure ranges. Upon freezing at atmospheric pressure, long-ranged order is established by completion of the tetrahedral coordination and the perfect hydrogen bonded network prevails over an entire system. It is the hydrogen bond that gives rise to a rich variety of ice crystalline forms in a wide range of pressure; at least 13 ice polymorphs are known in bulk phase [4]. Moreover, liquid (amorphous) water is believed to have two phases depending on temperature and pressure, (though it is still controversial)[5]. At least, several experimental and simulation studies support this view. This polyamorphism, in turn, accounts for anomalous properties of water mentioned above. If we turn our gaze upon mixtures of water with other molecules of comparable size to water, there exist three crystalline forms known as clathrate hydrates (structure I, II, and H) in which water molecules form a host lattice incorporating guest molecules [6, 7].

Some properties in a confined system may be far from those observed in a bulk phase such as the freezing point of water, which depends directly on the size of the free space. The confinement has also a close connection with hydration structure around macromolecules. Here we consider two types of confinement; either confinement of small hydrophobic molecules by water or confinement of water by hydrophobic environments. These confinements have been investigated via molecular dynamics (MD) and Monte Carlo (MC) computer simulations together with a newly developed method to evaluate the free energy of water and ice.

In the first type, confinement is by water. Water molecules constitute the principal component and mostly hydrophobic compounds are encapsulated in vacant space. The clathrate hydrate has a different structure from ice and is a kind of guest-host compound [6, 7]. The cages in structure I and II hydrates are pentagonal dodecahedron, tetrakaidecahedron, and hexakaidekahedron. Stability of a clathrate hydrate depends significantly on temperature and gas pressure of guest species to be encapsulated. The thermodynamic stability of these substances has long been estimated using empirical parameters [6, 7]. If those parameters can be obtained solely from intermolecular interactions, it is of great advantage to predict thermodynamic stability of clathrate hydrates that encage various kinds of guest species without invoking laboratory experiments or introducing empirical parameters. We show how to estimate thermodynamic stability more accurately by removing unphysical assumptions of the theoretical treatment thus far proposed. The free energy of those solids is calculated from various components separately; the interaction energy at temperature 0 K, the vibrational free energy, and the configurational entropy arising from desordering of protons and cage occupation [8]. This enables us to evaluate the thermodynamic stability of clathrate hydrates from only intermolecular potentials currently available.

Stable morphology of crystalline states at given temperature and pressure under some external fields has been predicted by means of MD simulation. This includes confinement by hydrophobic walls with variable cell size and shape, which was proposed by Andersen and by Parrinello and Rahman [9, 10]. These methods provide powerful tools to reproduce or predict phase transitions among various crystalline forms [11–13]. In the latter part of this chapter, we introduce new forms of stable ice under either lateral or axial pressure surrounded by two parallel planer walls and also in a carbon nanotube in advance of laboratory experiments. It is highly probable that ice forms under confinement in nano-scale space are quite different from those known under bulk conditions. The transitions observed in computer simulations correspond to the limit of mechanical stability. Thus, the true phase transition point should be evaluated by calculating the free energies of crystalline structures (and the liquid state) accurately in order to predict a phase diagram of various crystalline forms (freezing transition) on a temperature-pressure plane. We show, on the basis of computer simulations and free energy calculation, evidences of rich phase behavior of water – simultaneous occurrence of liquid-liquid and glass transitions in quasi two-dimensional water and new phases of ice and possible existence of a solid-liquid critical point in quasi one-dimensional water encapsulated in a carbon nanotube.

7.2 Clathrate Hydrates

7.2.1 Structure and the related properties

Gas hydrates comprise *guest molecules* encaged in a hydrogen bonded network of host water molecules. The clathrate hydrate structures, known as I and II [6, 7], differ from ice structures as shown in Fig. 7.1 and are stable only in the presence of guest molecules, which can be either hydrophobic or hydrophilic in nature. The unit cell of the structure I is cubic and contains 46 water molecules forming two kinds of cages; 2 smaller pentagonal dodecahedra and 6 larger tetrakaidecahedra. The unit cell of the structure II is also cubic and is composed of 16 smaller pentagonal dodecahedra and 8 larger hexakaidecahedra made from 136 water molecules. Some properties pertinent to the two kinds of clathrate hydrates are tabulated in Table 7.1 [7]. Those cages are combined together by sharing faces as displayed in Fig. 7.1. The other type, known as structure H, is less common [14]. We will not hereafter refer to the structure H clathrate hydrate but a similar treatment as proposed to the structure I or II can be made.

Table 7.1. Properties pertinent to the unit cells of clathrate hydrate I and II

structure	I		II	
number of water molecules	46		136	
cell dimension (Å)	12.03		17.31	
cage type	small	large	small	large
faces	5^{12}	$5^{12}6^2$	5^{12}	$5^{12}6^4$
number of cages	2	6	16	8
size of cage (Å)	7.82	8.66	7.80	9.37

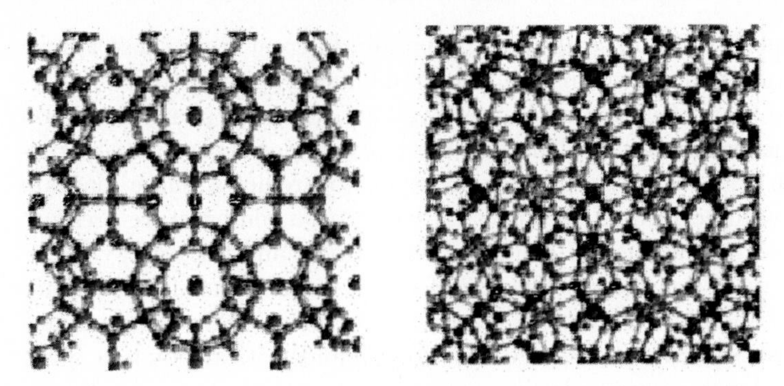

Fig. 7.1. Structure of clathrate hydrate I (left) and clathrate hydrate II (right).

Table 7.2. Diameters of guest molecules (Å) that form clathrate hydrate structure I and II, and the ratios relative to the effective cage sizes for smaller and larger cages. Effective cage size is defined as 'the cage size - 2.9Å'. A cage occupied by a guest molecule is marked with an asterisk.

guest	diameter	I(small)	I(large)	II(small)	II(large)
Ar	3.8	0.772	0.660	0.775*	0.599*
Kr	4.0	0.813	0.694	0.816*	0.619*
N_2	4.1	0.833	0.712	0.836*	0.634*
O_2	4.2	0.853	0.729	0.856*	0.649*
CH_4	4.36	0.886*	0.757*	0.889	0.675
Xe	4.58	0.931*	0.795*	0.934	0.708
H_2S	4.58	0.931*	0.795*	0.934	0.708
CO_2	5.12	1.041	0.889*	1.044	0.792
C_2H_6	5.5	1.118	0.955*	1.122	0.851
c-C_3H_6	5.8	1.178	1.007*	1.182	0.897*
$(CH_2)_3O$	6.1	1.240	1.059*	1.244	0.943*
C_3H_8	6.28	1.276	1.090	1.280	0.971*
iso-C_4H_{10}	6.5	1.321	1.128	1.325	1.005*

Clathrate hydrates are crystalline but nonstoichiometric compounds with respect to guest molecule arrangement; all the cages are not necessarily occupied and the cage occupancy depends on the temperature and pressure of the guest compound in equilibrium with the clathrate hydrate. Although there are no defects in hydrogen bonds of an empty clathrate hydrate structure, no clathrate hydrate without guest molecules has been found in nature. Clathrate hydrates are stable only when the interaction between guest and water molecules dominates over the sum of the two unfavorable terms: (1) the entropy decrease arising from confinement of guest molecules in small void cages, and (2) the free energy for formation of empty clathrate hydrate structure from ice or liquid water.

Since the cages are made from the firmly hydrogen bonded water molecules, the size of a cage is restricted to be distributed in a very narrow range as given in Table 7.1. Thus, the size of guest species must have an upper bound. Because the attractive interaction is responsible for stabilization of clathrate hydrates, an accommodated guest molecule is smaller than a butane molecule which has the critical size for balance between attractive and repulsive interactions. Some of the guest molecules for natural gas hydrates are listed in Table 7.2.

7.2.2 Van der Waals and Platteeuw (vdWP) theory

The thermodynamic stability of clathrate hydrates has been estimated by the vdWP theory [15]. This theory is applicable to any sort of hydrate, either type (I or II) and either simple or complex. (Simple means that there is only one kind of guest species in the clathrate hydrate.) Here, we describe only an essential part of it for convenience of the later argument, restricting the discussion to simple clathrate hydrates. Consider a system being in equilibrium with a gas phase of guest molecules. Each unit cell has $m_w = 46$ (136) water molecules and a maximum of 8 (24) guest molecules for structure I (structure II in parenthesis). The total number of smaller cages, N_s, is $2n_w$ ($16n_w$), and that of larger cages, N_l, is $6n_w$ ($8n_w$) for a system composed of n_w unit cells. If the number of the occupied larger and smaller cages are j_l and j_s, the relevant canonical partition function Z_{j_s,j_l} at temperature T is given by

$$Z_{j_s,j_l} = \binom{N_s}{j_s}\binom{N_l}{j_l} \exp(-\beta A_w^0) \exp(-\beta j_l f_l - \beta j_s f_s), \tag{7.1}$$

where A_w^0 denotes the free energy of the empty clathrate hydrate. In the above equation, β is $1/k_B T$ where k_B is the Boltzmann constant. Here, f_l and f_s are the free energy changes due to the introduction of a guest molecule in a larger and a smaller cage.

If the free energy arising from the water-water interactions does not change upon encaging, the free energy of cage occupancy by nonlinear ($l = 3$) or by linear ($l = 2$) molecule is given by

$$f = -k_B T \ln\{s^{-1} \prod_{j=1}^{l} (I_j k_B T/2\pi\hbar^2)^{1/2} (m k_B T/2\pi\hbar^2)^{3/2}\}$$

$$\int_v \int_\Omega \times \exp[-\beta w(\mathbf{r}, \boldsymbol{\Omega})] d\mathbf{r} d\boldsymbol{\Omega}, \tag{7.2}$$

where the integration spans the single cage v with respect to the position $\mathbf{r}$ and all orientations with respect to the angles, $\boldsymbol{\Omega}$, and s stands for the symmetry number of the guest molecule. The mass of the guest molecule and the j-th moment of inertia of the three (or two) principal axes are denoted by m and I_j respectively, and $w(\mathbf{r}, \boldsymbol{\Omega})$ stands for the interaction potential between water molecules and the guest inside the corresponding cage. The former part of the above equation is the contribution from the kinetic energy with $\hbar = h/2\pi$, where h is the Planck constant. (In the case of a spherical guest, the integration with respect to $\mathbf{\Omega}$ and the associated kinetic part are omitted.)

Transforming from the canonical to the grandcanonical ensemble with respect to guest species whose chemical potential is μ_g, the grand partition function, Ξ, is written as

$$\Xi = \exp(-\beta A_w^0)[1+\exp\{\beta(\mu_g - f_s)\}]^{N_s}[1+\exp\{\beta(\mu_g - f_l)\}]^{N_l}. \quad (7.3)$$

An averaged number of guest molecules in the hydrate, $\langle N \rangle$, is given by

$$\begin{aligned}\langle N \rangle &= \partial \ln \Xi / \partial(\beta\mu_g) \\ &= N_s \exp[\beta(\mu_g - f_s)][1+\exp\{\beta(\mu_g - f_s)\}]^{-1} \\ &+ N_l \exp[\beta(\mu_g - f_l)][1+\exp\{\beta(\mu_g - f_l)\}]^{-1}. \end{aligned} \quad (7.4)$$

The chemical potential of water, μ_c, can be calculated from

$$\begin{aligned}\mu_c &= -k_B T \partial \ln \Xi / \partial N_w \\ &= \mu_c^0 - \frac{k_B T}{m_w}\{m_s \ln[1+\exp\{\beta(\mu_g - f_s)\}] + m_l \ln[1+\exp\{\beta(\mu_g - f_l)\}]\}, \end{aligned} \quad (7.5)$$

where μ_c^0 is the chemical potential of the (hypothetical) empty hydrate. The most important part of the vdWP theory is described by (7.1)-(7.5). It is assumed in the vdWP theory [15] that (i) the cage structure is not distorted by the incorporation of guest molecules, (ii) the free energy is independent of the occupation of other cages, (iii) a guest molecule inside a cage moves in the force field created by water molecules fixed at lattice sites and there is no coupling between host and guest molecular motions, and (iv) that classical mechanics is adequate to describe these systems.

It seems that the coupling between guest and host water molecules is not negligible for a large guest species. A large guest molecule may give rise to modulation of host water vibrational frequency. Then, the free energy of cage occupation includes an extra contribution, which is not taken into account in the original vdWP theory. The assumption imposed on the vdWP theory can be eliminated by the following method. We assume that the free energy due to a large guest can be associated with harmonic vibrational motions and that the free energy can be approximated by (7.2) in the case of a smaller guest molecule.

Since the potential energy curve is well fitted to a quadratic function shown in Fig. 7.2(a), it is approximated to a harmonic oscillator. The free energy, g, for a harmonic oscillator is evaluated according to the classical mechanical partition function for a harmonic oscillator as

$$g = k_B T \int \ln(\beta\hbar\omega) h(\omega) d\omega, \quad (7.6)$$

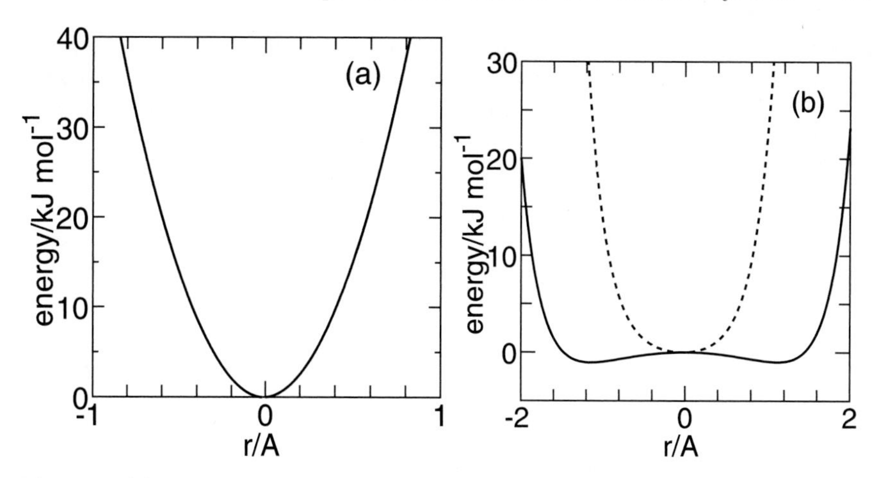

Fig. 7.2. (a) The potential energy of a guest propane molecule (approximated to a spherical LJ particle) in a larger cavity of the clathrate hydrate II. (b) The potential energy of a guest argon atom in a large cavity (solid line) of the clathrate hydrate II and in a smaller cavity (dotted line).

where $h(\omega)$ is the density of states normalized to the number of degrees of freedom per molecule for each system. In other words, the free energy of the system is given by the entropy arising from occupancy of guest molecules and the free energy due to the motions of guests inside the cages either by the intermolecular vibrational motions or the single particle integration as given in (7.2). In the former case, the free energy of cage occupation is given by

$$f = \Delta g + u_g, \tag{7.7}$$

where u_g is the minimum value of $w(\mathbf{r}, \boldsymbol{\Omega})$ in the integrand of (7.2) and Δg denotes the vibrational free energy difference per guest between empty and occupied hydrates [16–18].

It is appropriate to examine how the harmonic approximation works for occupation by a large guest molecule. Let us consider two extreme sizes, a large propane and a small argon. Here a propane molecule is tentatively approximated to a spherical Lennard-Jones (LJ) particle whose size, σ, and energy, ϵ, parameters are 5.6 Å and 2.0 kJ mol^{-1} [19] in order to extract a significant contribution from the vibrational frequency shift of the host lattice caused by guest molecules. The LJ parameters σ and ϵ for argon are set to 3.4 Å and 1.0 kJ mol^{-1}, respectively [19]. For the water-guest interaction, we assume the Lorentz-Berthelot (LB) rule with the LJ parameters for oxygen atoms set equal to those for TIP4P water; $\sigma_{\mathrm{OO}} = 3.2$ Å and $\epsilon_{\mathrm{OO}} = 0.65$ kJ mol^{-1} [19].

The potential energy of a guest propane molecule interacting with surrounding water molecules ($w(\mathbf{r})$ in (7.2)) is calculated along three axes in Cartesian coordinates and one of those energy curves is plotted in Fig. 7.2(a) as a function of displacement of the guest molecule from the minimum energy position. The potential surface of propane is well represented by a harmonic up to 20 kJ mol^{-1} from the minimum potential energy and therefore it is reasonable that the potential energy is expanded only to quadratic order.

The potential energy curves of guest argon in the structure II hydrate are shown in Fig. 7.2(b). Contrary to the propane, the potential energy curves are not quadratic even in the smaller cage. In the larger cage, the potential energy curve has two minima along each coordinate axis. Thus, a small guest molecule is only weakly coupled with the host water molecules and the guest motion is rather irrelevant to the condition as to whether the host water molecules are fixed or allowed to move. Therefore, use of (7.2) is justified for a smaller guest [17].

7.2.3 Dissociation pressure of clathrate hydrates

The equilibrium condition for formation and dissociation of a clathrate hydrate is given by (here we consider an equilibrium between ice and hydrate neglecting anharmonic vibrational free energy)

$$\mu_i = \mu_c \tag{7.8}$$

where μ_i is the chemical potential of ice at a given condition. By subtracting μ_c^0 from the both sides of (7.8), we obtain from (7.6)

$$\begin{aligned} \mu_i - \mu_c^0 = -\frac{k_B T}{m_w} \{ & m_s \ln[1 + \exp\{\beta(\mu_g - f_s)\}] \\ & + m_l \ln[1 + \exp\{\beta(\mu_g - f_l)\}]\}. \end{aligned} \tag{7.9}$$

The left hand side of (7.9), the chemical potential difference of water molecule between ice and the empty hydrate, $\mu_i - \mu_c^0$, is calculated from the sum of the differences in the vibrational free energy according to (7.6) and the configurational energy, which is rather independent of the gas pressure of the guest in the narrow range up to a few MPa. The right hand side is significantly dependent on the pressure through μ_g, though both f_s and f_l are insensitive to the pressure. In calculating μ_g, a gas phase can be regarded as an ideal one. If necessary, the second virial coefficient may be taken into account in calculating both the density and chemical potential of the gas phase guest in equilibrium with the hydrate.

Here, we show how the dissociation pressure at a given temperature is calculated for clathrate hydrates encaging guests such as propane and argon. As shown above, it is calculated via vibrational free energy for propane and via single particle partition function inside a cage for argon. The choice for argon is obvious; the effect of guest on the host lattice is small and the harmonic approximation is not good as plotted in Fig. 7.2(b). The free energies of occupation by argon thus calculated are −28.97 kJ mol^{-1} for smaller cage and −30.48 kJ mol^{-1} for larger cage, which lead to a correct dissociation pressure at temperature 273.15 K as shown below. On the other hand, the low frequency modes associated with translational motions, in the hydrate occupied by propane, shift toward higher frequencies in comparison with empty hydrates. The shift is thermodynamically unfavorable to stabilize the hydrate (see (7.6)). For the kinetic stability or melting of hydrate, however, this shift to higher frequency in the presence of guest molecules serves to prevent hydrates from collapsing, owing to reduction of amplitudes of vibrational motions. The vibrational free energy and the potential energy are given in Table 7.3. The largest portion of the free energy arises from the interaction between water and the guest propane molecules. The free energy arising from the vibrational motions of the guest molecules coupled with the host lattice, Δg, is negative. The free energy based on (7.2), f_{vdwp}, differs from $\Delta g + u_g$, the sum of the vibrational free energy difference and the interaction energy between water and guest molecules, (7.7). To evaluate the effect by a coupling of the host-guest term, the second derivative of the potential energy $\mathbf{V}$ is decomposed into 4 submatrices, $\mathbf{V}_{ww}$, $\mathbf{V}_{wg}$, $\mathbf{V}_{gw}$, $\mathbf{V}_{gg}$, where suffix w and g indicate the potential is differentiated with respect to water and guest coordinate. Here, $\mathbf{V}_{ww}$ and $\mathbf{V}_{gg}$ are diagonalized separately, setting $\mathbf{V}_{wg} = \mathbf{0}$. The (static) influence of the guest is incorporated into $\mathbf{V}_{ww}$ through the interaction between the guest and water. The free energies associated with $\mathbf{V}_{ww}$ for the fully occupied hydrate denoted by g'(host) and associated with $\mathbf{V}_{gg}$ denoted by g'(guest) are calculated. In Table 7.3, $\Delta g'$(host) and g'(guest) $+u_g$ are also given. The free energy difference in the host between occupied and empty hydrates, $\Delta g'$(host), is evaluated by neglecting the three lowest frequency modes which correspond to the whole host translation. The value of g'(guest)$+u_g$ is in good agreement with f_{vdwp} evaluated by (7.2). According to the original assumptions made in the vdWP theory, $\Delta g'$(host) should be zero. However, $\Delta g'$(host) is positive and large in a real hydrate. This arises from the fact that in the presence of guest molecules, some modes relevant to motions of water molecules shift to higher frequency regions although the guest molecules are fixed to the centers of cages. Moreover, this $\Delta g'$(host) accounts for most of the difference between the free energies evaluated by the mode analysis, (7.6), and by the single particle integration, (7.2). The free energy due to the coupling between host water and guest through $\mathbf{V}_{wg}$ is negative but very small as shown in Table 7.3.

We examine how the dissociation pressure of the (spherical) propane clathrate hydrate depends on the approximation in the free energy calculation. If ice is in equilibrium with hydrate at the given temperature, then the chemical potential of ice, μ_i, equals μ_c, as in (7.8), and therefore $\mu_i - \mu_c^0 = \mu_c - \mu_c^0$, assuming $\mu_i - \mu_c^0$ is independent of the guest gas pressure unless it is too high. The dissociation pressure, p_d, (Fig. 7.3) is obtained from the intersection between the chemical potential curve and the horizontal line corresponding to the difference in chemical potential between ice and empty hydrate, $\mu_i - \mu_c^0$, which is calculated to be -0.73 kJ mol^{-1}. The occupation number per unit cell is 7.97 at p_d= 0.56 MPa by a classical mechanical

Table 7.3. Free energy difference of intermolecular vibration, Δg, and potential energy, u_g, between empty and fully occupied hydrates at potential energy minimum structure. Energy is in kJ mol^{-1}. The free energy of a guest molecule, based on (7.2), is also given, which is denoted by f_{vdwp}. Primed values are evaluated by removing guest-host coupling terms $\mathbf{V}_{wg}$ and $\mathbf{V}_{gw}$

free energy	propane	argon (large)	argon (small)
$\Delta g + u_g$	−45.21		
u_g	−42.23	−11.43	−17.61
$\Delta g'(\text{guest}) + u_g$	−49.18		
$\Delta g'(\text{host})$	4.55		
f_{vdwp}	−48.93	−30.48	−28.97

partition function of harmonic oscillators. The occupation number calculated from (7.2) is 7.97 at p_d=0.10 MPa. Clearly, (7.2) gives a lower dissociation pressure compared to (7.7). Therefore, the influence of the guest molecule on the host lattice is fairly large and cannot be neglected [16–18]. Comparison with experiment will be made below. In the case of argon, guest atoms occupy both larger and smaller cages. The free energies of cage occupation are calculated according to (7.2) and listed in Table 7.3. The calculated dissociation pressure at 273.15 K is 10.4 MPa, which should be compared with the experimental one, 9.0 MPa.

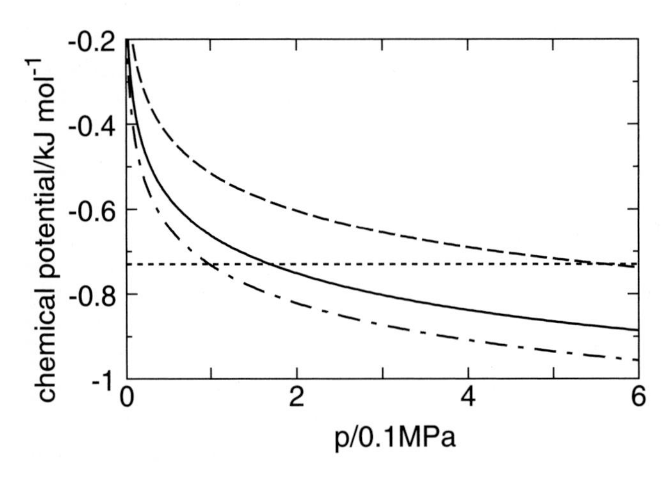

Fig. 7.3. Dissociation pressures of propane hydrate at 273.15 K. Solid line; non-spherical and anharmonic, dashed line; spherical and harmonic, dash-dot line; spherical and vdWP. Horizontal line; free energy difference between ice and empty clathrate hydrate II.

7.2.4 Further development

It is not expected that the anharmonic contribution to the free energy from the host molecules is negligible. This is especially serious when guest molecules are not spherical ones. This is because the rotational motions in the cages are essentially anharmonic. The anharmonic free energy can be calculated by several ways which turn out to be efficient. Here we show merely results from a revised method taking account of the anharmonic free energy to give a more accurate phase diagram, in which the anharmonic free energy is evaluated by MC simulations with the Gaussian statistics [21]. The chemical potential difference between occupied and empty hydrates is plotted in Fig. 7.3 for nonspherical (harmonic + anharmonic terms) and spherical (harmonic term) guest propane molecules together with that calculated from the original vdWP theory. The experimental dissociation pressure is 0.17 MPa [7], which should be compared to our present result, $p_d = 0.17$ MPa, to the harmonic oscillator approximation examined above, $p_d = 0.56$ MPa, and to the original vdWP theory $p_d = 0.10$ MPa. The occupation number of the cage per unit lattice is 7.9 in all the methods.

A similar approach has been applied to the ethane hydrate. The experimental dissociation pressure is 0.53 MPa [7], which is very close to our present result, $p_d = 0.50$ MPa, but is different from the harmonic oscillator approximation, $p_d = 0.24$ MPa, and from the original vdWP theory, $p_d = 0.16$ MPa. The occupation number of the cage per unit lattice ranges from 5.5 to 5.6 among the three methods. The present extension of the vdWP theory results in much better agreement with experiment than any other method previously proposed as far as dissociation pressures of propane and ethane guest molecules are concerned.

7.3 Water under extreme confinement

Now we consider water under extreme confinement by hydrophobic surfaces. It is meant by the term "extreme" that at least one dimension of the confined space is of the order of a few molecular diameters or of nanometers. Examples are found in biological systems (e.g., water between and inside proteins, membranes and cells), micropore systems (e.g., water in zeolites, fullerenes and porous glasses), and geological systems (e.g., water between mica crystals). It is meant by "hydrophobic" that no water molecules can form hydrogen bonds with the confining surfaces. When water is confined in hydrophobic pores, the hydrogen-bond network structure, which would extend infinitely otherwise, is forced to terminate at the surfaces. In such narrow and hydrophobic spaces, the entire hydrogen-bond network is so frustrated that the average potential energy per molecule rises by more than a few $k_B T$, and therefore properties of water would be significantly altered, rather than be merely perturbed. In particular in a low-temperature range where water displays properties uncommon to other simple liquids, water could respond to the extreme confinement quite differently from the way of simple liquids. Phase equilibria and phase transitions of water under such extreme confinement at low temperatures are reviewed here.

7.3.1 Water inside carbon nanotubes

The inner space of the carbon nanotubes is an ideal system for studying properties of confined substances since it has simple and well-defined geometry: the cylindrical quasi-one-dimensional space. Although experiments focusing on confined fluids or solids inside carbon nanotube have started recently, properties of confined water itself have not been reported at this moment. However, there are several computer simulation studies of water confined inside a carbon nanotube or a model cylindrical pore [22–25]. Here we focus on the structure and the phase behavior of pure water at low temperatures studied by computer simulations.

There are at least thirteen crystalline phases in the phase diagram of bulk water and remarkably all of them satisfy the ice rule (except for ice X in which a proton locates in the middle position between two neighboring oxygen atoms), i.e., every water molecule in each ice phase is fully-hydrogen bonded. This shows that water has a unique network-forming ability that is flexible (because the network structure can take different forms depending on pressure and temperature) and strong (because any form is a perfect network). We will see that the network-forming ability persists even under severe confinement.

Stability of ice nanotubes at 0 K in vacuum. Before we examine phase behavior of water confined in the quasi-one-dimensional space, let us ask a question if it is possible to construct a perfect hydrogen bond network in a quasi-one-dimensional space. The simplest form of a perfect network we can imagine is a prism formed by equally-spaced n-gonal rings (Fig. 7.4). The network is topologically identical to the 2-dimensional $n \times m$ square lattice network with n finite and under the periodic boundary condition in that direction. The standard ice rule is satisfied if every OH arm along each edge of the prism is oriented in one direction and every OH arm along each ring circulates in one direction. These quasi-one-dimensional hydrogen bond networks have topologically-perfect connectivity but suffer distortion from the ideal tetrahedral structure with a certain degree depending on n. It was shown, as we see later, that liquid water confined in a cylindrical space freezes into these ice forms with $4 \leq n \leq 7$, referred to as ice nanotubes due to the fact that their diameter is of nanometer scale.

Relative stability of the quasi-one-dimensional ice forms (as a function of their length) was examined by the classical intermolecular potential models of water and by the ab initio quantum calculation [26]. When the number of molecules is finite and small, the square ice form was found to be most stable,

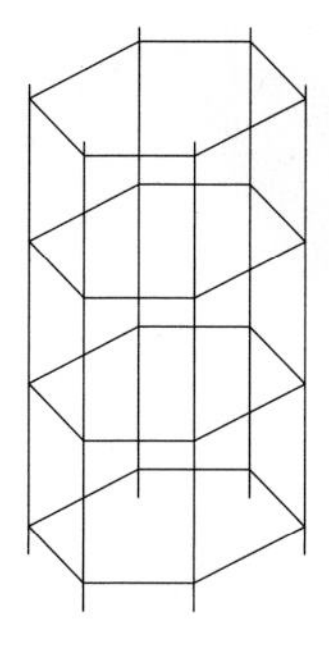

Fig. 7.4. Simplest form of perfect network in a quasi-one-dimensional space. This example is a hexagonal-prism type network ($n = 6$).

which is consistent with the previous studies of water clusters [27]. However, it was found that there is a crossover of stability from square to other ice forms,with increasing number of molecules; in the limit of infinite length the pentagonal or hexagonal ice nanotube was found to be most stable.

Of all configurations satisfying the bulk ice rule, ice nanotubes in which the (vertical) direction of hydrogen bonds along each edge alternates are significantly more stable than others. A consequence of the additional ice rule is that ice nanotubes with n even (i.e., square and hexagonal) have no net dipole but those with n odd (i.e., pentagonal and heptagonal) inevitably have a net dipole.

Formation of ices in carbon nanotubes. The phase behavior of water confined in zig-zag single-walled carbon nanotubes (SWCNs) was studied by MD simulations and free-energy calculations [24]. The diameter and helicity of a SWCN are uniquely defined by two integer indices, say (R,Q), and zig-zag SWCN's are those with the second index Q = 0; i.e, they are characterized by the single number R or by (R,0). Five different nanotubes with indices $R = 13 - 18$ were examined, whose diameter ranges from 11.1 to 13.4 Å. For each system, a series of MD simulation starts from an arbitrary arrangement of water molecules at a fixed high temperature (e.g., $T = 350$ K) under a fixed axial pressure (e.g., $P_{zz} = 50$ MPa). After equilibration, the phase behavior is examined by decreasing the temperature in steps. The hydrogen bond network, of liquid water confined in the hydrophobic environment, is less promoted than the bulk counterpart, as opposed to the conventional picture of hydrophobic effect seen in liquid water around a small hydrophobe. At sufficiently low temperatures, however, the liquid-like disordered structures turn into crystalline structures (Fig. 7.5), which are identified as ice nanotubes. Extremely-small or nearly-zero diffusivity of molecules confirms that the low-temperature phases are indeed solids.

The diameter of the carbon nanotube plays a crucial role in selecting a crystalline form of the ice nanotube. Under $P_{zz} = 50$ MPa, the square ice nanotube forms in (13,0) and (14,0) carbon nanotubes, pentagonal, hexagonal, and heptagonal ice nanotubes form in (15,0), (16,0), and (17,0) systems, respectively. It is an interesting coincidence that the $(R, 0)$ carbon nanotube

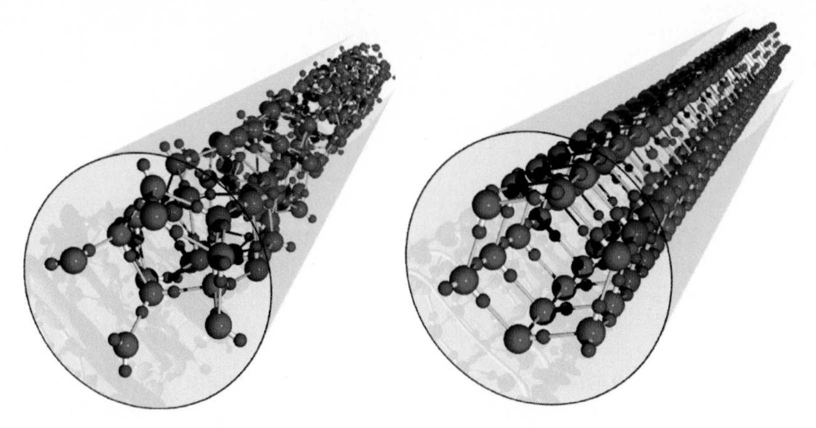

Fig. 7.5. Liquid (left) and solid (right) phases of TIP4P water confined in the (16,0) carbon nanotube. The solid phase has the hexagonal ice nanotube structure.

works as a template for $(R-10)$-gonal ice nanotube, except for R=13. In the widest (18,0) carbon nanotube, however, no crystalline structure has been found.

Nature of phase transformations in confined water. Two apparently-opposite views are possible for the freezing behavior of confined water in the carbon nanotube. One is that freezing into ice nanotubes must be a first-order phase transition, as there is no liquid-solid critical point. The other is that any phase change in the carbon nanotube must be continuous as no first-order transition is possible in one-dimensional systems. Thus, the result of simulation would contradict either view as to whether continuous or first-order. However, that is not necessarily the case because the basis of the first view is the experimental fact for bulk systems while the second one is rigorously proved only for one-dimensional systems with limited types of pair potentials. In the following we see that both continuous and first-order-like phase transitions are expected.

Figure 7.6 gives the potential energy of water inside six different SWCNs plotted as a function of temperature. When $R = 13$, 14, and 15 at $P_{zz} = 50$ MPa the potential energy decreases continuously with decreasing temperature while when $R = 16$ and 17 at the same pressure it drops abruptly at a certain point upon cooling and shows a large hysteresis upon heating. In each case the density and the structural order parameters exhibit the same continuous or discontinuous behavior. It was found that the phase behavior does depend on pressure too. Under a higher pressure, water in a (15,0) carbon nanotube freezes into the hexagonal ice nanotube via a first-order-like phase transition. (Note it freezes into a pentagonal form via a continuous change under 50 MPa.) More puzzling is the behavior of water confined in a (14,0) carbon nanotube. It undergoes a continuous phase change to a square

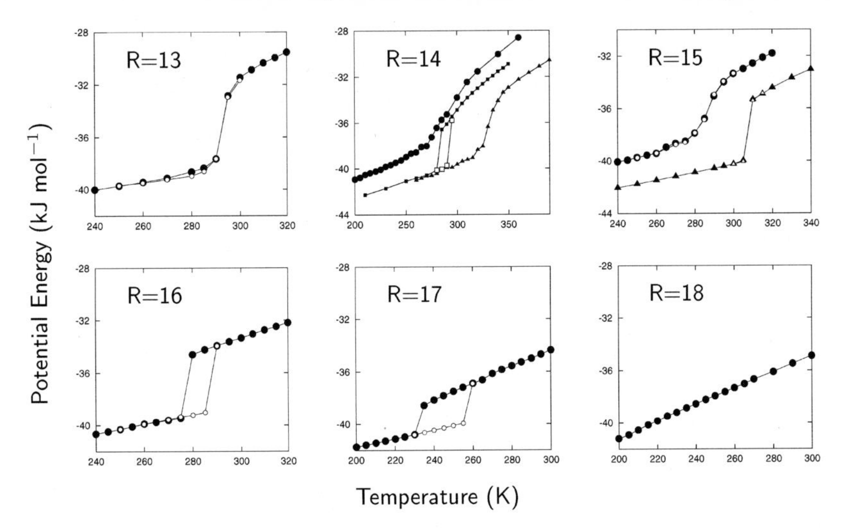

Fig. 7.6. Temperature dependence of the potential energy for TIP4P water confined in the carbon nanotubes. R is the index of carbon nanotube. The axial pressure is 50 MPa (circles), 200 MPa (squares), and 500 MPa (triangles). Filled and unfilled symbols indicate the cooling and heating processes, respectively.

ice tube at 50 MPa, a first-order transition to a pentagonal ice tube at 200 MPa, a continuous transformation to a pentagonal ice tube at 500 MPa. To understand the phase behavior of confined water, it is important to construct a phase diagram for each $(R, 0)$ carbon nanotube.

Phase diagrams for water in different carbon nanotubes. A set of isotherms for a (14,0) system in the pressure-volume plane shows that there exists a temperature below which the phase transformation proceeds discontinuously and above which it is continuous (Fig. 7.7). Such a temperature is known as a critical temperature, usually found for gas-liquid and liquid-liquid phase equilibria. This suggests that the first-order phase boundary extends from low T and low P_{zz} to high T and high P_{zz} and terminates at a critical point (Fig. 7.8). Let us call two phases separated by the boundary the square and pentagonal phases and examine the two phases along the coexistence curve from a very low temperature to the critical temperature.Then we find that both phases are solids in lower temperature region, i.e., square and pentagonal ice nanotubes but the square phase gradually turns to be liquid-like with increasing temperature while the pentagonal phase remains solid-like. Thus, it seems that the well-known fact that no liquid-solid phase boundary terminates at a critical point does not hold in this confined system.

For the (15,0) system, we find the phase boundary between pentagonal and hexagonal phase. The two phases are both solid phases at low temperatures but the pentagonal phase gradually turns into a liquid state with increasing temperature. No critical point is found for this system. For the

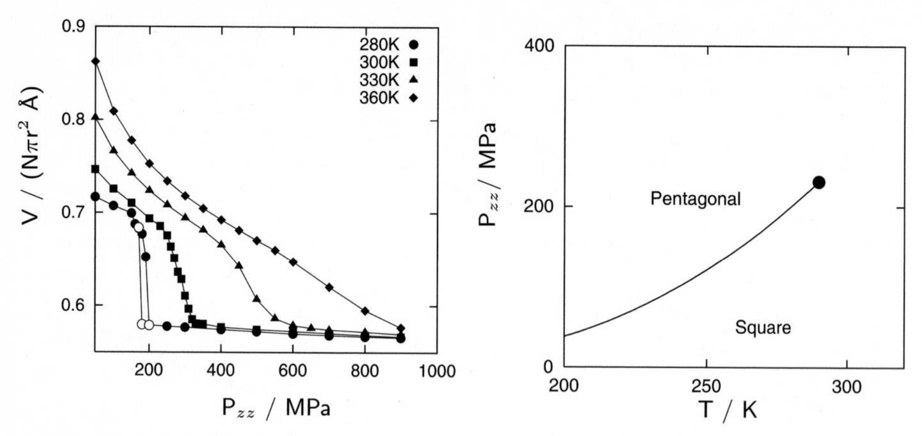

Fig. 7.7. (Left) Isotherms of water confined to the (14,0) carbon nanotube

Fig. 7.8. (Right) Schematic phase diagram for water confined to the (14,0) carbon nanotube.

(16,0) and (17,0) systems, the phase boundary seems always between liquid and solid phases. Again no critical point is found for these systems.

Global phase diagram of water in carbon nanotubes. More complete understanding of the phase behavior of confined water is made possible by a three-dimensional global phase diagram in the $TP_{zz}D$-space, where D is the diameter of the cylindrical pore. Here we assume D as a continuous variable although in practice it is discrete in the case of the carbon nanotube.

Consider a system with D slightly larger than that of the (14,0) system. Then one would find that the phase boundary between square and pentagonal phases is shifted slightly downward with respect to that in the (14,0) system. That is, the two-phase equilibrium P_{zz} decreases with increasing the radius of the carbon nanotube at given T. This is understood qualitatively by Le Chatelier's principle as follows. Increase in D decreases the normal pressure P_N on the cylindrical wall, which in turn induces a process having an opposite effect, i.e., a process driving the system to the pentagonal phase. (Note that the pentagonal phase is bulkier than the square phase and therefore has larger P_N.) More quantitative information is given by the Clapeyron equations for the quasi-one-dimensional system by which one can obtain the rates of the boundary shift: $(\partial P_{zz}^{\mathrm{eq}}/\partial D)_T$ and $(\partial T^{\mathrm{eq}}/\partial D)_{P_{zz}}$. Results of MD simulations at various D have verified the above discussion [33]. Moreover another phase boundary – a boundary between liquid and square phases – was found for the systems even narrower than the (13,0) system. Therefore, the reason that no first-order phase transition is found for the (13,0) system is not that the system is too narrow for any phase transition to be observed but that it is too narrow for the square-pentagonal phase transition and too wide for the liquid-square phase transition. In the global phase diagram, the phase boundaries are surfaces with very steep slopes with respect to the D axis. What we have seen in each system ($R = 14, 15$, etc) is the cross section of the 3-d global phase diagram at fixed D.

7.3.2 Phase behavior of water confined in slit nanopores

The strong ability of water to form a hydrogen bond network manifests itself not only in bulk and quasi-one-dimensional systems but also in quasi-two-dimensional systems. It was observed from MD simulations that water confined between two flat hydrophobic surfaces freezes into a bilayer ice crystal [29] or a bilayer of amorphous ice [30]. These are strong first-order transitions between phases with a highly-defective network and a (nearly) perfect network.

Here we review the quasi-two-dimensional confined water phases focusing on the equilibrium conditions, structures at low temperatures, similarity and difference between confined and bulk water, and possible discontinuity in the solvation force curve between hydrophobic surfaces due to the phase transition of confined water.

Thermodynamic description of confined systems. There are several ways to confine a fluid in a quasi-two-dimensional space. For example, one may confine a fluid in a rigid slit pore whose width is fixed or between two surfaces under a constant load. Each way of confining a fluid corresponds to a different thermodynamic "constraint" on the system. Thus it is important to use the thermodynamic description appropriate for a particular constraint when experimental or simulation results are analyzed There is such a case that a different constraint on the system (e.g., fixed width or fixed load) leads to different phase behavior of the same fluid.

Let us consider three ways of confining a fluid in a quasi-two-dimensional space as shown in Fig. 7.9. The first is a closed system with a fixed area density N/A (number of molecules per unit area) under a fixed load P_{zz} (normal pressure). This condition mimics the surface force apparatus (SFA) experiment [31] in which a thin film of liquid confined between two surfaces is under a fixed load. The second is a closed system with a fixed width H (separation distance between two surfaces) under a fixed lateral pressure P_{xx}. The lateral pressure refers to the pressure tensor components parallel to the surfaces. The third is an open system with a fixed width. Each system has a different equilibrium condition: the constant load in the first system is achieved by allowing volume change in the normal direction to the surfaces; the lateral pressure in the second system is kept constant by allowing volume change in the lateral direction; and the constant chemical potential μ in the third system is assured by making the confined system open to the bulk environment.

The fundamental differential for the first and second systems is given by

$$dU = TdS + \mu dN - P_{xx}HdA - P_{zz}AdH \tag{7.10}$$

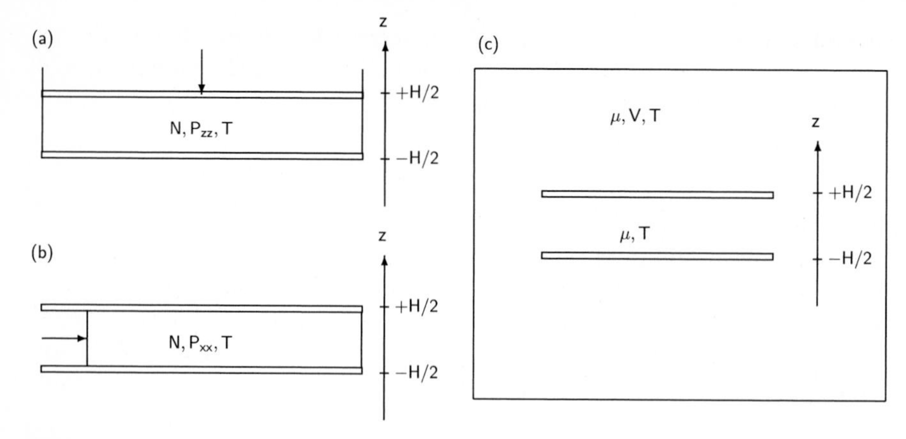

Fig. 7.9. Three different ways of confining a fluid in a quasi-two-dimensional space. (a) $NP_{zz}T$-constant system, (b) $NP_{xx}T$-constant system, and (c) μVT-constant system.

where U is the internal energy, T the temperature, S the entropy, and A the area of the confining surface. The appropriate thermodynamic potential for the constant-load system is given by

$$\Phi_z = U - TS + P_{zz}AH, \tag{7.11}$$

because any state of equilibrium is achieved under the condition of fixed N, T, A, and P_{zz}. Likewise, the thermodynamic potential for the constant-lateral pressure system is given by

$$\Phi_x = U - TS + P_{xx}AH. \tag{7.12}$$

One then finds the following differentials:

$$d\Phi_z = -SdT + H\Delta PdA + AHdP_{zz} + \mu dN \tag{7.13}$$

and

$$d\Phi_x = -SdT - A\Delta PdH + AHdP_{xx} + \mu dN \tag{7.14}$$

where $\Delta P = P_{zz} - P_{xx}$. The equilibrium conditions for the constant-load and constant-lateral pressure systems are $[\delta\phi_z]_{N,P_{zz},A,T} = 0$ and $[\delta\phi_x]_{N,H,P_{xx},T} = 0$, respectively.

The thermodynamic description of the open system in which two parallel plates are immersed in a bulk fluid focuses on the excess quantities – the differences between the inhomogeneous open system and the corresponding homogeneous open system with the same μ, V, and T [32]. The fundamental differential is

$$dU = -PdV + TdS + \mu dN + 2\sigma dA - FdH \tag{7.15}$$

where F is the total force exerted on each plate immersed in the fluid and σ is the fluid-plate surface tension. The thermodynamic potential is the grand potential $\Omega = U - TS - \mu N$, and the excess part is $\Omega^{\text{ex}} = \Omega - \Omega^b$, where superscript b denotes quantities of the homogeneous bulk system. The infinitesimal change in Ω^{ex} is then written as

$$d\Omega^{\text{ex}} = -S^{\text{ex}}dT - N^{\text{ex}}d\mu + 2\sigma dA - FdH \tag{7.16}$$

where the superscript "ex" denotes the excess quantity. The equilibrium condition is $[\delta\Omega^{\text{ex}}]_{\mu,T,A,H} = 0$

The global phase diagrams for these quasi-two-dimensional one-component systems are given in three-dimensional thermodynamic spaces as in the quasi-one-dimensional systems. Using intensive variables $v = V/N$, $s = S/N$, $a = A/N$, $\eta = S^{\text{ex}}/A$, $\Gamma = N^{\text{ex}}/A$, and $f = F/A$, slopes of the phase boundaries in the three-dimensional spaces are given by the Clapeyron equations [32, 33]: For the constant-load system

$$\begin{aligned} \frac{dT}{dP_{zz}} &= \frac{v^\alpha - v^\beta}{s^\alpha - s^\beta}, \quad \frac{dP_{zz}}{da} = -\frac{(H\Delta P)^\alpha - (H\Delta P)^\beta}{v^\alpha - v^\beta}, \\ \frac{da}{dT} &= \frac{s^\alpha - s^\beta}{(H\Delta P)^\alpha - (H\Delta P)^\beta}, \end{aligned} \tag{7.17}$$

for the constant-lateral pressure system

$$\begin{aligned} \frac{dT}{dP_{xx}} &= \frac{v^\alpha - v^\beta}{s^\alpha - s^\beta}, \quad \frac{dP_{xx}}{dH} = \frac{(a\Delta P)^\alpha - (a\Delta P)^\beta}{v^\alpha - v^\beta}, \\ \frac{dH}{dT} &= -\frac{s^\alpha - s^\beta}{(a\Delta P)^\alpha - (a\Delta P)^\beta}, \end{aligned} \tag{7.18}$$

and for the open system

$$\frac{dT}{d\mu} = -\frac{\Gamma^\alpha - \Gamma^\beta}{\eta^\alpha - \eta^\beta}, \quad \frac{d\mu}{dH} = -\frac{f^\alpha - f^\beta}{\Gamma^\alpha - \Gamma^\beta}, \quad \frac{dH}{dT} = -\frac{\eta^\alpha - \eta^\beta}{f^\alpha - f^\beta}. \tag{7.19}$$

Bilayer ice phase. It is much harder to observe a freezing transition of bulk water in MD simulation (unless under a strong electric field [34]) than in a refrigerator; freezing is a very slow process in the time scale of the molecular simulation [35]. In fact one can study many interesting properties of supercooled water by simulation without having apprehension of freezing. In contrast, it has been found that confined water in the constant-load system freezes quickly enough to be observed by simulation provided that the following conditions are satisfied. The first is hydrophobicity of the surfaces between which water is confined. The second is that the area density be chosen to 0.212 Å^{-2}. The third is that the load be not too small. This means that there are about two water molecules per each 3Å × 3 Å area, which corresponds to a bilayer film of water. The third condition is necessary for confined water to be a homogeneous thin film. Should these conditions be satisfied, it is straightforward to observe a crystallization of confined liquid water when the high-temperature phase is cooled stepwise in the MD simulation [29].

The phase change observed has common characteristics of a first-order phase transition as any other crystallization does. The freezing of confined water is accompanied by large and abrupt decrease in potential energy (4.6 kJ/mol) due to the formation of a perfect hydrogen bond network. Melting is observed at some temperature higher than the freezing temperature. It was also observed that depending on the constant load applied, confined water either contracts or expands as it freezes.

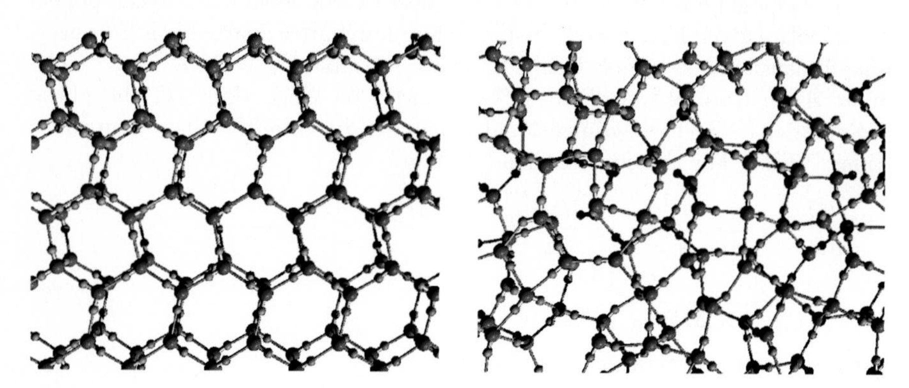

Fig. 7.10. Top views of the bilayer ice (left) and water (right).

The crystalline solid phase has a quasi-two-dimensional perfect hydrogen-bond network of bilayer structure – so the solid confined water is called bilayer ice. Figure 7.10 shows a view from a direction vertical to the bilayer ice surface. If we neglect orientation of each molecule (or, equivalently, configuration of hydrogen atoms), the two layers have an identical distorted honeycomb lattice structure, each node of which corresponds to an oxygen atom. Perfect connectivity of the hydrogen bond network is realized in the bilayer lattice as follows: every water molecule is hydrogen-bonded laterally to three neighbors in the same layer and vertically to the forth neighbor in the other layer. The bilayer ice satisfies, in addition to the common ice rules, two new ice rules, violation of which causes significant increase in potential energy and/or decrease in residual entropy. Due to this fact, the residual entropy of the bilayer ice is calculated exactly as $S^{\mathrm{bilayer}} = (Nk/4)\ln 2$, which is 43% of the residual entropy of ice Ih.

The phase boundary between liquid water and the bilayer ice can be examined by (7.17). Upon the freezing transition, the volume change $v^\alpha - v^\beta$ (where α and β denote the bilayer ice and liquid phase, respectively) is negative at 50 MPa and 150 MPa and slightly positive at 1 GPa. The corresponding entropy change $s^\alpha - s^\beta$ is always negative. Thus the slope is positive at low and medium P_{zz} and turns to be negative at high P_{zz}.

Bilayer amorphous phase. A phase transition in confined water is also observed when the slit-like pore is rigid (i.e., H is fixed) and the lateral pressure P_{xx} is kept constant. In this case, however, the resulting low-temperature phase is an amorphous phase rather than the crystalline phase [26].

Phase behavior very unique to this confined system is not that the low-temperature phase is amorphous but that the liquid-to-amorphous phase change is a first-order transition rather than a glass transition. Temperature dependence of the potential energy (at P_{xx} =0.1 MPa and H = 8.7 Å) shows two significant features of the first-order transition (Fig. 7.11): (i) there is an abrupt change in energy (at 270 K) within a small temperature range and (2) a hysteresis is observed in the reversed heating process. Discontinuous changes are observed not only for the potential energy but also for the density, the normal pressure, and the structure. The amount of the energy change is comparable to that for freezing of bulk water to ice Ih and that for freezing of confined water to the bilayer ice [29]. It is, however, much larger than that for the conjectured liquid-liquid transition of supercooled water [36].

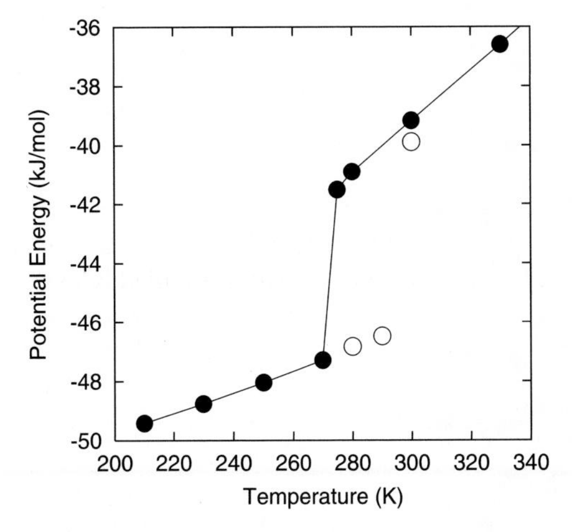

Fig. 7.11. Temperature dependence of the potential energy for water confined to a rigid slit pore at P_{xx} =0.1 MPa. Filled and unfilled circles indicate the cooling and heating processes, respectively.

It was shown from long-time simulations that the diffusivity before and after the large drop in energy (at 275 K and 270 K respectively) differs by four orders of magnitudes. This is again comparable to the change upon freezing of real water or TIP4P water to ice Ih. The drastic change in diffusivity is due to the phase change from liquid to solid (or solid-like) state.

Structure of the solid (or solid-like) phase was found to be very similar to that of the bilayer ice. Almost every water molecule is hydrogen-bonded to its four neighbors, it is a bilayer structure, and the spatial arrangement of oxygen atoms in one layer is superimposed on the other. It is clear that the substantial decrease in potential energy and the loss of fluidity at 270 K are due to the formation of the perfectly hydrogen-bonded network. The most important difference from the bilayer ice is its lack of periodicity in the two-dimensional structure (Fig. 7.12). That is, the low-temperature phase is an amorphous phase solid. It is interesting to see that the two-dimensional hydrogen-bond networks of the bilayer ice and the bilayer amorphous phase are exact analogues of the honeycomb structure and the continuous-random network – the well-known two-dimensional models for crystal and glass [37]. The two-dimensional network structure of the bilayer amorphous phase has a wide distribution in the bond angle and several kinds of n-gonal rings: 77% of hexagon (n =6), 11% of pentagon (n =5), 11% of heptagon (n =7) and 1% of the rest (n =4, 8, 9, 10). The significant fraction of pentagonal and heptagonal rings indicates that they are *not* exceptional imperfections but rather characteristic constituents of the amorphous phase.

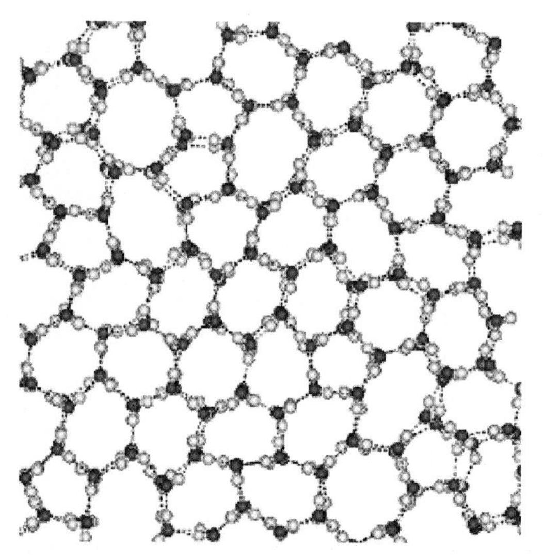

Fig. 7.12. Top view of the bilayer amorphous structure.

Existence of a phase boundary between two distinct amorphous or liquid phases of a pure substance is referred to as polyamorphism [44]. It has been reported [38] that silicon [38, 39], carbon [41, 42], and phosphorus [43] exhibit polyamorphic liquid-liquid transitions. Polyamorphism in water has also been observed as the solid-solid transition between the high density amorphous (HDA) and the low-density amorphous (LDA) phases by Mishima and co-workers [44, 45]. Although still controversial, the liquid-liquid transition of water has been implied by molecular simulation in a metastable supercooled region inaccessible by experiment [46–48]. The transition in confined water in the rigid slit pore is a liquid-solid polyamorphic transition and its first-order character is much clearer than that ever found in bulk water by simulation. This is also an uncommon example of the glass transition that a liquid freezes into an amorphous solid via a first-order transition.

Similarity of quasi-two-dimensional water and bulk water. It was pointed out by MD studies of confined water and bulk water that there are common features in the relationships among three phases: a crystalline phase and two phases (disordered phase I and II) involved in polyamorphism. Here let disordered phase I be the liquid phase stable in a temperature range between freezing and boiling points and let disordered phase II be the liquid or amorphous phase that is usually metastable at any thermodynamic state. Then disordered phase I and II for confined water correspond to the liquid and the bilayer amorphous phase. For bulk water, disordered phase I and II are high-density and low-density amorphous phases (observed by experiments at a high-pressure region) or high-density liquid and low-density liquid phases (observed by simulations at a low-pressure region).

Table 7.4. Properties characterizing the higher-energy amorphous phase (HEA), the lower-energy amorphous phase (LEA), and the crystalline solid phase in both quasi-two-dimensional system (Q2D) and three-dimensional bulk system (3D).

	Potential energy (kJ/mol)		4 H-bond species [a]		6-membered rings [a]	
Phase	Q2D	3D	Q2D	3D	Q2D	3D
HEA	−47.63	−53.5	73.1	80.6	35.4	46.4
LEA	−51.94	−54.9	99.7	88.3	76.7	62.8
Crystal	−52.98	−55.9	100.0	100.0	100.0	100.0

[a] - percentages

Potential energy and structural properties of the three phases for each system are summarized in Table 7.4. There are two significant features common to confined and bulk systems:
(i) The potential energy difference between disordered phase I and II is greater than that between disordered phase II and the crystalline phase.
(ii) Upon the polyamorphic transition from disordered phase I to II, the distribution of molecules not having coordination number 4 nearly vanishes or substantially decreases.
That is, concerning potential energy and structure of the hydrogen-bond network, disordered phase II is much closer to a crystalline phase than to disordered phase I. Whether or not this holds for other substances with polyamorphism remains to be examined.

7.3.3 Phase boundary between liquid and bilayer amorphous phases

The phase boundary between liquid and bilayer amorphous phases in TP_{xx}-plane for fixed H is examined using (7.18). Since $\Delta s < 0$ and $\Delta v > 0$ at the liquid-to-bilayer amorphous phase transition, the boundary has a negative slope in TP_{xx}-plane. The phase boundary extends into a metastable region of negative pressures ($P_{xx} < 0$), where the slope becomes positive. This trend was verified by simulations of isobaric paths at $P_{xx} = 0.1$, 100, 300 MPa. With increasing pressure the slope becomes smaller in magnitude (tends to be parallel to the T axis) and the transition is no longer observed along an isobaric path of 1 GPa, at least up to 190 K. The highest transition temperature of TIP4P water, confined in the rigid slit pore, should lie between 275 and 290 K at around 0.1 MPa. The negative slop of the phase boundary was also verified by observing the transition at around 300 MP upon decompression along an isothermal path of 250 K.

7.3.4 Phase transition and solvation force induced by confinement

So far we have seen the phase behavior of confined water along the isobaric path (fixed P_{zz} or P_{xx}) and along the isothermal path (fixed T), as we do in bulk system. Now we see the phase behavior along the isobaric *and* isothermal path by changing the wall-wall separation distance H [33].

For the closed system (see Fig. 7.4b) at $P_{xx} = 0.1$ MPa and $T = 280$ K, the force P_{zz} exerted on the walls oscillates with changing H, reflecting the structure of the confined water. At 270 K, however, the force curve exhibits discontinuous changes at several distances and a large hysteresis is observed when H is reduced and increased (Fig. 7.13a). Those discontinuous changes in force curve reflects phase transitions in confined water. For example, upon reducing H the system undergoes a freezing transition at around $H = 9$ Å, and then a melting transition at $H = 7.5$ Å. From structural analysis the solid phase was found to be the bilayer amorphous phase. Large hystereses were also observed upon increasing H.

The phase behavior of confined water in the open system (see Fig. 7.13b) was also examined by changing H at fixed P and T [33]. Also fixed is the chemical potential of confined water since $\mu(\text{bulk}) = \mu(P,T)$ and $\mu(\text{bulk}) = \mu(\text{confined})$. In this system the force exerted on the walls by water molecules is the solvent-induced part f of the solvation force. At 270 K and 0.1 MPa, the force curve exhibits oscillatory behavior when H is decreased. The two semifinite hydrophobic walls attract each other at around $H = 11$ Å by 30 MPa. With further decreasing H, the force turns to repulsive and increases continuously until sudden drainage of confined water takes place. The solvation force then becomes attractive. It was suggested recently that the drying transition of a confined fluid gives rise to an attractive solvation force between the two walls at large length scales [49]. In the timescale of this simulation,

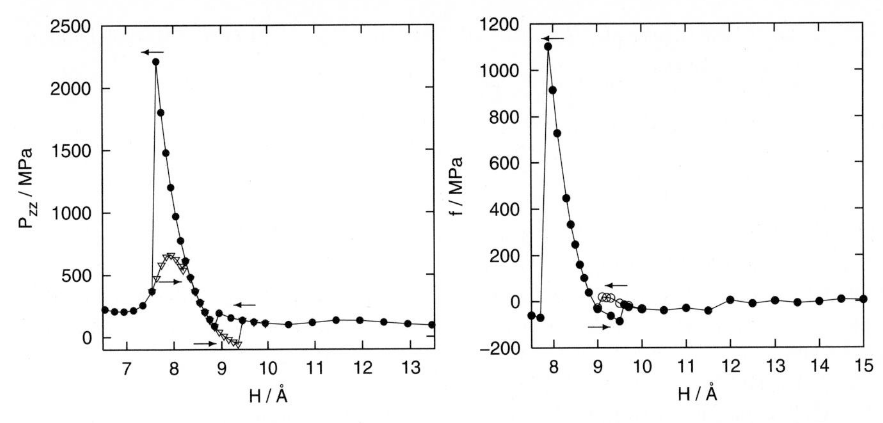

Fig. 7.13. Solvation force curves for (**a**) closed system at T =270 K and P_{xx} = 0.1 MPa and (**b**) open system at T =250 K and P = 0.1 MPa. The arrows indicate the direction that H is changed.

however, the drying transition was observed only at a small H. At 250 K (a temperature around the estimated melting point of the TIP4P water) the solvation force curve exhibits a sudden drop at 9 Å followed by a rapid increase (Fig. 7.13b). It was again confirmed that the sudden drop in f is due to the liquid-bilayer amorphous phase transition in confined water. When H is increased reversely, the force becomes attractive in a certain range of H, where the confined water in the solid state is under *negative* load, and then turns abruptly to be repulsive due to the melting transition in confined water.

Acknowledgments

The authors thank Professors X. C. Zeng and I. Ohmine, Drs. K. Kiyohara and G.T. Gao for a long term cooperative research work on water, ice and clathrate hydrates. This work is supported by Grant-in-Aid from the Ministry of Education.

References

1. D. Eisenberg, W. Kauzmann: *The Structure and Properties of Water* (Oxford University Press, London 1969).
2. I. Ohmine, H. Tanaka: Chem. Rev. **93**, 2545 (1993).
3. G. Dantl: Zeits. für Phys. **166**, 115 (1962).
4. C. Lobban, J.L. Finney, W.F. Kuhs: Nature, **391**, 268 (1998).
5. O. Mishima, H. E. Stanley: *Nature* **396**, 329 (1998).
6. D. W. Davidson: 'Clathrate Hydrates', in *Water - A Comprehensive Treatise*, Vol.5, ed. by F. Franks (Plenum, New York, 1973).

7. E. D. Sloan: *Clathrate Hydrates of Natural Gases*, 2nd edn. (Marcel Dekker, New York, 1998).
8. L. Pauling: J. Am. Chem. Soc. **57**, 2680 (1935).
9. H. C. Andersen: J. Chem. Phys. **72**, 2384 (1981).
10. M. Parrinello, A. Rahman: Phys. Rev. Lett. **45**, 1196 (1980).
11. P. M. Allen, D. J. Tildesley: *Computer Simulation of Liquids* (Oxford University Press, London, 1987).
12. D. C. Rapaport: *The Art of Molecular Dynamics Simulation* (Cambridge University Press, Cambridge, 1995).
13. D. Frenkel, B. Smit: *Molecular Simulation* (Academic Press, San Diego, 1996).
14. J. A. Ripmeester, J. A. Tse, C. I. Ratcriffe, B. M. Powell: Nature **325**, 135 (1987).
15. J. H. van der Waals, J. C. Platteeuw: Adv. Chem. Phys. **2**, 1 (1959).
16. H. Tanaka, K. Kiyohara: J. Chem. Phys. **98**, 4086 (1993).
17. H. Tanaka, K. Kiyohara: J. Chem. Phys. **98**, 8110 (1993).
18. B. Kvamme, H. Tanaka: J. Phys. Chem. **99**, 7114 (1995).
19. J. O. Hirschfelder, C. F. Curtiss, R. B. Bird: *Molecular Theory of Gases and Liquids* (Wiley, New York, 1954).
20. W. L. Jorgensen, J. Chandrasekhar, J. D. Madura, R. W. Impey, M. L. Klein: J. Chem. Phys. **79**, 926 (1983).
21. H. Tanaka: J. Chem. Phys. **101**, 10833 (1994).
22. M.C. Gordillo, J. Marti: Chem. Phys. Lett. **329**, 341 (2000).
23. I. Brovchenko, A. Geiger, A. Oleinikova: Phys. Chem. Chem. Phys. **3**, 1567 (2001).
24. K. Koga, G.T. Gao, H. Tanaka, X.C. Zeng: Nature **412**, 6849 (2001). For 'armchair' (R, R) in this reference read 'zig-zag' $(R, 0)$.
25. G. Hummer, J.C. Rasaiah, J.P. Noworyta: Nature **414**, 188 (2001).
26. K. Koga, R.D. Parra, H. Tanaka, X.C. Zeng: J. Chem. Phys. **113**, 5037 (2000).
27. C.J. Tsai, K.D. Jordan: J. Phys. Chem. **97**, 5208 (1993).
28. K. Koga, G.T. Gao, H. Tanaka, X.C. Zeng: in preparation
29. K. Koga, X.C. Zeng, H. Tanaka: Phys. Rev. Lett. **79** 5262 (1997).
30. K. Koga, H. Tanaka, X.C. Zeng: Nature **408**, 564 (2000).
31. J. N. Israelachvili: *Intermolecular and Surface Forces* (Academic Press, London, 1992)
32. R. Evans, U. Marini Bettolo Marconi: J. Chem. Phys. **86** 7138 (1987).
33. K. Koga: J. Chem. Phys. **116**, 10882 (2002).
34. I.M. Svishchev, P.G. Kusalik: J. Am. Chem. Soc. **118**, 649 (1996).
35. M. Matsumoto, S. Saito, I. Ohmine: Nature **416**, 409 (2002).
36. H. Tanaka: J. Chem. Phys. **105**, 5099 (1996).
37. R. Zallen: in *The physics of amorphous solids*. 63-67 (Wiley, New York, 1998).
38. C.A. Angell, R.D. Bressel, M. Hemmati, E.J. Sare, J.C. Tucker: Phys. Chem. Chem. Phys. **2**, 1559 (2000).
39. M.O. Thompson et al.: Phys. Rev. Lett. **52**, 2360 (1984).
40. C.A. Angell, S.S. Borick, M.H. Grabow: J. Non-Cryst. Solids **205**, 463 (1996).
41. M. Togaya: Phys. Rev. Lett. **79**, 2474 (1997).
42. J.N. Glosli, F.H. Ree: Phys. Rev. Lett. **82**, 4659 (1999).
43. Y. Katayama et al.: Nature **403**, 170 (2000).
44. O. Mishima, L.D. Calvert, E. Whalley: Nature **310**, 393 (1984).
45. O. Mishima, L.D. Calvert, E. Whalley: Nature **314**, 76 (1985).

46. P. H. Poole, F. Sciortino, U. Essmann, H.E. Stanley: Nature **360**, 324 (1992).
47. H.E. Stanley: Physica A **205**, 122 (1994).
48. H. Tanaka: Nature **380**, 328 (1996).
49. K. Lum, D. Chandler, J. D. Weeks: J. Phys. Chem. B **103**, 4570 (1999).

8 Thin Film Water on Insulator Surfaces

George E. Ewing, Michelle Foster, Will Cantrell, and Vlad Sadtchenko

8.1 Introduction

The study of thin water films on insulator surfaces started, as did many of the pioneering investigations in surface science, with Irving Langmuir [1]. In 1918 he measured film thicknesses on mica and glass. His procedure, elegant in its simplicity, involved taking many sheets of mica or cover glass slides from the ambient laboratory environment and stacking them in a small cell. The adsorbed molecules (principally H_2O) on these surfaces were driven off by heating to 300EC and captured in a trap cooled with liquid air. The number of water molecules caught, together with the known geometric area of the substrate surfaces, allowed a calculation of thin water film coverages: 2 molecular layers on mica and 4.5 on glass. If we view these insulator substrates as typical, then we come to expect any insulator surface to have a few molecular layers of water stuck to it under ambient conditions.

These are the types of questions we will address: What sort of intermolecular forces hold a water film to an insulator surface? What are the thermodynamic quantities that distinguish the film from the condensed phases of water? How do we characterize the hydrogen bonding network within the film? We shall explore the answers to these and other questions for three model insulator surfaces: NaCl(001), muscovite mica (001) and BaF_2(111).

This chapter unfolds as follows. The next section, **Overview**, first explores the reasons why thin films of water form on insulator surfaces. This is accomplished by analyzing the nature of the energies that constitute *physisorption*. Next, the principle experimental tool of our investigations, *infrared spectroscopy*, is described. Spectroscopic profiles are shown to contain information on the thickness of the films and the nature of their hydrogen bonding networks. Adsorption *isotherms*, extracted from the infrared data are then examined. The isotherm shapes are used to assess the balance between interactions within the films and bonding to the substrate surface. Temperature dependence of the isotherms reveals *thermodynamic quantities* of the adsorption process, for example, enthalpies of formation and absolute entropies of the thin films. With this background, we consider **Three Systems**, then end with a section we call **Afterward and Forward**.

8.2 Overview

Physisorption. Water is held near an insulator surface by energies that can be partitioned into five types: *electrostatic, dispersion, induction, repulsion* and *hydrogen bonding.* Taken together these energies account for the phenomenon of water physisorption. In the following qualitative discussion of physisorption, we will concentrate on the example of H_2O on NaCl(001) with asides that will encompass other substrates.

The electrical properties of the water molecule include its dipole, quadrupole and higher moments either located at a single site or distributed among a variety of sites [2]. The dominant electrostatic interaction of H_2O with NaCl is by way of these electric moments and the electric field near the surface generated by the ions of the substrate. To appreciate the magnitude of this type of energy, consider first an H_2O molecule in the field of a single Na^+ ion. The field strength, in volts per meter (Vm^{-1}), at distance z from the center of the ion is given by [3]

$$E = \frac{q}{4\pi\varepsilon_O z^2} \tag{8.1}$$

where q = 1.6 x 10^{-19} C is the elementary charge of Na^+ and ε_o = 8.85 x 10^{-12} $CV^{-1}m^{-1}$ is the vacuum permittivity. Consider water and the ion at a separation defined by the touching of their hard sphere radii. We take this distance to be the sum of the Na^+ radius (95 pm [4]) and the H_2O radius (140 pm [4]) and arrive at z = 235 pm. Application of (8.1) yields an electric field of 2.6 x 10^{10} Vm^{-1}. The energy of a dipole parallel to an electric field and favorably aligned [3] is

$$W(z) = -\mu E \tag{8.2}$$

and, with μ = 2.5 x 10^{-30} Cm for H_2O [3] , we arrive at W = -96 kJ molΓ^1. If we increase the distance of the molecule from the ion by the diameter of the water molecule we arrive at a separation, z = 515 pm, and find W = -20 kJ mol^{-1}, still a considerable energy of attraction.

The electric field directly over an ion at the (001) surface of the collection of ions that make up a cubic ionic crystal is given by the Lennard-Jones and Dent expression [5]

$$E(z) = \frac{8\pi q \exp(-\pi\sqrt{2}\, z/a)}{4\pi\varepsilon o a^2}. \tag{8.3}$$

The separation z is measured above an ion of charge q and the distance between nearest neighbor cations and anions is a = 282 pm for NaCl [6]. Comparison of (8.2) and (8.3) shows that while the decay of electric field from a single ion is quadratic with displacement, the decay is exponential from the surface of an ionic crystal. Using (8.3) we calculate E = 1.1 x 10^{10} V m^{-1} at z = 235 pm or roughly a factor of two smaller than for the same displacement from a single ion. The binding energy is comparably less, $W =$ -40 kJ mol^{-1}. However the big difference between interaction of the water with a single ion and a crystal surface comes about when we displace the water molecule again by its molecular diameter to $z = 515$ pm. This amounts to moving the molecule touching the surface, that is in the first layer, to the second layer. The exponential decay of the electric field results in a binding energy of only -0.5 kJ mol^{-1}, a decrease by two orders of magnitude over the molecule touching the surface. And while we have greatly simplified the electrical properties of water and ignored the details of its orientation on the surface, this numerical exercise has provided two lessons. The first is that the binding energy of a water molecule at the surface of the ionic crystal is large and, as we shall soon show, comparable to a hydrogen bond energy. The second is that if the molecule resides in the second layer its electrostatic binding to the ionic substrate is negligible. Only the first layer is directly affected by the electric field of the substrate.

An estimation of the dispersion energy between water and the substrate is not so straightforward, but we can begin by considering the interaction between a single water molecule and a single ion/atom within the substrate. The dominant attractive term in the two-body dispersion potential decays with displacement as z^{-6} [2]. If we now consider a water molecule against a surface, we must expand the number of participating ions/atoms by the substrate volume proportional to z^3 [4]. As a consequence, the net attractive energy now decays only as z^{-3} that might appear less severe than the exponential dependence of the surface electric field. And indeed it is, though the diminution of the dispersion energy when water moves from the first layer (z = 235 pm) to the second layer (z = 515 pm) is still an order of magnitude. So as in the consideration of electrostatic interaction, the dispersion energy between a water molecule and the substrate is only significant when it resides in the first layer.

Since both repulsion and induction energies also depend on high powers of the displacement [2], these two contributions to physisorption, like dispersion and electrostatics, will only influence the energetics of molecule-substrate interaction in the first layer.

In order to attach some quantitative numbers to the relative importance of the electrostatic, dispersion, induction and repulsion energies we refer to the calculations of Engkvist and Stone [7]. For a single water molecule atop the NaCl(001) surface, they find the electrostatic energy to be -57 kJ mol^{-1}, repulsion energy at 43 kJ mol$-^{-1}$, induction energy of -13 kJ mol^{-1} and a dispersion energy of -13 kJ mol^{-1}, for a net binding energy of -40 kJ mol^{-1}. The optimum structure they find, shown in Fig. 8.1, has the water molecule positioned nearly above Na^+ and lying almost flat against the surface. This orientation, qualitatively different from the one we supposed, considers many electric moments of the molecule and the intricate variation of the electric field over the surface to achieve a 50% increase in electrostatic binding energy over our crude estimate. The net binding energy, -40 kJ mol^{-1}, is a compromise among the electrostatic term and the three other energy types, all with comparable values.

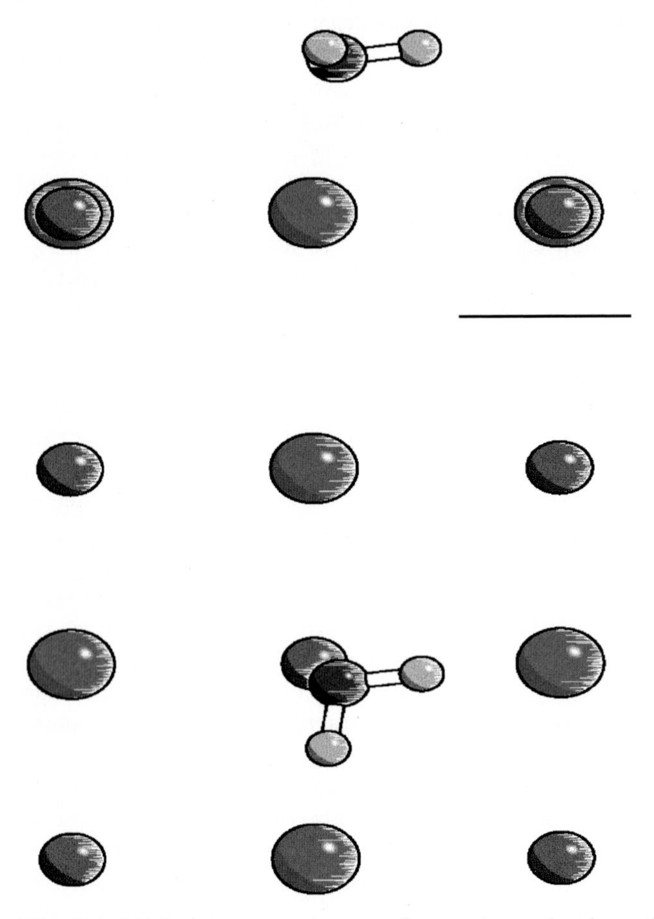

Fig. 8.1. Minimum energy structure for a water molecule on the NaCl(001) surface. This figure shows one view perpendicular to the surface and another parallel to it. From the calculations of Engkvist and Stone [7].

Before moving on to consider aggregations of water molecules on the surface of an insulator, it is helpful to consider the bonding within bulk liquid water. The enthalpy of condensation of water at 25 °C is -44 kJ mol^{-1} [3] and the formation of a single hydrogen bond in the liquid is estimated to be about -25 kJ mol^{-1} [8]. The numbers associated with hydrogen bonding energies are therefore comparable to the binding of water to an ionic substrate. We can anticipate then, that at least in the first adlayer of adsorbed water, that hydrogen bonding within the film will be as important as bonding of the film to the substrate. For the second (or higher layers), the water molecules will not be affected *directly* by the substrate because of the interactions fall off so rapidly with distance. However, water in the second layer will be affected by the water molecules in the first layer that are bound to the surface.

While hydrogen bonding, electrostatic, dispersion, and induction add complexity to the structure of thin film water, repulsion alone can account for molecular layering. This is even evident in the distribution function for hard spheres against a hard surface [9]. Near liquid densities, the distribution function achieves a maximum as the first layer of spheres touch the surface. The second layer, in its turn, is restricted in its approach to the surface by the first layer and also exhibits a distribution maximum. Third and higher layers exhibit distinct (but decreasing) maxima as well. Layering of water against a hydrophobic surface has been demonstrated through various calculations [10] and for water against NaCl(001) [11].

Infrared spectroscopy. The spectroscopic signature of water, particularly in the OH stretching region, provides two important pieces of information about thin films. The first is the absorbance band profile that contains information on the nature of the hydrogen bonding network [12]. The second is the extent of the absorbance from which the thickness of the film can be estimated.

There is a well documented red shift in the absorption feature associated with the OH stretching vibrations upon formation of hydrogen bonds [12]. This is evident upon a comparison of the water monomer with the spectra of the condensed phases. The myriad monomer vibration-rotation features have band centers at 3657 cm^{-1} (ν_1) and 3756 cm^{-1} (ν_3) [13] that collapse to a diffuse absorption centered near 3390 cm^{-1} in the liquid [14] and drop to 3220 cm^{-1} for ice [15]. This trend is indicated in Fig. 8.2. The red shift extends even further for water clusters where density functional calculations show vibrations of hydrogen bonded modes as low as 3000 cm^{-1} for the hexamer and octamer [16], and frequencies as low as 2935 cm^{-1} have been observed for the heptameter (Buck and Steinbach; Chapter 3 of this book).

The spectroscopic band shape can be an indication of the heterogeneity of the molecular environment of hydrogen bonded water. In general the more disordered the molecular environment the broader the band [8]. In addition to the broadening induced by disorder, the overall width is also a consequence of the coupling of the OH vibrations of all the molecules that make up the bulk sample [17, 18]. For clusters, the overall bandwidth is a measure of the spread of vibrational frequencies within the cluster. In the cyclic hexamer, the two stretching vibrations in each of the six isolated monomers become 12 normal modes in the complex. Calculations show that these range from a non-hydrogen bonded (dangling bond) vibration at 3726 cm^{-1} to 3058 cm^{-1} for the vibration of a hydrogen bonded mode [16]. Experimental measurement of the cyclic water hexamer finds overlapping features that extend from 3700 to 3075 cm^{-1} [19].

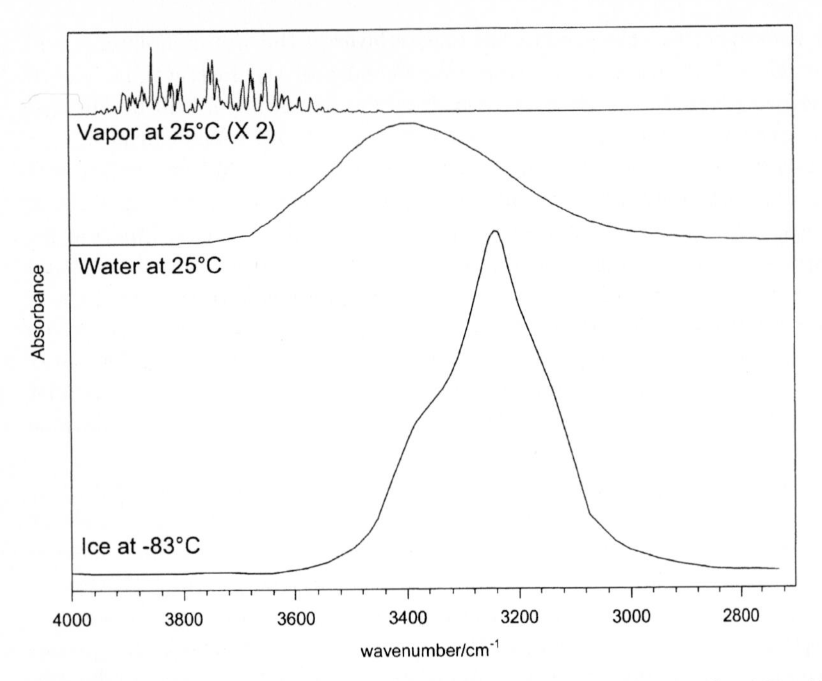

Fig. 8.2. Infrared spectroscopy of H_2O. The vapor spectrum is from David Weis [private communication], liquid from Downing and Williams [14], and ice from M. L. Clapp, R. E. Miller and D. R. Worsnop, *J. Phys. Chem.* **99**, 6317 (1995). The absorbances are scaled to the same number of H_2O molecules in the optical path for each phase. This figure was adapted from one prepared by David Weis [private communication].

The oscillator strength tends to increase with the hydrogen bond energy. For example the integrated cross section is larger by an order of magnitude for the stretching modes in going from the monomer to liquid water [14] and another factor of 1.5 on formation of ice [15]. The same trend is found in clusters with lowest oscillator strengths for the trimer with bent (strained and weak) hydrogen bonds and highest oscillator strengths in the cyclic, non-planar hexamer with strong linear hydrogen bonds [16].

As we shall see in the spectra of films to be shown, all of the profiles for the higher coverages resemble that of liquid water shown in Fig. 8.2. We shall therefore be emboldened to use the pure water optical constants to estimate coverage information.

We adopt the modified Beers law expression we have used previously [20],

$$\int_{band} A\mathrm{d}\tilde{\nu} = \frac{NS_{H_2O}}{\ln 10} \int_{\mathrm{band}} \sigma \mathrm{d}\tilde{\nu} \quad . \tag{8.4}$$

Here absorbance is defined by $A = \log_{10}(I_0/I)$, where I_0 is the intensity through the evacuated cell containing a stack of the insulator crystals, and I

is the intensity after water vapor has been admitted and the film formed. The wavenumber dependent liquid water cross section is σ [14], N is the number of faces exposed to vapor, and S_{H_2O} is the surface density of water in the thin film on each face. The integrations are over the OH stretching band profiles that we take to be from 2500 to 4000 cm^{-1}.

We can cast S_{H_2O} into a more easily interpretable form if we compare it with the surface density, S_{ref}, of a reference substance. The reference might be the density of sites on the surface to which the molecules in the thin film adhere. For example, if we take the binding sites to be Na^+ ions in the NaCl(001) surface, we use $S_{ref} = S_{NaCl} = 6.4 \times 10^{18}$ ions m^{-2} [20]. Alternatively we might take S_{ref} to be the number of H_2O molecules per unit area within a monolayer of the liquid. In this case we take the density of 25 °C liquid water of d = 1000 kg m^{-3} [6] to calculate the average monolayer thickness of 310 pm and $S_{water} = 10.4 \times 10^{18}$ molec m^{-2} within this layer. Selecting the appropriate reference surface density we arrive at a coverage,

$$\Theta = \frac{S_{H_2O}}{S_{ref}} \tag{8.5}$$

giving the average number of H_2O layers atop the selected surface binding sites of the substrate or the number of layers when the thin film is assumed to have the density of liquid water. The caveat here is that a coverage is not defined until the structure of the thin film, i.e. S_{ref}, is specified.

Uncertainty in our derived coverages also arises from photometric errors in the measured integrated absorbance and the assumption that the value of the integrated cross section is that of liquid water. These uncertainties vary as a function of coverage. The spectra at low coverages are noisier than those at high coverages and thus affect the measurement of the integrated absorbances. The cross section (either at a specified wavenumber or integrated over the band) also changes as a function of coverage since it is sensitive to the nature of the hydrogen bonding network. The cross section will change from some initial value representative of water bonded to the substrate at submonolayer coverages to an intermediate value for one or so layers where the water is affected by both substrate and other water molecules. At the highest coverages, the cross section should be close to that of the pure liquid water.

Isotherms. Since we can obtain coverages from the spectra of the thin films at selected values of water vapor pressure and temperature, we can construct isotherms.

The shapes of adsorption isotherms have been sources of information on adlayer structures since the interpretations provided by Brunauer *et al.*[21] in 1940. In the spirit of their analysis, we shall be comparing the isotherms of H_2O on NaCl(001), muscovite mica(001) and BaF_2(111) with isotherms of systems where adlayer structures have been provided by a variety of experimental or theoretical methods.

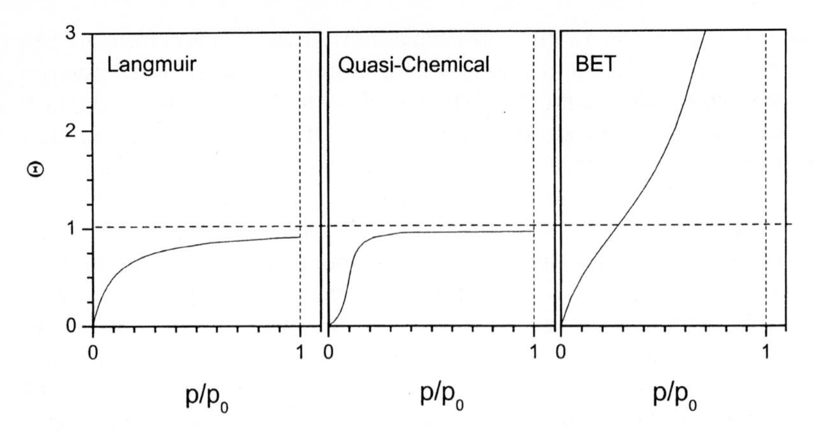

Fig. 8.3. Model adsorption isotherms. Reduced pressures are given by p/p_0, where p_0 is the equilibrium pressure of the bulk phase of the absorbant. Relationships to develop these isotherms are discussed in the text. From the study by Foster and Ewing [20].

The shapes of three model isotherms are given in Fig. 8.3. Here p_0 is the bulk vapor pressure of the adsorbant and p its pressure over the substrate. In these presentations we have arbitrarily taken $p/p_0 = 0.1$ for half monolayer coverage.

The simplest form, that from the Langmuir model [1], follows from the view that adsorption occurs randomly to available surface sites. When all sites are filled the surface is saturated by the monolayer and no further adsorption can occur. Lateral interactions are ignored and multilayer adsorption is denied. Physisorption of CO on NaCl(001) is accurately represented by the Langmuir isotherm [22]. For this particular system at 55 K, $\Theta = 0.5$ occurs at $p/p_0 = 2 \times 10^{-8}$ and monolayer coverage is effectively achieved by $p/p_o = 10^{-7}$. Increases in pressure by several orders of magnitude do not alter the adlayer structure or the extent of coverage as documented by polarized infrared absorption measurements [22]. Multilayer crystalline CO is formed at $p/p_0 = 1$. When the multilayer does form, the original monolayer structure (with CO molecules centered over Na^+ ions and each axis perpendicular to the (001) face [23]) is not appreciably altered [24]. Since the multilayer spectrum is identical to that of the α-phase of the crystal [25], the second and higher layers have CO molecules aligned along the diagonals of their cubic unit cell. This structural stability for the CO monolayer with multilayer formation on top is not found for H_2O on NaCl(001) under ambient conditions as we shall see.

While lateral interactions are not important to submonolayer and monolayer CO systems at 55 K, as implied by its accurate modeling with a Langmuir isotherm, they do become important below 30 K. This temperature signals a phase transition in which the CO axis becomes tilted [26]. To judge whether lateral interactions are important to an adlayer structure, it is helpful to compare nw with kT, where n is the number of nearest neighbors and w is the pairwise molecular interaction energy.

The quasichemical model treats nearest neighbor lateral interactions through the parameter nw/kT [27]. The form of the resulting isotherm for nw/kT = -4 (i.e., an attractive interaction), is shown in Fig. 8.3. This isotherm is concave for low coverages with an inflection point at $\Theta = 0.5$. Above half-coverage it becomes convex as it approaches saturation at $\Theta = 1$. The physical interpretation of the adlayer structure is that, at low coverages, molecules adsorb randomly and can be treated as a two-dimensional lattice gas. For high coverages lateral interactions are responsible for two-dimensional crystalline island formations. A coexistence region at $\Theta = 0.5$ finds both lattice gas and crystal phases in the adlayer. At cryogenic temperatures, several small molecules [28-30] including H_2O [31] are well represented by the quasichemical isotherm. For CO_2 on NaCl(001), molecular quadrupoles are the chief source of the lateral interaction [29]. Spectroscopic evidence for both lattice gas [30] and two-dimensional crystalline phases [29] of CO_2 near 100 K has been found. For H_2O on NaCl(001) near 100 K with the molecules nearly centered over Na^+ [7], they are separated too far for hydrogen bonds to form so dipole attraction is the major lateral interaction. For these cryogenic studies half-coverage occurs for $p/p_0 \ll 1$.

An *analytical* model that incorporates multilayer formation but ignores lateral interactions (e.g., $nw = 0$) is that of Brunauer, Emmett, and Teller (BET) [32]. The model contains, within its parameters, the enthalpy of adsorption. An example of a BET isotherm (type III from Brunauer) [21] is shown in Fig. 8.3. The nature of the model allows multilayer formation as revealed by the asymptotic approach of multilayer formation as $p => p_0$. Despite its denial of lateral interactions, many experimental systems are well represented by the form of the BET isotherm. As we shall see, the high coverage regions of the BET isotherm form of Fig. 8.3 is in apparent agreement with the high coverage regions of water isotherms on NaCl(001). However for water on BaF_2(111) and mica(001) the isotherm apparently terminates as $p => p_0$ to a finite coverage value.

The Langmuir and quasichemical monolayers at cryogenic temperatures that we have given as examples, retain stable structures well beyond pressures needed to saturate the monolayer. In the case of CO on NaCl(001), pressure increases of many orders of magnitude are required before multilayer adsorption is initiated. Thus in these systems clear distinctions among lattice gas, islands, monolayer and multilayer phases are realizable. This is possible because, at the cryogenic temperatures of these experiment, the free energy differences among the different phases are typically much greater than RT, so they are easily resolved. But as we shall see, distinctions among lattice gas, island, monolayer and multilayer descriptions are blurred when considering water adlayers on insulator surfaces under ambient conditions.

Thermodynamic quantities. Thermodynamic analysis of isotherms yields free energy, enthalpy, and entropy as a function of coverage. From these values, the changes in intermolecular bonding among the water molecules (the hydrogen bonding network) and bonding of water to the substrate can be assessed. In addition, it will be possible to gauge the ordering of water molecules on the substrate as a function of film thickness.

In order to extract thermodynamic quantities we make use of the expression [20]

$$\ln\left(\frac{p_{\mathrm{w}}}{p_{\theta}}\right) = \frac{\Delta H_w - \Delta H_\Theta}{RT} - \frac{S_w - S_\Theta}{R} \tag{8.6}$$

where ΔH_w is the enthalpy of adsorption of liquid water (-44 kJ mol^{-1} at 25C) [3], ΔH_θ is the enthalpy of adsorption of the thin film at coverage Θ. The entropy of liquid water is S_w (69.9 J mol^{-1} K^{-1} at 25 $^\circ$C) [3] and S_θ is the entropy of adsorbed water at a coverage of Θ. In applying (8.6) we plot $\ln(p_w/p_\theta)$ against $1/T$ using the appropriate value of the equilibrium bulk liquid water vapor pressure, p_w.

8.3 Three Systems

We have chosen three substrates as hosts for water adsorption studies: NaCl, muscovite mica and BaF_2. Our choices are not arbitrary. Each substrate is readily available in pure form. The NaCl and BaF_2 are grown from melts to form single crystals. Muscovite mica is mined and visually selected for its size and homogeneity. Each of the substrates is easily cleaved to reveal well defined faces: (001) for NaCl and mica and (111) for BaF_2. These faces are smooth to visual examination and at the atomic level. Each substrate offers windows of transparency to the infrared study of water absorption in the OH stretching region. The substrate surfaces have been examined by a variety of experimental and theoretical approaches [33-35]. The substrates do have distinctive characteristics that make their interfaces with water of particular interest. NaCl is water soluble and serves as a model for the solubilization process [36]. Mica is water insoluble, has its (001) face nearly comparable with the ice basal face and has been used as a nucleating agent [37]. Similarly, BaF_2 is water insoluble and its (111) face lattice constant is also a close match for the basal face of ice [38].

To obtain spectra of thin film water on these insulator surfaces, two obstacles need to be overcome. The first is that the absorbance of a thin film, consisting of a few monolayers or less, will be weak. The second obstacle is that the myriad spectroscopic vibration-rotation features of water vapor that accompanies the ambient system may easily overwhelm the absorbance of the thin film. These two obstacles are circumvented by stacking many substrate crystals close together, the strategy initiated by Langmuir [1]. This maximizes the number of films examined and minimizes the space between them that contains water vapor.

A schematic drawing of a cell containing many closely spaced crystals of NaCl with their (001) faces orthogonal to the direction of the interrogating infrared beam, is shown in Fig. 8.4 [20]. The same arrangement is used for BaF_2 crystals except that their (111) faces are orthogonal to the cell axis [39]. For mica, the crystalline sheets are canted with respect to the propagating radiation [40]. The temperature controlled cell is placed in the sample compartment of a Fourier transform interferometer (FTIR) spectrometer.

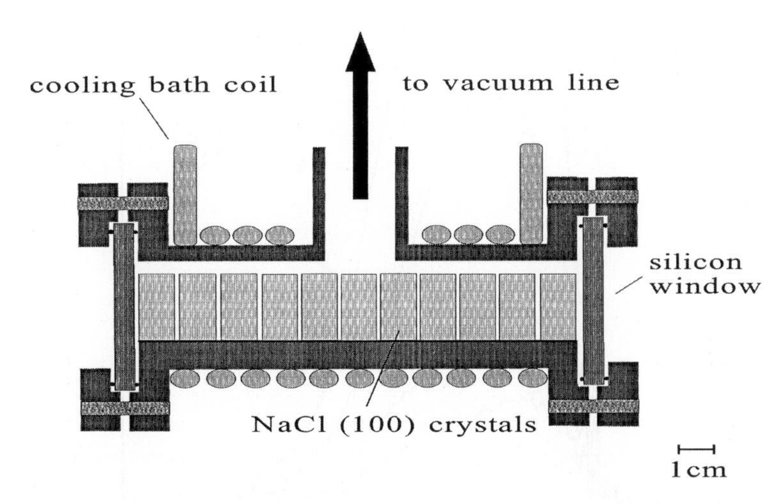

Fig. 8.4. The cell used for transmission FTIR experiments. From Foster and Ewing [20].

A. NaCl (001)

Spectroscopic Signatures. The absorption spectra of water adsorbed on 28 NaCl(001) faces (14 crystals) at 24 ° C and various pressures is shown in Fig. 8.5. These spectra by Foster and Ewing [20] are extensions of earlier work by Peters and Ewing [41]. At the highest pressure, 13 mbar, the band shape and bandcenter nearly match those of liquid water. We are emboldened to use the bulk water optical constants [14] together with (8.4) and (8.5) to determine that the coverage value is $\Theta = 2.2$. Here we have used $S_{NaCl} = 6.4 \times 10^{18}$ ions m^{-2}, the site density of Na^+ ions in the NaCl(001) surface [20], as reference. Based on its spectroscopic signature, thin film water at this coverage appears to be water-like. For lower pressures/coverages the thin film spectroscopic signature undergoes a change, in particular the band center shifts to higher wavenumbers. The property of the thin film is no longer

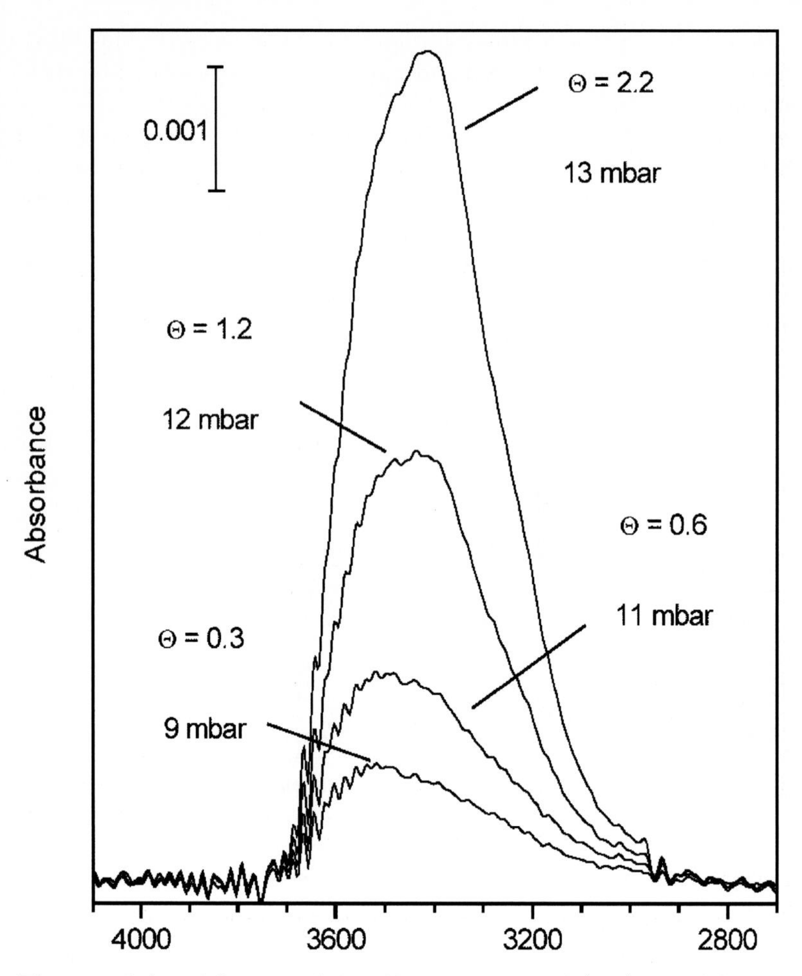

Fig. 8.5. Infrared Spectra of thin film water on NaCl(001) at 24 $^\circ$ C. The number of NaCl(001) faces is $N = 28$. From Foster and Ewing [20].

so water-like. Nevertheless if we continue to make use of the water optical constants we can construct an approximate isotherm throughout the pressure region.

Isotherms. Isotherms may be extracted from the pressure dependent adlayer spectra through the photometric analysis just described. Using the spectra shown in Fig. 8.5, and others as well, we present the 24 $^\circ$C isotherm in Fig. 8.6. Coverages obtained from spectra taken on ascending pressures are given by the closed circles while data taken on descending pressures are given by open circles. The slight coverage differences can be ascribed to random errors in the values of integrated absorbance obtained. Thus the isotherm exhibits little if any hysteresis.

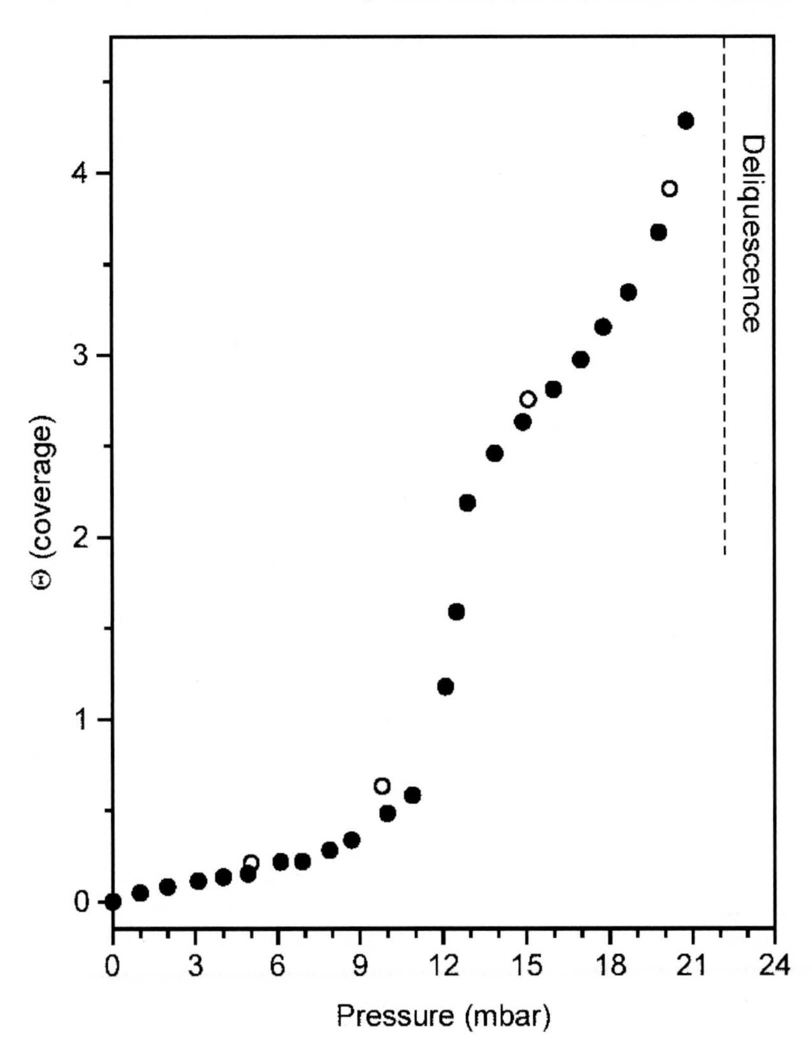

Fig. 8.6. The adsorption isotherms of H_2O on NaCl(001) at 24 °C. Data taken on ascending and descending pressures are given by closed and open circles, respectively. From Foster and Ewing [20].

Four separate regions can be distinguished in the isotherm of Fig. 8.6. The *low coverage region*, with $\Theta \leq 0.5$, is characterized by a linear increase in coverage becoming slightly concave at the end of this region. Later we shall identify the low coverage region with islands of water molecules bound to the surface and held together by hydrogen bonds. In the *transition region*, with $0.5 \leq \Theta \leq 2.5$, a sharp (near vertical) rise in coverage is accompanied by pressure increases with an inflection point near $\Theta = 1.5$. The transition region will later be proposed to consist of co-existing submonolayer islands and multilayer thin film water structures. The *high coverage region*, with $2.5 \leq \Theta \leq 3.5$, is characterized by a more gentle increase in coverage with pressure. We shall associate the high coverage region with a multilayer bound by a hydrogen bond network resembling that in liquid water. An inflection point near $\Theta = 3.5$ signals another upturn in coverage with increasing pressure as the deliquescence point is approached. We shall call coverages with $\Theta \geq 3.5$ the *presolution region.*

While the coverage in the isotherm of Fig. 8.6 appears to smoothly increase as the deliquescence pressure is approached, the dissolving away of the salt at the NaCl(001) face is actually a nucleated process [42]. The phase transition between NaCl(001) and water vapor to the saturated salt solution atop NaCl(001) and water vapor does not occur until the water vapor pressure exceeds its equilibrium value by 5 to 10% [42]. Dissolution of salt particles (rich in surface defects) with increasing water vapor pressure and the appearance of crystallites from brine droplets as a function of decreasing water vapor pressure have been studied [43].

Isotherms of water on NaCl crystallites under ambient conditions from other laboratories have been reported [44]. While these isotherm shapes resemble that in Fig. 8.6 for multilayer coverages, there are qualitative differences in the monolayer and submonolayer regions. These differences can be rationalized by the abundance of defects on the surfaces of the crystallites. Adsorption to defects is expected to be favored over adsorption to the smooth terraces of single crystal NaCl(001) faces [36,44-46]. This expectation is consistent with the observations that find a higher water adlayer coverage for a given pressure on NaCl crystallites than on the NaCl(001) faces of single crystals [46].

We now summarize comparisons between the water on the NaCl(001) isotherm at $24\,^{\circ}$C for Fig. 8.6 and the model isotherms of Fig. 8.3. The Langmuir isotherm and the water isotherm appear not to have common features. We conclude that H_2O molecules do not absorb randomly to NaCl(001) under ambient conditions. However, the concavity and the sharp rise in the quasichemical isotherm bear qualitative resemblance to the ambient water isotherm. This suggests the importance of lateral interactions in the H_2O adlayer. On the other hand, the quantitative mismatch of the inflection points, $\Theta = 0.5$ for the quasichemical isotherm and $\Theta = 1.5$ in Fig. 8.6, suggest a different nature of the coexistence phases for the model system and ambient H_2O on NaCl(001). Finally the BET isotherm and adlayer water reveal a similarity but only in the high coverage and presolution regions. Clearly attraction between layers is important. These qualitative comparisons of the model isotherms with that of adlayer H_2O on NaCl(001) appear to be at the limit to the interpretations of the nature of thin film water we can offer from this type of analysis.

The Monte Carlo calculations of Engkvist and Stone [11] allow us to deepen our understanding of the thin film molecular structure implied by the observed isotherms. While they provide quantitative information in the form of pair distribution functions, we shall draw from their calculations two configurations corresponding to coverages of $\Theta = 0.5$ and $\Theta = 3.0$. These are shown in Fig. 8.7. The exploded views into three distinct layers above the NaCl(001) surface, $0 \leq z \leq 360$ pm, $360 \leq z \leq 640$ pm, and $640 \leq z \leq 960$ pm, provide easier visualization of adlayer structures. These spacings have been guided by the layering revealed by the Monte Carlo distribution functions.

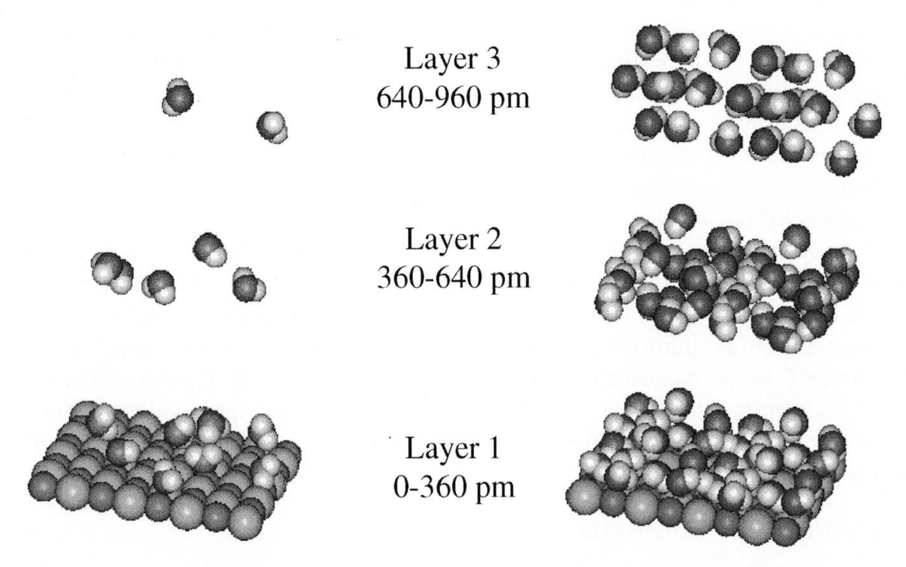

Fig. 8.7. Exploded views of typical configurations of water molecules near the NaCl(001) surface at $\Theta = 0.5$ and 3.0. The three layers have O atoms at $0 \leq z \leq 360$ pm; $360 \leq z \leq 640$ pm, and $640 \leq z \leq 960$ pm. These images were taken from the Monte Carlo calculations of Enkgvist and Stone [11].

Layer 1 for $\Theta = 0.5$ appears as a random assemblage of monomers and small clusters. However, the distribution functions actually indicate [11] that the molecules are not randomly distributed but favor an O-O separation consistent with hydrogen bond formation. Moreover H_2O molecules are more likely over Na^+ than Cl^{-1} ions. Examination of a number of Monte Carlo configurations suggest that molecules in layer 1 are rather anchored to the underlying ions. The few molecules in layer 2, some directly above molecules in layer 1, suggest the beginning of multilayer formation.

For a coverage of $\Theta = 3$, the water molecules are distinctly partitioned into three layers. This is clearly evident in the distribution function that explores molecular positions perpendicular to the (001) face [11]. The layering of water at a coverage of $\Theta = 3$, as revealed by the distribution of O centers above NaCl(001), is reminiscent of layering found for a liquid of hard spheres against a hard surface. However the details of the distribution for this real system - water molecules against the hydrophilic salt surface – are understandably more complicated. The molecules in layer 1 are now influenced by molecules in layer 2 as well as by the substrate below. Thus while water molecules in the uppermost layer and the layer closest to the surface are only partially surrounded by other molecules, the distribution function that explores the O-O separation finds a value near that of liquid water. Thus the adlayer $\Theta = 3$ resembles liquid water.

Thermodynamics. A series of isotherms, each like that in Fig. 8.6, spanning a range of temperatures from –30 ° to +30 ° C together with application of (8.6) allows the evaluation of enthalpy and entropy of the adsorption process. Values of ΔH_Θ for adsorption on NaCl(001) are always below that of the enthalpy of condensation of water of –44 kJ mol^{-1}. (The heat of condensation of vapor on neat water, is to two significant figures, the same as on brine [6]). For monolayer coverage the exothermicity of water condensation on NaCl(001), at ΔH_1 = -50 ± 3 kJ mol^{-1}, exhibits its greatest difference from neat water (or brine). The corresponding entropy ΔS_1 = 55 J mol^{-1} K^{-1} is considerably lower than the value for liquid water of 69.9 J mol^{-1} K^{-1} [3].This is to say that thin film water in the monolayer is more ordered than the bulk liquid. This conclusion is consistent with the ordering found in the Monte Carlo calculations of Engkvist and Stone [11]. At higher coverages, $\Theta = 3$, both enthalpy and entropy of adsorption achieve the values of bulk water. Thus thermodynamically, a multilayer film has come to resemble the properties of liquid water.

B.Muscovite Mica(001)

Spectroscopic Signatures. In the study of Cantrell and Ewing [40], absorption spectra of thin film water on mica were obtained in a cell that resembled that in Fig. 8.4 with two important differences. Since mica has a broad vibrational band near the OH stretching region of water, [47] thin sheets (~10–100 μm) were used as substrates to obviate absorption interferences. However, the thinness of the mica sheets if aligned perpendicular to the infrared interrogating beam, introduces etalon fringes. This fringing was effectively eliminated by canting the mica sheets at their Brewster angle of 57 ° [48].

Absorption spectra for water adsorbed to mica at high, $\Theta = 2.5$, monolayer, $\Theta = 1$, and low, $\Theta = 0.3$, coverages are shown in the three panels of Fig. 8.8. Although we have chosen thin films at 8.6 ° C for discussion, other temperatures explored reveal the same characteristic spectroscopic profiles. Each measured spectrum is shown with a simulated spectrum at the corresponding coverage of liquid water. The simulation was achieved with σ obtained from the extinction coefficient as a function of wavenumber listed in Downing and Williams [47]. The mismatches between the measured and simulated spectra are probably due to the differences in the hydrogen bonding networks within the thin film on mica and liquid water. Clearly, the mea-

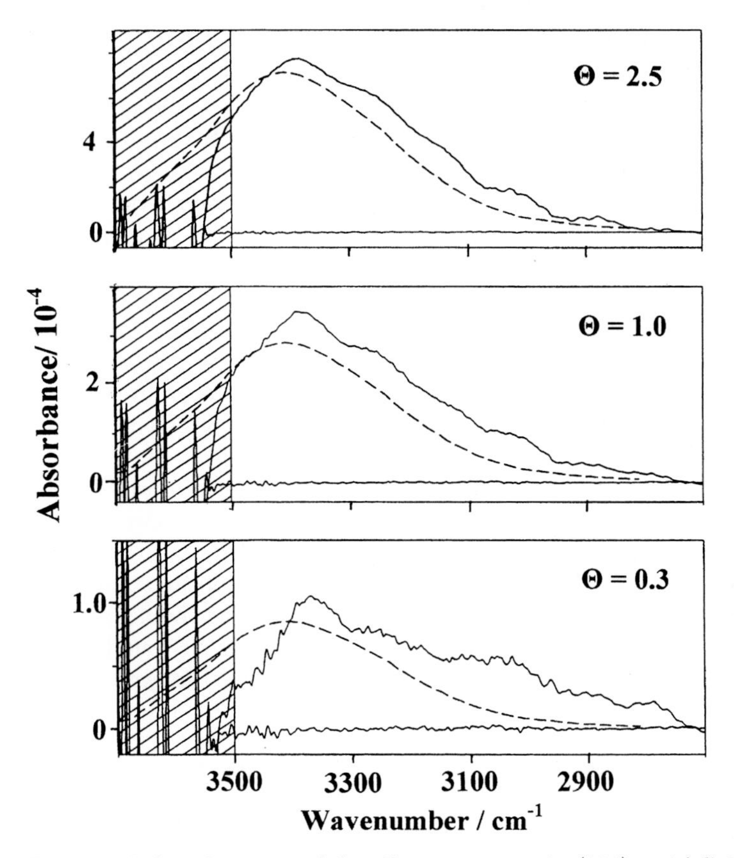

Fig. 8.8. Infrared spectra of thin film water on mica(001) at 9 ° C. The solid lines represent the measured spectra with $N = 16$ surfaces. The dashed lines simulate the spectra at the corresponding coverages using the optical constants of neat liquid water. Photometry in the cross hatched region is compromised because of the strong mica absorption in the region 3500 to 3700 cm^{-1}. From Cantrell and Ewing [40].

sured high coverage spectrum is most like liquid water and the low coverage spectrum suggests its own unique hydrogen-bonded network.

We start by examining the infrared spectra for low coverage. Although we have chosen to concentrate upon $\theta = 0.3$, other submonolayer coverage spectra down to $\theta = 0.1$ are qualitatively similar. Even a casual inspection will reveal that the measured spectrum does not match the spectroscopic profile calculated from the equivalent thickness of bulk water. In fact, the measured spectrum suggests a structured layer, a reasonable hypothesis considering the fact that the arrangement of the O^{-1} ions at the surface mark out a hexagonal lattice which approximates that of the basal plane of ice [49]. It seems logical therefore to assume that the adsorbed water molecules would be in a configuration favoring a hydrogen-bonded network mimicking that of ice. Previous theoretical and experimental work have argued for a structured, tightly bound first layer. Molecular dynamics calculations by Odelius et al. [50] indicate puckered hexagons of H_2O molecules locked to the underlying mica surface, while atomic force microscopy investigation by Miranda et al. [51] showed an ice-like layer for a coverage of about 1.

While it is true that the arrangement of the O^{-1} ions on the face of the mica resembles the lattice structure on the basal plane of ice, that similarity is disrupted by counter ions of K^+, left behind when the mica is exfoliated [49]. The formation of ice-like bilayers will be frustrated by these K^+ ions, though the water molecules may form strong hydrogen bonds to these ions, accounting for the low wavenumber features in the low coverage spectra. As we shall see later, the $BaF_2(111)$ face does present an uninterrupted hexagonal lattice. In that case, the low coverage spectra are consistent with an ice-like bilayer.

The thicker film corresponding to $\Theta = 2.5$ coverage, whose optical profile is given in the upper panel of Fig. 8.8, gives an understandable spectroscopic signature. Molecules in the upper layers are principally hydrogen bonded to neighboring water molecules. The band center of the composite spectrum is within 10 cm^{-1} of the liquid water spectrum. Poorly defined shoulders in the low wavenumber tail of the spectrum resemble those that we found at 0.3, suggesting that the layer of water next to mica has retained its well ordered structure.

Isotherms. Five isotherms for temperatures between 0.6 and 25.1 $^\circ$C are shown in Fig. 8.9. They are plotted against the reduced pressure, p/p_0, which is the ratio of the pressure of the vapor over mica, p, to its equilibrium value p_0, over liquid water at the specified temperature.

The shape of the isotherms is qualitatively the same. There is a near linear increase in coverage with p/p_0 in the submonolayer regime followed by a steeper rise in coverage beyond $p/p_0 \approx 0.8$. That the isotherms for increasing or decreasing pressures, track each other indicates no evidence of hysteresis. However, there *are* subtle differences in the isotherms as a function of temperature. The coverage at a given p/p_0 is greater at low temperatures, which is evident upon comparison of the 0.6 $^\circ$C and 25.1 $^\circ$C isotherms. These differences translate into the thermodynamic quantities we will obtain from the isotherms.

The shape of an adsorption isotherm is indicative of the types of interactions between the substrate and adsorbing species and among the adsorbed species as we have discussed. The shape of the isotherm also reveals how those interactions are altered with changing pressure, temperature and coverage. The smoothness of the composite isotherm, shown in Fig. 8.9, and the fact that it does not appear to diverge as p/p_0 approaches unity are evidence that there are no phase transitions within the film at these temperatures and that water wets mica incompletely.

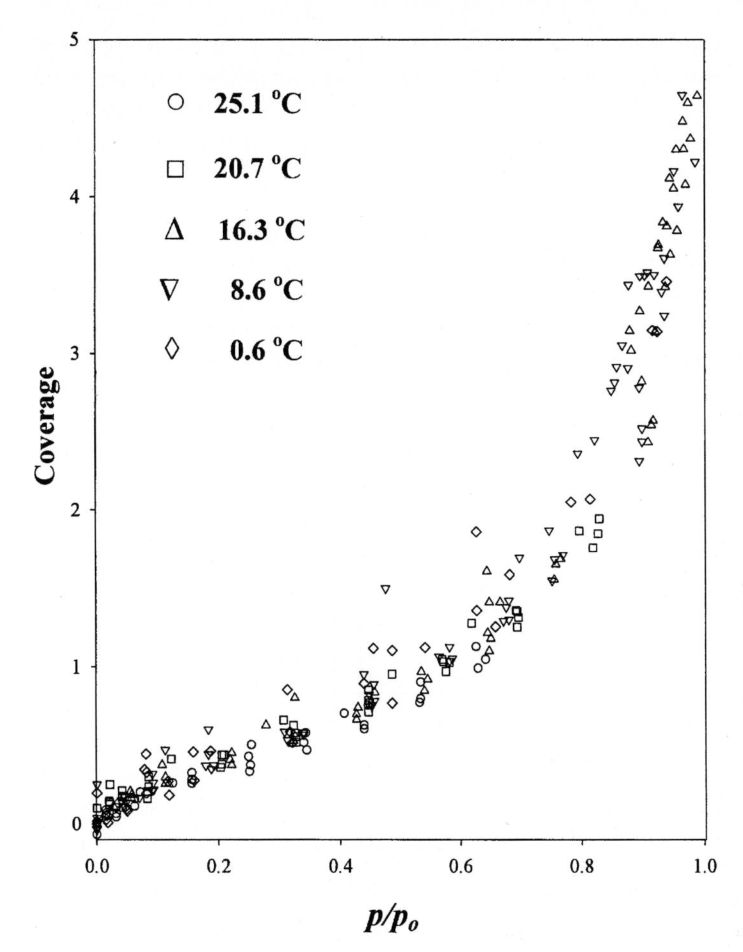

Fig. 8.9. Adsorption isotherms of water on mica(001) for 5 temperatures. From Cantrell and Ewing [40].

Changes in bonding networks are frequently marked by abrupt changes in film thickness with pressure as we have observed for water on the NaCl(001) surface in Fig. 8.6. The smoothness of our composite isotherm is evidence that the bonding network of water adsorbed to mica changes as a function of pressure in a continuous fashion.

Complete wetting is characterized by coverage (or film thickness) which diverges as coexistence between the bulk liquid and the vapor is approached. This is illustrated in the BET isotherm in Fig. 8.3. Partial wetting yields an isotherm that terminates at a finite coverage at $p/p_0 = 1$. Though straightforward in theory, it is not always a simple matter to distinguish a wetting isotherm from a partially wetting one as a practical matter [52]; the isotherm may approach saturation asymptotically. In our case, the coverage increases by a factor of 2 between $p/p_0 = 0.8$ and 0.95, however it does not appear to diverge but terminates near $\Theta = 5$. Our results are consistent with the work of Beaglehole et al. [53], who show limiting film thicknesses between 0.75 and 2.5 nm. Taking one layer of water to have a thickness of 0.31 nm, then our film thickness at saturation would be 1.5 nm.

Thermodynamics. As before we used p_Θ and p_w and (8.6) to obtain thermodynamic quantities. Figure 8.10 shows the enthalpy of adsorption and absolute entropy of the adsorbed water as a function of coverage. The dashed line in both panels of the figure delineates the values for liquid water.

The condensation of water onto mica is exothermic at all coverages with an extremum near $\Theta = 1$. In the model that describes intermolecular bonding of H_2O in ice [8], each molecule, in a tetrahedral arrangement, can donate two hydrogen atoms to hydrogen bond with a neighbor and can accept two

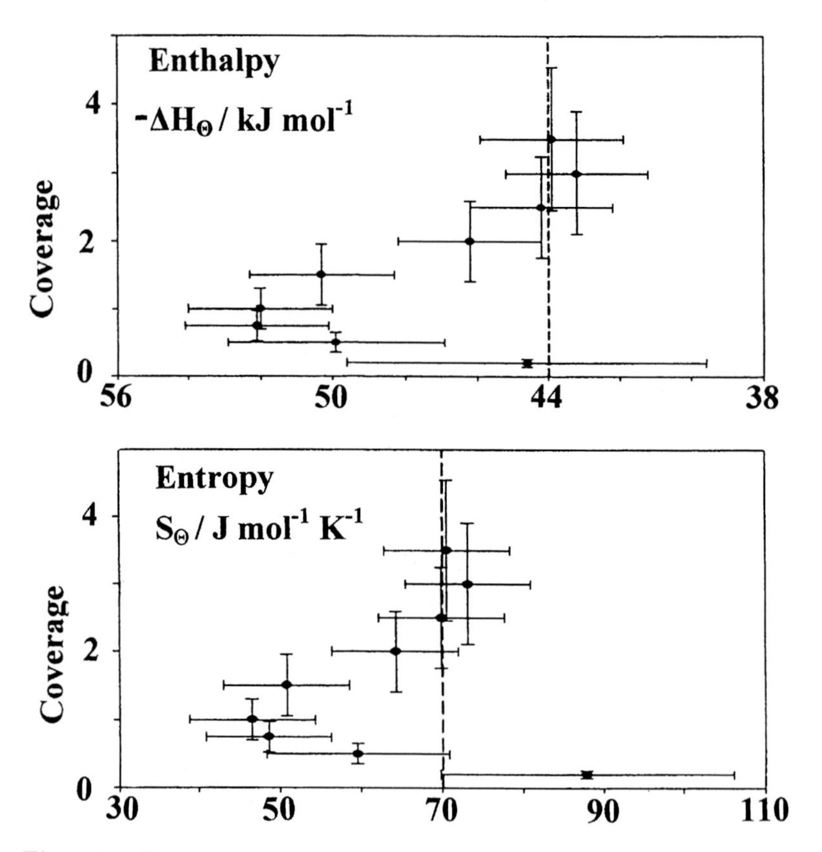

Fig. 8.10. Enthalpy and entropy of adsorption of water on mica(001). The dashed line in the upper panel marks the value of the enthalpy of condensation of liquid water from the vapor at 25C. The dashed line in the lower panel marks the absolute entropy of liquid water at 25 °C. From Cantrell and Ewing [40].

hydrogen bonds to its oxygen lone pairs. At submonolayer coverages, most molecules bonded to the surface are not fully tetrahedrally coordinated. We visualize this situation as islands of water molecules scattered across the mica. Water molecules on the periphery of these islands will lack H_2O or K^+ neighbors with which to form intermolecular bonds, i.e., they will have dangling bonds. As more water adsorbs, near $\Theta = 1$, the clusters begin to coalesce resulting in an increase in the enthalpy of condensation because water molecules can bond not only to the surface of the mica, but to neighbors laterally as well.

The entropy of adsorption mirrors the change in enthalpy. At the very lowest coverages, the water molecules are even more disordered than those in bulk water. This is demonstrated in the lower panel of Fig. 8.10 where the entropy of the thin film near $\Theta = 0.2$ actually exceeds that of the liquid. This may be a consequence of configurational entropy [27] because the islands containing small numbers of water molecules are randomly dispersed across the surface of the mica. As more water molecules are added to the surface, they are "locked" into position by their neighbors. The entropy decreases to a minimum near $\Theta = 1$ implying the most ordered adlayer arrangement.

We believe that the extrema in the enthalpy and entropy at $\Theta = 1$ are a consequence of the thin film structure on mica described by Miranda et al. [51] as an ice-like layer. This structure was observed to originate as polygonal islands that had an epitaxial relationship with the substrate. As the vapor pressure was increased, the islands merged into a single layer. Subsequent layers had liquid-like characteristics. Measurements of the conductivity of water adsorbed to mica have also suggested the presence of a structured first layer [51]. Significantly, the decrease in entropy between liquid water and the entropy for $\Theta = 1$ in Fig. 8.4 is 23 ± 7 J mol $^{-1}$ K^{-1}. The change in entropy for the liquid to crystalline phase transition for water at $0\,^{\circ}$C is 22 J mol^{-1} K^{-1} [3]. The absolute entropy of the adsorbed monolayer on mica at 47 ± 7 J mol^{-1} K^{-1}, within its low value extrema, approaches that of ice which we show later is 40 J mol^{-1} K^{-1}.

As the coverage increases, $\Theta > 2$, the enthalpy and entropy approach those of liquid water, indicating that the thickness of the film is increasing beyond the range of the interlayer forces [4]. As the influence of the mica is reduced, the water molecules behave more like molecules in the bulk liquid.

C. BaF_2(111)

Spectroscopic Signatures. In a study taken from Sadtchenko *et* al. [39], absorption spectra of H_2O adlayers on BaF_2(111) at $25\,^{\circ}$C for various pressures are shown in Fig. 8.11. At the lowest pressure, 2 mbar, the spectrum reveals a diffuse doublet with components at 3460 ± 5 cm^{-1} and 3220 ± 5 cm^{-1}. A pressure increase to 12 mbar produces a nearly three-fold increase in absorbance but only a subtle change in the doublet profile. Finally, in the upper spectrum of Fig. 8.11, a pressure increase by 10 mbar has produced an increase in absorbance by more than a factor of two. The resulting diffuse profile at 22 mbar, no longer a doublet, is centered at 3400 cm^{-1} with an ill-defined shoulder near 3200 cm^{-1}. If we disregard the shoulder, which significantly contributes to the overall band width of 500 cm^{-1} at half-height, the band center and high frequency side of this profile closely resembles that of liquid water (Fig. 8.2).

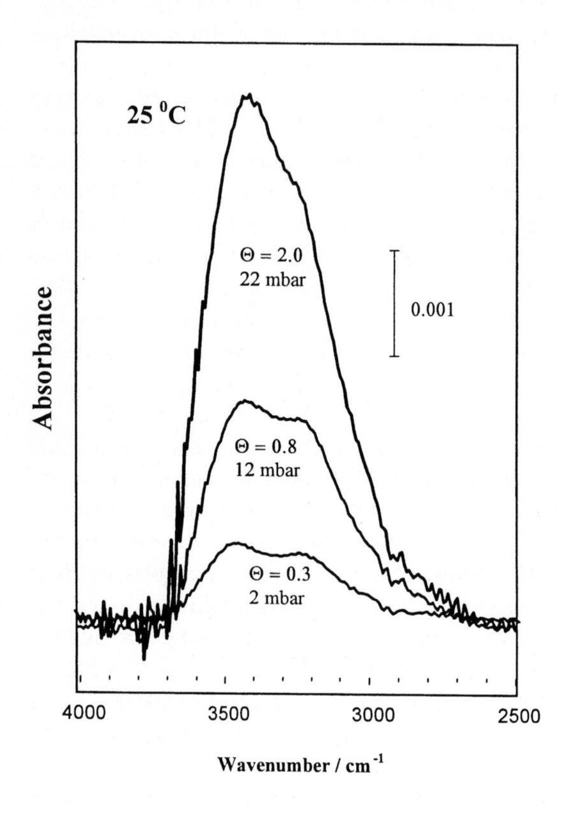

Fig. 8.11. Infrared spectra of thin film water on $BaF_2(111)$ at 25 °C. The number of $BaF_2(111)$ faces is $N = 46$. From Sadtchenko *et al.* [39].

We first consider the doublet profiles of the 25 °C lower-coverage spectra represented in Fig. 8.11. To begin these spectra do not resemble the diffuse singlet spectroscopic signature of liquid water. Nor do they resemble the spectra of submonolayer water on NaCl(001) represented by a broad asymmetric band located near 3420 cm^{-1} (Fig. 8.5) and associated with a disordered two-dimensional hydrogen bonded network (Fig. 8.7). While the spectra of low coverage water on mica(001) show several features, they are overlapping and poorly resolved (Fig. 8.8). Since there are two distinct features in the infrared spectrum of submonolayer water on $BaF_2(111)$, we will argue that, unlike thin films on NaCl(001) or mica(001), this adlayer is consistent with an ice-like bilayer even at ambient temperature.

One possibility for the doublet spectroscopic signature is that the water molecules are adsorbed to distinct sites at the surface (e.g. Ba^{2+} and F^{-1} ions) or reside in two different environments (e.g. liquid-like and solid-like H_2O). However, these arguments for distinct environments that could give rise to heterogeneous site splitting into a doublet are not consistent with the invariance of the spectroscopic profiles on temperature change [39]. The binding energies of H_2O molecules to two distinct sites would be different. Consequently the populations of molecules bonded to these sites would be temperature dependent. At higher temperatures the more tightly binding sites would be preferentially occupied. That the doublet features do not change in relative absorbances with temperature then speaks against heterogeneous site splitting. The invariance of spectroscopic profile shape with coverage from $\Theta = 0.3$ to $\Theta = 0.8$ as shown in Fig. 8.11 is also not consistent with heterogeneous site splitting. For two sites, the more strongly binding site will be covered first and then the less strongly binding site will be available for adsorption as pressure is increased. There *are* subtle changes in the doublet profile with coverage but we shall offer an explanation for these later.

We now consider an interpretation favored by other work [55], that leads us to suggest that the spectroscopic signature of H_2O on $BaF_2(111)$ is a consequence of the formation of an ice-like bilayer. For unlike the NaCl(001) face, $BaF_2(111)$ presents a face compatible with the basal surface of I_h ice.

Molecular dynamics (MD) calculations of Wassermann *et al.* [55] at $27\,^\circ$C have the bilayer bound to $BaF_2(111)$. The bilayer may be viewed as a lattice of buckled six-membered rings of H_2O molecules interconnected by hydrogen bonds. (The oxygen framework of the buckled six-membered rings of H_2O is of the same symmetry as the carbon framework in the chair form of cyclohexane.) We have shown one of the rings of this bilayer lattice in Fig. 8.12. Each oxygen, represented by the medium size ball, is associated with two lone pairs, the longer lines, and two hydrogen atoms, the small balls, to which they are chemically bonded. A lone pair of each H_2O molecule in the lower half of the bilayer is axial and directed downward toward a Ba^{2+} ion [55]. The other molecule lone pair is equatorial and accepts a proton from a neighboring water molecule in the upper half of the bilayer in either the ring shown or a neighboring ring not shown. The two protons in each of these lower H_2O molecules are also equatorial and are donated to the lone pairs of neighboring molecules in the upper half of the bilayer. Thus each oxygen in the lower half of the bilayer is four coordinated (to three H_2O molecules and one Ba^{2+} ion) in a tetrahedral arrangement. Devlin and Buch [56, 57] designate four-coordinated surface water molecules s-4. By contrast, each oxygen in the upper half of the bilayer is only three coordinated. There are two types of three coordinated water molecules. For one type, a non-hydrogen bonded proton (dangling hydrogen) is axial and directed upward. These molecules are designated d-H. The second type of molecule has an axial lone pair (dangling

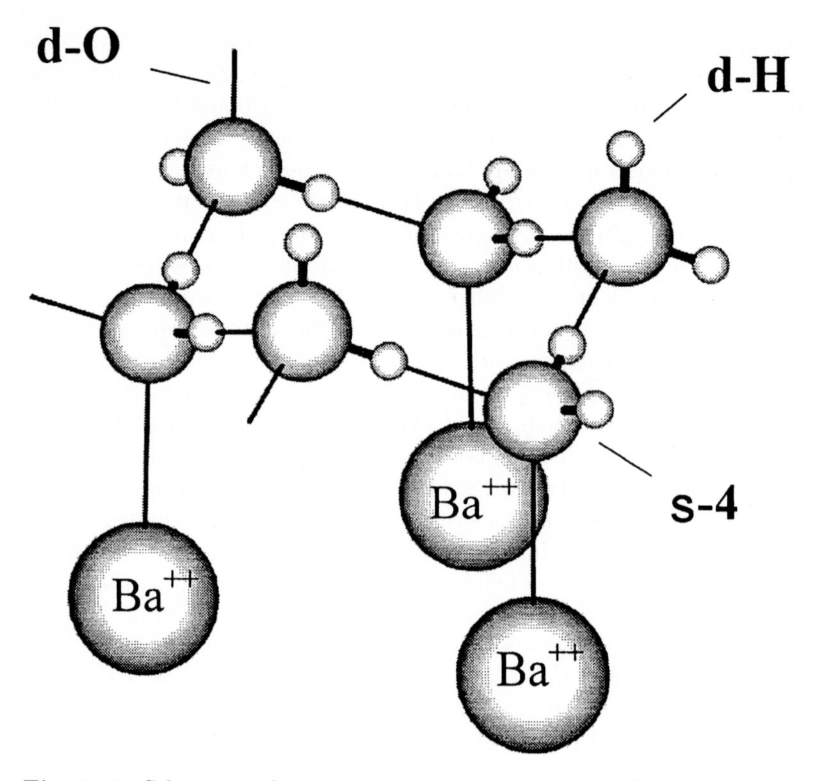

Fig. 8.12. Schematic of the bilayer ice structure on $BaF_2(111)$. From Sadtchenko *et al.* [39].

oxygen), that does not participate in hydrogen bonding, directed upward. These molecules are designated d-O.

Devlin and Buch have calculated the spectrum of the bilayer at the surface of an ice cluster [57]. The infrared profile of a bilayer of water molecules on a structurally matched substrate, *i.e.* $BaF_2(111)$, is likely to resemble the top bilayer on the basal face of an ice crystal. The principle difference for our case is that the lone pair in the lower layer is directed to a Ba^{2+} ion below, while in the ice case it is directed toward a proton to which it hydrogen bonds. The calculation of Devlin and Buch is for the infrared spectrum of a D_2O ice bilayer at the surface of a 450 molecule cluster. The three different types of molecules analogous to the H_2O bilayer are represented d-D, d-O and s-4. Each type of molecule gives rise to two vibrational modes corresponding to the D stretching motions against the two chemical bonds of a D_2O molecule. The highest frequency mode calculated is for the d-D molecule stretching vibration at 2720 cm^{-1}. In order to compare these calculations to the H_2O bilayer we are considering, each vibrational frequency is scaled by the factor 1.36 that brings the high frequency d-D molecule vibration into alignment with the corresponding d-H vibration at 3700 cm $^{-1}$ observed in experiments [58]. While each of the six vibrational stretching motions of the d-D, d-O, and s-4 molecules is associated with its own unique effective reduced mass, they all principally involve motions of D (or H) atoms. The single scaling factor we have employed glosses over the small reduced mass differences, but encompasses anharmonicity effects. The appropriately scaled calculation is compared to the $\Theta = 0.8$ at 25 $^\circ$ C spectrum in Fig. 8.13. The total theoretical spectrum, the sum of contributions from the three types of molecules in the bilayer, has two well resolved features in qualitative agreement with our observed adlayer spectrum. The principle contribution to the spectrum comes from the in-phase and out-of-phase vibrations of the s-4 molecules. The distinguishing vibration of the dangling-hydrogen, d-H, at 3700 cm^{-1} is presumably too weak to be observed. These observations at low coverages are consistent with the formation of an ordered ice-like adlayer on the $BaF_2(111)$ surface even at room temperature. An ice-like adlayer at temperatures above the bulk-ice melting point can be rationalized since the interactions of H_2O

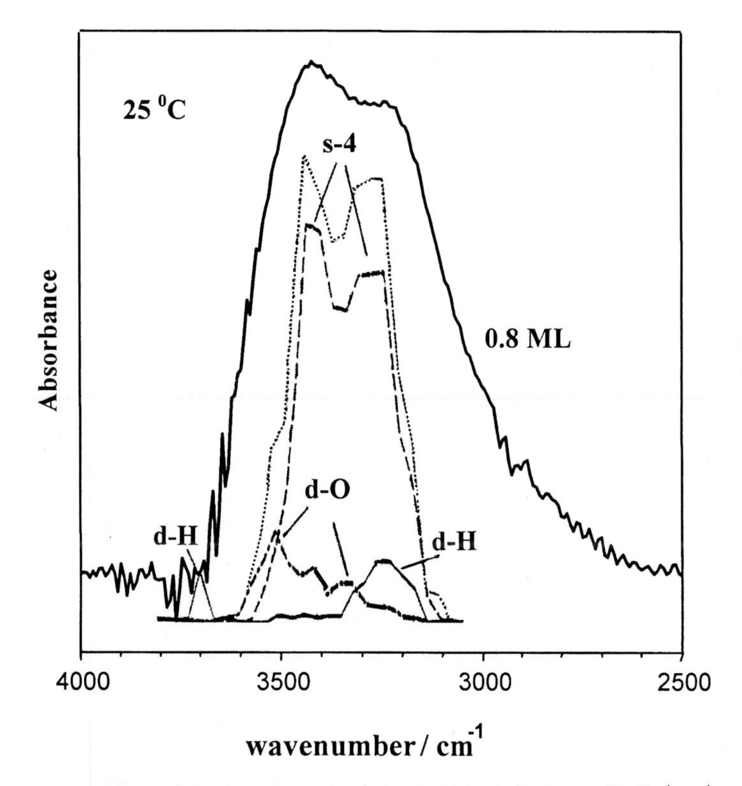

Fig. 8.13. Infrared spectrum of the ice-like bilayer on $BaF_2(111)$ near monolayer coverage at 25C. The broken curves are adapted from calculations of Devlin and Buch [54], and the experimental data from Sadtchenko *et al.* [39].

molecules with the substrate are strong as we shall show later. As a consequence, the order is preserved in the adlayer at temperatures above the ice melting point because H_2O molecules are immobilized at the adsorption sites.

Having argued that an ordered, ice-like bilayer of H_2O molecules forms on the $BaF_2(111)$ surface at submonolayer and monolayer coverages, it is easy to explain the spectral variation at $\Theta = 2$. The spectrum of such a liquid-like, disordered adlayer must be similar to the spectrum of bulk liquid water. The two-band structure, that is present in the spectra at low H_2O coverages disappears, and a broad asymmetric peak, centered at 3400 cm^{-1} is observed. As we have already mentioned, the profile at higher coverages is similar to the spectrum of neat water with one exception. An ill-defined shoulder near 3200 cm^{-1} contributes significantly to the overall band width. Some contrast between the spectrum of neat water and the spectra of high coverage H_2O adlayers, however, is expected. It is possible that unlike water, the multilayer H_2O films may not be completely disordered. Water molecules that are close to the surface of $BaF_2(111)$ must be influenced by the substrate in such a way that their entropy is slightly lower than the entropy of the molecules further from the substrate. The shoulder at 3200 cm^{-1}, whose frequency is the signature for ice (Fig. 8.2), may be a manifestation of the residual ordering of the surface bound H_2O adlayer in the case of high coverages.

Isotherms. We turn now to explore the isotherms of Fig. 8.14 to further interpret the nature of water adlayers on $BaF_2(111)$. Recalling the statistical nature of adsorption for systems under ambient conditions, we expect that for a coverage stated to be $\Theta = 1$ the bilayer actually will not be complete but consist of islands with their upper layers partially covered by adsorbed molecules.

We now consider interpretation of the low coverage region, $\Theta < 0.5$, of the isotherm below 5 mbar. The spectroscopic signatures have led us to identify small islands. Presumably these low coverages correspond to an equilibrium at the $BaF_2(111)$ surface between a two-dimensional gas of H_2O molecules and two-dimensional (bilayer) islands. The absorptions of isolated adsorbed H_2O molecules (the two-dimensional gas) is presumably too weak to be detected both because of their low concentration and low optical cross sections.

The adsorption and desorption isotherms are irreversible at coverages above $\Theta = 1$ as shown in Fig. 8.14. We have interpreted the variations in the spectra in Fig. 8.11 as an evidence for disordering of the first, ice-like adlayer by adsorption of water molecules on top of it. We would like to argue that the hysteresis in the isotherm at 25 $^\circ$C further supports this conclusion.

Hysteresis in the isotherm demonstrates that, depending on their history, two H_2O films of identical thickness may have different vapor pressures above their surfaces. Therefore, the observed isotherm hysteresis suggests that significant variations occur in the structure of the adlayer when the film thickness exceeds some critical value between one and two monolayers. As we propose, the variations in the structure of the H_2O adlayer may consist in

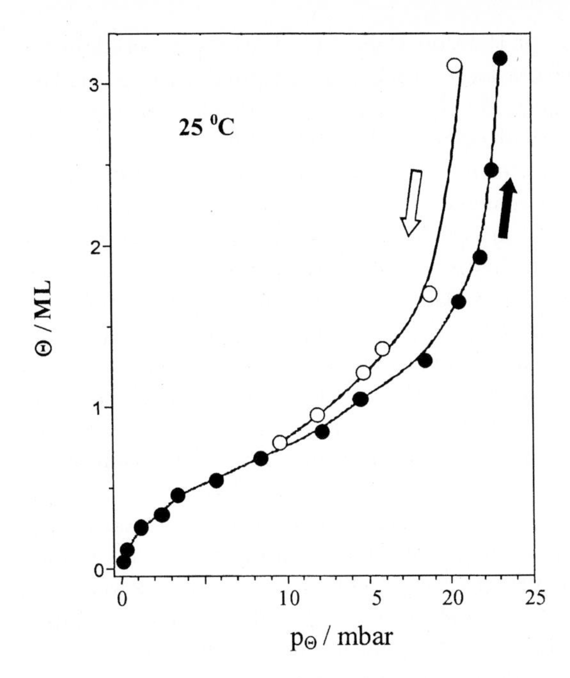

Fig. 8.14. The adsorption and desorption isotherms of water on $BaF_2(111)$ at 25 ° C. From Sadtchenko *et al.*[39].

transition from highly ordered hexagonal to liquid-like disordered structure as the coverage increases above $\Theta = 1$. The disordering of a thin ice-like adlayer covered by liquid-like film must be a kinetic process, which is probably complete on the time scale of our experimental measurement (10-20 min). Consequently, we were not able to study kinetics of such a transition from ordered to disordered H_2O adlayer. Miura et al. have studied the kinetics of water adsorption on $BaF_2(111)$ surface at room temperature by AFM near 22 mbar [35]. According to these results, the adsorption of a $\Theta = 1$ film was followed by formation of droplets with small contact angles. However the droplets eventually disappeared resulting in a multilayer film. This observation is in accord with our conclusion about the nature of variations in the adlayer structure during multilayer adsorption to which we attribute the spectrum changes in the range from 12 to 22 mbar. Indeed, a phase separation must occur initially after adsorption of H_2O molecules on the top of the ice-like adlayer that may manifest itself as liquid drops. The uniform liquid-like film is observed when the disordering of the ice-like layer is complete.

Thermodynamics. Enthalpy and entropy of H_2O adsorption on $BaF_2(111)$ surface have been derived from the isotherms spanning the temperatures – 1 to 25 ° C. The condensation of water onto $BaF_2(111)$ is exothermic at all coverages. At the lowest submonolayer coverages the adsorption enthalpy becomes the least exothermic. Indeed at $\Theta = 0.2$ the enthalpy is -38 kJ mol^{-1} which is a significant departure from the value for liquid water. As we suggested earlier, the adsorption of H_2O on $BaF_2(111)$ initially proceeds with formation of small islands with a large number of incompletely coordinated H_2O molecules on the periphery. At the island edges, the lack of neighbor molecules with which to form hydrogen bonds must result in a low value of adsorption bonding. As more water molecules adsorb to approach monolayer coverage, the islands begin to coalesce. The resulting adlayer is characterized by an extended network of lateral hydrogen bonds and, therefore, the adsorption enthalpy is more exothermic. Indeed, the adsorption enthalpy is most exothermic at $\Theta = 0.8$. The exothermic extremum of -58 kJ mol^{-1} is significantly lower than the enthalpy of condensation of either ice (-48 kJ mol $^{-1}$), or liquid water (- 44 kJ mol^{-1}). This value also suggests that the interactions of H_2O molecules with the $BaF_2(111)$ surface are strong. The enthalpy of adsorption measured at coverages close to $\Theta = 1$ supports the conclusion that H_2O adsorption on $BaF_2(111)$ leads to the formation of an ordered layer nailed to the substrate.

At H_2O coverages above $\Theta = 1$, the enthalpy of adsorption decreases significantly, and approaches the enthalpy of adsorption of bulk liquid water. The change in the adsorption enthalpy from –58 to – 44 kJ mol^{-1} is in agreement with the proposed transition from an ordered to disordered liquid-like structure of the H_2O adlayer on $BaF_2(111)$ surface at multilayer coverages.

The variation in entropy of adsorption with coverage mirrors the change in enthalpy. At low coverage, the entropy of H_2O adlayer on BaF_2 is actually higher than the entropy of bulk water. This may arise because many islands containing small numbers of water molecules contribute configuration entropy. Furthermore, the desorption of water molecules is likely to occur via a two-step mechanism resulting in a large number of mobile desorption precursors, essentially H_2O monomers, on the substrate surface [59]. As the coverage increases, the entropy of the adlayer decreases rapidly, achieving a minimum value of about 40 J K^{-1} mol^{-1} near $\Theta = 1$, which is significantly lower than the entropy of liquid water.

The low value of entropy near $\Theta = 1$ indicates that at this coverage the most ordered adlayer arrangement is achieved. It is interesting to compare entropy of the adlayer at this coverage to the entropy of I_h ice. As we have indicated, our value of entropy is appropriate to 25 $^\circ$C. Therefore the comparison must be made to the entropy of a hypothetical "overheated" ice at 25. This entropy is estimated by taking the value for ice at 0 $^\circ$C, 38.1 J K^{-1} mol^{-1} [6], its heat capacity of C_p = 40 J K^{-1} mol^{-1} and the use of the integrated form of the relationship $\mathrm{d}S = C_p dT/T$ [3]. The resulting value is 40 J K^{-1} mol^{-1} for "overheated" ice at 25 $^\circ$C. The entropy for neat "overheated" ice is compatible with the value for the entropy of monolayer water on BaF_2 (111) that we find to be 40 $\pm$ 5 J K^{-1} mol^{-1}. The close match of these entropies is consistent with the view that H_2O adsorption on $BaF_2(111)$ leads to the formation of a highly ordered ice-like structure even at 25 $^\circ$C. Of course, other tightly bound adlayer structures would also tend to result in a low entropy value. We note that the corresponding $\Theta = 1$ coverage entropy for water on mica(001) is somewhat higher (Fig. 8.10), suggesting again that this substrate does not favor an ice-like adlayer.

At H_2O coverages above $\Theta = 1$ the entropy of the adlayer approaches 68 J mol^{-1} K^{-1}, near the value for water [3]. The increase in entropy from 40 to 68 J mol^{-1} K^{-1} at multilayer coverages is in agreement with the proposed transition from ice-like to liquid-like structure of the H_2O adlayer on $BaF_2(111)$ surface.

Another important conclusion follows directly from these adsorption measurements. Since the enthalpy of adsorption at $\Theta = 3$ coverages is equal to the adsorption enthalpy of H_2O on the surface of liquid water, it is unlikely that the interactions of H_2O molecules with ions of the BaF_2 substrate propagate through a distance greater than a few H_2O molecular diameters. This conclusion is consistent with our findings for thick films of water on Na(001) and mica(001).

8.4 Afterward and Forward

Langmuir was right [1]. Mica has a film of water, a few molecular layers thick, adsorbed to its surface under ambient conditions. Our confirming spectroscopic experiments together with theoretical calculations are justified in that they add insight into the structure of the film, and provide isotherms, as well as yield thermodynamic values. For each of the substrate surfaces we have considered, NaCl(001), mica(001) and $BaF_2(111)$, the structures of the monolayer and submonolayer films are largely directed by the underlying ionic surfaces. This is confirmed by several theoretical studies and supported by the infrared absorption profiles that are significantly different from those of liquid water. The disruption of the liquid water hydrogen bonding network is most dramatically revealed for the monolayer on $BaF_2(111)$ that appears to have an ice-like arrangement even at room temperature. For multilayer coverages the thin film spectroscopic profiles approach those of liquid water for each of the substrate surfaces.

Both the shapes of the isotherms and the thermodynamic values they yield are consistent with ordered tightly bound monolayers. The enthalpy of adsorption and the absolute entropy approach liquid water values at multilayer coverages.

The above summary, in the form of an afterword, begs to looking forward to consider questions that need answering in future investigations.

What is the role of thin film water on the chemistry of insulator surfaces? This question has been answered in part for reactions of several oxides of nitrogen on NaCl, the subject of an extensive review by Hemminger [60]. Both this review and a recent one by Grassian [61], who considers a variety of substrates other than NaCl, point to the relevance of this surface chemistry on insulator particles to tropospheric processes. What is not clear is the relationship between the surface morphology of the substrates and the nature of the thin film. The spectra of thin water on defect-rich NaCl surfaces [46] and well defined NaCl(001) surfaces [20, 41] are qualitatively different. The chemistry that the thin-film water, frequently called the quasi-liquid layer [62], supports must be intimately related to the nature of these different forms of water.

Physical properties of insulator surfaces are also affected by thin-film water. This is of relevance to the mechanism of dissolution of water soluble insulators [36, 42]. Thin-film water also plays a role in the sintering of insulator particles. Sintering appears as the common experience of caking of granular salt or sugar. Commercially, sintering is of practical importance in the packaging and transport of many powders and particles from pharmaceuticals to fertilizers to detergents [63]. The role of thin-film water in the molecular interpretation of sintering is needed.

Thin-film water must also determine, at least in part, the ability of the insulator to act as an ice nucleating substrate. While mica, and other insulator substrates, have been long known as ice nucleating agents [37, 64], the mechanism for their effectiveness is not clear [65]. The exploration of $BaF_2(111)$ as an ice nucleation substrate, and the structure of thin water on its surface, is the subject of a recent investigation [39, 66]. More work needs to be done before the nature of thin film water as a precursor to ice formation on insulator surfaces is understood.

The most interesting example of thin film water on an insulator substrate is found on the surface of ice near its melting point. Michael Faraday set a standard in experimental design for the study of surface melting in 1860 [67]. A recent investigation on interfacial melting of ice [68] offers a sampling of references to previous work. Thin film water on ice provides the lubrication for skating and skiing [69], and plays a role in electrification of clouds and lightning [70]. Chemistry within the thin film water layer on snow pack and atmospheric ice particles is an active area of current research [71-74]. While it is common to treat the physical and chemical properties of thin film water on ice as ordinary water, further research will be required to establish its nature.

Acknowledgements

We wish to thank Zhenfeng Zhang for numerous discussions and experimental assistance by Charles McCrory and Peter Conrad, all at Indiana University. A continuing partnership with Dr. Anthony Stone has greatly deepened our understanding of these systems. The work was funded by the National Science Foundation.

References

1. I. Langmuir: J. Am. Chem. Soc. **40**, 1361 (1918).
2. A. J. Stone: The Theory of Intramolecular Forces, (Clarendon Press, Oxford,1997).
3. P. Atkins: Physical Chemistry, 6th ed; (W. H. Freeman, New York, 1998).
4. J. N. Israelachvili: Intermolecular and Surface Forces, (Academic Press, London, 1985).
5. J. F. Lennard-Jones, B. M. Dent: Trans. Faraday Soc. **24**, 92 (1928).
6. *National Research Council (US) International Critical tables* (McGraw-Hill, New York, 1926).
7. O. Engkvist, A. J. Stone: J. Chem. Phys. **110**, 12089 (1999).
8. D. Eisenberg, W. Kauzmann: The Structure and Properties of Water,(Oxford, New York, 1969).
9. D. Henderson, F. A. Abraham, J Barker: Mol. Phys. **31**, 1291 (1976).
10. C. Y. Lee, J. A. McCammon, P. J. Rossky: J. Chem.Phys. **80**, 4448 (1984).
11. O. Engkvist, A. J. Stone: J. Chem. Phys. **112**, 6827 (2000).
12. G. C. Pimentel, A. L. McClelland: The Hydrogen Bond, (Reinhold, New York, 1960).
13. T. Shimanouchi, Tables of Molecular Vibrational Frequencies, Part 1: NSRDS-NB86, (U. S. Government Printing Office, Washington, D. C., 1972).
14. H. D. Downing, D. Williams: J. Geophys. Res. **80**, 1656 (1975).
15. S. G. Warren: Appl. Opt. **23**, 1306 (1984).
16. E. Estrin, L. Paglieri, G Corongiu, E. Clementi: J. Phys. Chem. **100**, 8701 (1996).
17. S. A. Rice, M. S. Bergren, A. C. Belch, G. Nielson: J. Phys. Chem. **87**, 4295 (1983).
18. H. Witek, V. Buch: J. Chem. Phys. **110**, 3168 (1999).
19. K. Nauta, R. E. Miller: Science **287**, 293 (2000).
20. M. Foster, G. Ewing: J. Chem. Phys. **112**, 6817 (2000).
21. S. Brunauer, L. S. Deming, W. E. Deming, E. Teller: J. Am. Chem. Soc. **62**, 1723 (1940).
22. C. Noda, G. E. Ewing: Surf. Sci. **240**, 181 (1990).
23. A. W. Meredith, A. J. Stone: J. Chem. Phys. **104**, 3058 (1996).
24. H.-C. Chang, H. H. Richardson, G. E. Ewing: J. Chem. Phys. **89**, 7561 (1988).
25. G. E. Ewing, G. C. Pimentel: J. Chem. Phys. **35**, 925 (1961).
26. a) J. Heidberg, E. Kampshoff, M. Suhren: J. Chem. Phys. **95**, 9408 (1991). b) D. Schmicker, J. P. Toennies, R. Vollmer, H. Weiss: J. Chem. Phys. **95**, 9412 (1991).
27. T. L. Hill: Introduction to Statistical Thermodynamics, (Addison-Wesley, Reading, 1960).
28. a) K. R. Willian, G. E. Ewing: J. Phys. Chem. **99**, 2186 (1995), b) S. K. Dunn, G. E. Ewing: J. Phys. Chem. **96**, 5284 (1992).
29. O. Berg, G. E. Ewing: Surf. Sci. **220**, 207 (1989).
30. J. Heidberg, E. Kampshoof, R. Kühnemuth, O. Schömekäs: Can. J. Chem **72**, 795 (1994).
31. L. W. Bruch, A. Glebov, J. P. Toennies, H. Weiss: J. Chem. Phys. **103**, 5109 (1995); S. Fölsch, M. Henzler: Surf. Sci. **264**, 65 (1992).
32. S. Brunauer, P. H. Emmett, E. Teller: J. Am. Chem. Soc. **60**, 309 (1938).

33. a) G. Meyer, N. M. Amer: Appl. Phys. Lett. **56**, 2100 (1990),
b) A. L. Schuger, R. M. Wilson, R. T. Williams: Phys. Rev. B **49**, 4915 (1994),
c) Q. Dai, J. Hu, M. Salmeron: J. Phys. Chem. B **101**, 1994 (1997).
34. J. Hu, X.-D Xiao, D. Ogletree, M. Salmeron: Science **268**, 267 (1995).
35. K. Miura, T. Yamada, M. Ishikawa, S. Okita: Appl. Surf. Sci. **140**, 415 (1999).
36. M. Luna, F. Rieutford, N. A. Melman, Q. Dai, M. Salmeron: J. Phys. Chem. A **102**, 6793 (1998).
37. G. Bryant, J. Hallett, B. Mason: J. Phys. Chem. Solids **12**, 189 (1959).
38. A. Lehmann, G. Fahsold, G. Konig, K. H. Riedere: Surf. Sci. **369**, 289 (1996).
39. V. Sadtchenko, P. Conrad, G. E. Ewing: J. Chem. Phys. **116**, 4293 (2002).
40. W. Cantrell, G. E. Ewing: J. Phys. Chem. B **105**, 5435 (2001).
41. a) S. J. Peters, G. E. Ewing: Langmuir **13**, 6345 (1997). b) S. J. Peters, G. E. Ewing, J. Phys. Chem. B **101**, 10880 (1997).
42. W. Cantrell, C. McCrory, G. E. Ewing: J. Chem. Phys. **116**, 2116 (2002).
43. I. N. Tang, K. H. Fung, D. G. Imre, H. R. Munkelwitz, Aerosol Sci. Technol. **23**, 443 (1995).
44. a) P. B. Barraclough, P. G. Hall: Surf. Sci. **46**, 393 (1974), b) H. U. Walter, Z. Phys. Chem. (Frankfurt am Main) **75**, 287 (1971), c) R. A. Lad, Surf. Sci. **12**, 37 (1968), d) M. Kaiho, M. Chikazawa, T. Kanazawa: Nippon Kagaku Kaishi **8**, 1368 (1972).
45. G. E. Ewing, S. J. Peters: Surf. Rev. and Lettr. **4**, 757 (1997).
46. D. Dai, S. J. Peters, G. E. Ewing: J. Phys. Chem. **99**, 10299 (1995).
47. W. Vedder, R. S. McDonald: J. Chem. Phys. **38**, 1583 (1963).
48. G. Carson, S. Granick: Appl Spectrosc. **43**, 473 (1989).
49. F. H. Norton: Elements of Ceramics, (Addison-Wesley: Cambridge, 1952).
50. M. Odelius, M. Bernasconi, M. Parrinello: Phys. Rev. Lett. **78**, 2855 (1997).
51. P. Miranda, L. Xu, Y. Shen, M. Salmeron: Phys. Rev. Lett. **81**, 5876 (1998).
52. J. Dash: Phys. Rev. B **15**, 3136 (1977).
53. a) D. Beaglehole, H. K. Christenson: J. Phys. Chem. **96**, 3395 (1992), b) D. Beaglehole, E. Radlinska, B. Ninham, H. K. Christenson: Phys. Rev. Lett. **66**, 2084 (1991).
54. D. Beaglehole: Physica, A. **244**, 40 (1997).
55. B. Wassermann, J. Reif, E. Matthias: Phys. Rev. B **50**, 2593 (1994).
56. J. P. Devlin, V. Buch: J. Phys. Chem. **99**, 16534 (1995).
57. J. P. Devlin, V. Buch: J. Phys. Chem. **101**, 6095 (1997).
58. B. Rowland, M. Fisher, J. P. Devlin: J. Phys. Chem. **97**, 2485 (1993).
59. J. G. Davy, G. A. Somorjai: J. Chem. Phys. **55**, 3624 (1991).
60. J. C. Hemminger: Int. Rev. Phys. Chem. **18**, 387 (1999).
61. V. H. Grassian: Int. Rev. Phys. Chem. **20**, 467 (2001).
62. R. Vogt, B. F. Finlayson-Pitts: J. Phys. Chem. **98**, 3747 (1994).
63. G. I. Tardos, I. V. Nicolaescu, B. Ahtchi-Ali: Powder Handl. and Process. **8**, 7 (1996).
64. P. V. Hobbs, Ice Physics, (Clarendon Press, Oxford, 1974).
65. H. R. Pruppacher, J. D. Klett: Microphysics of Clouds and Precipitation, 2nd ed. (D. Reidd, Dordrecht, 1997).
66. V. Sadtchenko, G. E. Ewing, D. Nutt, A. J. Stone: Langmuir (to be published, 2002).
67. M. Faraday: Proc. Roy. Soc. [London] **10**, 152 (1860).
68. V. Sadtchenko, G. E. Ewing: J. Chem. Phys. in press (2002).

69. V. F. Petrenko, R. W. Whitworth: Physics of Ice, (Oxford University Press, Oxford, 1999).
70. M. Baker, J. G. Dash: J. Cryst. Growth **97**, 770 (1989).
71. S. M. Clegg, J. P. D. Abbatt: J. Phys. Chem. A **105**, 6630 (2001).
72. T. Huthwelker, D. Lamb, M. Baker, B. Swanson, T. Peter: J. Colloid and Interface Sci. **238**, 147 (2001).
73. D. R. Hanson and A. R. Ravishankara: J. Phys. Chem. **96**, 2682 (1992).
74. M. Zondlo, S. B. Barone, M. A. Tolbert: Geophys. Res. Lett. **24**, 1391 (1997).

74. M. Sugisaki, H. Suga and S. Seki: Bull. Chem. Soc. Jpn. 41, 2591 (1968)

9 Protein Hydration Water

Douglas J. Tobias, William I-Feng Kuo, Ali Razmara, and Mounir Tarek

9.1 Introduction

Proteins are linear heteropolymers of amino acids connected by amide (peptide) linkages. There are twenty naturally occurring amino acids whose side chains span a wide range of polarity (e.g. aliphatic, aromatic, neutral polar, acidic, and basic). The polypeptide backbone is largely organized into regular "secondary" structures (α-helices, β-sheets, reverse turns) by networks of hydrogen bonds, along with unstructured regions (loops). Interactions between side chains and of side chains with the solvent determine how the secondary structural elements are uniquely packed together in the functional (native) state of the protein.

Proteins may be broadly classified in terms of the primary components of their solvent environment: soluble proteins reside in aqueous solution, while membrane proteins reside in the amphipathic milieu of a lipid bilayer, which consists of a non-polar, hydrocarbon interior sandwiched between polar groups that are in contact with aqueous solution. Although water molecules are intimately associated with both soluble and membrane proteins, in this chapter we will restrict our attention to the soluble proteins, which are completely surrounded by water molecules. To a first approximation, soluble proteins are "folded" into an architecture that sequesters the nonpolar side chains from the solvent in a "hydrophobic core, and places the polar side chains in the aqueous solution on the surface of the protein.

Water plays a vital role in determining the structures and dynamics, and hence, the function of globular soluble proteins. Water molecules in protein solutions may be broadly classified into three categories [1]: strongly bound, internal water molecules that occupy internal cavities and deep clefts; water molecules that interact with the protein surface; and bulk water. Internal waters, which can be identified crystallographically and are conserved in homologous proteins, are extensively hydrogen bonded and comprise an integral part of the protein structure. They have residence times ranging from about 10 ns to ms, and their exchange with the bulk solvent requires local unfolding to occur. Surface water molecules are much less well-defined structurally than internal water molecules (in the sense that surface binding sites identified crystallographically are not highly conserved among different crystal forms of the same protein), and are much more mobile, with residence times on the order of tens of ps. In addition to being important for protein stability, and in the energetics and specificity of ligand binding, surface waters also have a profound influence on the dynamics of a protein molecule as a whole [2]. Proteins require a threshold level of hydration in order to function [3]. Although the details of the connection between protein hydration and function have not yet been worked out, it has become clear that surface water is required for the activation of fast conformational fluctuations that appear to be important in protein folding and function [4]. The observation of enzyme activity in partially hydrated powders [3], where the amount of water present is far less than sufficient to completely cover the protein surface, suggests a crucial role for the water molecules in the first solvation shell, the so-called "protein hydration" water.

This chapter reviews the structural, energetic, and dynamical properties of protein hydration water. This topic has been the subject of numerous experimental and theoretical investigations. Space limitations preclude this from being an exhaustive review with a complete set of references. Thus, we have provided an overview, citing review articles wherever possible. We apologize to authors of original work that is not cited here.

9.2 Structural and Energetic Aspects of Protein-Water Interactions

The surface presented to the aqueous solvent by a soluble protein is chemically heterogeneous and rough. This is evident in Fig. 9.1, where we have colored the surface of the protein barnase according to the polarity of the amino acid side chains. The surface can be described as predominantly polar, but a substantial portion ($\approx$ 25 % of the total surface area) is nonpolar, and within the polar patches of the surface there is considerable variability in the charge state of the side chains across areas corresponding to a few water molecules. Thus, we anticipate that the details of protein-water interactions will vary substantially from site to site on the protein surface. It is also clear from Fig. 9.1 that the protein surface is very rough on the length scale of a water molecule. The packing of the side chains on the surface produce a number of highly curved protrusions around which water molecules must arrange themselves, as well as numerous nooks and crannies of varying extent and shape. In light of the chemical heterogeneity and roughness depicted in Fig. 9.1, we anticipate that the details of protein-water interactions will vary substantially from site to site on the protein surface.

The snapshot from a molecular dynamics (MD) simulation of barnase in solution shown in Fig. 9.2 illustrates qualitatively how the chemical heterogeneity and roughness of the protein surface produces an inhomogeneous first hydration shell (defined here as water molecules with oxygen atoms within 4 Å of any protein heavy atom, for reasons discussed below). Note that there

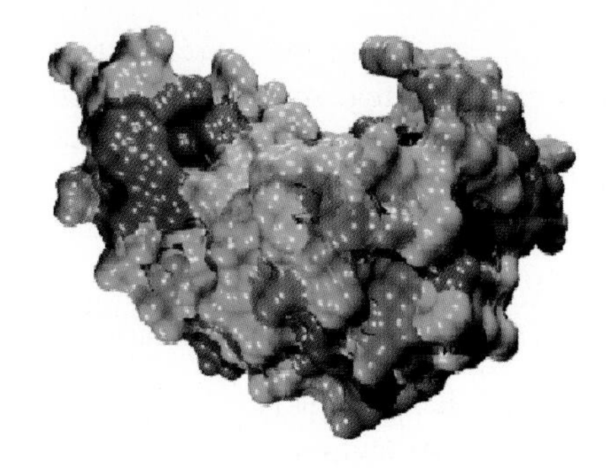

Fig. 9.1. Solvent accessible surface representation of the protein barnase with the protein surface colored according to amino acid type: non-polar, gray; neutral polar, green; anionic (at neutral pH), orange; cationic (at neutral pH), blue.

are not enough of these water molecules that are the most intimately associated with the protein to coat the entire surface. Water molecules in the first shell tend to avoid the highly curved protrusions, preferring instead to embed themselves in crevices, which may be regarded as water binding sites or "hydration sites" [5]. Fig. 9.2 is useful for visualizing how individual water molecules might arrange themselves on the surface of a protein at a given instant in time, but it is a bit misleading because it is a single snapshot from a MD simulation.

Fig. 9.2. Solvent accessible surface of barnase colored magenta with hydration water molecules (those whose O atom is within 4 Å of any protein heavy atom) colored blue.

Although some water molecules appear to be sufficiently well immobilized on a protein surface to be identified crystallographically, most are disordered. Hence, it is more appropriate to discuss the hydration of a protein in terms of probabilistic (ensemble averaged) measures, such as solvent distribution functions [5]. Using X-ray and neutron diffraction, solvent density (electron and neutron scattering length) distributions can be determined in protein crystals, although technical difficulties have limited the number of such measurements to a few. Three-dimensional solvent density distributions are readily computed from MD simulations [5]. In Fig. 9.3 we show water isodensity surfaces from the barnase solution simulation.

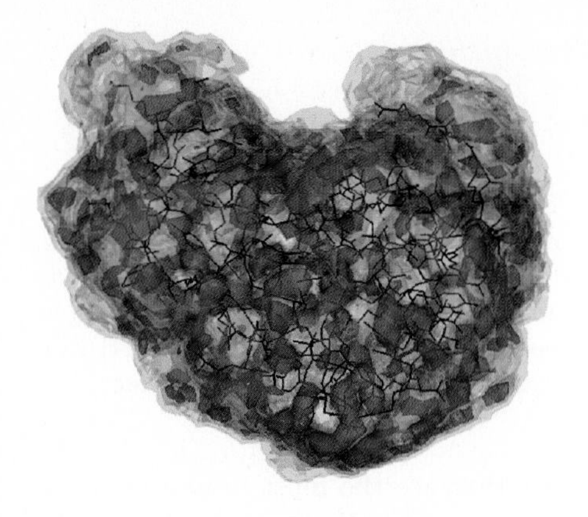

Fig. 9.3. Water isodensity surfaces constructed from a MD simulation of barnase in solution. The protein is drawn as a black wireframe model; the surfaces are contoured at densities relative to bulk of 0.5 (blue), 1.0 (green), and 1.5 (red).

Inspection of solvent density distributions around proteins reveals that there are, roughly speaking, four distinct regions [5]. These regions become readily apparent when the three-dimensional solvent density distribution is decomposed into radial distributions referred to the closest protein atom, and are averaged for each protein atom type. Examples of such distributions determined from analysis of X-ray diffraction data from a protein crystal [6] are shown in Fig. 9.4. Starting from the interior of the protein and moving outward, one passes from a largely solvent-free region inside the protein ($r <$ 1.5 Å), through a transition region where there is a significant degree of interpenetration of protein and solvent (1.5 Å $< r <$ 2.5 Å), into a region where the solvent density is higher than the bulk (2.5 Å $< r <$ 4 Å), and, finally, into the bulk solution ($r >$ 4 Å).

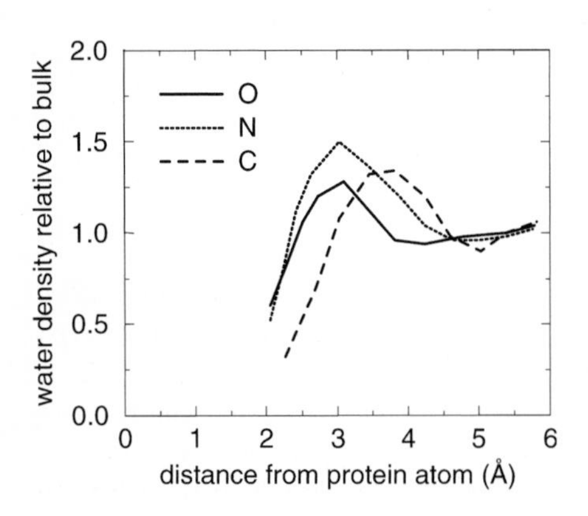

Fig. 9.4. Radial solvent density profiles extracted from analysis of X-ray diffraction data from rat mannose binding protein A [6], plotted versus distance from the closest protein heavy atom, separately for each protein atom type: O (solid curve), N (dashed curve), and C (dotted curve).

The first hydration shell of the protein, which may be identified as the region where the water density is higher than the bulk, does not extend outward from the protein surface beyond roughly 4 Å on average (this is the basis of the operational definition, used throughout this chapter, of hydration waters as those within 4 Å of protein heavy atoms). Essentially the same picture of the first hydration shell of proteins in solution has been obtained from MD simulations, and small angle X-ray and neutron scattering data [5, 7, 8].

Fig. 9.5. Solvent accessible surface of barnase colored gray with "bound" water molecules (those whose O atom is within 4 Å of any protein heavy atom) colored according to the energy of their interaction with the protein, with blue being the most negative (favorable) and red being the most positive (unfavorable).

The diffuse nature of the first hydration shell is a manifestation of the roughness and chemical heterogeneity of the protein-water interface . These features are also manifested as a pronounced heterogeneity in the strengths of protein-water interactions (Fig. 9.5). The distribution of protein-water interaction energies for water molecules in the first hydration shell of barnase, computed from a MD simulation (using the CHARMM22 force field for the protein [9] and the TIP3P water model [10]), is plotted in Fig. 9.6. The protein-water interaction energies range from -36 kcal/mol to +8 kcal/mole. The distribution is peaked at -1.5 kcal/mole, and has an average of -7.4 kcal/mol.

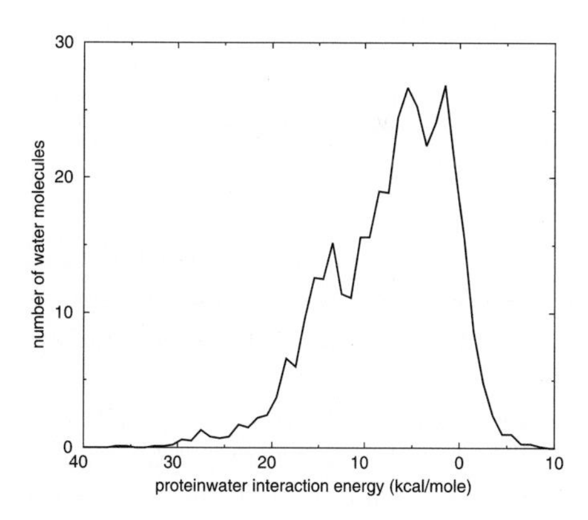

Fig. 9.6. Histogram of protein-water interaction energies for hydration water molecules (those within 4 Å of protein heavy atoms) from a 1 ns MD simulation of barnase in solution.

9.3 Water Dynamics Near Protein Surfaces

The results presented in the previous section paint a picture of a rough and chemically heterogeneous protein surface, which produces an inhomogeneous first solvation shell with a broad range of protein-water interaction energies. The surface roughness and/or heterogeneity of hydration sites is also manifested in the anomalous dynamics of protein hydration water.

Inelastic incoherent neutron scattering experiments on partially hydrated protein powders in which the hydration level is sufficiently low ($h \approx 0.3$ g H_2O/g protein) that all of the water molecules can be considered in contact with the protein, have clearly demonstrated that both the vibrational and diffusive motion of protein hydration water is anomalous [11]. In particular, the dynamical susceptibility spectrum (essentially energy-weighted dynamical structure factor) indicate that the diffusive motion of protein hydration water is strongly suppressed relative to bulk water, and the single-particle density correlation function (i.e. the intermediate scattering function, obtained by Fourier deconvolution of the dynamical structure factor) displays a secondary (structural) relaxation that is described well by a stretched exponential function, rather than an exponential as expected for Brownian motion. Fourier inversion of the intermediate scattering functions revealed that the (MSDs) of protein hydration water molecules follow a power law (fractal) time dependence , $\langle r^2(t) \rangle \propto t^{0.4}$, in contrast to the linear (Brownian) time dependence, $\langle r^2(t) \rangle \propto t$, exhibited after 5 ps by bulk water at room temperature.

The signatures of anomalous hydration water dynamics evident in neutron scattering data have been qualitatively reproduced by numerous MD simulations (see [12, 13] for reviews), but quantitatively accurate simulations of hydration water dynamics have been scarce [13, 14]. We have recently shown that neutron scattering data can only be quantitatively reproduced if the environment in the powder samples employed in the experiments is realistically modelled [13], and we have found that a protein crystal is a good model for a powder at an equivalent hydration level [16]. In Fig. 9.7, we illustrate the anomalous translational diffusion of protein hydration water with a comparison of the MSDs of water molecules in the bulk and in a crystal of the protein Ribonuclease A (RNase), where most of the water molecules are in contact with protein molecules.

The anomalous character of hydration water dynamics is also reflected in the rotational motion, which can be described by orientational correlation functions that have been computed from several MD simulations [12, 14], but have only been measured directly in experiments [17]. Here we discuss water rotational motion based on the second-rank rotational correlation functions, $C(t) = \langle P_2 [\mathbf{u}(0) \cdot \mathbf{u}(t)] \rangle$, where $\mathbf{u}$ is a molecule-fixed unit vector (the results presented here are for the H–H vector, but other choices, like the O–H bond and the dipole moment, give similar results), and $P_2(x)$ is the second Legendre polynomial of argument x. The correlation functions computed from

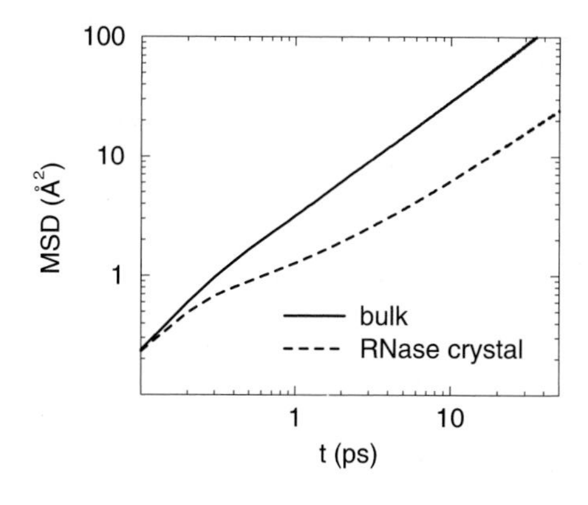

Fig. 9.7. Water center-of-mass mean-squared displacements computed from MD simulations of bulk water (solid curve) and water in a RNase crystal (dashed curve).

MD simulations of bulk D_2O and D_2O in contact with surface of barnase in solution are shown in Fig. 9.8 (we used the SPC/E water model [18] for these calculations). The contribution of a given bound water to the $C(t)$ was included only while that water was near the protein surface (within 4 Å of any protein heavy atom). The rotational correlation function for bulk water decays much faster than that of the protein hydration water, indicating that the rotational motion of water molecules in contact with the protein is significantly slowed relative to the bulk.

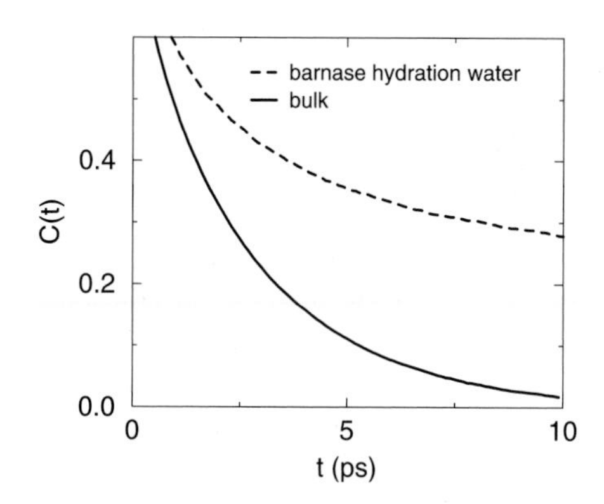

Fig. 9.8. Water second-rank rotational correlation functions computed from MD simulations at 300 K of bulk D_2O (solid curve) and D_2O on the surface of barnase in solution (dashed curve).

The correlation functions in Fig. 9.8 do not exhibit an exponential decay, as expected for rotational Brownian motion, but rather are well-described by a stretched exponential, $exp\left[-(t/\tau)^{\beta}\right]$, which is a signature of a distribution of relaxation times. Individual water molecules interacting with different parts of a rough, chemically heterogeneous protein surface have different reorientation rates [19]. Although the rotational correlation function for each water molecule may exhibit an exponential decay with a characteristic correlation time, when all of the correlation functions with different correlation times are averaged to give the overall $C(t)$, the result is a stretched exponential [19].

Average water rotational correlation times have been deduced from a quantity, $N_S \rho_s$, determined by magnetic relaxation dispersion (MRD) measurements on protein solutions [20]. Here, N_S is the number of water molecules on the surface of the protein, $\rho_S = \langle\tau\rangle_S / \langle\tau\rangle_B - 1$, and $\langle\tau\rangle_S$ and $\langle\tau\rangle_B$ are the average rotational correlation times of water molecules on the surface of the protein and in the bulk, respectively. Estimating N_S from the solvent-accessible surface area of proteins with known structures, values of $\rho_S \approx$ 4-5 have been obtained for several globular proteins [20, 21]. We have computed estimates of ρ_s from our simulations to check for consistency with the MRD data. To this end, the average correlation times are computed from $\langle\tau\rangle = (\tau/\beta)\Gamma(1/\beta)$ [11], where τ and β are parameters from fits to the stretched exponential function, and $\Gamma(x)$ is the gamma function of argument x. From fits to the data in Fig. 9.8, we obtained $\langle\tau\rangle_B = 2.4$ ps for bulk D_2O, and $\langle\tau\rangle_S = 14.4$ ps for water on the surface of barnase. The bulk result is in reasonable agreement with the experimental value of 3.1 ps [22]. Taking N_S to be the average number of water molecules in the first hydration shell (i.e. within 4 Å of a protein heavy atom), we obtain $\rho_S = 4.9$ for the hydration water, which is in the range of values extracted from MRD data on native proteins [20, 21].

The results presented so far in this section have demonstrated that water mobility is perturbed (specifically, slowed down) near a protein molecule relative to in the bulk. It is therefore conceivable that protein hydration water could play a role in the kinetics of protein folding, and protein-ligand and protein-protein association. A quantity that is perhaps more directly relevant to these processes than the MSDs and rotational correlation functions is the residence time of a water molecule on the protein surface. The residence time *per se* is not directly accessible from experimental measurements, but it can be readily computed from a MD trajectory [12]. In Fig. 9.9 we show the distribution of residence times for water molecules on the surface of barnase in solution. The average residence time is $\approx$ 20 ps, but the distribution is very broad, underscoring again the heterogeneity of protein-water interactions. There does not appear to be a strong correlation between residence time and the chemical nature of the protein moieties defining a hydration site [23]. Rather, it appears that the residence time is simply related to the accessibility of a hydration site, with longer times associated with the less accessible sites (i.e. deeper nooks and crannies in Fig. 9.2).

9.4 Protein Coupling to Solvent Motion

At low temperatures (< 200 K) proteins exist in a glassy state. As the temperature is increased, the atomic motional amplitudes increase linearly, as

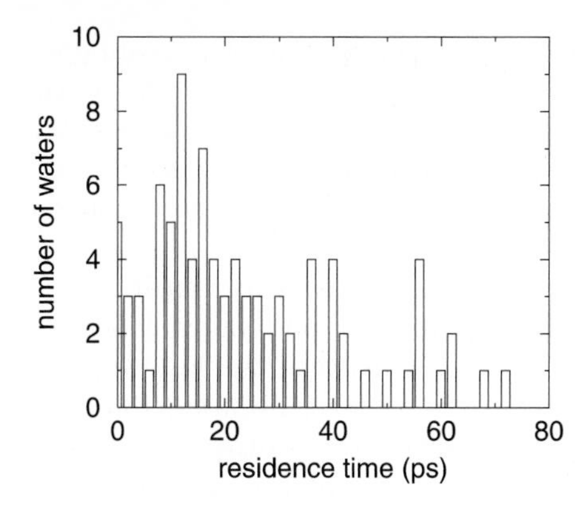

Fig. 9.9. Distribution of water residence times on the surface of barnase from a MD simulation in solution.

in a harmonic solid. In hydrated proteins, at approximately 200 K, the amplitudes suddenly increase, signaling the onset of additional anharmonic and diffusive motion. This so-called "dynamical transition," which is depicted by the neutron scattering data shown in Fig. 9.10, has been observed for atoms distributed throughout proteins over a wide range of length and time scales [24–26], and it has been correlated with the onset of the function of several proteins [27, 28]. The dynamical transition temperature and the amplitudes of the motion above the transition are sensitive to the solvent environment of the protein. As can be seen in Fig. 9.10, the transition is suppressed in a dehydrated protein [28, 29]. Moreover, increasing the solvent viscosity increases the transition temperature and decreases the amplitudes of the motion above the transition [29].

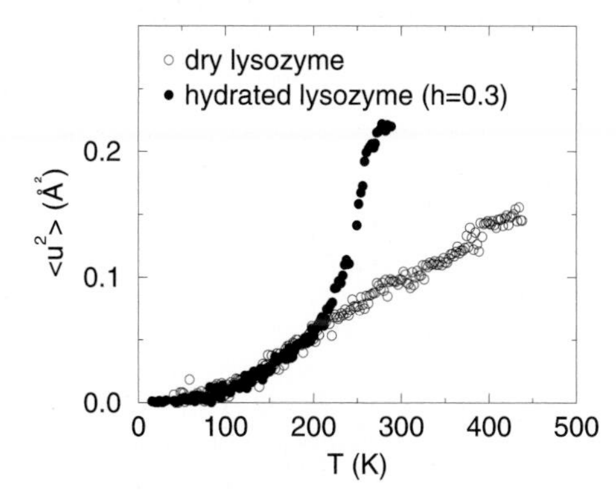

Fig. 9.10. Temperature dependence of the mean-squared displacements of non-exchangeable H atoms in lysozyme measured by elastic incoherent neutron scattering experiments (probing a time scale of tens of ps) [29]. Filled circles: hydrated powder; open circles: dehydrated powder.

In analogy with supercooled liquids, the additional motion above the dynamical transition has been associated with a structural relaxation, or α process [30]. The observation that protein structural relaxation is suppressed by dehydration suggests that water molecules participate in some sort of bond breaking process on the surface of the protein on a time scale that is shorter than that of the structural relaxation [26]. Specifically, we suppose that the surface of a dehydrated protein is rigidified by strong (electrostatic and hydrogen bonding) interactions between polar side chains. Protein-water hydrogen bonds break up these interactions, and water mobility is expected to facilitate the protein conformational fluctuations involved in the structural relaxation.

To elucidate the role of solvent in the protein dynamical transition, we have recently analyzed the temperature dependence of protein-water hydrogen bond dynamics [31]. In order to distinguish between the fast ($\approx$1 ps) formation and break up of hydrogen bonds due to water libration/rotation, and slower (tens of ps) relaxation of the protein-water hydrogen bond network due to diffusion of water molecules between sites on the protein surface and/or exchange with bulk water, we employed two measures of hydrogen bond lifetime, following a recent analysis of fast and slow hydrogen-bond dynamics in supercooled water [32]. The fast hydrogen bond lifetime, τ_{HB}, is simply defined as the average time that a given protein-water hydrogen bond remains intact. The slow hydrogen bond network relaxation time is defined in terms of the decay of the bond correlation function, $c(t) = \langle h(0)h(t) \rangle / \langle h \rangle$ [32, 33]. Here $h(t)$ is a hydrogen bond population operator, which is equal to one if a given donor-acceptor (D-A) pair is hydrogen bonded at time t, and zero otherwise, and the angular brackets denote an average over all D-A pairs. The function $c(t)$ is the probability that a random D-A pair that is hydrogen bonded at time zero is still bonded at time t, regardless of whether or not the bond was broken at intermediate times. Thus, beyond an initial transient period, the decay of $c(t)$ is not determined by fast hydrogen bond breaking by water rotation/libration, but rather by rearrangement of the protein-water hydrogen bond network. The hydrogen bond network relaxation time, τ_R, is defined as the time at which $c(t)$ decays to $1/e$, i.e. $c(\tau_R) = e^{-1}$ [32].

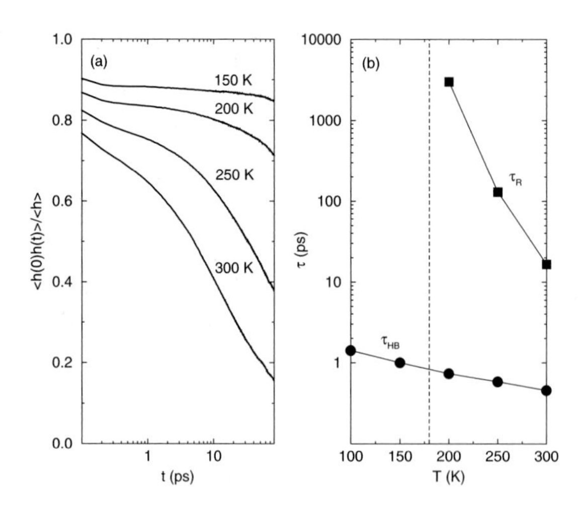

Fig. 9.11. Temperature dependence of the average protein-water hydrogen bond lifetime, τ_{HB}, and network relaxation time, τ_R. The dynamical transition temperature (180 K) is indicated by the vertical broken line.

The temperature dependence of the fast hydrogen bond lifetime, τ_{HB}, and slow network relaxation time, τ_R, of protein-water hydrogen bonds in MD simulations of the RNase crystal is shown in Fig. 9.11. The lifetime τ_{HB} shows a smooth variation with temperature over a range of temperature that includes the protein dynamical transition. In contrast, as the temperature is decreased from 300 K, the relaxation time τ_R appears to diverge at the protein dynamical transition temperature. The coincidence of structural arrest in the protein and the protein-water hydrogen bond network at the same temperature suggests a role for the onset of restructuring of the protein-water hydrogen bond network in the protein dynamical transition. The most plausible mechanism for this network relaxation is water translational diffusion.

To investigate the role of water translational diffusion in protein structural relaxation, we performed a simulation of the RNase crystal at 300 K in which the positions of the water O atoms were restrained by a harmonic potential [31]. The restraints had the effect of inhibiting water translational diffusion, while preserving nearly complete librational/rotational freedom. In light of the small impact on water rotational motion, it is not surprising that the restraints hardly affected the protein-water hydrogen bond lifetime, τ_{HB}, which was 0.45 ps in the free simulation and 0.47 ps in the restrained simulation. Moreover, as expected, by inhibiting water translational diffusion the restraints significantly slowed the relaxation of the protein-water hydrogen bond network: the network relaxation time (τ_R) was 65 ps in the restrained simulation, which is about 3.6 times longer than the 18 ps obtained from the free simulation. Thus, by imposing restraints, we have to a large extent decoupled the protein dynamics from the water restructuring.

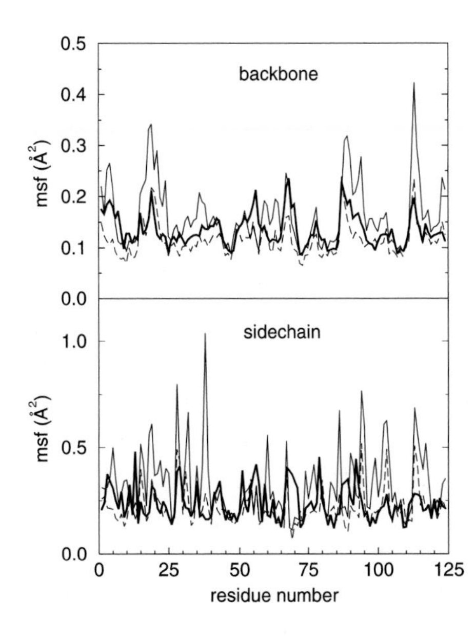

Fig. 9.12. Comparison of protein dynamical properties at 300 K during simulations of the hydrated RNase crystal in which the water oxygen atoms were free (solid curves) and harmonically restrained (broken curves), and a dehydrated RNase powder (heavy solid curves): mean-squared fluctuations of protein (a) backbone and (b) side chain heavy atoms. The msfs were computed for several blocks of 100 ps and averaged over blocks and protein molecules.

To gain more insight into specific protein motions affected, we calculated the mean-squared fluctuations (msfs) of protein heavy atoms on the 100 ps time scale. It is evident in Fig. 9.12 that restraining water translational motion reduces the protein atomic fluctuations throughout the protein, both in the backbone and side chains, and that the extent of the reduction is similar to that of a dehydrated system. Averaged over all the protein residues, the backbone msfs in the restrained crystal and dehydrated powder, 0.12 $Å^2$ and 0.13 $Å^2$, respectively, are 20 to 25% lower than the 0.16 $Å^2$ in the unrestrained crystal, and the side chain msfs in the restrained crystal and dehydrated powder, 0.22 $Å^2$ and 0.23 $Å^2$, are about 30% lower than the 0.32 $Å^2$ in the unrestrained crystal. Overall, the reduction of the motion of the side chains is significantly greater than that of the backbone. The effects appeared to be greatest in the more mobile regions of the protein structure (i.e. loops and solvent exposed side chains).

9.5 Summary

This chapter has given an overview of the structure and dynamics of water molecules near the surface of a globular protein, from the viewpoint of both experimental measurements and molecular dynamics simulations. In summary, the protein surface is rough and chemically heterogeneous. These properties give rise to strong inhomogeneities in the structure and anomalies in the dynamics of water near the surface. The range of the perturbation of water properties by the protein surface does not extend much beyond the first solvation shell. The dynamics of a protein is intimately coupled to the dynamics of the solvent. The time scale of the structural relaxation in protein molecules that is necessary for their function is tuned to that of the reorganization of the network of protein-water hydrogen bonds. This dynamical matching appears to be a key aspect of the special role that water plays as a solvent for biological molecules.

Acknowledgements

D.J.T. acknowledges support of his research by a grant (MCB-0078278) from the National Science Foundation. We thank Amos Tsai for providing the neutron scattering data plotted in Figure 9.10.

References

1. V.P. Denisov, B. Halle: Faraday Disc. **103**, 227 (1996)
2. C. Mattos: Trends Biochem. Sci. **27**, 203 (2002)
3. J.A. Rupley, G. Careri: Adv. Prot. Chem. **411**, 37 (1991)

4. L.D. Barron, L. Hecht, G. Wilson: Biochemistry **88**, 13143 (1997)
5. V. Makarov, B.M. Pettitt, M. Feig: Accts. Chem. Res. **35**, 376 (2002)
6. F.T. Burling, W.I. Weis, K.M. Flaherty, A.T. Brunger: Science **271**, 72 (1996)
7. D.I. Svergun, S. Richard, M.H.J. Hoch, Z. Sayers, S. Kuprin, G. Zaccai: Proc. Natl. Acad. Sci. USA **95**, 2267 (1998)
8. F. Merzel, J.C. Smith: Proc. Natl. Acad. Sci. USA **99**, 5378 (2002)
9. A.D. MacKerell Jr. et al.: J. Phys. Chem. **102**, 3586 (1998)
10. W.L. Jorgensen, J. Chandrasekhar, J.D. Madura, R.W. Impey, M.L. Klein: J. Chem. Phys. **79**, 926 (1983)
11. M. Settles, W. Doster: Faraday Disc. **103**, 269 (1996)
12. A.R. Bizzarri, S. Cannistraro: J. Phys. Chem. B **106**, 6617 (2002)
13. M. Tarek, D.J. Tobias: Biophys. J. **79**, 3244 (2000)
14. M. Marchi, F. Sterpone, M. Ceccarelli: J. Am. Chem. Soc. **124**, 6787 (2002)
15. C. Arcangeli, A.R. Bizzarri, S. Cannistraro: Chem. Phys. Lett. **291**, 7 (1998)
16. M. Tarek, D.J. Tobias: J. Am. Chem. Soc. **121**, 9740 (1999)
17. S.K. Pal, J. Peon, A.H. Zewail: Proc. Natl. Acad. Sci. USA **99**, 1763 (2002)
18. H.J.C. Berendsen, J.R. Grigera, T.P. Straatsma: J. Phys. Chem. **91**, 6269 (1987)
19. R. Abseher, H. Schreiber, O. Steinhauser: Proteins **25**, 366 (1996)
20. V.P. Denisov, B. Halle: J. Mol. Biol. **245**, 682 (1995)
21. V.P. Denisov, B.H. Jonsson, B. Halle: Nature Struct. Biol. **6**, 253 (1999)
22. B. Halle, H. Wennerström: J. Chem. Phys. **75**, 1928 (1981)
23. A. Luise, M. Falconi, A. Desideri: Proteins **39**, 56 (2000)
24. H. Frauenfelder, G.A. Petsko, D. Tsernoglu: Nature (London) **280**, 558 (1979)
25. E.W. Knapp, S.F. Fischer, F. Parak: J. Phys. Chem. **86**, 5042 (1982)
26. W. Doster, S. Cusack, W. Petry: Nature (London) **337**, 754 (1989)
27. B.F. Rasmussen, A.M. Stock, D. Ringe, G.A. Petsko: Nature (London) **357**, 423 (1992)
28. M. Ferrand, A.J. Dianoux, W. Petry, G. Zaccai: Proc. Natl. Acad. Sci. USA **90**, 9668 (1993)
29. A.M. Tsai, D.A. Neumann, L.N. Bell: Biophys. J. **79**, 2728 (2000)
30. W. Doster, S. Cusack, W. Petry: Phys. Rev. Lett. **65**, 1080 (1990)
31. M. Tarek, D.J. Tobias: Phys. Rev. Lett. **88**, 138101 (2002)
32. F. Starr, J.K. Nielsen, H.E. Stanley: Phys. Rev. Lett. **82**, 2294 (1999)
33. A. Luzar, D. Chandler: Phys. Rev. Lett. **76**, 928 (1996)

10 Computational Studies of Liquid Water Interfaces

Liem X. Dang and Tsun-Mei Chang

Summary. A series of classical molecular dynamics simulations were carried out to study the molecular properties and mass transfer processes at the CCl_4-H_2O liquid/liquid interface. We found the computed interfacial properties to be unique and markedly different from the bulk properties. The potential of mean force and the transport mechanics of a $CHCl_3$ molecule across the CCl_4-H_2O interface were investigated using constrained molecular dynamics. The resulting free energy profiles show a monotonic decay from the aqueous phase into the nonpolar CCl_4 liquid. The computed density profile indicates that the presence of the $CHCl_3$ molecule essentially exerts no perturbation to the interface. By examining the solvation structures, we found that the transport of the $CHCl_3$ molecule involves a smooth change of the composition of the solvation shells around the solute molecule. We also calculated the transport free energy profile of a Cl^-/Cs^+ ion across the CCl_4-H_2O interface and found that the equilibrium transfer free energy of the ion increases from the bulk H_2O phase into the nonpolar CCl_4 phase, which is opposite to the case for organic solutes. By decomposing the free energy into contributions corresponding to each molecular potential parameter, we show that the contribution associated with the atomic charges accounts for the sharp rise in the free energy from liquid H_2O to liquid CCl_4.

10.1 Introduction

The study of molecular properties and chemical processes at liquid/liquid interfaces is essential to many fields of science including chemistry, physics, and biology [1-5]. For example, the nature of solvent-solvent and solvent-solute interactions dictates the thermodynamic, structural, and dynamical properties and the assembly of amphiphilic molecules adsorbed at the interface. Characterization of the behavior of molecules at interfaces may have a significant impact on areas such as biological membrane phenomena, detergency, and oil recovery. Transfer of solutes or ions across two immiscible liquids is fundamental to many chemical processes such as liquid chromatography, phase transfer catalysis, electrochemistry, and ion extraction [6-9]. The study of interfacial transfer also is crucial in processes used for remediating environmental problems (e.g., separation chemistry). These processes often are performed in binary solvent systems (i.e., organic and aqueous phases) or involve interactions of contaminated organic solvents with groundwater.

For many decades, significant research efforts have focused on investigating the thermodynamic properties and chemical processes at liquid interfaces. Although these previous investigations provided a macroscopic picture of the interfaces and were able to determine the important factors that affected interfacial chemical reactions, a detailed understanding of liquid interfaces remains limited because of the complicated nature of interfacial phenomena. Nevertheless, with the advent of the new experimental techniques, more powerful computational resources, and advanced statistical mechanical theories, we can now address these interfacial processes at the molecular level [10 - 14].

Computer simulation techniques such as Monte Carlo (MC) and molecular dynamics (MD) methods have contributed a great deal to our knowledge of bulk liquids [15–17], liquid/vapor interfaces [18–20] and liquid/solid [21, 22] interfaces. Since the late 1980s, several research groups have used these approaches to examine the equilibrium properties of neat liquid/liquid interfaces [23 - 29]. The results reported by these groups have provided valuable information on the structures, dynamics, thermodynamics, and conformational equilibria of liquid interfaces. Recently, this approach has been extended to study the transfer of ions and organic solutes across liquid/liquid interfaces [3, 4, 30–32]. With the use of these simulation techniques, it is now possible to directly probe the mechanisms and dynamics of mass transport processes across liquid/liquid interfaces.

In this study, we focused on the equilibrium properties and mass transport reactions at the CCl_4-H_2O liquid/liquid interface. This particular liquid/liquid interface is of great environmental importance. Chlorinated hydrocarbons such as CCl_4 and $CHCl_3$ have been used extensively as solvents in the processing of radioactive materials. Contaminated solutions containing these solvents were discharged into the soil in the past, and their interaction with the groundwater is considered a major problem in environmental remediation efforts. We felt that a detailed investigation of the interfacial properties and transport processes across the CCl_4-H_2O liquid/liquid interface from a molecular perspective would further the scientific understanding of ion and solute extraction from CCl_4 and, thus, help solve related environmental problems.

Our chapter is organized as follows. The polarizable model potentials and computational details of the MD simulations are briefly described in Section 2. Equilibrium properties of the CCl_4-H_2O interface including density fluctuations, structures, and electrostatic properties are briefly summarized in Section 3. In Section 4, we describe our investigations of the free energy profiles and transfer mechanisms governing the transfer of organic solutes and ions across liquid/liquid interfaces. Finally, our conclusions and description of future work are presented in Section 5.

10.2 Potential model and computational details

10.2.1 Polarizable potential model

In this study, we used non-additive polarizable potential models to describe the molecular interactions of H_2O, CCl_4, and $CHCl_3$ molecules. For the CCl_4-$CCl_4$and H_2O-H_2O intermolecular potentials, we used data previously developed in our laboratory and reported in the literature [33, 34]. To represent the water molecule, we used a rigid three-site potential model with a fixed O-H bond length of 1.00 Å and an H-O-H bond angle of 109.5^o. Lennard-Jones parameters associated with the O atom, and fixed charges were assigned to both the O and H atoms. In addition, we placed point polarizabilities on both the O and the H atoms to describe the non-additive, induced polarization energies, and we used a rigid, all-atom potential model to describe the CCl_4 potential. Each CCl_4 molecule was fixed at the experimental tetrahedron geometry with a C-Cl bond length of 1.77 Å and a Cl-C-Cl bond angle of 109.47^o. In this model, Lennard-Jones parameters, partial charges, and point polarizabilities are associated with each carbon and chlorine atom. Both potential models were constructed to reproduce accurately the thermodynamic, structural, and dynamical properties of the bulk liquid of H_2O and CCl_4 [33, 34]. The CCl_4 potential also has been applied to the study of the CCl_4 liquid/vapor interface and has yielded a reasonable estimate of the surface tension.For completeness, the potential parameters are summarized in Table 10.1. The Lorentz-Berthelot combining rule was used to determine the cross interactions between CCl_4, H_2O, CH_4, and $CHCl_3$.

Table 10.1. Potential parameters for CCl_4, H_2O, CH_4, $CHCl_3$,and Cs^+ used in the MD simulations. σ and ε are the Lennard-Jones parameters, q is the atomic charge, and α is the atomic polarizability.

Molecule	Atom type	σ(Å)	ε(kcal/mol)	q (e)	α(Å^3)
CCl_4	C	3.41	0.10	-0.1616	0.878
	Cl	3.45	0.26	0.0404	1.910
H_2O	H	0.00	0.00	0.365	0.170
	O	3.205	0.16	-0.730	0.528
CH_4	C	3.40	0.1094	-0.4176	0.878
	H	2.65	0.0157	0.1044	0.235
$CHCl_3$	C	3.41	0.137	0.5609	0.978
	Cl	3.45	0.275	-0.1686	1.910
	H	2.81	0.020	-0.0551	0.135
Cs^+	Cs	3.831	0.10	1.000	2.440
Cl^-	Cl	4.410	0.10	-1.000	3.690

The total interaction energy of the system can be decomposed into pairwise additive and non-additive components [35].

$$U_{tot} = U_{pair} + U_{pol}. \tag{10.1}$$

$$U_{pair} = \sum_i \sum_j \left(4\varepsilon_{ij} \left[\left(\frac{\sigma_{ij}}{r_{ij}} \right)^{12} - \left(\frac{\sigma_{ij}}{r_{ij}} \right)^{6} \right] + \frac{q_i q_j}{r_{ij}} \right) \tag{10.2}$$

and

$$U_{pol} = -\sum_{i=1}^{N} \mu_i \bullet E_i^0 - \frac{1}{2} \sum_{i=1}^{N} \sum_{j=1, i\#j}^{N} \mu_i \bullet T_{ij} \bullet \mu_j + \sum_{i=1}^{N} \frac{|\mu_i|^2}{2\alpha_i} \tag{10.3}$$

Here, r_{ij} is the distance between sites i and j, q is the charge, and σ and ε are the Lennard-Jones parameters. The term E_i^0 represents the electric field at site i produced by the fixed charges in the system, μ_i is the induced dipole moment at atom site i, and T_{ij} is the dipole tensor. The first term in (10.3) represents the charge-dipole interaction, the second term describes the dipole-dipole interaction, and the last term is the energy associated with the generation of the dipole moment μ_i. During MD simulations, a standard iterative, self-consistent field procedure was used to evaluate the induced dipoles.

10.2.2 Simulation details

The MD simulations for mass transfer processes were performed on a system consisting of 714 H_2O molecules, 140 CCl_4 molecules and a single chloroform molecule with linear dimensions of 28x28x56.3 Å. The same procedure was used to create and equilibrate the liquid/liquid interface. A schematic representation of the simulation cell for transferring a solute molecule across the CCl_4-H_2O liquid/liquid interface is depicted in Fig. 10.1. The interface was chosen to be perpendicular to the z-axis. The H_2O molecules stay mostly in the region of z < 28 Å, while the CCl_4 molecules occupy roughly the region of z > 28 Å.

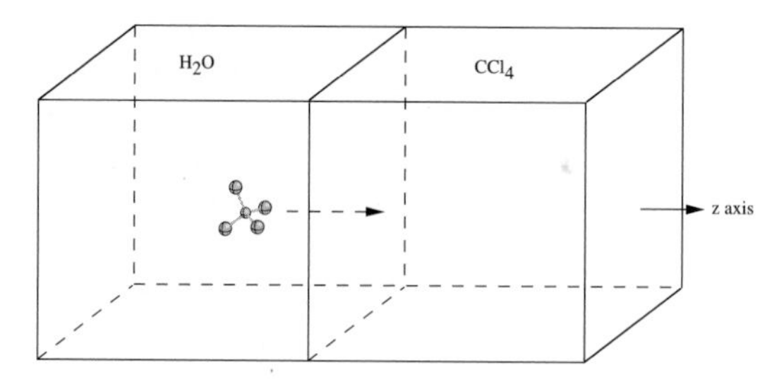

Fig. 10.1. Schematic representation of the simulation box for the mass transport processes across the CCl_4-H_2O liquid/liquid interface. The interface is at the middle of the cell, H_2O is at the left side, and CCl_4 is at the right side. The Z-axis is chosen to be perpendicular to the interface.

To evaluate the free energies associated with the transfer of an ion across the liquid/liquid interface, we used constrained MD techniques similar to the approach used for ionic association [18]. In this approach, the reaction coordinate for ion transfer can be considered as the z_s position of the ion. The Helmholtz free energy difference, $\Delta F(z_s)$, between a state where the ion is located at z_s, F (z_s), and a reference state, where the ion is at z_o, F_o, is simply

$$\Delta F(z_s) = \Delta F(z_s) - F_0 = \int_{z_0}^{z_s} \langle f_z(z_s^{'}) \rangle dz_s^{'}, \tag{10.4}$$

where $f_z(z_s')$ is the z component of the total force exerted on the ion at a given z position, z_s', averaged over the canonical ensemble. Here, F_0 is chosen as the free energy of the system with a solute particle located in the bulk water region. During the simulation, the z position of the solute (chloroform) molecule is reset to the original value after each dynamical step, and the average force along the z-direction on the center of mass of the solute is evaluated. No restriction is applied in the xy direction; therefore, the solute molecule is free to diffuse in the xy plane. The average forces subsequently are integrated to yield the free energy profile. For our calculation, the positions of the chloroform molecules ranged between z=16 to 38 Å, with a position increment of 0.5-1.0 Å. The total simulation time at each solute location was at least 70 ps.

10.3 Molecular properties of the liquid/liquid interface

10.3.1 Density profiles

We evaluated the CCl_4 and H_2O liquid densities within 1-Å-thick liquid slabs along the interfacial normal direction as a function of the their z coordinate across the CCl_4-H_2O liquid/liquid interface (see Fig. 10.2). The H_2O density stays at the experimental value in the bulk water region and then falls off quickly to zero into the bulk CCl_4 region.The same behavior also is observed for the CCl_4 density profile. These results indicate that there are two well-defined interfaces, and the bulk CCl_4 and H_2O regions remain separated. There is a slight overlap between the CCl_4 and H_2O densities at the interface regions, suggesting that the interfaces are not flat at a molecular level, which may be a result of thermal fluctuations [23, 25]. Interestingly, we found that the density profile of H_2O is very smooth, while the CCl_4 density profile exhibits some oscillations. A similar trend also was observed for other liquid/liquid interfaces [23,25-28]. The oscillatory behavior in the CCl_4 density profile may be due to the thermal capillary broadening or the smaller number of CCl_4 molecules used in the simulation as compared to that of water. Simulations using a larger system size and longer trajectory are required to resolve this point. Generally, the density profiles can be fitted into a hyperbolic tangent function [20, 37], which yields estimates of the "10-90" surface thickness of 3.0 and 2.2 Å for H_2O and CCl_4, respectively. The interfacial thickness of CCl_4 was found to be much smaller at the CCl_4-H_2O liquid/liquid interface than at the CCl_4 liquid/vapor interface, possibly due to unfavorable CCl_4-H_2O hydrophobic interactions.

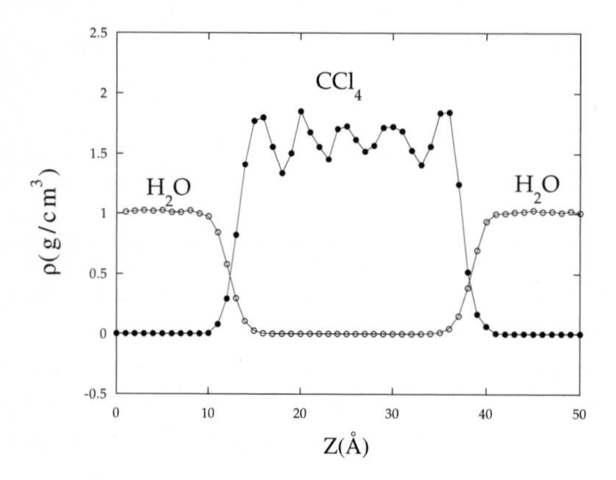

Fig. 10.2. Computed density profiles of CCl_4 (circles) and H_2O (open circles) at room temperature from a 300-ps MD simulation.

10.3.2 Molecular structures and orientations

In this section, we discuss the detailed structures of the CCl_4 and H_2O molecules, which we analyzed via the pair correlation functions and orientational probability distributions; the local solvation structures of H_2O and CCl_4, which we investigated as a function of their distance to the interface; and the preferred structures between the CCl_4 and H_2O molecules at the interface. We examined the change in the water local structures from the bulk region to the interface in terms of the z-dependent H_2O atomic radial distribution functions. The peak positions of these pair correlation functions are almost identical regardless of the z coordinate of the H_2O molecules, indicating that the local structure of H_2O remains unchanged. However, the presence of the interface has a profound effect on the molecular orientations of the H_2O molecules. We calculated the probability distributions for angles between the water molecules and the interface normal direction as a function of their z coordinates. The angular distributions in the interfacial region showed significant deviations from those in the bulk liquids, indicating that the interfacial forces indeed induce an orientational order in H_2O near the interface. Similar behavior has also been observed at the liquid/vapor and other liquid/liquid interfaces.

A unique feature of H_2O is its ability to form hydrogen bonds. We examined the average number of hydrogen bonds and the degree of hydrogen

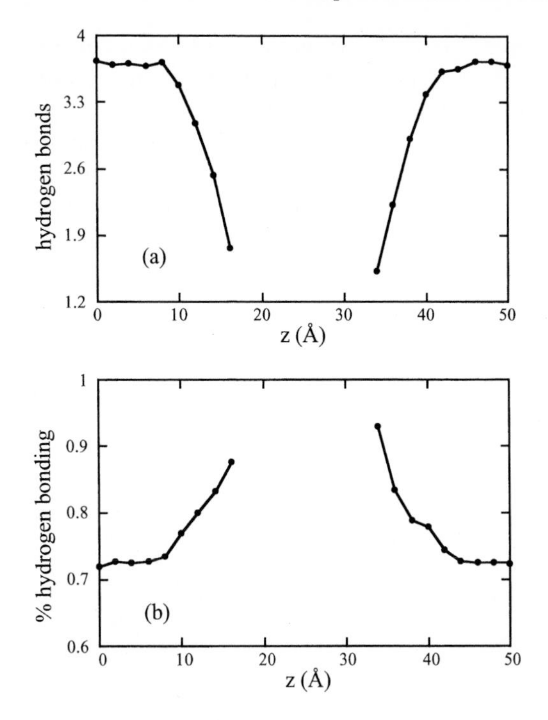

Fig. 10.3. Hydrogen bonding analysis at the interface and the bulk: (a) presents the average number of hydrogen bonds per H_2O molecule as a function of the z coordinate of the H_2O molecules; (b) shows the degree of hydrogen bonding as a function of the z coordinate of the liquid slabs, which is defined as the number of hydrogen bonds divided by the number of H_2O molecules in the first solvation shell per H_2O molecule.

bonding per water molecule as a function of the distance perpendicular to the interface (see Fig. 10.3). As expected, because of the smaller density of water molecules at the interface, the average number of hydrogen bonds per water molecule decreases from the bulk value of 3.7 to 1.9 as H_2O approaches the liquid/liquid interface. Interestingly, an opposite trend has been observed for the degree of hydrogen bonding per H_2O molecule, which is defined as the ratio between the average number of hydrogen bonds and the average number of water molecules in the first solvation shell. It has been found that the degree of hydrogen bonding actually increases from the bulk liquid into the interface. Consideration of the energetics of the system explains this feature. By strengthening the hydrogen bonding, the interfacial energy of water can be reduced. A similar conclusion also has been reached for other liquid/liquid interfaces [23, 26]. In addition, the preferred orientational order of the water molecules near the interface may be caused by the reinforcement of the hydrogen bonding in the interfacial regions.

Similarly, we studied the CCl_4 radial distribution functions as a function of their z coordinates. Again, the peak positions of these distribution functions remained unchanged from the bulk region to the interface, suggesting that the local structure of CCl_4 is not modified by the presence of the interface. The molecular orientation of the CCl_4 molecules was characterized by the z-dependent probability distributions for the angle between the CCl_4 molecules and the z-axis. A weak orientational order was found in the region close to the interface. The relatively weak orientational order of CCl_4 compared to H_2O is a direct consequence of the high molecular symmetry and the fairly isotropic molecular interaction of the CCl_4 molecules.

To characterize solvation structures of the H_2O-CCl_4 molecules at the interface, we calculated the C-O, C-H, Cl-O, and Cl-H radial distribution functions. The pair correlation functions were obtained for water molecules in the interface region. Well-defined features were observed in these atomic radial distribution functions, thus clearly indicating the existence of a preferred local solvation structure between the CCl_4 and H_2O molecules at the interface. However, because of the weak H_2O-CCl_4 molecular interactions, we found that these features were rather shallow, as would be expected. This result suggests that the structure correlation between CCl_4 and H_2O is fairly weak at the interface.

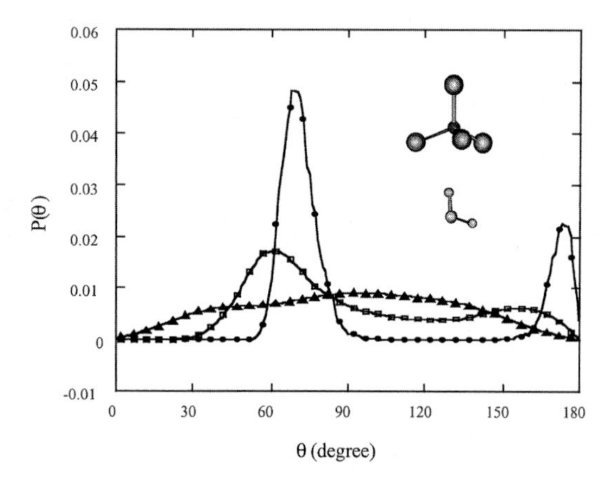

Fig. 10.4. Probability distributions for the angle between the intramolecular C-Cl bond and the vector pointing from C to the O atom as a function of the C-O separation. The data points correspond to C-O distances between 0-3.5 (circles), 3.5-4.5 (squares), and 4.5-5.5 Å (triangles), respectively. A uniform angular distribution function is described by a sine function.

By examining the angular distribution functions between the H_2O and CCl_4 molecules near the interface, we gained a better understanding of the CCl_4-H_2O solvation structures (see Fig. 10.4). For small CCl_4-H_2O separations, the computed angular distribution functions deviated significantly from the sine curve that describes a uniform distribution. This behavior suggests that there is a strong orientational order between the H_2O and the CCl_4 molecules at short distances. Further analysis of the angular probability distribution revealed that the CCl_4-H_2O dimer structure at the interface was very similar to the optimized gas-phase structure [29]. As the CCl_4-H_2O distance increased, the angular probability distribution became a sine function, indicating the disappearance of the orientational correlation between the CCl_4 and H_2O molecules.

10.3.3 Electrostatic properties at the interface

Using the polarizable potential model allows us to describe more realistically changes in molecular electrostatic properties in an inhomogeneous environment. Here, the total dipole moments of the water molecules were evaluated as a function of their coordinates in the interface normal direction. We found that when the water molecules were far away from the interface, they had an average dipole moment of 2.6 D, which was very close to the bulk value. However, as the H_2O molecules approached the interface, the dipole moments decreased monotonically to values close to their gas phase values. This effect might be caused by changes in the hydrogen-bonding network and/or the molecular orientations of water molecules near the interface. We also evaluated the induced dipole moments of both the H_2O and CCl_4 molecules as a function of the distance to the interface. As expected, the average induced dipole moments of the H_2O molecules decreased near the interface. Interestingly, an opposite trend was observed for the induced dipole moments of the CCl_4 molecules. The CCl_4-induced dipole moments were enhanced significantly near the interface as compared to those in the bulk CCl_4 liquid. This change is caused by the strong local electric field induced by the H_2O molecules at the interface, and it might have pronounced implications in liquid/liquid-interface transport processes. The results of this investigation are depicted in Fig. 10.5.

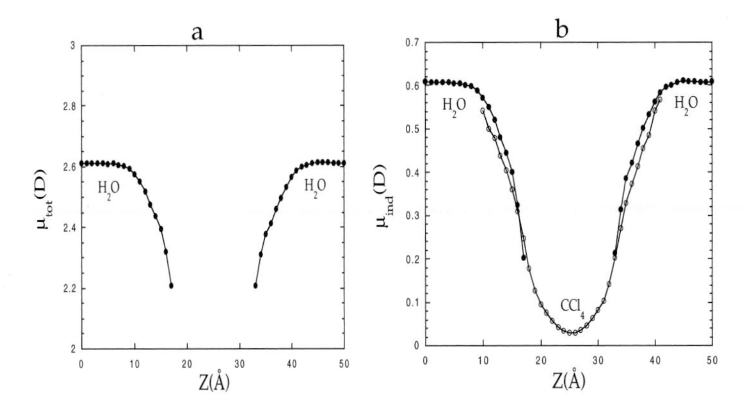

Fig. 10.5. Total dipole moment of water molecules as a function of the Z-axis (a). The induced dipole moments of H_2O (filled circles) and CCl_4 (circles) molecules as a function of the Z-axis normal to the interface (b).

10.4 Mass transport across liquid/liquid interfaces

10.4.1 Transport of organic solutes

Computer simulation techniques provide a useful and practical approach to the study of solute transfer processes across liquid/liquid interfaces. With the use of MD techniques, Pohorille and Benjamin obtained the free energy of adsorption of p-n-pentylphenol across a water liquid/vapor interface [38]. Marrink and Berendsen carried out simulations to study the transport of water molecules through phospholipid/water systems [2]. Through use of a Lennard-Jones liquid/liquid interface and MD methods, Hayoun et al [1] described solute transfer as an activated process that reflects the changes in the degree of solvation by the two solvents. More recently, Pohorille and Wilson [4] investigated the excess chemical potentials of a series of small solutes across the water-membrane and water-hexane interfaces. Their results showed that the shape of the free energy profiles depends on the magnitude of the dipole moments of the solute molecules [4].

In the following discussion, we examine the solvation structures, free energy profile, and transport mechanism associated with the transfer of a small organic solute – a single chloroform across the CCl_4-H_2O liquid/liquid interface. The mechanisms and dynamics of mass transport processes are governed by the free energy profiles (potential of mean force). We have examined the free energy profiles associated with transferring a single methane molecule [40] and a chloroform molecule from the liquid water phase to the liquid CCl_4 phase at 298 K. The free energy change between two different solute locations in the free energy profiles represents the reversible work necessary to bring the solute from one position to the other. The region through which the free energy undergoes changes extends over roughly 10 Å, which yields an estimate of the interfacial width. In both cases, the organic solute exhibits free energy stability in the bulk CCl_4 relative to that in the bulk H_2O phase. The free energy difference is 6.5 kcal/mol for the transfer of a chloroform molecule. The decrease in free energy in the non-aqueous phase is consistent with the fact that $CHCl_3$ is more soluble in CCl_4 than in H_2O.

In Fig. 10.6, we show the free energy profile for the transfer of chloroform across the CCl_4-H_2O interface at 298 K as a function of its distance to the interface. The free energy profile exhibits a monotonic decrease from the aqueous phase into the organic phase. Interestingly, the dipolar solute, chloroform, is not surface active, as indicated by the fact that no interfacial minimum is observed in the free energy profile. This result disagrees with the conclusion of Wilson and Pohorille's study [4] on the mass transfer of methane and floromethanes across the water/hexane interfaces. In their study, they found that the excess chemical potentials exhibit deep interfacial minima for the dipolar solutes and much more shallow minima for the nonpolar solutes. The differences in these free-energy profiles may be due to the nature of

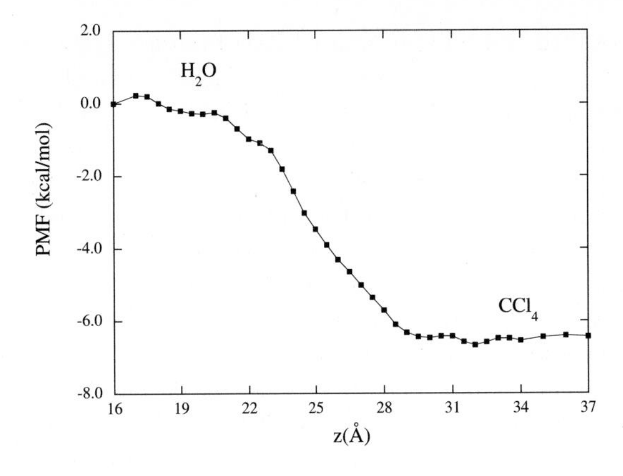

Fig. 10.6. Computed free energy profiles of transferring a single chloroform molecule across the CCl_4-H_2O liquid/liquid interface at room temperature.

the nonaqueous solvent or the potential models employed in describing the intermolecular interactions.

To further characterize the transfer of free energy across the liquid/liquid interface, the potential of mean force (PMF) can be decomposed into contributions that correspond to the individual solute-solvent potential interactions. This decomposition can be achieved by separating the force into the Lennard-Jones (f^{L-J}), Coulombic (f^{Coul}), and the polarization (f^{Pol}) components.

$$f = f^{L-J} + f^{Coul} + f^{Pol}, \tag{10.5}$$

$$\Delta F(z_s) = \Delta F_{L-J}(z_s) + \Delta F_{Coul}(z_s) + \Delta F_{Pol}(z_s) \tag{10.6}$$

$$= -\int_{z_0}^{z_s} [\langle f_z^{L-J}(z_s^{'})\rangle + \langle f_z^{Coul}(z_s^{'})\rangle + \langle f_z^{Pol}(z_s^{'})\rangle] dz_s^{'}. \tag{10.7}$$

Figure 10.7 shows the Lennard-Jones contribution to the transfer free energy for moving chloroform across the CCl_4-H_2O interface. We found that the PMF decreased as the uncharged, nonpolarizable, Lennard-Jones solute

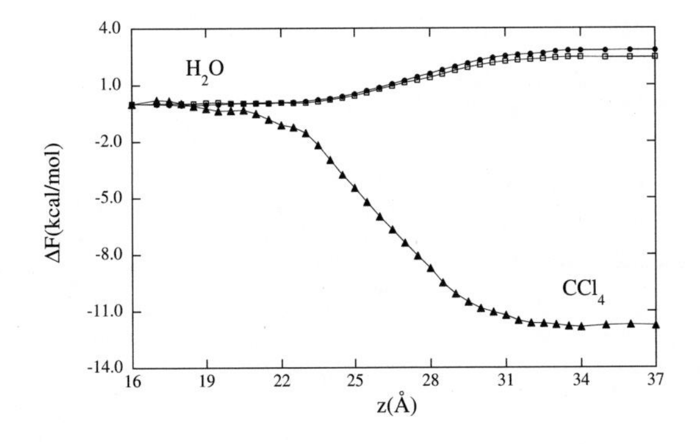

Fig. 10.7. Lennard-Jones (triangles), Coulombic (squares), and polarization (circles) contributions to the solvation free energies for transferring a chloroform molecule across the CCl_4-H_2O liquid/liquid interface.

moves from the aqueous phase into the organic phase. This finding suggests that a considerably larger free energy is required to create a cavity in H_2O to accommodate this Lennard-Jones solute than in the CCl_4 liquid. A consideration of the system's energetics may yield an understanding of this finding. Because water molecules form hydrogen bonds with each other in the liquid phase, more energy is required to disrupt the hydrogen bonding network to create a large enough cavity in H_2O to accommodate the solute than is required in a more weakly interacted CCl_4 liquid phase.

The Coulombic and polarization contributions to the transfer free energies for the chloroform molecule near the interface also are shown in Fig. 10.7. Clearly, these contributions increase monotonically as the $CHCl_3$ molecules approach the CCl_4 liquid from the aqueous phase. This effect is a direct consequence of the more favorable electrostatic interactions that exist between water and $CHCl_3$ than between the nonpolar CCl_4 solvent and the chloroform. Interestingly, the trend from these contributions is the opposite of the trend from the Lennard-Jones contributions. For the cases in which small organic solutes are transferred across the CCl_4-H_2O interfaces, the Lennard-Jones contributions are the dominating component, and the resulting total

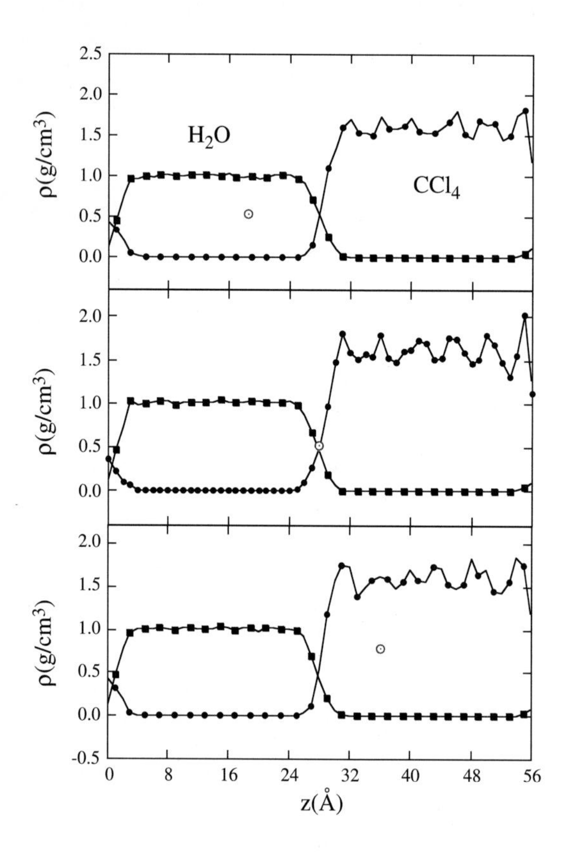

Fig. 10.8. Computed density profiles of CCl_4 (circles) and H_2O (squares) at room temperature for the $CHCl_3$ molecule located at z=18, 28, and 36 Å, respectively.

free energy shows a monotonic decrease from the aqueous phase into the non-aqueous phase.

To gain insight into the transport mechanism of organic solutes across the liquid/liquid interfaces, we performed a detailed examination of the solvation structures. The computed density profiles for CCl_4 and H_2O in thin liquid slabs as a function of their z coordinate in the interface normal direction are shown in Fig. 10.8. The density profiles were obtained for $CHCl_3$ located at z=20 (the bulk aqueous phase), 28 (the interfacial region), and 36 Å (the bulk CCl_4 region). Clearly, regardless of the chloroform positions, the bulk H_2O and CCl_4 phase remained well separated. The small overlap of the density profiles in the interfacial region was due largely to the thermal broadening. In general, the CCl_4 and H_2O density profiles exhibited similar features, indicating that this liquid/liquid interface was not broadened by the presence of the chloroform molecule. A similar behavior was observed for methane near the CCl_4-H_2O interface [40]. It is interesting to note that recent studies on ion transport across liquid/liquid interfaces suggested that the ion significantly perturbed the interface and caused the formation of a water finger that extended into the organic phase. Such a phenomenon was not observed for moving organic solutes across the liquid/liquid interface, which might indicate that the transport of ions and organic solutes is governed by different mechanisms.

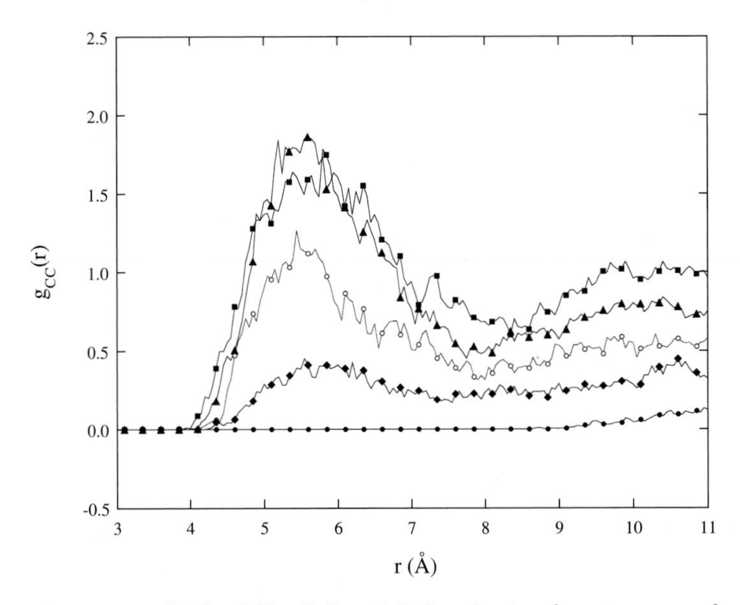

Fig. 10.9. $CHCl_3$-CCl_4 C-C radial distribution functions as a function of the z coordinate of the $CHCl_3$ molecule. The solid curves from the top to the bottom correspond to the $CHCl_3$ molecule located at z = 36 (squares), 32 (triangles), 28 (open circles), 24 (diamonds), and 20 Å (circles), respectively.

In addition to the density profiles, we further examined the atomic pair correlation functions between $CHCl_3$ and CCl_4 or H_2O as a function of the chloroform position. In Fig. 10.9, we show the computed C-C radial distribution functions between $CHCl_3$ and CCl_4 for chloroform located at Z=18-36 Å. The first peak positions of all the radial distribution functions are almost at the same C-C distance, suggesting that the local solvation structures of CCl_4 molecules around $CHCl_3$ remain unaltered regardless of the chloroform locations. The decrease in the magnitudes of the pair correlation functions simply reflects the decreasing number of CCl_4 solvent molecules available to solvate the chloroform near the interface. A similar trend also was observed for the chloroform-H_2O radial distribution functions. In a previous study on methane transfer across the CCl_4-H_2O interface, we reached the same conclusion [40]. This result suggests that the mechanism of transferring small organic solutes across the liquid/liquid interface involves a smooth transition of the solvent composition in the solvation shell around the solute molecule.

10.4.2 Ion transport

Interest in computer simulation of ion transport has intensified since Benjamin's pioneering work on the mechanism and dynamics of transferring a single Cl^- across a H_2O/1,2-dichloroethane interface [3]. Because of approximations made on the electrostatic free energy calculation, Benjamin suggested that, due to the presence of a free energy minimum (~5 kcal/mol) at the liquid-liquid interface, ion transfer into the aqueous phase is an activated process. Schweighofer and Benjamin recently studied transport of the ammonium ion from H_2O to the nitrobenzene liquid-liquid interface and found no barrier when the system was fully equilibrated [50]. In a paper describing their MD study on the structural and energetic characteristics of ion-assisted transfer between H_2O and chloroform, Lauterbach et al. reported that the cesium ion diffuses spontaneously from the interface to H_2O and displays apparently no free energy minimum [51]. Recently, Fernandes and co-workers reported a series of MD simulations on ion transfer processes from H_2O to organic solvents. Their computed ion transfer free energies are in reasonable agreement with the experimental data, and no minima were observed at the liquid-liquid interface [52].

We recently carried out an extensive study in which we used MD techniques and polarizable potential models to study the mechanism for transporting Cs^+or Cl^- across a H_2O/CCl_4 liquid-liquid interface [53]. The results obtained in these studies provided new insight into both the free energies and solvent structures as the ions moved across the interface. The rigid-body intermolecular polarizable interaction models for H_2O and CCl_4 were used The $Cs^+$and Cl^- potential parameters also were developed so that the minima and structures and energies were in good agreement with high-level electronic structure calculations and experimental measurements [54]. A summary of

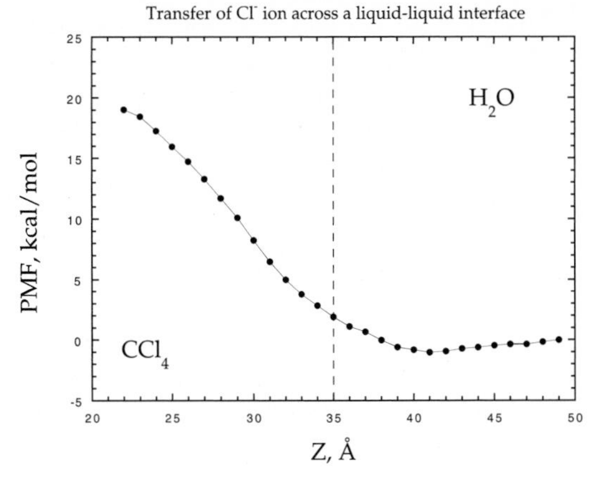

Fig. 10.10. Computed free energy profile of transferring a Cl^- ion across a H_2O - CCl_4liquid/liquid interface.

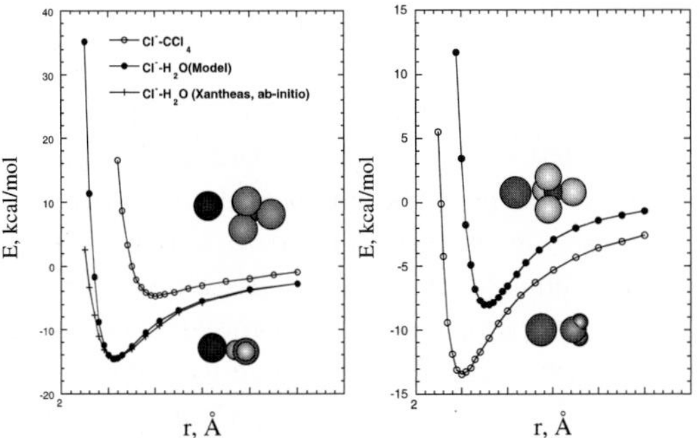

Fig. 10.11. Left: gas-phase Cl^-- H_2O and Cl^- - CCl_4 interaction energies as a function of ion-molecule separations. Insets are the lowest-energy structures. The *ab initio* data are taken from the work of Xantheas [54]. Right: same for Cs^+.

the Cs^+- CCl_4, Cl^-- CCl_4, Cl^-- H_2O, and Cs^+- H_2O interaction energies as a function of ion-molecule separation is shown in Fig. 10.10.

In Fig. 10.10, the average PMF for transferring a Cl^-across the H_2O/CCl_4 liquid-liquid interface at 298 K as a function of Z-axis normal to the interface is shown. Upon examining the free energy profile, we found that it exhibited a monotonic increase from the aqueous phase into the organic phase. The computed free energy undergoes major changes as the ion begins to cross the interface. No barrier was found at the liquid-liquid interface. The change in free energy is positive and it can be understood by comparing the Cl^--H_2O and Cl^--CCl_4 dimer potential energy surfaces shown in Fig. 10.11. It is clear that the preferred phase for the Cl^- is the aqueous phase rather than the non-aqueous phase. The PMF was decomposed into energy contributions that correspond to the individual ion-solvent interactions. These energy contributions indicate that the computed free energy of ion transfer is the result of a competition between the Lennard-Jones and the electrostatic interactions (Coulombic and polarization). This finding is significantly different from the transferring of a neutral solute molecule (methane or chloroform) across this

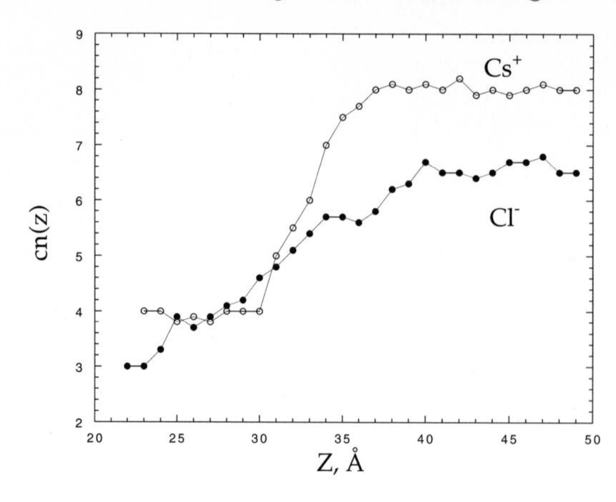

Fig. 10.12. Computed hydration numbers of the ions as a function of z-axis normal to the interface.

interface, where the free energy required creating a cavity is a major contribution [40].

During the transfer process, we monitored the coordination number for the Cl^-as a function of the Z-axis normal to the interface. As the Cl^- moved across the interface, the first hydration shell of the ion began to be reduced as shown in Fig. 10.12. The characteristic features shown in this figure are clearly similar to the computed free energy profile. For instance, the free energy profile moves upward while the coordination number moves downward. Thus, an ion transfer mechanism that involves changes in the hydration shell of the ion has been demonstrated. This finding is in excellent agreement with a recent experimental study by Osakai et al. [55], who measured the Gibbs free energy of ion transfer between water and nitrobenzene interface for various ions. They also measured the water content in nitrobenzene and found a fraction of water associated with the ion in the nitrobenzene liquid phase. For example, the coordination number of Na^+ is approximately 6 in water. The coordination number deceased to approximately 4 when the ion is transferred to nitrobenzene. In Fig. 10.13, a snapshot was taken from the MD simulation when the Cl^- was located in the CCl_4 liquid phase, 13 Å away from the interface. As can be seen from the features of Fig. 10.13, our results very closely agree with the model suggested by Osakai et al. [55]. We examined the interface for various Cl^- positions along the Z-axis and found some distortion to exist as the Cl^- moves from the aqueous phase to the non-aqueous phase. The most severe case of distortion occurs when the ion is located deep inside the non-aqueous phase. At short distances (<12 Å), fingering effects were observed as the Cl^- moves into the non-aqueous phase. Due to the strong hydrogen-bonding network between H_2O molecules, these fingering effects persist for at least a few nanoseconds in our simulations. The fingering effects disappear as the Cl^- moves further into the CCl_4 liquid phase.

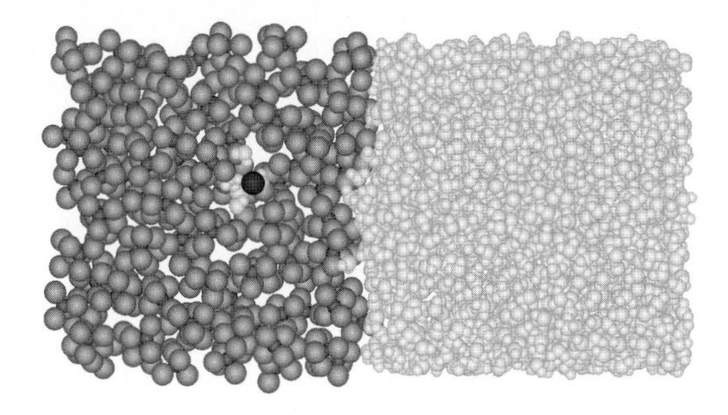

Fig. 10.13. Snapshots taken 6 ns into the MD simulations when the Cl^- was pulled away from the interface in the CCl_4 liquid phase.

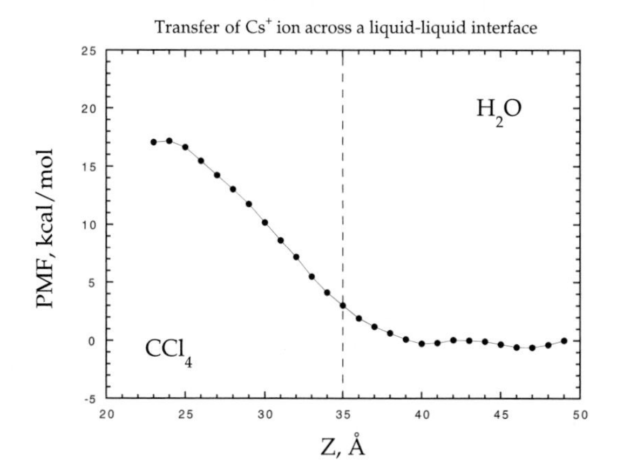

Fig. 10.14. Computed free energy profile of transferring Cs^+ across a H_2O - CCl_4 liquid/liquid interface.

In addition to the Cl^- simulations, a detailed study of the transport properties for Cs^+ across a H_2O/CCl_4 interface was also completed. The PMF of transferring Cs^+across the H_2O/CCl_4 interface at 298 K as a function of Z-axis normal to the interface is shown in Fig. 10.14. As one can see, the PMF is very similar to that for Cl^-. For instance, the free energy profile exhibits a monotonic increase from the aqueous phase into the organic phase, and no minimum was found at the interface. The increase in the free energy is expected because the Cs^+ interacts more strongly with H_2O than with CCl_4. We also have performed a reverse transfer for the $Cs^+/H_2O/CCl_4$ system and found the computed PMFs for both directions are nearly identical. Thus, the results presented in this paper represent the equilibrium free energy of the system.

The estimated free energy of transfer is 17 ± 1 kcal/mol. There is experimental data available for the Gibbs transfer energy of Cs^+ from the aqueous phase into a mixed organic phase (nitrobenzene [NB] - CCl_4) [56]. For example, the net free energy of transfer from water into a mixed organic phase (*i.e.*, 60% NB and 40% CCl_4) is about 5 kcal/mol. Because of complications in experimental techniques, electrochemical measurements on Gibbs transfer energy from the aqueous phase into neat CCl_4 have not been carried out yet.

The results on mixed organic phase systems have indicated that with decreasing dielectric permitivity the value of the Gibbs transfer energy of Cs^+ increases. Thus, we can expect that the value of the Gibbs transfer energy of $Cs^+$from water to neat CCl_4 will be greater than 5 kcal/mol.

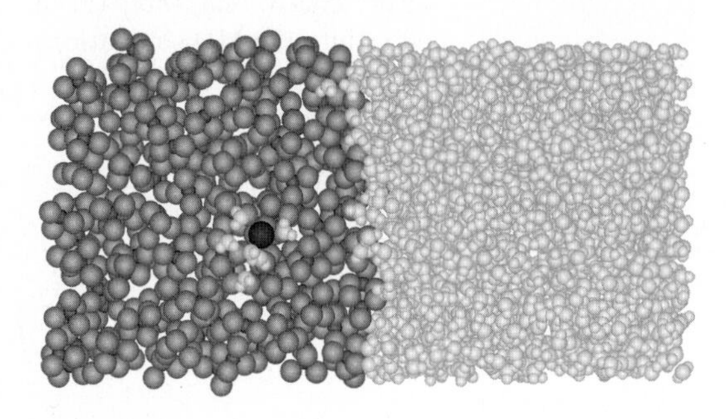

Fig. 10.15. Snapshots taken 6 ns into the MD simulations when the Cs^+ was pulled away from the interface in the CCl_4 liquid phase.

Similar to the Cl^- transfer, as the Cs^+moves across the interface, the first hydration shell of the Cs^+ as a function of Z-axis normal to the interface is significantly reduced (i.e., 8 to 4) as shown in Fig. 10.12. In Fig. 10.15, a snapshot was taken when the Cs^+ was located deep in the CCl_4 liquid phase. It is clear that the Cs^+ was completely solvated even though its first solvation shell was reduced by a half. A similar observation was reported in previous studies of aqueous clusters [54]. The computed density profiles for various ion positions along the Z-axis were much less distorted when compared to the corresponding results for Cl^-.

10.5 Conclusions and future work

In this study, we carried out a series of MD simulations to investigate the equilibrium properties of the neat CCl_4-H_2O liquid/liquid interface and the free energy profile and mechanisms of mass transport across this interface. Non-additive polarizable potential models were used to describe certain molecular properties of the interface.

We found that the interfacial properties were unique and were markedly different from the bulk properties. For example, the interface induced an orientational order of the H_2O and CCl_4 molecules near the interface that was lacking in the bulk liquids. The change in the molecular orientation also modified the molecular interactions near the interface as supported by the hydrogen bonding analysis. The interfaces were found to be dynamical and rough on a molecular scale. Furthermore, we observed that the induced dipole moments of the H_2O molecules decreased significantly in the interfacial region, while the CCl_4 induced dipole moments were greatly enhanced near the interface.

Using the constrained MD techniques, the mass transport of a $CHCl_3$ molecule across the CCl_4-H_2O interface was investigated. The transfer free energies were found to decrease monotonically from the aqueous phase into the nonpolar CCl_4 liquid. The presence of the solute exerted essentially no perturbation to the interface as revealed by the density profile analysis. Upon closely examining the solvation structures, we found little change in the peak positions of the solute-solvent atomic radial distribution functions as a function of the distance to the interface. This observation suggests that transport of the CH_4 or $CHCl_3$ molecule involves a smooth transition of the solvent composition in the solvation shells around the solute molecule.

The solvation free energies of a Cl^-/Cs^+ ion across the CCl_4-H_2O interface were evaluated using the PMF approach. The ion transport free energy increased as the ion moved from the aqueous phase into the nonpolar CCl_4 phase. The solvation PMF was decomposed into contributions corresponding to the individual interaction potential parameter. The Lennard-Jones contribution decreased across the interface from H_2O to CCl_4 with a shallow minimum near the interface. On the other hand, the polarization and charge contributions increased as the ion moved from the aqueous phase into the liquid CCl_4. This result might be caused by the stronger electric field produced by the H_2O molecules as is supported by the dipole moment analysis.

In summary, we present unique molecular properties of liquid/liquid interfaces and discuss the transfer free energies and mechanisms of organic solutes across the liquid/liquid interfaces. We investigated the ion transfer processes across the interface by examining the equilibrium solvation free energy profile. However, we suspect that the nonequilibrium effects may be important in ion transport reactions.

Future work will include characterization of crown ether-ion interactions in water using MD techniques. That research will involve the development of polarizable potential models for water-crown ether and ion-crown ether because pair potentials significantly underestimated the gas-phase, ion-crown ether binding energies. Extensive MD simulations will be carried out to predict the mechanism and thermodynamics for ion selectivity in macrocylic crown ethers in aqueous solution. These results will be compared to our previous work and to experimental data on the same systems. The results obtained in these simulations will provide both energetic and structural insights and also a means for evaluating the polarization effects in ion-crown ether interactions.

Acknowledgments

This work was performed in the Environmental Molecular Sciences Laboratory (EMSL) at Pacific Northwest National Laboratory under the auspices of the Division of Chemical Sciences, Office of Basic Energy Sciences, U.S. Department of Energy. Pacific Northwest National Laboratory is operated by

Battelle for the Department of Energy. Computer resources were provided by the Division of Chemical Sciences and by the Scientific Computing Staff, Office of Energy Research, at the National Energy Research Supercomputer Center (Berkeley, California). Operation of EMSL is supported by DOE's Office of Biological and Environmental Research.

References

1. M. Hayoun, M. Meyer, P. Turq: J. Phys. Chem. **98**, 6626 (1994).
2. H. Marrink, J. C. Berendsen: J. Phys. Chem. **98**, 4155 (1994).
3. K. J. Schweighofer, I. Benjamin: J. Phys. Chem. **99**, 9974 (1995).
4. A. Pohorille, M. A. Wilson: J. Chem. Phys. **104**, 3760 (1996).
5. M. C. Messmer, J. C. Conboy, G. L. Richmond: J. Am. Chem. Soc. **117**, 8039 (1995).
6. T. Wandlowski, V. Marecek, Z. Samec, J. Electroanal: Chem. **331**, 765 (1992).
7. R. B. Gennis: Biomembranes (Springer, New York, 1989).
8. C. M. Starks, C. L. Liotta, M. Halpern: Phase Transfer Catalysis (Chapman & Hall, New York, 1994).
9. R. P. W. Scott: Liquid Chromatography Column Theory (Wiley, New York, 1992).
10. S .G. Grubb, M. W. Kim, T. Rasing, Y. R. Shen: Langmuir **4**, 452 (1988).
11. K. B. Eisenthal: Annu:Rev. Phys. Chem. **43**, 627 (1992).
12. R. M. Corn, D. A. Higgins: Chem. Rev. **94**,107 (1994).
13. C. A. Croxton: Statistical Mechanics of the Liquid Surface (Wiley, New York, 1980).
14. J. K. Percus, G. O. Williams: Fluid Interfacial Phenomena (Wiley, New York, 1986).
15. J. Caldwell, L. X. Dang, P. A. Kollman: J. Am. Chem. Soc. **112**, 9144 (1990).
16. M. P. Allen, D. J. Tildesley : Computer Simulation of Liquids (Oxford University Press, Oxford, 1987).
17. G. King, A. Warshel: J. Chem. Phys. **93**, 8682 (1990).
18. J. K. Lee, J. A. Baker, G. M. Pound: J. Chem. Phys. **60**, 1976 (1974).
19. C. A. Croxton: Physica **106A**, 239 (1981).
20. J. Alejandre, D. J. Tildesley, G. A. Chapela: J. Chem. Phys. **102**, 4574 (1995).
21. F. F. Abraham: J. Chem. Phys. **68**, 3713 (1978).
22. D. A. Rose, I. Benjamin: J. Chem. Phys. **100**, 3545 (1994).
23. P. Linse: J. Chem. Phys. **86**, 4177 (1987).
24. M. Meyer, M. Mareschal, M. Hayoun: J. Chem. Phys. **89**, 1067 (1988).
25. J. Gao, W. L. Jorgensen: J. Phys. Chem. **92**, 5813 (1988)
26. I. Benjamin: J. Chem. Phys. **97**, 1432 (1992).
27. I. L. Carpenter, W. J. Hehre: J. Phys. Chem. **94**, 531 (1990).
28. A. R. van Buuren, S.-J. Marrink, H. J. C. Berendsen: J. Phys. Chem. **97**, 9206 (1993).
29. T.-M. Chang, L. X. Dang: J. Chem. Phys. **104**, 6772 (1996).
30. C. Chipot, M. A. Wilson, A. Pohorille: J. Phys. Chem. **101**, 782 (1997).
31. I. Benjamin: Science **261**, 1558 (1993).
32. I. Benjamin: Acc. Chem. Res. **28**, 233 (1995).

33. T.-M. Chang, K. A. Peterson, L. X. Dang: J. Chem. Phys. **103**, 7502 (1995).
34. L. X. Dang: J. Chem. Phys. **97**, 2659 (1992).
35. P. Ahlström, A. Wallqvist, S. Engström, and B. Jönsson, Mol. Phys. **68**, 563 (1989).
36. J. Ryckaert, G. Ciccotti, M. Ferrario, H. J. C: J. Comput. Phys . **23**, 327 (1977).
37. G. C. Lie, S. Grigoras, L. X. Dang, D.-Y. Yang, and A. D. McLean : J. Chem. Phys. **99**, 3933 (1993).
38. A. Pohorille, I. Benjamin: J. Phys. Chem. **97**, 2664 (1993).
39. D. E. Smith, L. X. Dang: J. Chem. Phys. **100**, 3757 (1994).
40. T.-M. Chang, L. X. Dang: Chem. Phys. Lett. **263**, 39 (1996)
41. G. J. Hanna, R. D. Noble: Chem. Rev. **85**, 583 (1985).
42. A. A. Kornyshev, A. G. Volkov: J. Electroanal. Chem. **180**, 363 (1984).
43. Y. Marcus: Ion Solvation (John Wiley and Sons Ltd., London, 1985).
44. . Lin, Z. Zhao, H. Freiser: J. Electroanal. Chem. **210**, 137 (1986).
45. P. E. Smith, W. F. van Gunsteren: J. Phys. Chem. **98**, 13735 (1994).
46. S. Boresch, M. Karplus, J. Molec: Bio. **254**, 801 (1995).
47. T. P. Straatsma, H. J. C. Berendsen, J. P. M. Postma: J. Chem. Phys. **85**, 6720 (1986).
48. S. H. Fleischman, C. L. Brooks III: J. Chem. Phys. **87**, 3029 (1987).
49. R. M. Levy, M. Belhadj, D. B. Kitchen: J. Chem. Phys. **95**, 3627 (1991).
50. K. Schweighfer, I. Benjamin: J. Phys. Chem. A . **103**, 10274 (1999) .
51. M. Lauterbach, E. Engler, N. Muzet, L. Troxler, G. Wipff: J. Phys. Chem. B **102**, 245 (1998).
52. P. Fernandes, A. Natalia, M. Cordeiro, D. S. Gomes, J. A. N. F : J. Phys. Chem. *B* **103**, 6290 (1999); ibid. B **104**, 2278 (2000).
53. L. X. Dang: J. Phys. Chem. B **103**, 8195 (1999).
54. S. Xantheas: J. Phys. Chem. **100**, 9703 (1995).
55. T. Osakai, A. Ogata, K. Ebina: J. Phys. Chem. B **101**, 8341 (1997).
56. A.Paulenova, V. Svec, R. Kopunec: J. Radioanalytical and Nuclear Chem. **150**, 303 (1991).

[illegible], J. Chem. Phys. 103, [illegible] (199[illegible])
[illegible] Dang, J. Chem. Phys. 97, [illegible] (1992)
[illegible], Mol. Phys. [illegible] (199[illegible])
[illegible], J. Comput. Phys. 23, 327 (1977)
[illegible], A. D. McLean, J. Chem. Phys. [illegible] (19[illegible])
[illegible], J. Chem. Phys. [illegible]
[illegible], J. Chem. Phys. 108, [illegible]
[illegible] Chang, [illegible] 204, [illegible]
[illegible], Chem. Rev. [illegible]
[illegible]

11 Water Confined at the Liquid-Air Interface

Mary Jane Shultz, Steve Baldelli, Cheryl Schnitzer, and Danielle Simonelli

Summary. The liquid interface of aqueous solutions is of central importance to numerous phenomena from cloud processing of combustion generated oxides to corrosion degradation of structural materials to transport across cell membranes. Recently, the nonlinear spectroscopic method, sum frequency generation (SFG), has been applied to investigate the structure of liquid interfaces and alteration of that structure by materials in solution. This chapter focuses on two categories of materials in solution: inorganic ionic materials that are nonvolatile – H_2SO_4, HNO_3, alkali sulfates and bisulfates, NaCl, and $NaNO_3$ – and soluble molecules that are volatile – HCl and NH_3. Ionic materials influence the structure of water at the interface through an electric double layer that arises from the differential distribution of anions and cations near the interface. Two models for the effect of the double layer are discussed. Soluble molecular materials of lower surface tension partition to the interface and displace surface water molecules. Ammonia is a rather unique probe of water at the surface. At low concentrations, ammonia merely docks to the dangling-OH groups. At intermediate concentrations, the surface changes little as the bulk concentration increases and at higher concentrations, ammonia blankets the surface and displaces water at the surface.

11.1 Introduction

The influence that ions have on the aqueous interface is an issue not only of long standing interest but also of tremendous practical importance. Inorganic ions in aqueous solutions play a fundamental role in areas as diverse as determining the stability of large biomolecules, corrosion, and heterogeneous ozone destruction. Since the beginning of physical chemistry, it was recognized that small, inorganic ions in solutions alter the surface tension via a depletion layer in which the ion concentration is lower than that of the bulk solution. Greater detail about the ion distribution, however awaited more refined experimental and theoretical techniques. This contribution focuses on experimental measurements of the configuration of water at the interface, a configuration determined by the details of the ion interfacial distribution.

Understanding the mobility and configuration of water in the hydration sphere around a charged or polar species is an important first step in the study of reactions involving these species. Since the Coulombic field extends over a long range, this is a challenging task. The challenge increases as the hydration sphere approaches the interface confining the water molecules by the asymmetric environment at the interface. Indeed even at the neat water/air interface, those molecules in the interface are restricted, caught between the bulk solution and the less polar environment on the other side. The net result for neat water is a slight preference for the molecular dipole to point into bulk water (with the dipole defined as pointing from the oxygen atom toward the hydrogen atoms) [1]. This contribution concentrates on these confined water molecules at the interface, with Section 3 focusing on the response of these water molecules to inorganic ions in solution. Section 4 presents results from using the polar ammonia molecule as a test-probe of the surface, producing a more detailed picture of surface water. The interaction between ammonia and small water clusters is further explored in Section 5. The primary experimental technique is described in Section 2.

The theoretical challenge for treating any but dilute solutions is daunting. Ions, particularly small inorganic ones, necessarily have a Coulombic field that extends over distances that are many times the diameter of the ion. The long-range nature of the Coulombic field influences the configuration of water molecules over several hydration layers. For all but very dilute solutions, there arises a correlated motion for the counter charged ions that alters properties such as diffusion rates, polarizability and viscosity. Furthermore, the aqueous solution can no longer be treated as a dielectric continuum – the structure and configuration of the water molecule must be considered. Recent molecular dynamics simulations [2-7] have begun to surmount these difficulties, though higher concentration solutions still present a challenge.

On the experimental side, techniques with the capability of measuring molecular-level detail at the aqueous/air interface have recently become available. The primary technique capable of yielding such detail at this soft and dynamic surface is the nonlinear laser spectroscopic technique, sum frequency generation (SFG). This contribution concentrates on SFG results at the aqueous/air interface, focusing specifically on water at these interfaces. Initially SFG was applied to investigate the structure of surface active solutes, and soon after to explore the configuration of water at the neat interface [11]. Since small, inorganic ions are generally depleted in the surface region, these solutions have been investigated only recently [12-21]. Though ions are depleted at the interface relative to the bulk, ions affect the configuration of interfacial water due to the differential distribution of anions and cations with respect to the surface. At concentrations for which ion pairing is significant, the ion pairs displace water even from the topmost layer. The polar ammonia molecule serves as a test probe of the interface. At low concentration, ammonia appears to simply dock to dangling-OH groups on the surface. For higher concentrations, ammonia binds the dangling-OH groups into an extended, hydrogen-bonded network. This extended network is observed both with SFG and with FTIR studies of ammonia-water clusters.

11.2 Experimental technique: sum frequency generation

Sum frequency generation is a nonlinear laser spectroscopy that derives its surface sensitivity from the lack of inversion symmetry at the surface. As the name implies, sum frequency generation takes two input frequencies and produces the sum of these. If one of the two input frequencies is in the infrared region, the result is a vibrational spectrum of the surface. Currently, SFG is the only technique capable of producing a vibrational spectrum of a neat liquid interface, a liquid-air solution interface or a buried liquid-liquid interface [22-25]. It is the vibrational signature that imparts species specificity. Configuration or orientation information is contained both in vibrational frequency shifts and in polarization information.

The nonlinear response of a medium to the incident electromagnetic waves is determined by the hyperpolarizability [26 - 31]. To second order, an incident electric field induces a polarization, $\boldsymbol{P}$, in the medium as

$$\mathbf{P} = \alpha^{(\mathbf{1})} \cdot \mathbf{E_1} + \chi^{(\mathbf{2})} : \mathbf{E_1 E_2} \tag{11.1}$$

where $\alpha^{(1)}$ and $\chi^{(2)}$ are the first and second order polarizabilities respectively. The first order response, $\alpha^{(1)}$, is a matrix that describes Rayleigh and Raman scattering. The second order response is a tensor which is the focus of this discussion. The observed intensity is proportional to the absolute square of the polarization, and the square is the source of interferences that can complicate spectra. Equation (11.1) implies that there are two relationships that must be considered to understand this second order response. One is the relationship of the electric fields, $\boldsymbol{E}_1$ and $\boldsymbol{E}_2$ in the nonlinear medium to the incident fields. The other is the relationship of the second order polarizability tensor, $\chi^{(2)}$, to the molecules at the surface. Each of these relationships is described below.

11.2.1 Optical factors and electric fields

The electric fields in the medium are related to the incident fields by Fresnel factors that reflect the efficiency for coupling into the nonlinear medium. There is a similar optical factor for coupling the generated sum frequency out of the medium. These factors result from the boundary condition that the tangential component of the electric field is continuous across a boundary [32]. The Fresnel factors reduce the electric fields and, for the *ppp* polarization combination, interference between components results in alteration or cancellation of expected intensity.

Since SFG is a coherent process, the direction of the sum frequency beam is determined by momentum matching at the interface. For incident angles of θ_1 and θ_2 at frequency ω_1 and ω_2 respectively, the sum frequency angle,

θ_{SF}, is

$$n_{SF}^2\omega_{SF}^2 \sin^2\theta_{SF} = \\ n_1^2\omega_1^2 \sin^2\theta_1 + n_2^2\omega_2^2 \sin^2\theta_2 + 2n_1n_2\omega_1\omega_2 \sin\theta_1 \sin\theta_2. \tag{11.2}$$

The coherence enables spatial filtering of the sum frequency signal from the much more intense reflected fundamentals.

11.2.2 Nonlinear susceptibility

The relationship between the surface nonlinear susceptibility, χ_{IJK}, and the molecular hyperpolarizability, β_{abc}, depends on the molecular orientation. The link between molecular orientation and surface susceptibility is commonly analyzed using Euler angle relationships among four coordinate systems: the laboratory system (XYZ) that defines the plane of incidence, the surface system (xyz) that defines the surface normal, the molecular Cartesian system (abc), and the molecular normal coordinates (ABC). The general results of these transformations are given in the literature [28,29]. The Cartesian susceptibility elements are related to the normal coordinate elements as

$$\beta_{\mathrm{abc}} = \frac{1}{n} \sum_{orientations} \beta_{\mathrm{ABC}}. \tag{11.3}$$

where n is the number of equivalent orientations. The Cartesian coordinate susceptibilities are similarly projected onto the surface coordinate system and averaged over the equivalent surface orientations

$$\chi_{\mathrm{IJK}} = N \left\langle \beta_{\mathrm{IJK}} \right\rangle \tag{11.4}$$

where N is the number of molecules and $\langle \cdots \rangle$ denotes the orientational average. For a two-dimensional, isotropic surface the orientational average results in zero intensity for *sss*, *spp*, *psp*, and *pps* polarizations. Finally, the SFG signal intensity is proportional to the polarization squared

$$Signal\ intensity \propto |P|^2 \propto \left|\chi^{(2)}\right|^2 = N^2 \left|\left\langle \beta^{(2)} \right\rangle\right|^2 \tag{11.5}$$

resulting in an SFG intensity that is proportional to the square of the number of molecules contributing to the signal modulated by the average orientation. Specifically for the work discussed below, with the *ssp* polarization combination, the SFG intensity is strongly affected by the projection of the infrared dipole onto the surface normal.

11.2.3 Experimental layout

A schematic of the experimental arrangement is shown in Fig. 11.1. The objective is to overlap pulsed visible and infrared beams in space and time

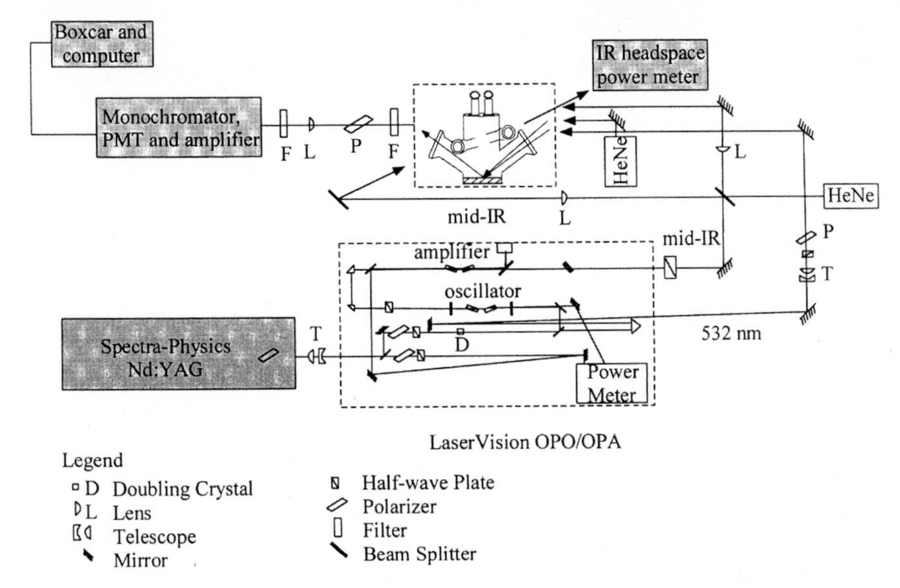

Fig. 11.1. Schematic of experimental apparatus for sum frequency generation. The master clock is the 1064-nm pulse from the Nd:YAG laser. Nonlinear processes in the optical parametric oscillator/optical parametric amplifier generate the infrared beam. In the OPO stage, the 532-nm pump beam is parametrically split into a signal and an idler beam the frequency of which is angle dependent. Angle tuning the oscillator stage results in an idler beam of the desired frequency to difference frequency mix with a 1064-nm pump from the YAG laser to produce tunable mid IR in the amplifier stage. The 532-nm visible beam for SFG is generated in a KTP crystal. The power of the infrared and visible beams are controlled with a combination of a half-wave plate and a polarizer. Both beams are brought to a gentle focus at a point beyond the liquid surface. The generated sum frequency is spatially filtered, filtered through a short-pass filter and focused on the entrance slit of a monochromator. The SFG photons are detected by a photomultiplier tube and the signal is sent to a boxcar averager and computer for analysis. All signals are normalized to the infrared intensity reaching the surface, which is determined by sampling the headspace.

on the liquid interface. Here, the source of the pulsed visible beam is a 1064 nm pulse from a *n*sec. Nd:YAG laser (Spectra-Physics GCR150) doubled in a KTP crystal. The infrared beam is generated in the optical parametric oscillator, optical parametric amplifier (LaserVision OPO/OPA) by parametrically splitting the 532-nm beam into a signal and an idler beam in the OPO stage and difference frequency mixing the idler with 1064 nm in the OPA stage. The polarization of both the visible and infrared beams is rotated as required. The two beams are colinearly focused to a spot beyond the liquid interface. Energy density is 100 mJ cm^{-2} for the infrared and 400 mJ cm^{-2} for the visible light. To ensure nonsaturation, the signal is checked for linearity in both the infrared and visible power density, and spectra are collected well within the linear regime. A portion of the infrared beam is split off prior to the sample and sent through the headspace to monitor the infrared intensity reaching the surface. All spectra discussed here are normalized to the infrared intensity and referenced to the free-OH stretch, assigned unit intensity.

11.3 Ions in solution modify water at the surface

Discussion concerning modification of the structure of water at the aqueous-air interface begins with an examination of the spectrum of the neat water-air interface. The spectrum shown in Fig. 11.2, obtained by collecting the *s*-polarized sum frequency signal upon excitation with an *s*-polarized visible beam and a *p*-polarized infrared beam (labeled *ssp*) is consistent with the first SFG spectrum of water published by Shen *et al.* [11]. Three major features are apparent: a sharp peak at 3700 cm^{-1} and two broader peaks centered at approximately 3400 cm^{-1} and 3150 cm^{-1}. Although details of the water spectrum vary somewhat with acquisition parameters, e.g. visible excitation wavelength and incident angles, all share these three major features. Interpretation of the 3700 cm^{-1} peak is the most straightforward and non-controversial. Midway between the symmetric and antisymmetric gas-phase absorptions of water, 3700 cm^{-1} corresponds to the decoupled OH oscillator frequency. It is thus assigned to an OH stretch of surface water molecules with a hydrogen free of hydrogen bonding, dubbed the 'free-OH' or 'dangling-OH' stretch. To be free of hydrogen bonding, the hydrogen atom must protrude from the surface. Significantly for the work discussed here, in order for the hydrogen atom to be free of hydrogen bonding, these OH bonds must be in the topmost monolayer. The intensity of the free-OH peak therefore serves as an indicator of perturbation of water in the very top layer.

Interpretation of the remaining two hydrogen-bonded peaks remains controversial. The lower frequency peak is variously termed the structured peak, the 'ice-like' peak, or the symmetric stretch of symmetrically bonded water. These characterizations stem from the similar characterization of Raman spectra of water, ice and aqueous solutions [33-39]. Conversely, the peak at 3400 cm^{-1} is attributed to disordered water at the surface in analogy to the prominent feature in the Raman spectrum of so called 'structure breaking' ions. An alternate characterization is based on theoretical calculations for ice [40-44] and water clusters [45] . The lower frequency peak is attributed to those water molecules that are strongly hydrogen bonded and include those for which the other O-H bond is dangling. In this interpretation, the 3400 cm^{-1} peak is due to more weakly hydrogen-bonded O-H bonds. These alternate interpretations present somewhat different pictures of water. Resolution of the controversy over the interpretation awaits further experimental and theoretical investigation. In this discussion, the lower frequency peak is referred to as the stronger hydrogen-bonded peak and the 3400 cm^{-1} peak as

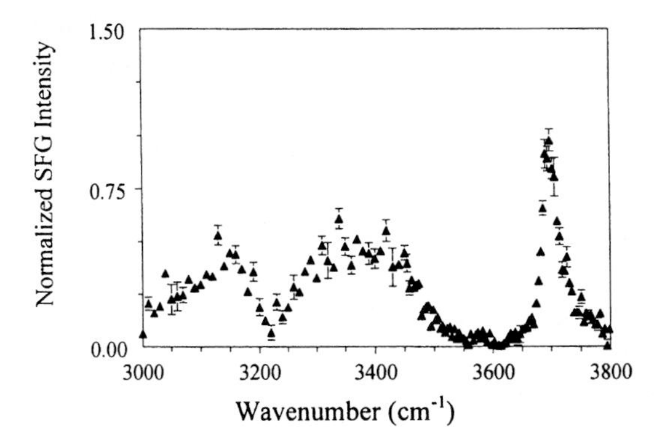

Fig. 11.2. Sum frequency spectrum of the neat air/water interface in the range of 3000-3800 cm^{-1} at 0 °C. The polarization combination is *s*-polarized sum frequency, *s*-polarized visible, and *p*-polarized infrared, labeled *ssp*. The sharp feature at 3700 cm^{-1} is assigned to an O-H bond free of hydrogen bonding, referred to as the 'free OH'. The broad features between 3000 and 3550 cm^{-1} are assigned to O-H bonds involved in hydrogen bonding.

the weaker hydrogen bonded peak. Both hydrogen-bonded peaks originate with water molecules in the asymmetric environment of the interface and the depth probed extends as far as this asymmetry persists.

11.3.1 The first layer – free OH

As indicated above, any perturbation of the free-OH resonance indicates that the influence of the solute has reached the topmost layer of the solution. With the exception of ammonia (discussed below) all solutes investigated to date uniformly diminish the intensity of the free-OH resonance. Specifically, the free-OH resonance is not observed to shift in frequency or broaden. The absence of a shift supports the model that the intensity decrease is due to free water being displaced from the surface rather than these three-coordinate water molecules becoming directly bonded to subsurface ions or ion pairs. Spectra of sulfuric acid solutions, shown in Fig. 11.3 are typical. With as little as 1 mol% solute, the free-OH intensity diminishes. The decrease continues as the H_2SO_4 concentration increases, and with 20 mol% acid (not shown) the free-OH intensity is below the noise level [12, 46, 47].

As indicated in (11.5), the SFG intensity is proportional to the square of the number of scatterers. So, for the aqueous surface, the square root of the free-OH intensity is proportional to the monolayer coverage of free-OH groups. The result, shown in Fig. 11.4, is based on a free-OH coverage on neat water of 25% [11]. For dilute acid solutions, sulfuric acid is totally dissociated into SO_4^{2-} ions and hydrated protons. As the acid concentration increases, a portion of the protons associate with the anion to form bisulfate ions diminishing the hydrated proton concentration. At yet higher concentrations,

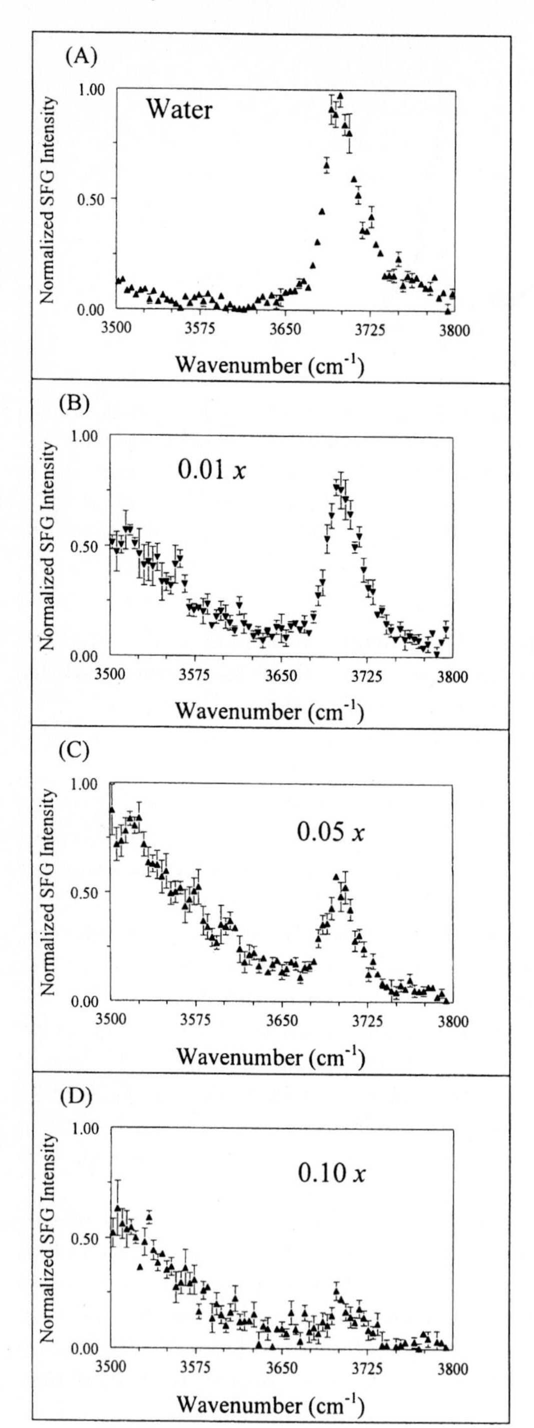

Fig. 11.3. The free-OH resonance of water from aqueous sulfuric acid solutions at 0 °C. (A) Neat water, (B) 0.01 x, (C) 0.05 x, and (D) 0.10 x sulfuric acid. All spectra are obtained with the *ssp* polarization combination, normalized to the incident infrared intensity and referenced to the free-OH resonance of neat water.

hydrated protons associate with bisulfate ions forming neutral, molecular sulfuric acid. Molecular sulfuric acid forms strong hydrogen bonds to water, a property that accounts for the increase in surface tension of water on forming solutions of less than 13 mol% acid (Fig. 11.5). The decrease in free-OH intensity correlates well with the decrease in degree of dissociation of sulfuric acid (Fig. 11.4).

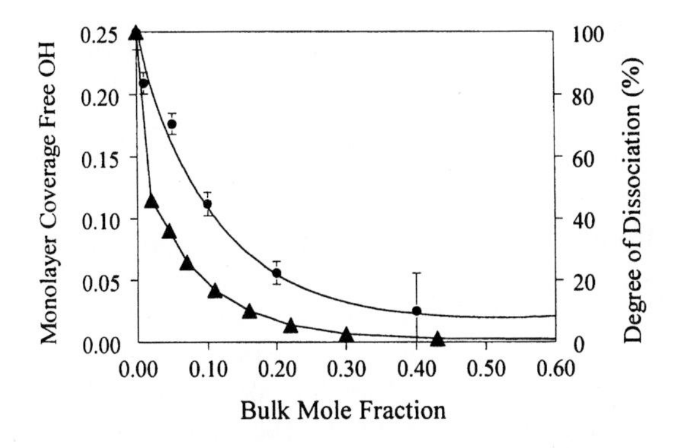

Fig. 11.4. The monolayer free-OH coverage (○) on the surface of sulfuric acid solutions is correlated with the degree of dissociation (△) of the acid in solution.

Note that the SFG intensity is proportional to the square of the number of scatterers. As a result, SFG is particularly sensitive to even a small decrease in the normal coverage of free-OH groups on neat water. The intensity at 1 mol% acid, corresponds to a 10% decrease in dangling-OH groups. This magnified effect indicates that the hydration sphere around the $H_2SO_4/HSO_4^-/SO_4^{2-}$ is both extensive and, at this concentration, penetrates to the very top monolayer of water. Using the dangling-OH coverage as a gauge of unperturbed water at the surface, indicates that at 1 mol% sulfuric acid, 90% of the surface water is unaffected by the presence of the acid. The lowest acid concentration found in stratospheric aerosols is 4 mol% in acid. At this concentration, 70% of the surface water is unperturbed. For the higher concentrations, over 10 mol%, all the surface water is perturbed. These conclusions are similar to those drawn by Phillips [48] from an analysis of the vapor pressure of water over sulfuric acid solutions. In a more direct examination of the surface, Somorjai *et al.* [49] used Auger Electron Spectroscopy (AES) and X-Ray Photoelectron Spectroscopy (XPS) to determine the sulfur to oxygen ratio in the first three layers of the solution. The data indicate that the S:O ratio is the same on the surface and in the bulk. Taken together, these results indicate that water in the topmost monolayer is involved in the hydration sphere of the various sulfate species present. Since ions in the very top layer would necessarily be missing the upper half of the hydration sphere, these ions are more stable in the lower layers. Molecular sulfuric acid can, however, penetrate to the very top layer. Hence, the free-OH intensity decrease lags behind the degree of association.

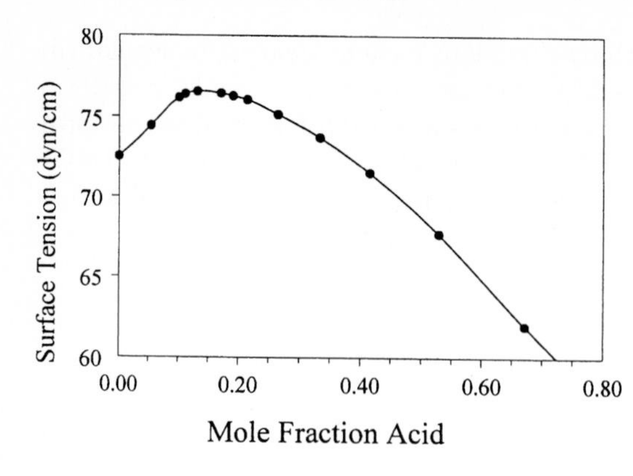

Fig. 11.5. Surface tension versus mole fraction acid for aqueous sulfuric acid solutions.

An examination of data from bisulfate and sulfate salts supports the model that associated ion pairs penetrate to the surface and displace free water (Fig. 11.6). Association between the bisulfate anion and alkali metal cations increases as the size of the cation increases. So $LiHSO_4$ is less associated, more ionic, than is $CsHSO_4$. $CsHSO_4$ therefore penetrates to the topmost layer more readily than $LiHSO_4$. Accordingly, the intensity of the free-OH resonance from a $CsHSO_4$ solution is diminished relative to that of a $LiHSO_4$ solution of the same concentration. The $CsHSO_4$ complex with its associated hydration sphere displaces free surface water.

A similar picture applies to the bisulfate vs. sulfate salt of a given cation. For example, Li^+ forms a neutral complex with HSO_4^- while two Li^+ cations are required to form a neutral complex with SO_4^{2-}. Once formed, the neutral complex can penetrate to the surface region. The free-OH resonance of the $LiHSO_4$ solution is therefore diminished relative to the Li_2SO_4 solution. Correspondingly, the free-OH resonance intensity for $CsHSO_4$ is less than that of Cs_2SO_4.

From the perspective of water on the surface, ions in solution far from the surface do not affect water. The surface of very dilute solutions looks the same as neat water and molecules impinging on such solutions encounter the same environment as for pure water. This picture changes as the ion concentration increases and the counter ions associate. The associated ions along with the hydration sphere penetrate the very top layer and displace free water. In the next subsection, hydrogen-bonded water is examined to augment this picture of water at these interfaces.

11.3.2 Hydrogen-bonded water

At low concentrations, less than 1 mol%, the topmost monolayer of the aqueous surface is unperturbed by the presence of inorganic ionic solutes as measured by the full free-OH resonance intensity. The picture for hydrogen-bonded surface water is quite different, however. Figure 11.7 shows the vibrational spectrum of the surface of a 1 mol% sulfuric acid solution compared

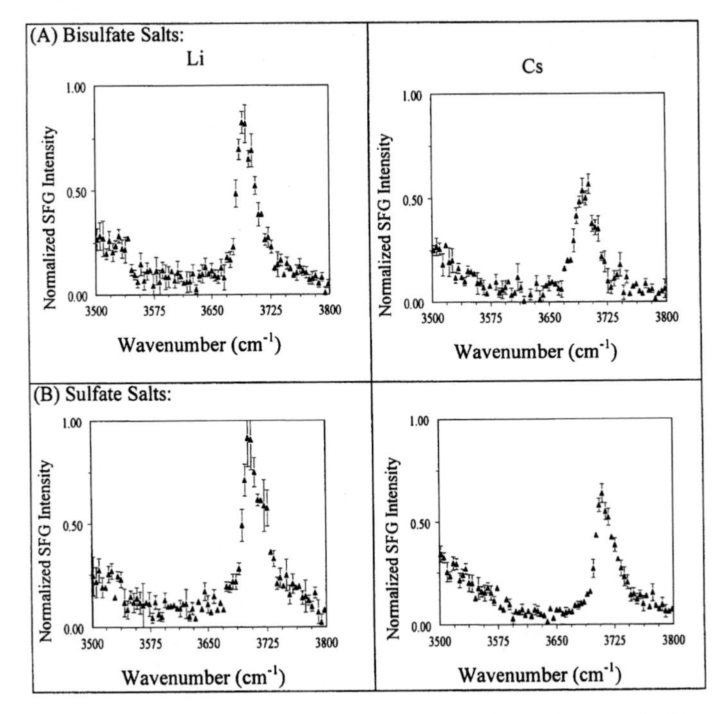

Fig. 11.6. The free-OH SFG intensity from the surface of bisulfate (A) and sulfate (B) solutions. Concentration 0.01 x, 0 $^\circ$C, *ssp* polarization combination.

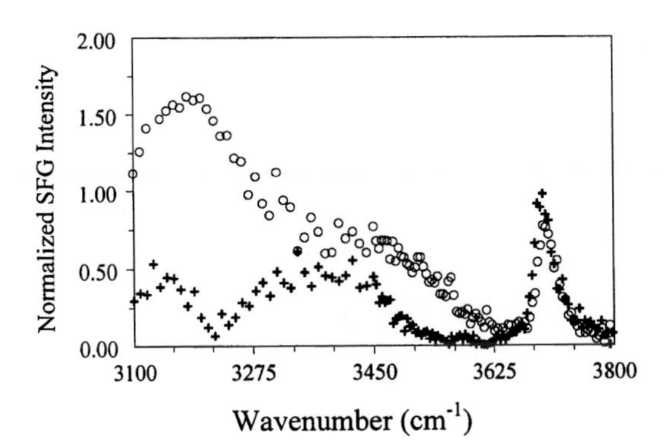

Fig. 11.7. SFG spectrum (*ssp*) of (+) water-air interface and (o) 0.01x sulfuric acid at 0 $^\circ$C. (Error bars have been eliminated for clarity.)

with that of neat water. Particularly, the strongly hydrogen bonded peak, centered around 3150 cm^{-1} for the neat air-water interface, increases in peak intensity by more than a factor of three. Two models have been proposed to explain this enhanced intensity. Both are based on the distribution of anions versus cations near the surface. In a 1 mol% solution, sulfuric acid is dissociated, at least to HSO_4^- and the hydrated proton. Since the HSO_4^- ion is both larger and more polarizable than the hydrated proton, the tail of the bisulfate ion distribution extends closer to the interface than the tail of the hydrated proton distribution. This differential distribution creates a double layer in the immediate subsurface region. The cations need not penetrate the topmost monolayer for this double layer to exist. It is merely necessary for the average distance from the top monolayer to be somewhat larger for the cations than for the anions. The double layer field is oriented with the negative pole pointing to the surface and the positive pole pointing into the bulk.

In one model, called a rotation model, water responds to this double layer field as follows. At the neat air-water interface, water is oriented with its dipole pointing slightly into the bulk [1], i.e. hydrogen atoms pointing slightly toward the bulk. Hydrogen atoms with their slightly positive charge are attracted toward the negative field near to the surface and the water molecules rotate relative to neat water. The rotation produces a larger projection of the water dipole onto the surface normal. Understanding how this rotation generates an enhanced SFG intensity requires examining the factors that affect the SFG intensity.

The species dependent terms that affect the SFG intensity are the number of scatterers, discussed above, and the orientation of those scatterers. The polarization combination used for these aqueous surface studies is *ssp*. In particular, the infrared beam is *p* polarized. So if the infrared transition dipole is aligned more vertically, parallel with the polarization of the infrared excitation, the macroscopic transition is enhanced. This enhancement results in an augmented sum frequency intensity. Within this model, the lower frequency hydrogen-bonded peak intensity is enhanced since the infrared dipole moment for the low frequency transition and the water dipole are coincident. In contrast, the infrared dipole for the higher frequency peak, corresponding to the antisymmetric stretch of water, is coincident with the rotation axis. Hence the higher frequency infrared transition is essentially unaffected by the rotation. Similarly, to a first approximation, the SFG intensity is also unaffected.

Exact deconvolution of the two hydrogen-bonded peaks in the sum frequency spectrum requires modeling the transitions giving rise to the two peaks since SFG is a nonlinear spectroscopy. The nonlinearity results in interferences that can lead to deceptive results. None-the-less, the spectrum in Fig. 11.7 indicates a significant enhancement of the lower frequency ($\sim$3150 cm^{-1}) peak relative to the higher frequency peak ($\sim$3450 cm^{-1}). Similarly, interpretation of what appears to be a slight blue-shift in the 3150 cm^{-1} peak for the sulfuric acid solution should be approached with caution until further theoretical and experimental work discerns the details of these resonances.

The alternate interpretation, called a global enhancement model, also begins with the differential distribution of anions and cations in the near-surface region. The alternate model does not require rotation of the water molecules. Instead, the model is based on the macroscopic second order polarizability, decomposing the polarizability into an infrared and a Raman contribution. The bulk of neat water generates no SFG intensity due to the isotropic distribution of dipoles. The Raman scattering intensity from symmetric environments is strong. However, due to mutual exclusion, the infrared transition in a totally symmetric environment is zero. At the surface, the inversion symmetry of the symmetric environment is broken and this gives rise to the SFG signal from neat water. The double layer field of ions enhances this symmetry breaking, thus enhancing the infrared contribution to the SFG intensity. In this model, the higher frequency peak corresponds to a less symmetric motion that has a substantial infrared transition intensity already in neat water. Enhancement of the asymmetry due to the double layer field, therefore has little affect on the intensity.

The two alternate models have very different implications for water at aqueous solution interfaces as well as for the hydrogen bonding environment for the two transitions. In the rotation model, incoming reactants are presented with a different face of water when ions are dissolved compared with neat water. In the global infrared enhancement model, the molecular configuration at the interface is the same in neat water and in aqueous solutions. Only the presence of the double layer distinguishes the two interfaces. Since the implications for the molecular-level picture of interactions and reactions at the aqueous surface are quite different, further experimental and theoretical efforts aimed at differentiating these models are currently underway.

Sulfuric acid is not unique in its affect on the aqueous surface (Fig. 11.8 and Fig. 11.9). The spectrum from the surface of 0.01 x sulfuric (A) and hydrochloric (B) acid solutions are quite similar. The lower frequency hydrogen-bonded peak for the H_2SO_4 solution is enhanced relative to that of the HCl solution, consistent with the presence of both HSO_4^- and SO_4^{2-} ions generating a more pronounced double layer field for aqueous sulfuric acid. The intensity of the free-OH peak for aqueous HCl is somewhat greater than that of the sulfuric acid solution when both are 0.01 x, indicating that the hydrated complex containing HCl does not perturb the topmost monolayer as much as those of H_2SO_4. HCl, however, is considerably more volatile than H_2SO_4. For HCl, associated pairs in the near surface region escape into the gas phase rather than engaging in hydrogen bonded complexes with water. Indeed, the search for molecular HCl on a variety of surfaces has met with failure. The only liquid surface on which molecular HCl has been observed is the surface of neat, liquid HCl [50]. Nitric acid is somewhat less volatile than HCl. Consistently, the concentration at which nitric acid perturbs the free-OH intensity is lower than the nonperturbing HCl concentration. Fig. 11.8 (C) shows the spectrum from 0.005 x HNO_3. At this concentration, the free-OH intensity is unperturbed and the hydrogen-bonded peaks bear the signature enhancement of the 3150 cm^{-1} peak, typical of a variety of strong acids.

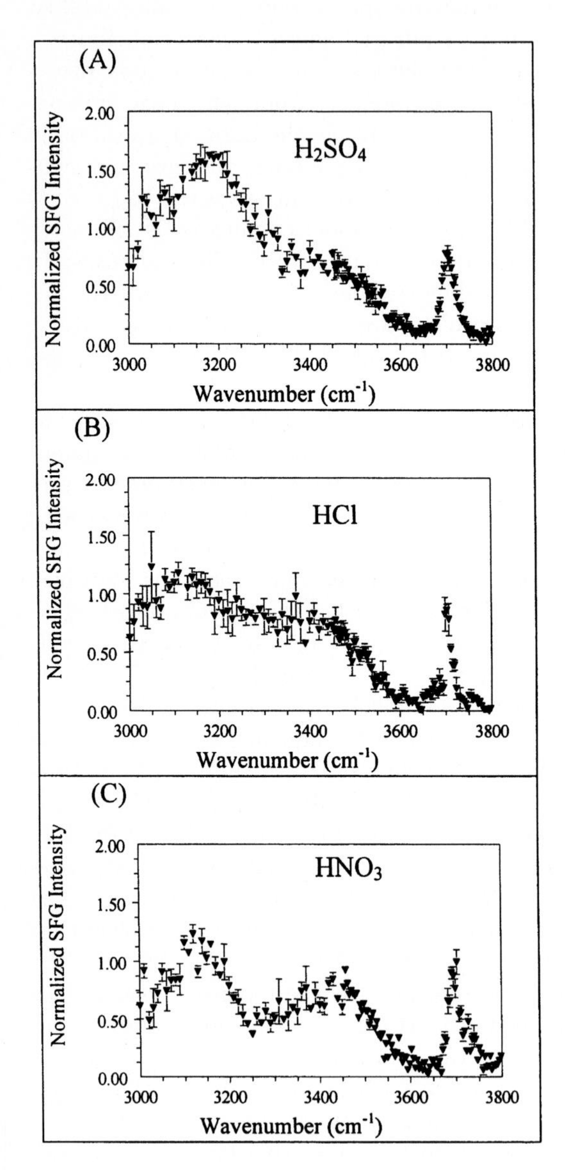

Fig. 11.8. SFG spectra of water at the surface of aqueous acid solutions at 0 °C. (A) 0.01 x H_2SO_4, (B) 0.01 x HCl, (C) 0.005x HNO_3. All spectra collected with the *ssp* polarization combination.

Replacing the proton with the small cation, Li^+ also results in enhancement of the lower frequency, hydrogen-bonded peak as shown in Fig. 11.9. For a salt concentration of 0.01 x, the free-OH intensity from $LiHSO_4$ solution is nearly unperturbed, indicating that associated ion pairs do not displace water in the top monolayer at this concentration. However, enhancement of the lower frequency, hydrogen-bonded peak is similar to that observed for acids, though less pronounced. Continuing this trend, the spectra of bisulfate salts of larger cations (Na^+, K^+, or Cs^+) show even less intensity enhancement. This correlation of decreasing enhancement with larger cation size is consistent with the expected ion distributions. On approaching the surface, the tail of the cation distribution extends nearer to the interface for the larger, more polarizable cations. When the anion and cation distribution functions are similar, the double layer field diminishes along with the intensity of the lower frequency hydrogen-bonded peak.

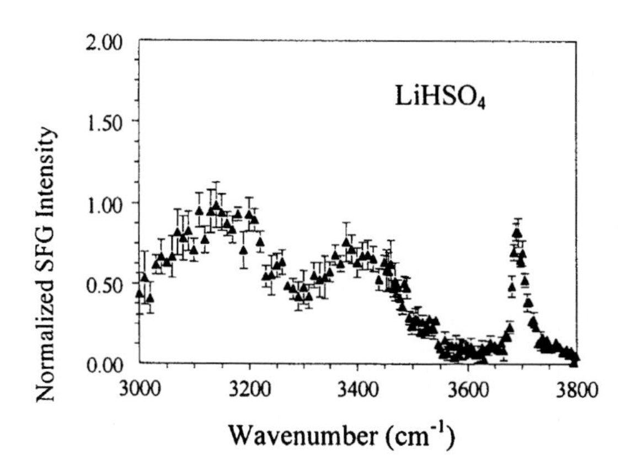

Fig. 11.9. SFG spectra of lithium bisulfate solution at 0 °C taken with *ssp* polarization. The SFG intensity is normalized to the infrared intensity reaching the interface and referenced to the free-OH intensity of neat water, assigned a unit intensity.

Within either the rotation or the field-enhanced-infrared-transition-dipole models, water at the interface exists in a dipole field that affects the macroscopic polarizability of the interface. Reactants approaching the interface also are influenced by this field. In the next section, the interaction of ammonia with the aqueous interface is examined.

11.4 A test-probe of the surface – ammonia

11.4.1 Orientation analysis

The aqueous ammonia interface is very dynamic. At 0 °C, the vapor pressure of water (~4.5 Torr) results in exchange of about a monolayer every 300 ns. With a comparable ammonia vapor pressure, the surface would seem to be very chaotic. However, surface tension [51, 52] and uptake measurements

[53–55] indicate the presence of an ammonia-water complex at the interface. This complex is of some importance to atmospheric chemistry since ammonia is essentially the only basic, gas-phase species present to any appreciable concentration in the atmosphere. Field measurements [56] indicate that ammonia is incorporated into tropospheric aerosols only to the level of neutralizing acid in the droplets. To model the rate of atmospheric processes, it is of interest to elucidate the structure of this complex and to unravel the mechanism of incorporation of ammonia into heterogeneous material in the atmosphere [54, 55, 57, 58].

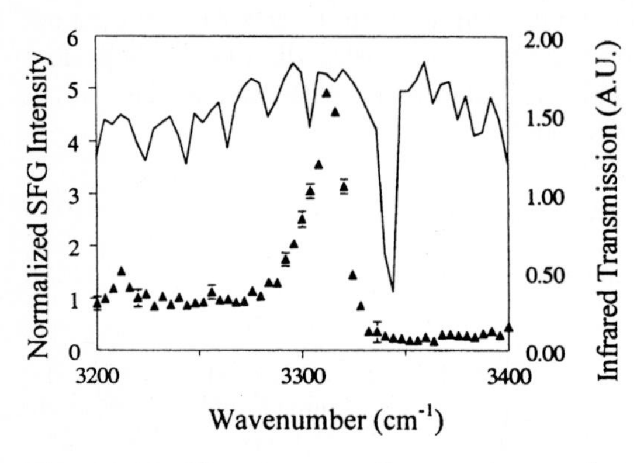

Fig. 11.10. The spectrum of ammonia, gas phase absorption (—) and SFG (△) from concentrated ammonia. The SFG spectrum is normalized to the infrared intensity reaching the surface as determined by a simultaneous infrared probe through the headspace of the cell. The infrared absorption is determined through this headspace port. Temperature 0 °C. SFG polarization is *ssp*.

Sum frequency generation is a nearly ideal method for determining the distribution and orientation of material at the aqueous ammonia interface. In addition to the free-OH and hydrogen-bonded resonances referred to above, ammonia has a very narrow, intense SFG resonance (Fig. 11.10) [59]. This resonance at 3312 cm^{-1} is assigned to the ν_1 symmetric stretch of ammonia. The slight red-shift of the ammonia NH_3 symmetric stretch from the gas-phase value ($\sim$20 cm^{-1}) indicates that the N-H bonding is only slightly perturbed by interaction with the aqueous surface. It should be noted that these SFG results provide the first, direct experimental observation of the ammonia-water complex, first suggested by Rice in 1928 [51] and more recently modeled by Donaldson [60]. The molecular-level detail that SFG can provide becomes apparent with the results from ammonia-water. As indicated in (11.5), the SFG intensity is proportional to the square of the macroscopic polarizability. It is the macroscopic polarizability that carries the molecular orientation and excitation polarization information.

Due to the vibrational resonance, the orientation analysis involves determining the average projection of the molecular vibrational dipole and Raman polarization onto the excitation electric fields. Or equivalently, determining the average projection of the excitation electric fields onto the molecular infrared and Raman transition moments. A general analysis of generated intensity as a function of orientation and polarization is given in the literature [28, 29, 61]. Because liquid interfaces are isotropic in the surface plane, the general equations simplify. In addition to the surface plane isotropy, the symmetry of the adsorbate can further reduce the general relationships. Specifically for ammonia, an important feature is the link between the symmetric and antisymmetric stretch modes [62]. The infrared dipole of the symmetric stretch normal mode is coincident with the molecular C_3 axis. The antisymmetric stretch mode, ν_3, is doubly degenerate, and importantly for locating the orientation, the ν_3 infrared dipole is in the plane perpendicular to the molecular C_3 axis. Thus, the ν_1 and ν_3 transition dipoles are mutually perpendicular. Probing with the *ssp* polarization combination uses the mutual perpendicularity to resolve the orientation as follows. For *ssp* polarization, the generated intensity is proportional to the average projection of the molecular vibrational transition dipole onto the surface normal. Since the ν_1 and ν_3 resonance frequencies are well separated, the intensity generated from each can be determined. For ammonia on water, the *ssp* intensity from the antisymmetric stretch is very weak resulting in a limit to the tilt of the C_3 axis with respect to the surface normal of less than 38 $^\circ$ [62].

The orientation of the C_3 axis can be narrowed yet further by examining the intensity resulting from excitation with other polarization combinations. Specifically, if the C_3 axis was perpendicular to the interface, as is the case for ammonia on a variety of solid surfaces, there would be a significant intensity for the anitsymmetric stretch with *sps* polarization. The *sps* polarization detects vibrational transition moments in the surface plane for nonsymmetric modes. The aqueous surface generates a nondetectable intensity with the *sps* polarization combination while aqueous ammonia generates some intensity. The polarization combination that restricts the tilt is the *ppp* combination. Analysis of the *ppp* intensity is challenging since several macroscopic polarizabilites contribute to this combination, and interferences among these complex quantities can result in some unexpected cancellations. However, the intersection of the intensity from *ssp*, *sps*, and *ppp* locates the tilt angle to between 25 $^\circ$ and 38 $^\circ$. This result is similar to the orientation of ammonia on the surface of water from the *ab initio* calculations [60].

Although SFG can determine the tilt angle, it does not determine the phase of the ammonia dipole, i.e. whether the dipole points into or out of the bulk. However, the minor red shift and the known weakness of donor hydrogen bonds formed by ammonia [63] indicate that ammonia is oriented with the hydrogen atoms pointing away from the bulk. Further, the antisymmetric stretch remains a single resonance at essentially the same frequency as the gas-phase resonance. If ammonia were oriented with the hydrogen atoms pointing toward the bulk, the differing interactions would broaden this resonance, as observed in liquid ammonia [64-67].

The orientation analysis supports a picture of an ammonia-water complex on the surface of aqueous ammonia in which the ammonia molecule bonds to the surface via the nitrogen lone pair. The next section contains an analysis of the free-OH and ammonia symmetric stretch intensities as a function of ammonia bulk concentration that is consistent with this picture.

11.4.2 Concentration dependence

Ammonia is highly soluble in water, yet uptake and surface tension measurements indicate that ammonia forms a surface complex with water [51, 60]. The structure of this surface complex is of interest for unraveling the mechanism of incorporation of ammonia into atmospheric aerosols. The orientation analysis discussed above provides a picture of the structure of the ammonia-water complex. The concentration dependence of ammonia on water enriches this picture and indicates a more intricate behavior of this surface complex.

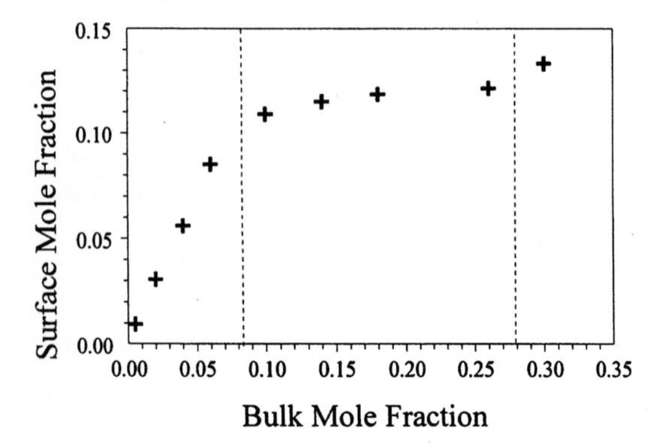

Fig. 11.11. Surface versus bulk mole fractions for ammonia on aqueous ammonia. Three regions are apparent: a low concentration region with a linear relationship in which the surface concentration exceeds that of the bulk, a middle region in which the surface concentration remains nearly constant, and a high concentration region in which the surface concentration again increases. In the high concentration region, the surface concentration is lower than that of the bulk.

Figure 11.11 shows the results of determining the surface mole fraction versus the bulk mole fraction for ammonia. Three regions are apparent. In the low concentration region, the surface mole fraction exceeds that of the bulk by a factor of 1.4. The concentration relevant to most atmospheric processes is in this region. The surface excess is consistent with the measured resistance to incorporation into the bulk [54] and suggests that ammonia is surface active, consistent with ammonia lowering the surface tension of water. In the middle region, the surface concentration of ammonia is nearly constant, indicating that the surface sites are saturated. Solid surfaces frequently show such saturation, but this is the first example of a soluble inorganic species saturating sites on such a dynamic surface. Surface tension measurements show a slight indication of this saturation with a slower decrease versus concentration in this middle region. The SFG results suggest that the change in surface tension in this middle region is a result of changes in the structure of the bulk solution as the bulk water accommodates ammonia molecules. Note that in the middle concentration region, the surface concentration has switched from exceeding that of the bulk to being less than that of the bulk. Finally, the surface concentration again rises as the bulk solution becomes saturated with ammonia and any further ammonia escapes into the gas phase.

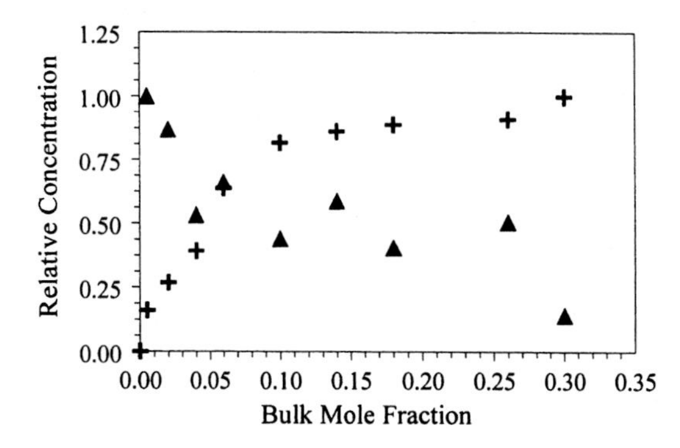

Fig. 11.12. Relative surface concentration of ammonia (+) and free water (△) as a function of the bulk mole fraction of ammonia. The free water concentration refers to the relative coverage by free-OH groups. For ammonia, the relative coverage is measured relative to ammonia on concentrated aqueous ammonia.

A detailed picture of water at the aqueous surface is obtained by examining the correlation between the increasing ammonia concentration and the diminishing free-OH resonance. For the data shown in Fig. 11.12, SFG intensity has been converted to relative coverage. This conversion involves dividing the water free-OH intensity by that for neat water and taking the square root. Similarly, the ammonia relative concentration is referenced to the signal intensity for concentrated ammonia, taking the root of this ratio. In region one, a bulk mole fraction of less than 0.08, there is a linear trade off of water free-OH coverage for ammonia coverage, suggesting that ammonia merely titrates the dangling-OH bonds for this concentration range. When about one-third of the dangling-OH bonds are decorated with ammonia, the surface is saturated. Increasing the bulk concentration then has no affect on the amount of ammonia on the surface. Saturation is a common phenomena on solid surfaces and often corresponds to the filling of favorable bonding sites with adsorbate molecules. Saturation on dynamic surfaces such as aqueous solutions at 0 °C has not been observed directly previous to this result. Titration of the dangling-OH bonds suggests that the hydrogen-bonded network on the aqueous surface is quite robust, remaining intact not only during the dynamic exchange of molecules with the gas, but also as ammonia molecules attach to the dangling-OH groups. The robustness of the hydrogen-bonded network is consistent with the well-defined vibrational resonances observed on the surface despite the very dynamic nature of the surface.

Taken together, the surface tension [51, 60], uptake measurements [54], and the present SFG results [59, 62] indicate that the ammonia-water complex on the surface consists of a single ammonia molecule with the C_3 axis tilted about 35 $^\circ$ off the normal, docked onto a dangling-OH bond of a surface water molecule, as illustrated in the cartoon picture shown in Fig. 11.13. Surface crowding and the substantial dipole moment of ammonia, inhibits adsorption of more than one ammonia for every three dangling-OH groups. The resistance to adsorption noted in the uptake experiments suggests that even at substantially lower concentration, decoration of dangling-OH groups by ammonia molecules inhibits further adsorption. It is likely that protonation of ammonia in acidic solutions, incorporates the resulting NH_4^+ ion into the bulk solution, freeing the dangling-OH groups for further ammonia docking. Upon neutralization of the acid, the surface saturates, inhibiting further uptake of ammonia.

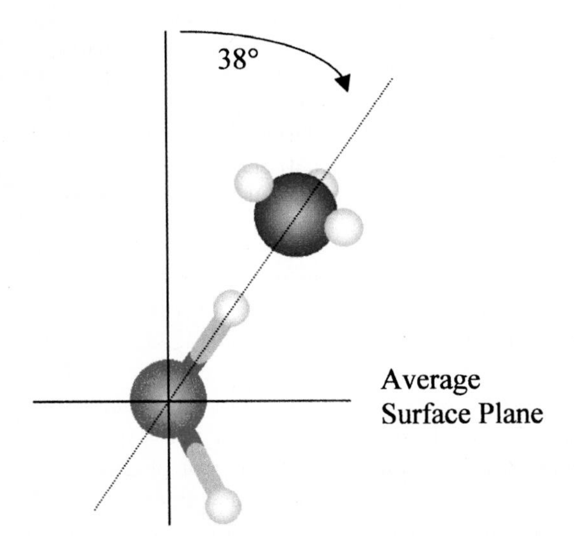

Fig. 11.13. Cartoon rendition of the ammonia-water surface complex. The C_3 axis of ammonia is tilted 25-38 $^\circ$ from the normal, docking onto the dangling-OH bond of a water molecule in the surface. On the molecular level, the surface normal is the perpendicular to the average surface plane. The aqueous surface is very chaotic on the molecular level, with exchange of a monolayer between the surface and the gas phase every 300 nsec.

From the point of view of water confined to the surface region, it appears that ammonia lowers the surface tension of water by tying up the dangling-OH bonds, making surface water more like bulk water. Furthermore, the hydrogen bonded network on the surface of water is quite robust since ammonia on the surface saturates well below the solubility limit.

11.5 Water in a liquid matrix

11.5.1 Water in CCl_4

Water is not very soluble in CCl_4. A previous literature report that water forms small clusters in CCl_4 [68] suggests that incorporating water into CCl_4 provides the potential for probing the interaction of a variety of molecules with small water clusters. However, it appears that water forms only isolated monomers in CCl_4. Figure 11.14 shows the infrared absorption spectrum of 0.02 M water in CCl_4. This spectrum was obtained with a commercial FTIR spectrometer in transmission mode as follows. A cell was constructed of glass with IR quartz windows (path length, 10 cm). Since the surface of quartz and glass are typically covered with OH groups that readily hydrogen bond to water, the entire cell was silanized prior to use in this experiment. Silanization is known to coat the surface with organic residues that render the surface hydrophobic. This key step was absent in the previous work with water in CCl_4. The success of this procedure is verified by an absence of any absorption in the water hydrogen-bonded region, even with 0.02 M water. Indeed, no hydrogen-bonded absorption is observed to saturation of CCl_4 with water. Only two peaks are observed. The feature at 3620 cm^{-1} is assigned to the symmetric stretch and that at 3712 cm^{-1} to the antisymmetric stretch. The broad absorption under these two sharper features are due to restricted motion of water within the CCl_4 cage. Carbon tetrachloride, thus serves as a room-temperature matrix isolation material.

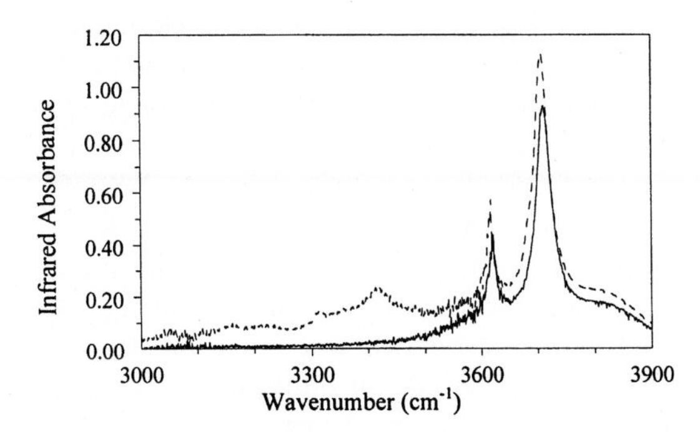

Fig. 11.14. Infrared absorbance of 0.02 M water in CCl_4 (solid line) and 0.008M in concentrated ammonia in CCl_4 in the region 3000 cm^{-1} to 3900 cm^{-1}.

Replacing water with concentrated aqueous ammonia results in a very different absorption spectrum, as indicated by the dotted line in Fig. 11.14. Notice that the concentration of water in the aqueous ammonia-CCl_4 mixture is more than a factor of two lower than that of water in CCl_4, yet the water vibrations have greater intensity for aqueous ammonia in CCl_4. This intensity increase is consistent with the well known increase of the transition moment of water vibrations upon hydrogen bonding [69]. In addition to the intensity increase, the intensity in the hydrogen bonded region indicates that ammonia acts like a surfactant, hydrogen bonding to water molecules.

11.6 Conclusions

Inorganic ions in aqueous solution are relevant to many phenomena from corrosion to atmospheric chemistry to biological fluids. Often, it is the interface between the aqueous solution and another phase – a metallic solid or air or a cell wall – that is of particular interest, yet is the least understood. Until recently, the major hurdle to probing these interfaces was the availability of a technique capable of specifically probing the interface. Specificity is desired so that the signal from the interface is not swamped or masked by the often much larger signal from the bulk phases on either side of the interface. When water is the object of the investigation, surface specificity is particularly important to distinguish water in the confined geometry of the interface from the more mobile water in the bulk.

This contribution describes results of using the nonlinear spectroscopic probe, sum frequency generation (SFG) to probe water at the aqueous interface focusing on the aqueous-air interface. A large variety of inorganic ions in solution influence water at the interface at low concentration through an electric double layer that forms due to the differential distribution of anions and cations with respect to the interface. The systems investigated to date share the characteristic of a large anion and small cation. In this case, the tail of the anion distribution extends closer to the interface than that of the cation. Two models have been proposed to explain the effect of this double layer on the intensity of the sum frequency. In one model, water molecules rotate in response to the field so that the water dipole has a larger projection onto the surface normal. In the other model, the oscillator strength of the infrared transition dipole is increased due to the enhanced symmetry breaking of the interface due to the field. These alternate models have very different implications for the configuration and mobility of water at the interface. In addition, distinguishing these models will provide insight into the vibrational modes of water in the hydrogen bonded water because the two hydrogen bonded peaks respond differently to ions in solution. The primary intensity enhancement is in the lower frequency peak with a much smaller perturbation of the higher frequency peak. If the interfacial water molecules actually rotate in response to the field, the rotated molecules present a different face to incoming reactants, potentially altering the rate of reactions at the interface.

For higher concentrations of ions, the Coulombic field extends sufficiently far that the anions and cations become associated into ion pairs or molecular complexes. These neutral species can penetrate to the very top monolayer of an aqueous solution. Water at the interface then is either tied up into hydrogen bonded complexes, or is displaced from the interface. In either case, the availability of water is reduced. If water is displaced it is unavailable for reaction until an adsorbate becomes incorporated into the bulk. If water is bound in hydrogen-bonded complexes, a potential reactant must pull water out of this complex, again reducing availability of water.

The latter portion of this contribution focuses on ammonia as a probe of the aqueous interface. At low concentration, ammonia titrates the dangling-OH bonds saturating the surface when approximately one-third of the dangling bonds are decorated with ammonia. Similarly, ammonia hydrogen bonds to water molecules isolated in a CCl_4 liquid environment. Without ammonia, water exists in CCl_4 as isolated water molecules. Addition of ammonia induces hydrogen bonding in the water molecules and increases the oscillator strength of the vibrational transitions. Isolated water molecules in CCl_4 show the effect of confinement in the liquid environment by a coupling of low frequency modes with the symmetric and antisymmetric stretch modes that is not observed in nonrestrictive environments.

References

1. M. Wilson, A. Pohorille: J. Chem. Phys. **95**, 6005-6013 (1991).
2. D. Michael, I. Benjamin: J. Chem. Phys.**107**, 5684-5693 (1997).
3. I. Benjamin : Acc. Chem. Res. 28, **233** (1995).
4. D. J. Tobias, P. Jungwirth, M. Parrinello: J. Chem. Phys. **114**, 7036-7044 (2001).
5. P. Jungwirth, D. J. Tobias: J. Phys. Chem. **B**, 10468-10472 (2001).
6. P. Jungwirth, D. J. Tobias: J. Phys. Chem. B **104**, 7702-7706 (2000).
7. P. Jungwirth: J. Phys. Chem. A **104**, 145-148 (2000).
8. X. D. Zhu, H. Suhr, Y. R. Shen: Phys. Rev. B **35**, 3047 (1987).
9. J. H. Hunt, P. Guyot-Sionnest, Y. R. Shen: Chem. Phys. Lett. **133**, 189-192 (1987).
10. Y. R. Shen: Nature. **337**, 519 (1989).
11. Q. Du, R. Superfine, E. Freysz, Y. R. Shen: Phys. Rev. Lett. **70**, 2313-2316 (1993).
12. C. Radüge, V. Pflumio, Y. R. Shen,: Chem. Phys. Lett. **274**, 140-144 (1997).
13. S. Baldelli, C. Schnitzer, M. J. Shultz, D. Campbell: J. Phys. Chem. B **101**, 10435-10441 (1997).
14. S. Baldelli, C. Schnitzer, M. J. Shultz, D. Campbell: J. Chem. Phys. Lett. **287**, 143-147 (1998).
15. S. Baldelli, C. Schnitzer, M. J. Shultz: Chem. Phys. Lett. **302**, 157-163 (1999).
16. S. Baldelli, D. Campbell, C. Schnitzer, M. J. Shultz: J. Phys. Chem. B. **103**, 2789-2795 (1999).
17. C. Schnitzer, S. Baldelli, D. J. Campbell, M. J. Shultz: J. Phys. Chem. A. **103**, 6383-6386 (1999).
18. C. Schnitzer, S. Baldelli, M. J. Shultz: Chem. Phys. Lett. **313**, 416-420 (2000).
19. C. Schnitzer, S. Baldelli, M. J. Shultz: J. Phys. Chem. B **104**, 585-590 (2000).
20. M. J. Shultz, C. Schnitzer, D. Simonelli, S. Baldelli: Int. Rev. Phys. Chem. **19**, 123-153 (2000).

21. M. J. Shultz, S. Baldelli, C. Schnitzer, D. Simonelli: J. Phys. Chem. B **106** (2002).
22. J. C. Conboy, M. C. Messmer, G. L. Richmond: J. Phys. Chem. **100**, 7617 (1995).
23. D. E. Gragson, G. L. Richmond: J. Chem. Phys. **107**, 9687-9690 (1997).
24. D. E. Gragson, G. L. Richmond: J. Phys. Chem. B **102**, 569-576 (1998).
25. M. C. Messmer, J. C. Conboy, G. L. Richmond: J. Am. Chem. Soc. **117**, 8039 (1995).
26. B. Dick, A. Gierulski, G. Marowsky, G. A. Reider: Appl. Phys. B **38**, 107-116 (1985).
27. B. Dick: Chem. Phys. **96**, 199-215 (1985).
28. C. Hirose, N. Akamatsu, K. Domen: Appl. Spec. **46**, 1051-1072 (1992).
29. C. Hirose, N. Akamatsu, K. Domen: J. Chem. Phys. **96**, 997-1004 (1992).
30. R. E. Muenchausen, R. A. Keller, N. S. Nogar: J. Opt. Soc. Am. **4**, 237-241 (1987).
31. N. Akamatsu, K. Domen, C. Hirose: Appl. Spec. **46**, 1051-102 (1992).
32. N. Bloembergen, P. S. Pershan: Phys. Rev. **128**, 606-622 (1962).
33. H. Chen, D. E. Irish: J. Phys. Chem. **75**, 2672-2681 (1971).
34. D. E. Irish, H. Chen: J. Phys. Chem. **74**, 3796-3802 (1970).
35. D. E. Irish, M. H. Brooker: Raman and Infrared Spectral Studies of Electrolytes, Clark, R. J. H. and Hester, R. E., Ed.; Heyden & Son: London, 1981, pp 212-311.
36. C. I. Ratcliffe, D. E. Irish: J. Phys. Chem. **86**, 4897-4905 (1982).
37. C. I. Ratcliffe, D. E. Irish Can: J. Chem. **63**, 3521-3525 (1985).
38. J. R. Scherer: The Vibrational Spectroscopy of Water, Clark, R. J. H. and Hester, R. E., Ed.; Heyden: Philadelphia, **5**, 149-216 (1978).
39. J. R. Scherer, M. K. Go, S. Kint: J. Phys. Chem. **78**, 1304-1313 (1974).
40. V. Buch, J. P. Devlin: J. Chem. Phys. **110**, 3437-3443 (1999).
41. J. P. Devlin, V. Buch: J. Phys. Chem. **99**, 16534-16548 (1995).
42. J. P. Devlin, V. Buch: J. Phys. Chem. B. **101**, 6095-6098 (1997).
43. J. P. Devlin, C. Joyce, V. Buch: J. Phys. Chem. A **104**, 1974-1977 (2000).
44. B. Rowland, N. S. Kadagathur, J. P. Devlin, V. Buch, T.Feldman, M. J.Wojcik: J. Chem. Phys. **102**, 8328-8341 (1995).
45. C. J. Tsai, K. D. Jordan: J. Phys. Chem. **97**, 5208-5210 (1993).
46. S. Baldelli, C. Schnitzer, M. J. Shultz, D. J. Campbell: *Sum Frequency Generation Study of Water at H_2SO_4 and Cs_2SO_4 Solutions*: Las Vegas, NV, 1997.
47. S. Baldelli, C. S. Schnitzer, M. J. Shultz, D. J. Campbell: *Probing H_2O Molecules At the Interface of H_2SO_4/H_2O Solutions Using Sum Frequency Generation*: Las Vegas, NV, 1997.
48. L. F. Phillips: Aust. J. Chem. **47**, 91-100 (1994).
49. D. H. Fairbrother, H. Johnston, G. Somorjai: J. Phys. Chem. **100**, 13696-13700 (1996).
50. S. Baldelli, C. Schnitzer, M. J. Shultz: J. Chem. Phys. **108**, 9817-9820 (1998).
51. O. K. Rice: J. Phys. Chem. **32**, 583-592 (1928).
52. N. G. McDeffie: Langmuir. **17**, 5711-5713 (2001).
53. G. Nathanson, P. Davidovits, D. Worsnop, C. Kolb: J. Phys. Chem. **100**, 13007 (1996).
54. Q. Shi, P. Davidovits, J. T. Jayne, D. R. Worsnop, C. E. Kolb: J. Phys. Chem. A **103**, 8812-8823 (1999).

55. E. Swartz, Q. Shi, P. Davidovits, J. T. Jayne, D. R. Worsnop, C.E.Kolb: J. Phys. Chem. A **103**, 8824-5533 (1999)
56. B. J. Finlayson-Pitts, J. N. Pitts Jr: Chemistry of the Upper and Lower Atmosphere; Academic Press: San Diego,(1999).
57. C. E. Kolb, D. R. Worsnop, M. S. Zahniser,P. Davidovits, L. F. Keyser, M. T. Leu, M. J. Molina, D. R. Hanson, A. R. Ravishankara: Laboratory Studies of Atmospheric Heterogeneous Chemistry; Barker, J., Ed.; World Scientific: Singapore, 771-875 (1995).
58. T. S. Bates, B. J. Huebert, J. L. Gras, F. B. Griffiths, P. A. J. Durkee: Geophys. Res. **103**, 16 (1998).
59. D. Simonelli, S. Baldelli, M. Shultz: J.Chem. Phys. Lett. **28**, 400-404 (1998).
60. D. J. Donaldson: J. Phys. Chem. A **103**, 62-70 (1999).
61. C. D. Bain, P. B. Davies, T. H. Ong, R. N. Ward: Langmuir. **7**, 1563 (1991).
62. D. Simonelli, M. J. Shultz: J. Chem. Phys. **112**, 6804-6816 (2000).
63. D. D. Nelson Jr, G. T. Fraser, W. Klemperper: Science. **238**, 1670-1674 (1987).
64. O. S. Binbrek, A. Anderson: Chem. Phys. Lett. **15**, 421 (1972).
65. J. J. Lagowski: The Chemistry of Non-Aqueous Solvents, II Acidic and Basic Solvents; Academic Press: New York, 1967; Vol. II.
66. W. B. Fischer, H. H. Eysel,: J. Mol. Struct. **415**, 249 (1997).
67. C. A. Plint, R. M. B. Small, H. L. Welsh: Can. J. Chem. **32**, 653 (1954).
68. L. B. Magnusson: J. Phys. Chem. **74**, 4221-4228 (1970).
69. M. Falk, E. Whalley: J. Chem. Phys. **34**, 1554 (1961).

85. [illegible] G. [illegible], J. T. [illegible], G. R. [illegible]: [illegible] Chem. A 108, [illegible] (1998)
86. [illegible] Chemistry of the Upper and Lower Atmosphere (Academic Press, San Diego 1999)
87. G. [illegible], D. [illegible], M. A. [illegible], J. Kaiser, [illegible] [illegible] M. [illegible] [illegible] [illegible] (1999)
88. [illegible] Phys. [illegible] 104 [illegible]
89. [illegible] Chem. Phys. Lett. 286, [illegible] 101 (1998)
90. [illegible] J. Phys. Chem. A 103, [illegible] (1999)
91. [illegible] (1991)
92. [illegible]
93. [illegible] Science 248, [illegible] (1990)
94. [illegible] Phys. Lett. [illegible] (1977)
95. [illegible] (1992)
96. [illegible]
97. [illegible]
98. [illegible] Phys. [illegible] (1982)

Part III

Atmospheric and Astrophysical Water and Ice

12 Physical Properties and Atmospheric Reactivity of Aqueous Sea Salt Micro-Aerosols

Pavel Jungwirth

Summary. Aqueous sea salt micro-aerosols play an important role in the heterogenous chemistry of the lower marine troposphere both in polluted and in remote areas. In particular, reactions with gases such as ozone or OH radicals leading to the release of molecular chlorine have been intensely studied, both experimentally and theoretically. Moreover, thin layers of sea water deposited on Arctic ice packs have been discovered to be a major source of reactive bromine species which destroy the surface ozone layer during polar sunrise. There is increasing evidence that the air-water interface is of a key importance in these chemical processes. Despite this, little has been known about the structure and physical properties of aqueous sea salt aerosols at a detailed, molecular level. Here, we summarize results of classical molecular dynamics, Car-Parrinello molecular dynamics and *ab initio* quantum chemistry calculations on concentrated aqueous sodium chloride and bromide solutions confined to cluster and slab geometries. The main questions addressed by the simulations concern the onset of NaCl ionic solvation in water clusters, transition from clusters to slabs, structure of solvation layers and degree of ion pairing in concentrated solutions with confined geometries. A key result of the simulations is the observation that polarizable halogen anions (chloride and bromide) are present at the air-water interface of bulk solutions in amounts sufficient for the heterogenous atmospheric chemistry to take place. The calculations also reveal that bromide actually exhibits surfactant activity, i.e. its concentration at the interface is higher than in the bulk. This is in accord with the observed enhanced atmospheric reactivity of aqueous bromide compared to chloride and with SEM experiments on wetting and re-drying of NaCl/NaBr co-crystals.

12.1 Introduction

Despite impressive progress in our understanding of atmospheric chemistry [1, 2] heterogenous processes involving rain or aqueous sea salt droplets, ice particles, and other aerosols are still in general poorly understood. The importance of including heterogenous chemistry in atmospheric modelling can be demonstrated using the spectacular failure of earlier computer models to predict the Antarctic ozone 'hole' due to the omission of heterogenous reactions occurring on polar stratospheric clouds [2]. Recently, a lot of attention has been directed to elucidation of the role of aqueous sea salt aerosols in processes leading to the release of chlorine into the lower marine troposphere

[3-11]. While the (ozone destructive) role of chlorine in the stratosphere is notoriously known its tropospheric role is less well established.

Field measurements revealed chlorine concentrations in the lower marine troposphere up to 150 ppt [3], which are more than an order of magnitude higher than predictions based on standard atmospheric models [12]. In the search for the heretofore unrecognized chlorine source, aerosol chamber experiments have been performed [4, 6]. In the experiment dry sea salt or NaCl microparticles have been humidified above the deliquescence point in the presence of ozone as polluting gas and UV radiation; subsequently, production of molecular chlorine has been observed.

Consequently, a chemical kinetic model has been suggested, involving the following four steps: i) ozone UV photolysis in the presence of water vapor leading to creation of OH radicals, either directly or via hydrogen peroxide, ii) uptake of OH into the aqueous sea salt (or rock salt) aerosol, iii) creation of a Cl^-...OH complex followed by creation of a Cl radical (by charge transfer to the naturally present hydronium cation) and, subsequently, Cl_2^-, iv) self-reaction of two Cl_2^- anions leading to formation of molecular chlorine, which diffuses and is eventually released from the aerosol. This kinetic model successfully reproduced the initial photochemical depletion of ozone; however, it underestimated the chlorine production by three orders of magnitude. The bottleneck turned out to be the charge-transfer step due a too low natural concentration of hydronium cations in roughly pH-neutral aerosols. This model could provide the observed chlorine concentration only when unrealistic acidities of the aerosols were assumed.

Another example of the role of aqueous sea salt in a confined geometry in atmospheric processes has been discovered recently. Indeed, thin layers of sea water deposited on ice packs have been found to be a major source of reactive halogen species during polar sunrise. Sea spray carried by winds settles on polar ice leading to an accumulating interfacial concentration of chloride and bromide during the polar winter. At the polar sunrise, a vigorous outburst of photochemistry is observed, leading to an almost complete destruction of ozone in the boundary layer on a timescale of hours [13-15]. Bromide, released from the snow packs in the form of an oxide has been shown to play a key role in polar tropospheric ozone destruction.

In the light of the above findings a new model, where chloride ions at the air-water interface play a prominent role, has been suggested [6]. It was, however, somehow disturbing that the new model is in conflict with standard theories of simple electrolytes and their surfaces [16]. Indeed, it has been a generally accepted textbook knowledge (albeit not supported by any direct experimental evidence) that atomic ions are repelled from the air-water interface by the electrostatic image forces; therefore, the interfacial layer is effectively devoid of ions [16-18]. In contrast, photoelectron spectroscopy measurements [19-21] and molecular dynamics (MD) simulations [22-28] performed in the last decade indicate that polarizable atomic anions such as chloride, bromide, or iodide reside at the surface of water clusters with several to several tens of solvent molecules.

The main purpose of the present study has been to investigate theoretically aqueous solutions of simple electrolytes such as sodium chloride or bromide in confined geometries corresponding to cluster or slab arrangements. In order to validate the basic assumption of the new atmospheric model (i.e., that heavier halides are present at the air-water interface of aqueous sea salt aerosols) we have in particular investigated the composition and structure of the interfacial layer. At the same time we have also addressed, by means of a pragmatic combination of *ab initio* calculations, Car-Parrinello and classical MD simulations, the following questions: i) how many solvent molecules are necessary for NaCl solvation in water clusters, ii) what structural changes accompany the transition from cluster to slab geometry of simple electrolytes, iii) what is the structure of solvation shells around ions at the interface compared to the bulk, iv) what is the degree of ion pairing in concentrated solutions with confined geometries and at what concentrations is the linear Debye-Hückel regime no longer valid, and, finally, v) what is the role of minor substituents, such as bromide, in aqueous sea-salt particles.

12.2 Systems and Computational Strategies

Typical sizes of aqueous sea salt aerosols range from 0.1 to 1 μm. The salt concentration in the aerosols is higher than that in sea water due to rapid water evaporation. As a matter of fact saturation is often reached or even exceeded, leading eventually to dry salt microparticles. In the present study we have investigated aqueous sodium chloride solutions (in some cases doped also with bromide) in confined geometries. The atmospheric micro-aerosols possess far too many particles to allow for a direct computational modelling at a molecular level. As a first step, we have investigated much smaller clusters, containing from six to several hundreds of water molecules. However, a much more realistic computational model with molecular representation of the atmospheric reality corresponds to a slab geometry with two roughly planar air-water interfaces. Therefore, most of our simulations were performed for slabs covering the whole salt concentration range from infinite dilution all the way to saturation.

The basic computational strategy consists in performing extended MD simulations with the quality of the force field checked and validated by Car-Parrinello and *ab initio* quantum chemical calculations for smaller systems of the same composition. The use of a polarizable potential model turned out to be essential for any realistic predictions. We have employed the SPC/POL model of water [29] and accounted explicitly for polarizabilities of the solute ions [30, 31]. MD simulations have been run at 300 K employing a time step of 1 fs. The OH bonds have been kept frozen using the SHAKE algorithm [32]. Typically, the systems have been equilibrated for 500 ps after which a relatively long sampling period of 0.5 - 1 ns followed. The slab condition has been realized by confining 864 water molecules and an appropriate number of ions into a box of 30×30×100 Å. Periodic boundary conditions have been applied and the particle mesh Ewald method has been used to calculate the electrostatic energies and forces [33]. The van der Waals interactions and the real space part of the Ewald sum have been truncated at 12 Å. All the classical MD simulations have been performed using the Amber code [34].

Car-Parrinello MD simulations have been performed for a water hexamer containing a single chloride anion. The electronic equations of motion [35] have been integrated using a time step of 0.121 fs and a fictious electron mass of 800 a.u. The electronic structure has been computed with the gradient corrected BLYP exchange-correlation functional [36, 37], treating the valence electrons explicitly, while the core electrons have been described using a pseudopotential [38]. The cluster has been placed in a periodic cubic box with a box length of 13.2 Å and an energy cutoff of 70 Ry for the plane-wave basis has been applied. Sampling has been based on 5 ps microcanonical trajectories corresponding to an average temperature of 250 K. Calculations have been performed using the CPMD program [39].

Quantum chemical *ab initio* calculations have been aimed at the investigation of NaCl ionic solvation in small clusters with up to six water molecules. Geometry optimizations have been performed at the MP2/6-31_{+Cl}G* level, while single point calculations for the allocated stationary points have been done at MP2/6-311_{+Cl}G(2d,p) and CCSD(T)/6-31_{+Cl}G* levels of theory. Zero-point energy as well as entropy effects at ambient conditions have been accounted for within the harmonic-oscillator/rigid-rotor model. The Gaussian98 program [40] has been employed.

12.3 Results and Discussion

12.3.1 NaCl Solvation in Small Water Clusters

An obvious question concerning salt solvation in confined aqueous systems is how many water molecules are necessary to dissolve a single rock salt molecule, i.e., to create a stable solvent-separated Na^+/Cl^- ion pair [41]. A direct extrapolation from the bulk saturated concentration leads to nine water molecules [42]. However, it is by no means obvious that such an extrapolation is justified since the molecular structure of small water clusters and the bulk differ significantly from each other. Calculations based on a continuum dielectric model and employing a potential of mean force between aqueous Na^+ and Cl^- from bulk liquid MD simulations predicted that at least twelve water molecules are necessary for ionic solvation of sodium chloride [41]. More recent electronic structure calculations on NaCl in water clusters with up to ten solvent molecules using the effective fragment potential method did not lead to location of a solvent-separated ion pair [43], possibly since the relevant part of the potential energy landscape has not been explored in sufficient detail.

In our study, we have launched an extensive *ab initio* search of potential energy surfaces of clusters with up to six water molecules containing a single NaCl species with special emphasis on geometries corresponding to a possible solvent-separated ion pair [44]. For clusters with five and less waters we have not been able to observe ionic solvation, the only stable structure corresponding to the contact ion pair, i.e., to an unsolvated NaCl molecule. However, for the water hexamer we have located as stable minima both the contact and solvent-separated ion pairs, the latter corresponding to ionic solvation. The two minima are roughly isoenergetic (within 2 kcal/mol) and separated by a relatively high barrier (3-5 kcal/mol).

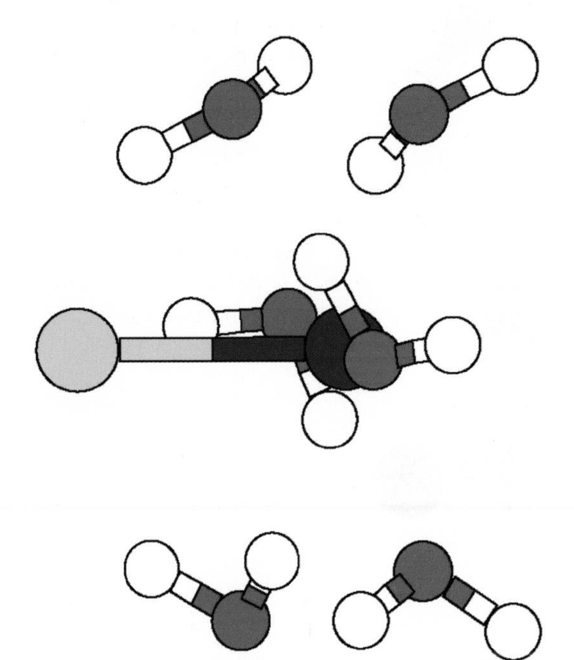

Fig. 12.1. Optimized structure of the $NaCl(H_2O)_6$ cluster with contact Na^+/Cl^- ion pair. O - red, H - white, Na^+ - brown, and Cl^- - yellow.

Figures 12.1, 12.2 and 12.3 depict the geometries of the three stationary points pertinent to NaCl solvation in the water hexamer. The unsolvated NaCl molecule (the contact ion pair) is shown in Fig. 12.1. The bond between sodium and chloride is well developed, although it is extended by some 10 % with respect to the gas phase (i.e. to 2.66 Å). The hydrogen bonded water structure is only moderately disturbed by the contact ion pair. The situation is very different for the almost isoenergetic solvent separated ion pair with Na-Cl distance of 4.43 Å (see Fig. 12.3). Here, all six waters are strongly

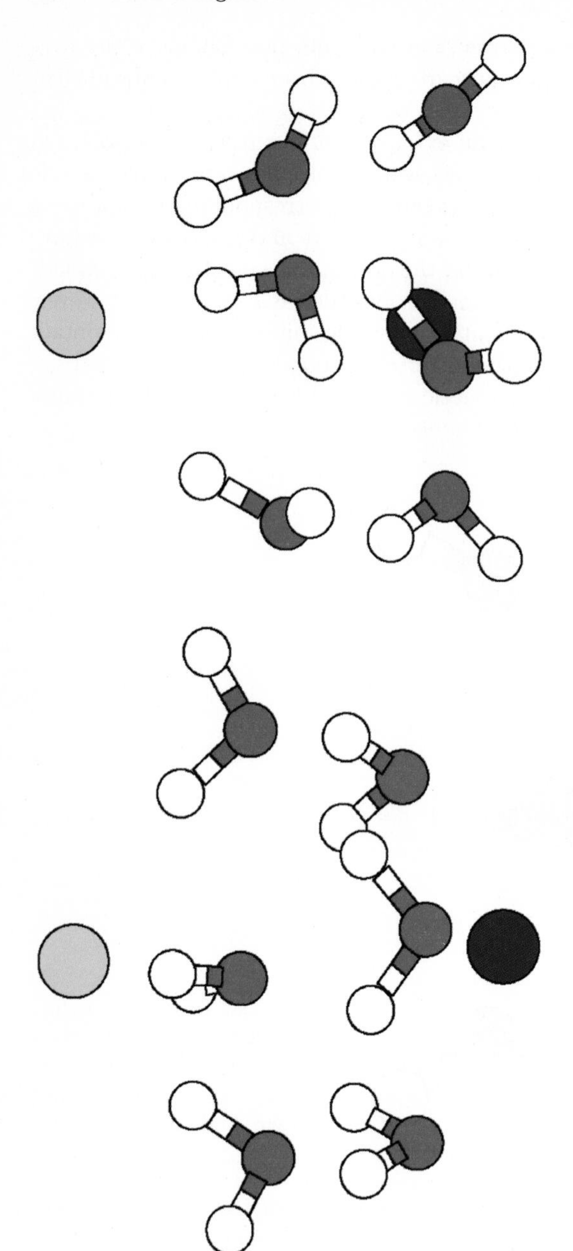

Fig. 12.2. Optimized saddle point, i.e., the barrier between the contact and solvent separated ion pairs. O - red, H - white, Na^+ - brown, and Cl^- - yellow.

Fig. 12.3. Optimized solvent separated ion pair corresponding to ionic solvation. O - red, H - white, Na^+ - brown, and Cl^- - yellow.

oriented in the electric field created by the two separated ions to the extent that virtually nothing is left from the water-water hydrogen bonding. Finally, the barrier separating the two stable minima corresponds to a Na-Cl distance of 3.61 Å and water geometries closer to that of the contact ion pair structure (see Fig. 12.2).

The principal result, i.e. the fact that only six waters are necessary for NaCl solvation, is not accidental. Indeed, water hexamer is the smallest water cluster which forms stable three-dimensional (non-cyclic) structures, which can then help in stabilizing a solvent separated Na^+/Cl^- ion pair. It is maybe not surprising that this result cannot be reached by simple extrapolation from the bulk. Finally, we can conclude, that ionic rock salt solvation occurs already at water cluster sizes much smaller than those pertinent to the atmosphere.

12.3.2 Water Clusters Saturated with NaCl

In a first attempt to bridge the gap between the single solvated NaCl molecule on one side and atmospheric micro-aerosols on the other side we have launched a series of studies of water clusters saturated with sodium chloride (salt:water number ratio of 1:9) containing 18 to 288 water molecules. For such systems rigorous quantum chemistry methods can no longer be applied since the cluster sizes are too large and the potential energy landscapes too complex. A remedy lies in employing empirical force fields within molecular dynamics simulations. It has been, however, demonstrated in earlier calculations that simulations of aqueous ions are rather sensitive to the quality of the force field, namely to the use of polarizable potentials [22, 26].

In our study, we have verified the force field against Car-Parinello simulations employing the BLYP density functional for chloride anion in/on water hexamer [45]. Among others, we have found a very good agreement for cluster geometries and chloride surface properties in particular between Car-Parrinello results and those obtained from classical MD with polarizable potentials. On the other hand, a non-polarizable force field yields results qualitatively different from those following from the Car-Parrinello simulations. Therefore, in all MD simulations discussed below we have employed a polarizable force field described in the previous section.

Our cluster results show that the polarizable chloride anion behaves very differently from the almost non-polarizable sodium cation. Indeed, while sodium ions are well solvated within the water clusters, chloride ions tend to spend a significant time at the air-water interface of the clusters. This observation is in accord with earlier simulations of a single chloride anion or sodium cation in medium-sized water clusters [22]. It is also important to note that the propensity of chloride ions for the interface is strong for all cluster sizes under investigation.

In the view of possible atmospheric surface reactivity we have found it very useful to quantify the amount of the ions at the air-water interface by

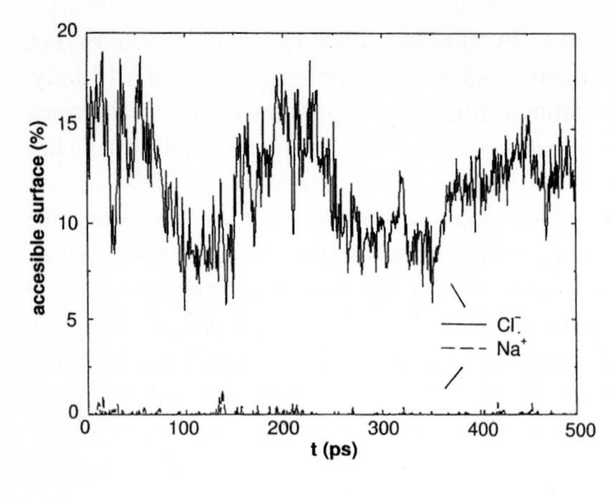

Fig. 12.4. Accesible surface (Percentage of total) of the chloride anion compared to that of sodium cation for the $(NaCl)_4(H_2O)_{36}$ cluster.

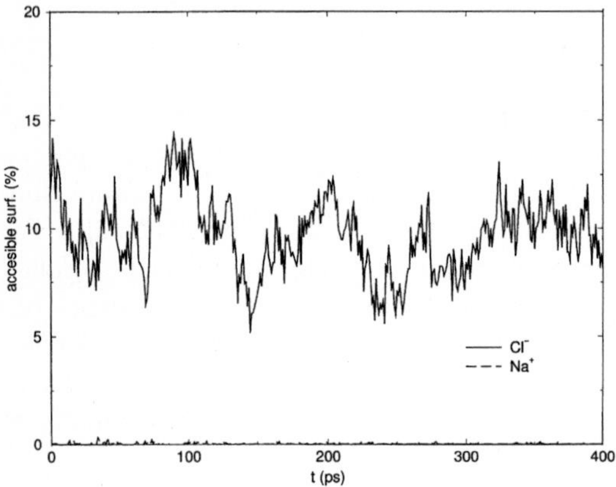

Fig. 12.5. Accesible surface of the chloride anion compared to that of sodium cation for the $(NaCl)_{32}(H_2O)_{288}$ cluster.

evaluating the averaged accessible surface of the two ionic species. There is a well established computational method for this based on rolling a spherical probe (of a diameter of 1.7 Å) over the interface at each time step of the simulation [46]. Figures 12.4 and 12.5 show the accessible surfaces of the sodium and chloride ions along extended MD trajectories for a small cluster with 36 water molecules and a large cluster possessing 288 water molecules. In both cases chloride anions occupy a significant portion of the surface (about 12 %), while sodium is almost absent from the air-water interface. The propensities for the interface for the two ionic species differ by two orders of magnitude and the affinity of chloride to the surface is primarily due to its large value of polarizability. Indeed, by switching to a non-polarizable force field the chloride accessible surface drops dramatically, by a factor of three.

12.3.3 Aqueous slabs containing NaCl

As mentioned before, an appropriate computational model of atmospheric aqueous sea salt aerosols of a micrometer size is a slab containing a sufficient amount of water molecules in order to simulate faithfully both the bulk and the interface (864 waters in our case), and the appropriate amount of ionic solute species. Similarly as in the laboratory experiment [6] we have in the first study included only the dominant solute, i.e. sodium chloride. Other minor impurities such as bromide will be discussed in the next section. We have explored the whole range of salt concentrations from infinite dilution to saturation [47], the latter being close to the laboratory as well atmospheric conditions.

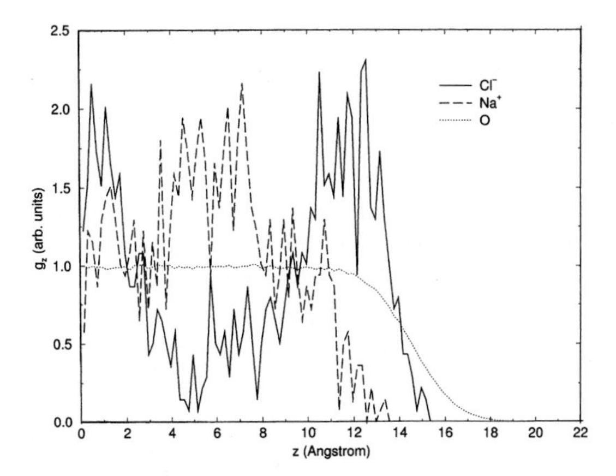

Fig. 12.6. Density profiles of chloride anions, sodium cations, and water oxygens in a water slab with NaCl at infinite dilution.

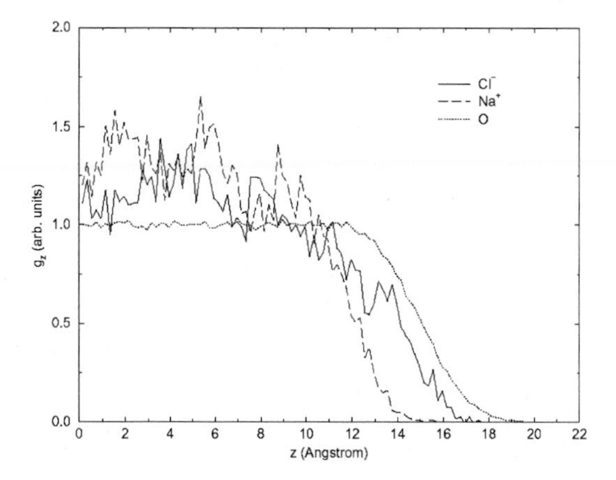

Fig. 12.7. Density profiles of chloride anions, sodium cations, and water oxygens in a water slab with NaCl at 1.2 M saturation.

We have employed two methods to quantify the abundance of the solvated ions in the water slabs. First, we have evaluated density profiles of sodium and chloride ions across the slab from the center to the surface by averaging over nanosecond MD simulations. Figures 12.6, 12.7 and 12.8 depict such density profiles on the background of the density of water oxygens for three

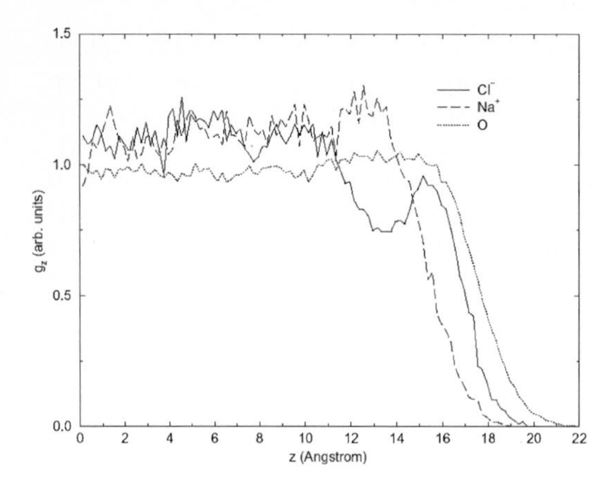

Fig. 12.8. Density profiles of chloride anions, sodium cations, and water oxygens in a water slab with NaCl at 6.1M saturation.

concentrations: "infinite dilution" (only one NaCl species), 1.2 M, and 6.1 M (saturation). In the whole concentration range a different behavior of the two ionic species is immediately apparent. While sodium cations are always fully solvated in the interior of the water slab, chloride anions penetrate significantly towards the air/water interface. The reason the signal for low salt concentrations is more noisy turns out to be purely technical - there are simply less ions to sample over.

From the point of view of a possible atmospheric chemical reactivity of ions at the surface of aqueous sea salt micro-aerosols it is better to quantify the propensity to the interface by evaluating the accessible surface of the ions along the MD trajectory (as demonstrated for cluster systems in Figs. 12.4 - 12.5). Figure 12.5 shows the dramatic, two orders of magnitude difference in accesible surfaces between Cl^- and Na^+. While at 6.1 M chloride occupies some 12 % of the surface (i.e. exhibits almost the same surface activity as an average water molecule) sodium is practically absent from the air water interface. Most of the difference between the two species can be attributed to the fact that chloride is a soft polarizable ion, while sodium is a hard non-polarizable ion. Indeed, MD simulations employing a non-polarizable force field show a threefold reduction of the propensity of chloride for the surface (see Fig. 12.9).

An important question with respect to the atmospheric reactivity, but having also broader consequences for our understanding of concentrated salt solutions, is what is the degree of ion pairing. In other words, what is the ratio between solvated and unsolvated NaCl species at a given concentration, and how does this ratio differ at the surface from the bulk value. Figure 12.10 shows the percentage of chloride anions paired with sodium cations along 500 ps MD trajectories for concentrations of 1.2 M and 6.1 M (saturation). We see that at 1.2 M the degree of ion pairing, reaching about 5 % both in the bulk and at the air/water interface (Fig. 12.10). At this concentration the aqueous solution dominantly contains separated solvated ions, in accord

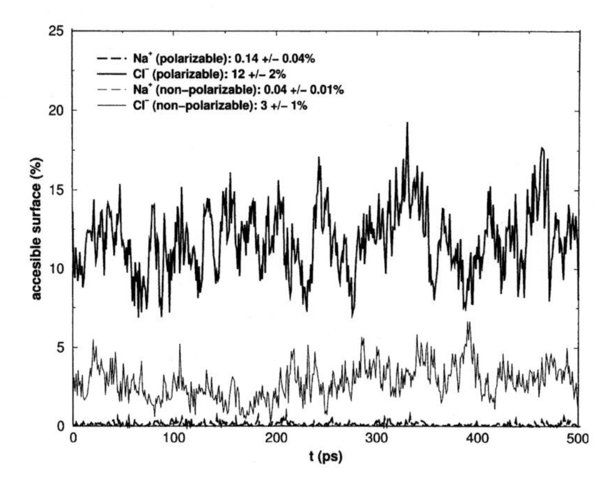

Fig. 12.9. Accesible surface of the chloride anion compared to that of sodium cation for a water slab saturated with NaCl. Comparison of Cl^- results using a polarizable (top line) and non-polarizable (middle line) force fields.

with the Debye-Hückel model of strong electrolytes. However, the situation is dramatically different when the concentration approaches saturation. At 6.1 M, three out of four chloride anions are paired with Na^+ in the bulk of the solution (see Fig. 12.11). Interestingly, ion pairing of chloride drops to 54 % at the interface, a clear consequence of the much lower concentration of sodium cations in the interfacial layer.

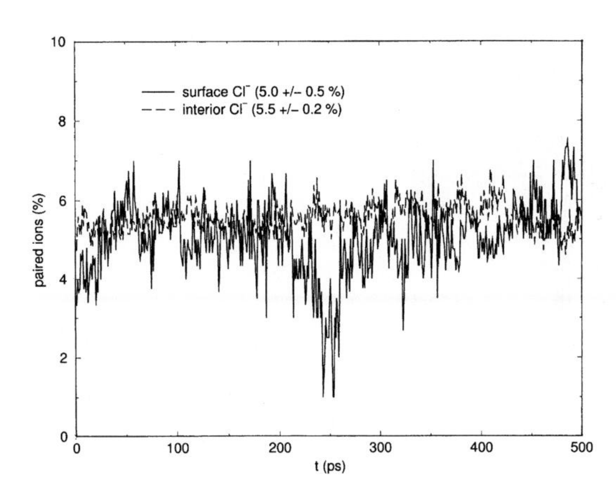

Fig. 12.10. Degree of ion pairing of chloride anions with sodium cations in a water slab with NaCl at 1.2 M.

Another structural question concerns the solvation number, i.e. the number of waters in the first solvation shell, for chloride in the bulk vs. at the interface as a function of salt concentration. At infinite dilution the number of water molecules in the immediate vicinity of a bulk solvated chloride anion amounts roughly to six. We see from Fig. 12.12 that already at 1.2 M concentration the chloride solvation number drops to 5.5. At saturation the solvation number further decreases to 4.5 (see Fig. 12.13). As expected, the solvation number for chloride at the interface is always smaller than in the bulk. However, the difference is rather small, indicating that, even in the

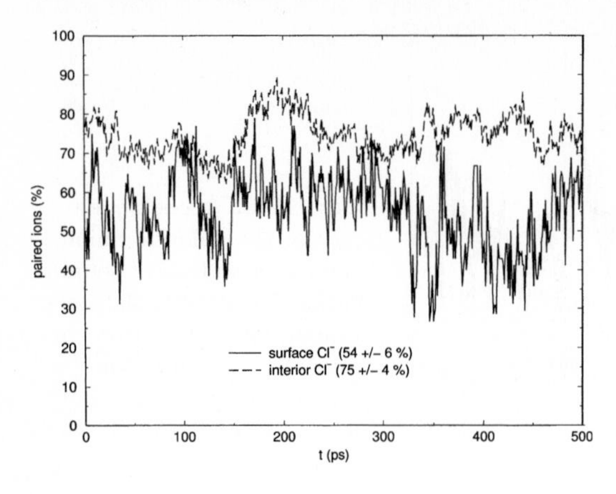

Fig. 12.11. Degree of ion pairing of chloride anions with sodium cations in a water slab with NaCl at 6.1 M (saturation).

vicinity of the surface, chloride anions form an almost complete water solvation shell. There is also a clear connection between solvation numbers and the degree of ion pairing - upon pairing of chloride, water molecules in the first solvation shell are being replaced by sodium cations.

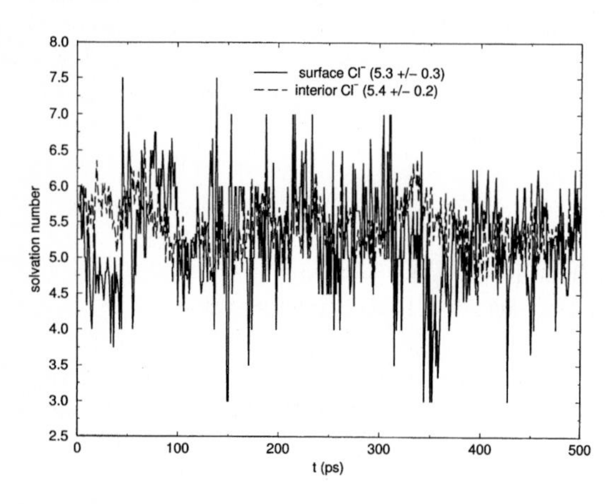

Fig. 12.12. Aqueous chloride solvation numbers in a water slab with NaCl at 1.2 M.

12.3.4 Aqueous slabs with a mixture of NaCl and NaBr

Sodium chloride is the dominant solute contained in sea water and aqueous sea salt aerosols. However, other solvated species also exhibit important atmospheric reactivity. Among these, bromide anions, the natural concentration ratio of which with respect to chloride is 1:650, play a prominent role. For example, it has been discovered recently that the destruction of the surface layer of ozone in the Arctic during polar sunrise is due to bromine chemistry [13-15]. Interestingly, it seems that, relative to chlorine, bromine reactivity is

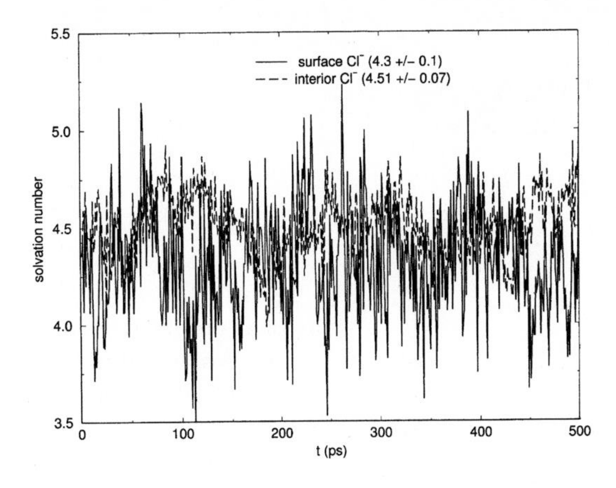

Fig. 12.13. Aqueous chloride solvation numbers in a water slab with NaCl at 6.1 M (saturation).

enhanced more than would follow from the direct comparison of the relevant rate constants. Since our previous results suggested that ionic polarizability is responsible for the propensity for the surface, we have speculated that bromide relative (with respect to chloride) and absolute concentration at the air/water interface could be enhanced due to its larger value of polarizability.

In order to mimic the atmospherically relevant situation we have taken an aqueous slab saturated with NaCl from our previous simulations and "transmuted" one of the chlorides into bromide. We have actually performed two simulations; one with initial position of the bromide anion at the interface and another with bromide initially near the center of the slab. Results of these two simulations, i.e. the time evolution of the distance of bromide from the center of the slab, are presented in Fig. 12.14. The emerging picture is quite clear. If bromide is initially placed at the interface, it tends to stay there, while bulk solvated bromide finds its way to the surface within 500 ps. Bromide and chloride differ only marginally in their values of the ionic radii, however, bromide is significantly more polarizable (soft). It is obviously the difference in polarizabilities and the related energy gain from asymmetric solvation, which drives bromide to the interface.

In order to obtain better statistics and to quantify the difference between surface propensities of the two halides we have performed two sets of simulations; first for a slab doped with 0.6 M of NaCl and 0.6 M of NaBr, and second for a more concentrated slab with 3 M of sodium chloride and 3 M of sodium bromide. Figure 12.15 shows representative snapshots (top views) of the surfaces of the slabs from the two simulations. Qualitatively, the emerging picture is clear already from these snapshots. Namely, both halides exhibit propensity for the interface; however, that of the more polarizable bromide is much stronger. Quantitative analysis shows that bromide is actually a surfactant, it's interfacial concentration being roughly twice the bulk value [48]. At lower concentrations of the salt mixture a linear regime prevails, i.e. the

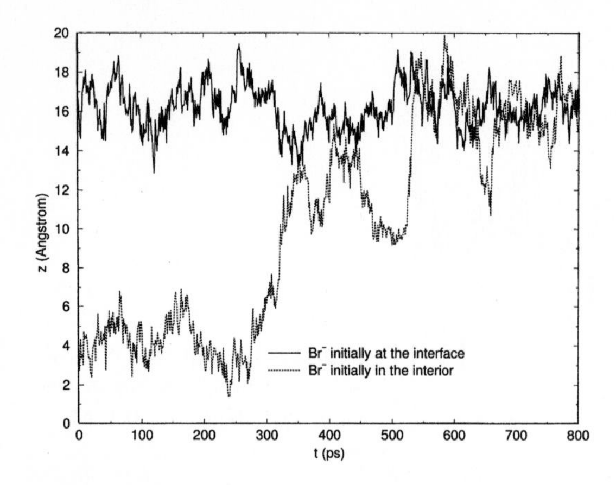

Fig. 12.14. Time evolution of the position of a bromide anion in a water slab saturated with NaCl, placed initially at the surface or near the center of the slab.

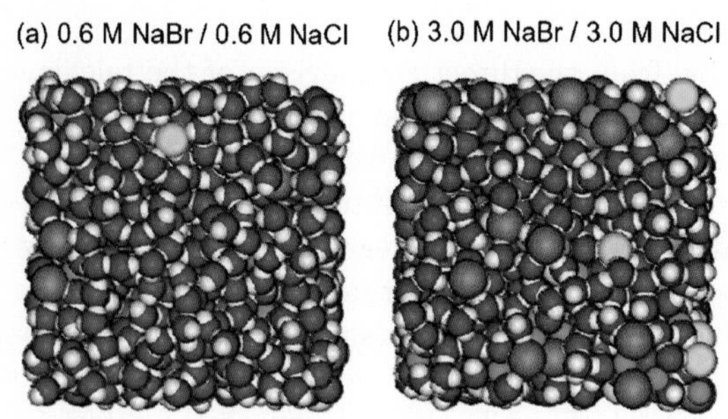

Fig. 12.15. Top views of representative snapshots from MD simulations of mixed NaBr/NaCl aqueous slabs at two different concentrations. Br^- ions are red in color.

ionic interfacial concentrations are roughly the same as one would obtain from separate simulations of single salt (NaCl or NaBr) solutions at the appropriate concentration. However, the simulation of the highly concentrated slab shows non-linear effects, namely that bromide further replaces chloride at the interface and it's surface concentration exceeds three times that in the bulk. Also, due to a significant accumulation of negative charges at the interface, a larger amount of sodium cations is dragged towards the surface than in the low concentration solution.

Finally, Fig. 12.16 shows the accessible surfaces of the three ionic species along the MD trajectory for the concentrated mixed solution. We see again that bromide is a clear winner occupying more than 30 % of the air/water interface. In contrast, chloride covers less than 5 % and sodium cations less than 1 % of the surface area. The high absolute and relative (to chloride) propensity of bromide for the interface is in accord with recent SEM experiments, which show strongly increased surface concentration of sodium bromide upon wetting and re-drying of NaCl/NaBr co-crystals [49].

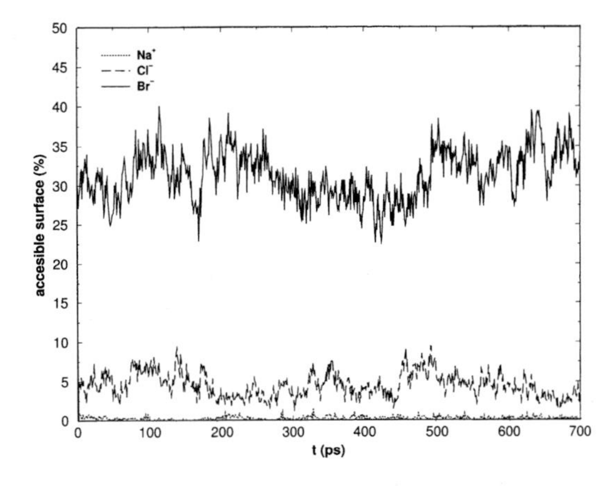

Fig. 12.16. Accessible surfaces of bromide and chloride anions, and sodium cations for a mixed aqueous slab containing 3 M of NaBr and 3 M of NaCl.

12.4 Conclusions

The principal result of the present computational study is that the propensity for the air/water interface of polarizable ions (such as heavier halides) confirmed both experimentally and theoretically for aqueous clusters pertains also to surfaces of extended salt solutions in the whole concentration range. The determining factor is the value of the polarizability of the solvated ion - the larger the polarizability, the more the ion can profit from asymmetric solvation at the air/water interface. Our findings, which are indirectly supported by several experiments [6, 7], contradict the standard Onsager model of the surface of simple electrolytes [16], which has been developed for solvated point charges and does not take polarization effects into account.

Results of quantification of the propensity of chloride anions for the surface of concentrated bulk salt solutions suggest a new kinetic model of photochemical reactivity of salt aerosols in the presence of ozone leading to a release of molecular chlorine [6]. Predictions based on this model, which involves heterogenous processes at the air/water interface of the aerosol, are in quantitative accord both with laboratory and field measurements [6, 50, 51]. Present simulations of mixed NaCl/NaBr aqueous solutions, which show a strongly enhanced interfacial propensity of bromide compared to chloride, have also direct implications for ozone-destructing bromine chemistry above Arctic snow packs during polar sunrise [52].

Acknowledgments. Most of the calculations have been carried out together with Prof. Douglas J. Tobias from UC Irvine. The author acknowledges a fellowship from the Lady Davis Trust and the hospitality of Profs. R. Benny Gerber and Victoria Buch during his sabbatical at the Hebrew University of Jerusalem. The Center for Complex Molecular Systems and Biomolecules in Prague is supported by the Czech Ministry of Education via a grant No. LN00A032.

References

1. F. Pitts, B. J. Pitts, J. N. Jr: Chemistry of the upper and lower atmosphere, Academic Press: San Diego, (2000).
2. R. P Wayne: Chemistry of atmospheres, Oxford University Press: New York, (2000).
3. C. W. Spicer, E. G. Chapman, B. J. Finlayson-Pitts, R. A. Plastridge, J. M. Hubbe, J. D. Fast, C. M. Berkowitz: Nature. **394**, 353 (1998).
4. K. W. Oum, M. J. Lakin, D. O. DeHaan, T. Brauers, B. J. Finlayson-Pitts: Science. **279**, 74 (1998).
5. J. Hirokawa, K. Onaka, Y. Kajii, H. Akimoto: Geophys. Res. Lett. **25**, 2449 (1998).
6. E. M. Knipping et al.: Science. **288**, 301 (2000).
7. F. Pitts, B. J., J. C. Hemminger: J. Phys. Chem. A. **104**, 11463 (2000).
8. V. Glasow, R. R. Sander: Geophys. Res. Lett. **28**, 247 (2001).
9. M. O. Andreae, P. J. Crutzen: Science. **276**, 1052 (1997).
10. R. Sander, Y. Rudich, V. Glasow, R. Crutzen, P.J: Geophys. Res. Lett. **26**, 2857 (1999).
11. A. R. Ravishankara: Science. **276**, 1058 (1997).
12. V. R. Crutzen, P. J., R. Sander: Nature. **383**, 327 (1996).
13. H. Boudries, J. W. Bottenheim: Geophys. Res. Lett. **27**, 517 (2000).
14. B. Ramacher, J. Rudolph, R. Koppmann : J. Geophys. Res. **104**, 3633 (1999).
15. P.A. Aryia, B. T. Jobson, R. Sander, H. G. W. Niki, J. F. Hopper, K. G. Anlauf: J. Geophys. Res. **103**, 13 (1998).
16. L. Onsager, N. Samaras, N. T: J. Chem. Phys. **2**, 528 (1934).
17. N. K Adam: The Physics and Chemistry of Surfaces, Oxford University Press: London,(1941).
18. J. E. B. Randalls: Phys. Chem. Liq. **7**,107 (1977).
19. G. Markovich, R. Giniger, M. Levin, O. Cheshnovsky: J. Chem. Phys. **95**, 9416 (1991).
20. G. Markovich, S. Pollack, R. Giniger, O.Cheshnovsky: Reaction dynamics in clusters and condensed phases, Jortner, J. (ed.), Kluwer: Amsterdam, 13 (1994).
21. G. Markovich, S. Pollack, R. Giniger, O. Cheshnovsky: J. Chem. Phys. **101**, 9344 (1994).
22. L. Perera, M. L. Berkowitz: J. Chem. Phys. **95**, 1954 (1991).
23. L. Perera, M. L. Berkowitz: J. Chem. Phys. **99**, 4222 (1993).
24. L.Perera, M. L. Berkowitz: J. Chem. Phys. **100**, 3085 (1993).
25. L. S. Sremaniak, L. Perera, M. L. Berkowitz: Chem. Phys. Lett. **218**, 3779 (1994).
26. L. X. Dang, B. C. Garrett: J. Chem. Phys. **99**, 2972 (1993).
27. L. X. Dang, D. E. Smith : J. Chem. Phys. **99**, 6950 (1993).
28. L. X. Dang: J. Chem. Phys. **110**, 1526 (1999).
29. J. Caldwell, L. X. Dang, P. A. Kollman: J. Am. Chem. Soc. **112**, 9144 (1990).
30. G. Markovich, L. Perera, M. L. Berkowitz, O. Cheshnovsky: J. Chem. Phys. **105**, 2675 (1996).
31. Jr. MacKerell et al.: J. Phys. Chem. B **102**, 3586 (1998).
32. J. P. Ryckaert, G. Ciccotti, H. Berendsen, J. C: J. Comput. Phys. **23**, 327 (1977).

33. U. Essmann, L. Perera, M. L. Berkowitz, T. Darden, L. G. Pedersen: J. Chem. Phys. **103**, 8577 (1995).
34. D. A. Case et al.: AMBER 6 Univ. of Calif., San Francisco. (1999).
35. R. Car, M. Parrinello: Phys. Rev. Lett **55**, 2471 (1985).
36. A. D. Becke: Phys. Rev. A **38**, 3098 (1988).
37. C. Lee, W. Yang, R. C. Parr: Phys. Rev. B **37**, 785 (1988).
38. N. Troullier, J. Martins: Phys. Rev. B **43**, 1993 (1991).
39. J. Hutter, P. Ballone, M. Bernasconi, P. Focher, E. Fois, S. Goedecker, M. Parrinello, M. Tuckerman CPMD, Version 3.0 (MPI für Festkörperforschung, Stuttgart, 1997).
40. M. J. Frisch, et al.: Gaussian Inc. Pittsburgh, PA, (1998).
41. G. Makov, A. Nitzan: J. Chem. Phys. **96**, 2965 (1992).
42. G. Gregoire, M. Mons, C. Dedonder-Lardeux, C. Jouvet: Eur. Phys. J D **1**, 5 (1998).
43. C. P. Petersen, M. S. Gordon: J. Phys. Chem. A **103**, 4162 (1999).
44. P. Jungwirth: J. Phys. Chem. A **104**, 145 (2000).
45. D. J. Tobias, P. Jungwirth, M. Parrinello: J. Chem. Phys. **114**, 7036 (2001).
46. B. Lee, F. M. Richards: J. Mol. Biol. **55**, 379 1971.
47. P. Jungwirth, D. J.Tobias: J. Phys. Chem. B **104**, 7702 (2000).
48. P. Jungwirth, D. J. Tobias: J. Phys. Chem. B **105**, 10468 (2001).
49. S. Ghosal, A. Shbeeb, J. C. Hemminger: Geophys. Res. Lett. **27**, 1879 (2000).
50. O. W. Wingenter, D. R. Blake, N. J. Blake, B. C. Sive, F. S. Rowland: J. Geophys. Res. **104**, 21819 (1999).
51. F. B. Griffiths, T. S. Bates, P. K. Quinn, L. A. Clementson, J. S. Parslow: J. Geophys. Res. **104**, 21649 (1999).
52. T. Koop, A. Kapilashrami, L. T. Molina, M. J. Molina: J. Geophys. Res. **105**, 26393 (2000).

13 Interactions and Photochemistry of Small Molecules on Ice Surfaces: From Atmospheric Chemistry to Astrophysics

Andrew B. Horn and John R. Sodeau

13.1 Introduction

Ice is ubiquitous in the environment, from solid ice at the planet surface to small particles in the atmosphere. The recently identified reactions which occur on the surface of polar stratospheric clouds (PSCs) in the Antarctic winter show that far from being inert, ice is an important surface for heterogeneous atmospheric reactions [1]. Indeed, the majority of studies performed in the laboratory indicate that the availability of accessible water for reaction and product stabilisation is the key to understanding heterogeneous processes on surfaces ranging from sea-salt to sulfuric acid aerosols. Over the last 10-15 years, a wide range of laboratory methods, broadly divided into those which probe surface reaction kinetics and those which probe surface reaction mechanisms [1,2], have been used to study processes relevant to atmospheric ice chemistry. Surprisingly, there are few examples where both mechanism and kinetics have been directly obtained simultaneously, and much of the original ideas concerning the nature of the ice/gas interactions were made by inference from kinetic trends with temperature and pressure. In the first part of this chapter, the nature of the interaction between atmospherically relevant molecules and ice is described, as seen through the application of infrared spectroscopy to the study of thin films of condensed ice at low temperatures. This provides the opportunity to create controlled ice surfaces and to trap and characterise intermediates and fundamental processes whilst reducing or eliminating many of the problems which arise in the presence of competing adsorption processes in complex systems. Following on from this, a discussion of the effect of ice upon the heterogeneous photochemistry of small molecules is presented. This is an area driven by two rather widely separated areas: atmospheric pollution processes and astrophysical interactions on interstellar dust.

Heterogeneous processes in the atmosphere on ice are currently receiving intense interest from laboratory studies, modellers and field measurement groups because it is now clear that they produce significant perturbations to many of the long-established gas-phase cycles. Of particular interest to stratospheric chemistry are the reactions of the principal halogen and nitrogen reservoir species with/on ice. Much is now known about the processes whereby essentially stable molecules such as chlorine nitrate ($ClONO_2$) and hydrogen chloride (HCl) and their bromine and iodine analogues are recycled into active species such as ClO and Cl, as well as the mechanisms by which the nitrogen oxides which normally sequester these active species are removed. Lower down in the atmosphere, there is much more uncertainty in the role and reactivity of the large amounts of ice present in the upper troposphere as cirrus clouds and aircraft contrails. Work in this area is currently being driven by a significant amount of evidence from field studies that there may be a heterogeneous component adding to the gas phase cycles involving small oxygen-containing volatile organic molecules such as acetone, acetic acid, methanol, ethanol and formaldehyde adsorbed on ice surfaces [3].

The interactions of small molecules such as those described above with ice surfaces are readily studied using infrared spectroscopy. The sensitivity of the technique along with its capability for *in situ* determination of surface composition and environment gives unprecedented access to the nature of ice surface interactions through fragment specific vibrational fingerprints and absorption band shifts. Bulk and surface processes can be characterised and detailed mechanistic models can be constructed and tested over a range of conditions. To date, this methodology has been applied by us for a number of fundamental studies of the interaction of atmospherically relevant molecules including chlorine nitrate, dinitrogen pentoxide, hypochlorous and hydrochloric acids (and to a lesser degree, their bromine analogues) with ice surfaces. To set these results in context, we first give an outline of the experimental techniques required to produce and support ice films with known and controllable surface properties, along with a simple description of the optical requirements and limitations for the observation of surface features. Studies of the binding of small molecules through hydrogen bonding to an ice surface is used to give insight into the physical processes which control initial sticking probabilities, surface accommodation coefficients, surface reaction mechanisms (*i.e.*, direct or precursor-mediated adsorption) and the desorption of surface reaction products. This treatment is further extended to cover the hydrolysis of atmospherically relevant adsorbates by ice surfaces and the role of the ice surface in both facilitating the reaction and the stabilisation of reaction products and intermediates. The latter effect is particularly important given the ionic nature of many of the observed reaction mechanisms. Reactions between adsorbed species and gaseous reactants are also discussed. This is followed by an examination of ice-surface mediated photochemistry, an area which is little studied but of potentially high relevance to areas of the atmosphere ranging from snowpack NO_x emission to stratospheric ozone chemistry. In a wider context, we also touch upon the potential effect of photochemistry in the production or loss of species on cometary material and ice in the interstellar medium.

13.2 Experimental aspects of the study of ice surface interactions using infrared spectroscopy

13.2.1 Optical considerations

For any material, the production of appropriate substrates for the spectroscopic study of surface chemistry requires a careful consideration both of the optical properties of the materials involved and of the optical geometries available if one is to attain sub-monolayer sensitivity. In the case of ice, this is something of a challenge, since water itself has a strong absorption spectrum in the mid-IR. Transmission and reflection are both possible and have been applied with varying degrees of success. For example, Tolbert et al. used IR transmission at normal incidence through thin ice films supported on a cooled silicon wafer to investigate the formation of various nitric acid hydrates of atmospheric relevance using films of between 1 – 10 μm to obtain optimal absorbance values for the major bands [4,5]. However, even on a 1 μm thick film, the ratio of the number of molecules in the surface compared to the bulk is of the order of 0.002, making surface specific features difficult to identify. The poor sensitivity as a result of the small number of surface species contributing to the overall signal in this configuration is compounded by the unfavourable optical properties of Si, which with a refractive index of 3.44 at 2000 cm^{-1} has a reflectivity of ca. 30% at normal incidence (although this limitation can be reduced by transmission through the Si substrate at the Brewster angle, ca. 75°) [6]. Improvements in surface specificity can be made by the use of thinner films on substrates with more favourable IR transparency, but sensitivity to sub-monolayer concentrations in transmission remains low, even with highly sensitive modern FTIR spectrometers. Additionally, there are potential problems in the use of ultrathin (1-5 monolayer) films as mimics as the surface structure is likely to be considerably different to that of the normal surface termination of the predominant bulk phases.

An alternative approach to transmission is to use a reflected IR beam to sample the surface of a condensed film in a manner analogous to that employed in the spectroscopic observation of species adsorbed on the surface of metals. For highly reflective surfaces, reflection-absorption infrared spectroscopy (RAIRS) has contributed significantly to our fundamental understanding of heterogeneous catalysis by metals [7]. RAIRS relies upon both the high underlying reflectivity of metal surfaces and an interaction between vibrating dipoles and the metal conduction electrons at grazing incidence to give a significant signal enhancement, permitting its use in the observation of sub-monolayer concentrations of surface species. An adaptation of the RAIRS method in which a composite substrate consisting of a flat metallic surface upon which an ultrathin (20 nm) ice film has been deposited provides an ideal method for studying ice surface chemisty [2]. Ice films of 20 nm thick have a surface to bulk ratio of *ca.* 0.1, and in conjunction with high light through-

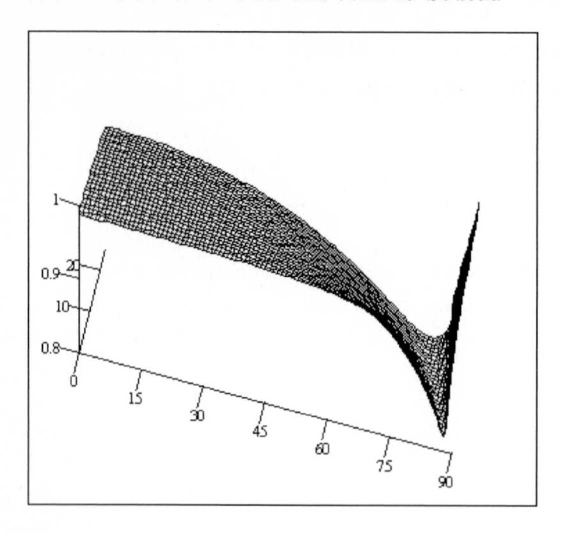

Fig. 13.1. Calculated infrared reflectivity of a composite ice/metal substrate (vertical axis) as a function of angle of incidence (0 – 90° on horizontal axis) and ice film thickness (0 – 25 nm into page). Optical constants used were Au = 3.0 + 30i, H_2O = 1.5 + 0.1i at 2000 cm^{-1} (5 μm).

put and metal-enhanced spectroscopic sensitivity, this yields a submonolayer sensitivity for species with reasonably strong absorption bands.

This gain in surface sensitivity is accompanied by an optical effect, the so-called "metal surface selection rule", which dictates that only vibrating dipoles with a component perpendicular to the surface (p-polarised) are observed and those parallel to the surface are suppressed [7]. For isotropic films such as condensed ice in which the component molecules have a range of orientations with respect to the surface plane, this is not a significant drawback. In RAIRS on single crystal metal surfaces, the maximum enhancement of p-polarised IR is observed to be sharply peaked at grazing incidence. The presence of a dielectric such as ice perturbs this behaviour and in general reduces the angle at which maximum enhancement occurs and also leads to a reduction in the magnitude of the enhancement. These factors are strongly dependent upon the ice film thickness and are readily calculated from classical thin film optics [8]. Typically for an ice film of 20 nm thickness upon a gold substrate, the optical angle shifts to ca. 78° (compared to 88° for pure gold). Figure 13.1 shows a sample calculation of this effect on the p-polarised reflectivity of ice on Au. Further complications arise for films thicker than ca. 100 nm as a result of the wavelength and distance-dependence of the optical factors which give rise to the metal surface selection rule. Above such thicknesses (comparable to the wavelength of the IR radiation used), the p-polarised response is attenuated whereas the s-polarised response is enhanced. This leads to complicated lineshapes (i.e. each individual absorption band is a composite of the s- and p-polarised response) which make detailed vibrational mode analysis of resonances arising from either the substrate or any absorbates rather less than straightforward in the region of the OH bands [9,10].

13.2.2 Ice film deposition

Producing appropriate mimics for atmospheric ice surfaces provides a further challenge for the study of reactive uptake on its surface. Ice has a wide range of phases and secondary structures, and although the atmospheric forms of ice are reasonably well documented, reproducing them in the laboratory is far from straightforward. The need for a reproducible and well-characterised ice surface for experimentation is highlighted in recent work by Rossi et al. [11], who have demonstrated that the rate of diffusion of HCl in ice (after reactive uptake) is strongly dependent upon the form of the ice and the method by which it is deposited in the experimental apparatus. Similarly, it has also been demonstrated that aspects of the physical binding of small molecules to ice is strongly dependent upon the nature of the ice surface, which in turn depends upon the nature of the ice film used (*i.e.* crystalline or amorphous). In the light of these observations, it is therefore essential in any experiment to produce a reliably reproducible ice surface with known bulk and surface structure. There are a number of recent experimental [12-15] and computational [16-18] studies of the surface structure of laboratory-prepared ice films which provide a guide to the conditions necessary for the formation of suitable mimics, and related studies of the effects of both substrate morphology [19-21] and interactions between adsorbed water oligomers (dimers, trimers, hexamers *etc.*) [22-24] further serve to increase our understanding of how to produce well-defined ice films for experimentation. In our studies, *in vacuo* vapour deposition from a gaseous H_2O source onto a cooled gold substrate is utilised. The studies described above show the structure of films deposited in this manner is effectively controlled by the deposition temperature, H_2O partial pressure and subsequent annealing procedures. In the majority of the thin film experiments from our laboratories discussed below, exposure of the gold substrate to $1\text{x}10^{-7}$ mbar H_2O at 110 K results in the formation of an amorphous film with an open, snow-like structure. Annealing this to *ca.* 145 K causes a transformation to a polycrystalline thin film. Reactive and non-reactive (physical) uptake onto these surfaces are described below in Section 3.

13.2.3 Experimental apparatus

In order to support thin condensed ice films for uptake studies in a manner suitable for spectroscopic observation and to expose them to appropriate concentrations of atmospheric trace gases, the apparatus shown schematically in Fig. 13.2 is utilised in our experiments. It consists of a cylindrical stainless steel vacuum chamber pumped by an oil-vapour diffusion pump to better than $1\text{x}\ 10^{-9}$ mbar. The chamber is optically coupled to an FTIR spectrometer using transfer optics and vacuum compatible KBr windows to focus the infrared beam onto a flat, metal substrate (either polycrystalline Au foil or single crystal Pt(111)) located at the centre of the chamber. The optics are

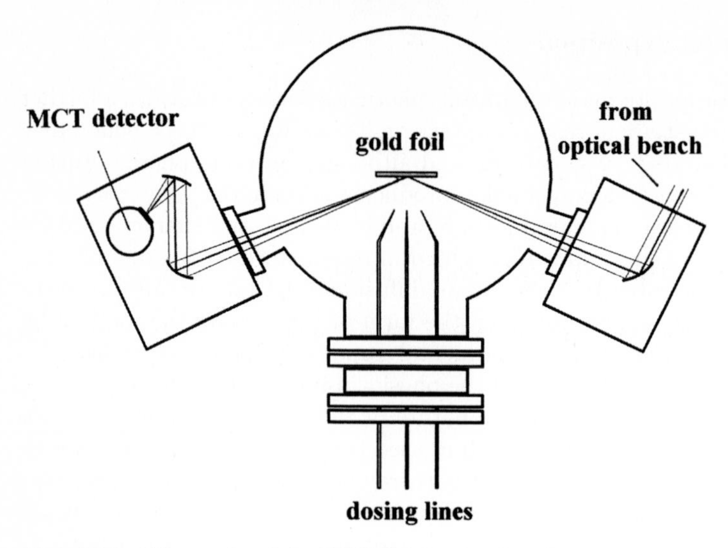

Fig. 13.2. Schematic of the RAIRS vacuum chamber.

aligned for RAIRS at an angle of *ca.* 75° as discussed in section 13.2.2, and the reflected beam is focussed onto a liquid nitrogen cooled wideband MCT detector. Substrate temperature control is effected by resistively heating the substrate whilst it is in contact with a liquid nitrogen reservoir. The substrate temperature is measured by a calibrated thermocouple. The H_2O and sample gases are deposited through piezoelectric valves from external 10 dm^3 reservoirs. Inside the vacuum chamber, the gases are directed onto the substrate by glass guide tubes to prevent any reaction in the gas phase prior to their encounter with the surface. Pressures inside the vacuum chamber and the total exposure of the substrate to reactive gases and H_2O are measured by a cold cathode ionisation gauge and a quadrupole mass spectrometer.

13.3 Uptake and reaction of small molecules on ice surfaces

There are a number of ways in which atmospherically relevant organic and inorganic molecules can interact with ice surfaces, ranging from weak forces such as van der Waals interactions and hydrogen bonding through to the strong forces associated with chemical bond formation. As for all surface processes, the nature and strength of these forces controls the thermodynamics and kinetics of adsorption and desorption as well as influencing the potential energy surface upon which subsequent reactions occur. Many ice surface reactions are observed to obey precursor-mediated reaction mechanisms in which these physical interactions control the surface concentration and presentation of reactant molecules prior to their reaction with either co-adsorbed species, incoming gas molecules or the ice substrate itself. In this section, the effect of the nature of the ice surface upon the binding of small molecules at low temperatures is briefly discussed prior to an examination of the mechanisms by which specific adsorbed molecules and incoming gases react.

13.3.1 Binding through physical interactions

As described in section 2.1 above, a number of experimental and theoretical methods have been applied in order to develop a detailed understanding of the nature of the exposed ice surface. Single crystal ice at low temperatures is known to have a repeated bilayer structure which shows a significant degree of reconstruction in the surface layer [15]. As a result, a general feature of all types of ice surface is the presence of pendant O and H atoms in the surface layers which are not fully coordinated (as they would be in the bulk). These pendant features are often referred to as "dangling bonds" [25-26], although this functionality should not be confused with uncoordinated orbitals (as opposed to atoms) present on the surface of semiconductors such as Si with the same name. What these dangling bonds represent is a highly polarised surface with a strong propensity to hydrogen bond. When incident upon the ice surface, any molecule possessing a hydrogen-bond receptor or donor group will as a result of this favourable potential show a greater tendency to stick to the surface as a result of H-bond formation (whatever its subsequent fate). Although amorphous ice has a lower surface density of dangling bonds compared to crystalline ice, films of amorphous ice have a rather more open structure and significantly higher number of dangling bond sites per unit geometric area, which potentially affects its ability to adsorb incoming molecules by hydrogen bonding.

If present in significant concentrations, the OH-stretching mode of dangling bonds can be observed in vibrational spectra since they occur at a higher frequency than the comparable vibrations of fully-coordinated bulk H_2O. Dangling bond vibrations of different types (with either H or O free above the surface, denoted d-H or d-O) are readily identified in vibrational spectra of ice clusters and nanoparticles as a result of the high surface/bulk ratios, and are discussed at length in chapters 3 and 17. In thin film spectra, dangling bond absorptions of the d-H type are typically seen as a small feature on the high frequency tail of the broad OH stretching band of ice. During film growth, the dangling bond intensity grows to a saturation value, as would be expected from a largely surface feature. It should be noted however that rapid deposition of a thick film does lead to a slightly higher saturation intensity for the dangling bond because of the larger surface area associated with such a disordered film. A number of studies have reported this sort of film as being microporous with a surface area of the order of $20 m^2 g^{-1}$ by BET analysis [27]. IR spectroscopic studies also confirm the significantly reduced

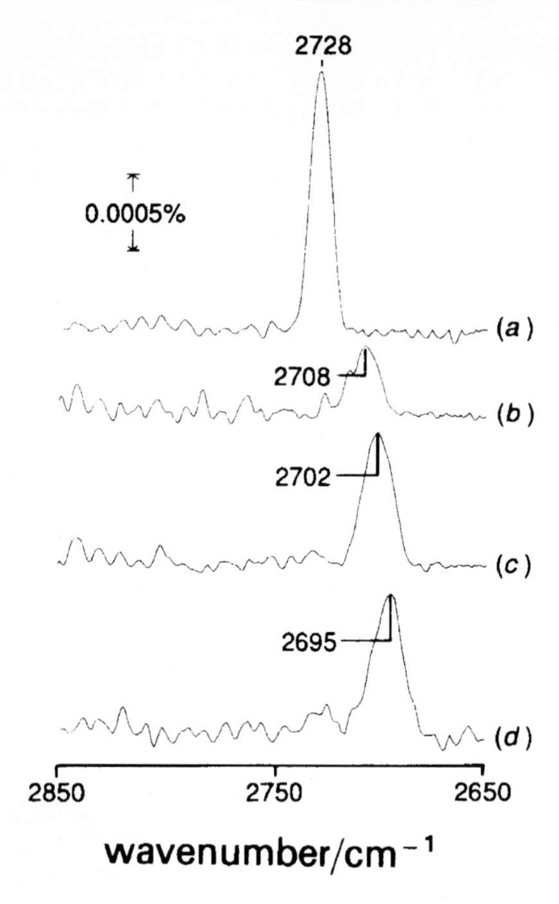

Fig. 13.3. Baseline-corrected RAIR spectra of adsorbates on a thin heavy-water ice film on Pt(111) at 100 K: (a) pure heavy water ice; (b) surface(a) exposed to 3 L CF_2Cl_2;(c) surface (a) exposed to 3 L CCl_3F; (d) surface (a) exposed to 3 L CCl_4. Copyright of the Royal Society of Chemistry, 1992.

number of dangling bonds in crystaline ice compared to amorphous ice as a result of the higher surface area of the latter.

The strength of the H-bonding interaction with different types of adsorbed molecule can be probed through vibrational frequency shifts of these surface modes. As an example of weak interactions with an ice surface, Fig. 13.3 shows baseline corrected RAIR spectra at 100 K of the dangling OD region of amorphous D_2O ice films both before and after exposure to small doses of a range of chlorofluorocarbons with increasing H-bonding affinity [28]. The ice dangling bond (d-D) for D_2O can be seen clearly at 2728 cm^{-1}. Typical CFC exposures were of the order of 3 L (1 L = $1x10^{-6}$ Torr s), sufficient to provide a monolayer coverage. Upon adsorption, a shift of between 22 and 32 cm^{-1} occurs, corresponding to perturbation of the dangling bond by H-bond interaction with the adsorbate. The shifts follow a regular trend of chlorine substitution. For CF_3Cl (not shown), no shift occurs because no sticking occurs at 100 K. The magnitude of the shift shows a strong correlation with a number of physicochemical properties of these compounds, most specifically the molecular polarisability and the C-Cl bond dissociation energy. Comprehensive studies of the infrared spectra of a series of twenty-three halocarbon compounds co-deposited onto water-ice held at 12 K show that perturbations to the water (D_2O) d-D sites which are not saturated in the absence of guests exhibit distinct trends depending on the type and extent of halogen substitution in the occluded molecules. A high level of correlation is found between the shift in the dangling bond and both the charge density and the molecular polarisability within each separate series for compounds of the formula CF_nH_{4-n}, CCl_nH_{4-n}, and CBr_nH_{4-n} (n = 1-4). When a combined polarisability/charge interaction parameter is plotted against the dangling bond shift a straight line (with predictive capabilities) results [29]. Similar trends are observed for the adsorption of species as diverse as N_2, CO_2, CF_4 and SO_2 on ice clusters studied by IR and Raman spectroscopy, which also reveal details of the coordination of adsorbates with the various types of surface-localised H_2O molecules. Shifts in the OH-stretching frequencies also show a significant degree of predictability by computational methods [15,30]. Of both atmospheric and astrophysical relevance are spectroscopic and isotherm measurements of the adsorption of CO onto ice surfaces, which are similarly illuminating as to the physical interaction between ice and adsorbate [31-33]. These studies reveal strong yet reversible adsorption onto a range of surface sites with a typical type I BET isotherm, whilst the observed CO stretching frequency also shows a dependency upon the nature of the local environment on the surface (or subsurface). Again, the fact that the magnitudes of these effects are accessible computationally is significant since it demonstrates that the nature of the relatively weak interactions involved are reasonably well defined. A wide range of examples of the adsorption onto ice of other atmospherically relevant trace gases such as O_3 and NO_3 are also to be found, again demonstrating that weak physical adsorption is possible and may therefore play a role in influencing their heterogeneous reaction kinetics [34-36].

In order for adsorbates to be retained on ice surfaces at the temperatures prevalent in the stratosphere and troposphere, much stronger adsorption enthalpies are required. Experimental uptake measurements on both clusters and thin films reveal that oxygen-containing organic molecules containing the alcohol (OH), carbonyl (C=O) and carboxylic acid (CO_2H) functional groups have a strong affinity for ice surfaces [37-39]. When such stronger hydrogen bonds are formed with the surface water molecules on ice, the dangling bond modes shift down in frequency to such an extent that they overlap the strong bulk fundamentals and are no longer clearly discernable. This behavior is exemplified in a study of the adsorption of acetone on ice between 100 – 150 K by Schaff & Roberts [40,41], where C=O stretching modes are evident as a result of adsorption but the weak surface water features are not observable once a full monolayer coverage has been achieved. Also observed is a change in the vibrational frequency of the C=O functional group of the ice-adsorbed acetone, which shifts from its condensed phase value of 1717 cm^{-1} down to 1703 cm^{-1}. This clearly demonstrates that the hydrogen bonding occurs through a d-H surface group to the carbonyl oxygen. Schaff & Roberts also observed a significant difference in adsorption behaviour between amorphous and crystalline ice, noting two distinct thermal desorption states from amorphous ice compared to a single state from crystalline ice. They attribute the two states on amorphous ice to hydrogen-bonded and bulk condensed acetone, and on the basis that they only observe desorption from the bulk state on crystalline ice conclude that the absence of a hydrogen-bonded state reveals that such a surface has a substantially lower number of dangling bond states (below their detection limits). This is certainly reasonable on the basis of the smooth morphology of the crystalline surface. A range of other organic molecules have been studied in recent years, arriving at many of the same conclusions as for acetone. In low temperature work, alcohols are observed to stick readily to ice surfaces through both H-bonding and direct condensation whilst at higher temperatures, kinetic studies of the partitioning between the gaseous and adsorbed phases reveal a small influence of H-bonding in uptake rates on ice [39].

The implication of these observations is that although the crystalline surface has a reduced dangling bond density compared to amorphous ice, it should still be capable of holding a significant number of impacting gas molecules of the type typically found in the atmosphere, either in a weakly-held precursor state or in a strongly bound physisorption potential well. Inorganic molecules such as $ClONO_2$ and HCl certainly have the potential to interact with surface dangling bonds in this way through their halogen atoms as well as the H-atom in the case of HCl, whilst many of the relevant organic species possess both donor and acceptor functionalities. Furthermore, in cases where multiple interactions between the surface and different parts of the same molecule can occur, it is possible that the H-bonding interactions influence surface presentation of the adsorbate, with obvious ramifications for surface-mediated photolysis or further reactions with the substrate and/or incoming gases.

13.3.2 Chemical interactions: ionisation and hydrolysis of HCl, N_2O_5 and $ClONO_2$

Once incoming molecules have equilibrated with the ice surface (even for a short time), a number of further channels are available in competition with direct desorption. Weak trapping of an adsorbed molecule may provide sufficient time for the adsorbed state to either react with other adsorbates or form stronger bonds with the surface (*i.e.* chemisorption). In the following sections, we examine the fate of three atmospherically important molecules on ice surfaces, both separately and in the presence of other potential reactants.

Hydrogen Chloride. In terms of heterogeneous chemistry, one of the most widely studied atmospherically relevant molecules is the important chlorine reservoir HCl. HCl is formed in the stratosphere by the reaction between reactive, ozone depleting Cl atoms and CH_4, and consequently represents an important route by which this key ozone destroying species is removed. It is relatively inert in the gas phase under upper atmosphere conditions, being removed principally by reaction with OH [1]. Heterogeneously, it is known to react with $ClONO_2$ and with HOCl to produce molecular chlorine, which goes on to participate in gas-phase cycles which destroy ozone:

$$HCl + ClONO_2 \longrightarrow HONO_2 + Cl_2 \tag{13.1}$$

$$HCl + HOCl \longrightarrow H_2O + Cl_2 \tag{13.2}$$

Given its important role in the Antarctic ozone hole phenomenon, the kinetics of its uptake onto ice surfaces and reactions with $ClONO_2$ under stratospheric conditions were first measured using coated wall flow tube methods [42-44]. The natural solubility of HCl in ice is small [45,46], and the observation of significant reactivity of HCl-doped ice towards other incoming molecules is therefore strong evidence for a surface process involving a significant surface excess of adsorbed HCl. In such kinetic measurements, HCl is observed to be readily taken up by ice surfaces and the uptake is observed to saturate at temperatures below 180 K for atmospherically realistic partial pressures. However, there are aspects of the temperature and HCl partial pressure dependence of the reaction which suggest that a number of competing processes are likely to be involved. At high HCl partial pressures for example, surface melting is observed, accompanied by the formation of a quasi-liquid layer into which continuous uptake can occur (as for the bulk liquid)[47,48]. Uptake is also observed to be dependent upon ice surface structure and surface areas, and is also influenced in some temperature regimes by the formation of stable surface hydrates [49]. Such quantitative measurements generally lead to the conclusion that solid ice surfaces are able to take up a maximum of 1 monolayer of HCl under stratospheric conditions [42,47,50]. However, the mechanism whereby this uptake occurs is not directly accessible from kinetic data. To address this problem, a number of experimental and theoretical studies have attempted to elucidate the fundamental mechanisms associated with HCl adsorption with varying degrees of success.

Of particular interest is the nature of the species adsorbed on the ice surface, a subject upon which there has been a considerable degree of controversy. Early debate centered on the physical nature of the adsorbate, i.e. whether the adsorbate is molecular or ionic. Thermal desorption measurements by Graham and Roberts [51] presented evidence for both species, whilst computational studies of the concentration of intact molecular HCl by Kroes and Clary predicted equilibrium values of ca. 10^{-7} monolayers at stratospheric conditions [52]. Vibrational spectroscopic observations have been particularly helpful in exploring this paradox. Fingerprints of molecular (intact) HCl and its many ionic hydrates obtained over a range of conditions provide a comprehensive database for the analysis of equilibrium surface composition from IR spectra. Below ca. 60K, molecular H_2O:HCl complexes are readily formed, which upon heating ionise to form amorphous ionic hydrates. At temperatures above 140 K, IR spectra are dominated by the characteristic features of hydrated protons, indicating ionisation of HCl into $H^+(H_2O)_n$ and Cl^-. At appropriate temperatures and stoichiometries, stable crystalline hydrates of HCl with characteristic spectral features are formed [53-55]. Generally, these types of measurement are reasonably conclusive in identifying ionic species as the majority surface species at stratospheric conditions. This is borne out by physical methods such as secondary ion mass spectrometry, which show a high surface density of ions after exposure to HCl [56,57]. The nature of adsorbed HCl has also been explored theoretically using a range of methods, and most computational studies now seem to agree that the most stable adsorbed form of HCl on an ice surface at stratospheric temperatures is indeed an ion pair[58-63], whilst calculations of H_2O:HCl clusters show that at least three water molecules are needed for the reaction to proceed efficiently [64-68]. In all of these calculations, the H_2O cage surrounding the intact HCl molecule serves to facilitate the ionisation by (a) providing a cyclic transition state in which proton transfer can occur with a low barrier; and (b) by solvating the product ions. Experimentally however, there seems to be a situation below 125 K where the relative interplay of these various competing processes appears to favour the molecular form ([69] and section 3.5 of chapter 17), and significant amounts can be trapped and observed in IR spectra. This temperature dependence of the ionic-molecular ratio has been effectively demonstrated by the reactive ion scattering experiments of Kang et al. [70]. Temperature-dependent diffusion measurements further suggest that the motion of HCl and its ionised components along the surface and into the bulk plays a role in determining the rate at which the surface is refreshed for further adsorption events to occur [71,72] or for further surface reactions to proceed [11]. Furthermore, Uras-Aytemiz *et al.* have observed that the rate of HCl diffusion through the hydrate crust on ice nanoparticles is likely to be the dominant factor in controlling the kinetics of hydrate formation [73].

Experiments and calculations such as those described above provide an excellent insight into the equilibrium composition and structure of HCl-doped ice surfaces. However, what they do not clearly reveal are the dynamics of the surface interactions whereby the various components are formed. Recent molecular beam scattering experiments have revealed considerably more detail of the fundamental trapping and ionisation events. For example, Andersson et al. observed three channels for incoming HCl molecules: (a) direct scattering; (b) trapping de-sorption; or (c) sticking to the surface [74]. At low initial coverages, the sticking channel dominates with a sticking coefficient close to unity (in agreement with the low coverage flow-tube and Knudsen cell results discussed above). At higher coverages, where fewer surface water molecules are available to participate in the ionisation step, a trapping de-sorption channel is clearly observed with a surface lifetime of the (intact) HCl molecule of less than 30 μs, in agreement with the calculations of Kroes and Clary for non-dissociating HCl [52]. Under similar experimental conditions, Haq *et al.* have observed the surface products of the adsorption of HCl from molecular beams to produce ionised HCl hydrates and have measured the difference in reaction barrier height for dissociative chemisorption compared to desorption for a range of coverages and temperatures [75]. They also observe strong evidence for the disruption of the ice lattice as adsorption proceeds and for the formation of stable surface hydrates when appropriate stoichiometry is achieved.

On the basis of all of the above observations, the uptake of HCl onto ice can simplistically be viewed as follows. An initial HCl molecule impinging on the ice surface from the gas phase first encounters a weak physical adsorption potential well. The equilibrium population of this state is low as a result of the shallow trap, but since trapping and equilibration with the surface is efficient, barrier-less ionisation rapidly ensues for the adsorbed HCl when sufficient water is available in the vicinity of the adsorption site. Changes in the orientation of the reaction components (including the solvent cage surrounding the ion products) as the reaction proceeds also affect the subsequent reactivity of the remaining ice surface to additional incoming gaseous HCl. At higher exposures, stable phases ranging from the monohydrate to the trihydrate may nucleate, upon which subsequent HCl adsorption becomes reversible. Transport rates also increase markedly with temperature, eventually leading to near-continuous uptake above 230 K, which may also be a result of surface melting when the HCl partial pressure exceeds the saturated vapour pressure of HCl above the ice/HCl mixture (hydrate). The variation of the rates of each of the above processes with temperature leads to a rather complex overall behaviour.

Dinitrogen Pentoxide. Another atmospherically significant molecule which shows an unforeseen heterogeneous reactivity on ice is N_2O_5. This molecule is the main night-time reservoir of nitrogen oxides, forming in the reaction between NO_3 and NO_2. During the daytime, it is rapidly photolysed back to its components. NO_2 plays an important role in the deactivation of halogen-based ozone destruction as it sequesters ClO to form $ClONO_2$, so any heterogeneous interaction which reduces N_2O_5 concentration concomitantly affects this process [1]. Field measurements have shown significant seasonal reductions in the amount of N_2O_5 correlated with the appearance of PSCs [76], and laboratory kinetic measurements have shown that the reaction is particularly efficient on ice [77]. As for HCl, the uptake of N_2O_5 appears to saturate as the ice surface becomes covered in the reaction product, indicating that the reaction is with the ice surface rather than catalysed by it. The uptake shows far less complexity as a function of temperature and pressure than HCl, and

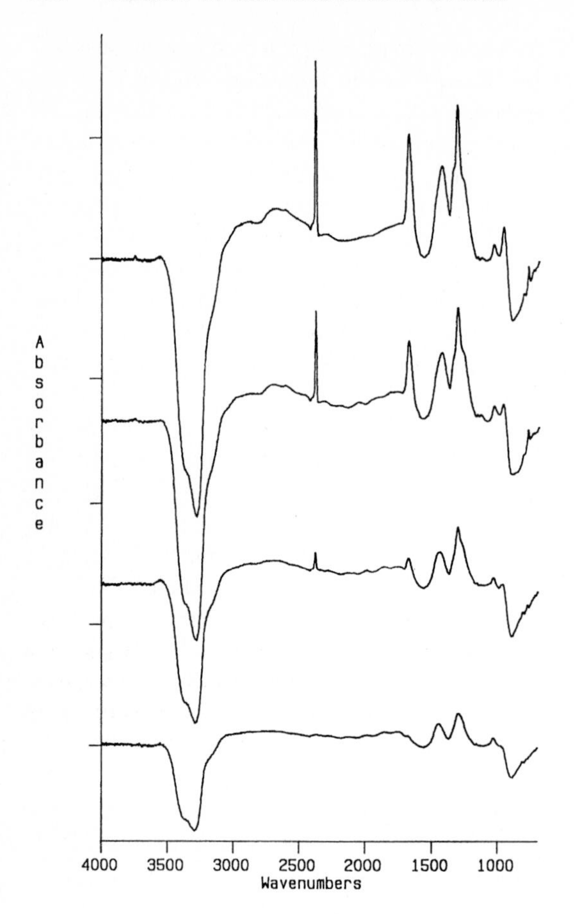

Fig. 13.4. RAIR spectra of the interaction of gas-phase N_2O_5 with a crystalline ice film at 140 K. From the bottom up: low exposure; medium exposure; high exposure; excess N_2O_5. Copyright of the American Chemical Society, 1994.

the major factors which control its hydrolysis are discussed below alongside experimental observations of reaction products and intermediates.

The product of the reaction of N_2O_5 with ice is expected to be $HONO_2$. However, given the low vapour pressure of $HONO_2$ over ice at low temperature, the (surface) reaction product cannot be detected simultaneously with the loss of reactant in flow tube or Knudsen cell measurements. Consequently, mechanistic inferences from kinetic data are necessarily vague. In conjunction with computational studies of reaction pathways and the stability of intermediates, the fragment-specific nature of infrared spectroscopy provides the necessary detail to complete the picture of the mechanism of heterogeneous ice/N_2O_5 interactions. A reasonably comprehensive database of vibrational fingerprints of N_2O_5 in its molecular and covalent forms [78-81], molecular $HONO_2$ and its dissociated ionic hydrates [82-87] and a wide variety of nitrate and related ion spectra [88] exists in the literature. To illustrate this point, Fig. 13.4 shows the IR spectra obtained during the interaction of gas-phase N_2O_5 with a condensed crystalline ice film at 140 K [89]. These are absorbance-difference spectra, such that positive bands show gained species and negative bands show lost species. At low exposure, the bands gained can be identified as being due to H_3O^+ and NO_3^- ions, whilst those lost are due to H_2O. As exposure continues, additional features due to molecular nitric acid and ionised N_2O_5 (as nitronium nitrate, $NO_2^+NO_3^-$) are seen along with the further loss of water. These spectra alone do not contain enough information for an accurate identification of the reaction mechanism, since it cannot be determined directly if molecular $HONO_2$ is formed and subsequently ionises or whether it is directly formed as H_3O^+ and NO_3. Similarly, it is not possible to identify whether covalent N_2O_5 or $NO_2^+NO_3^-$ is the reactant, or indeed, whether either form is directly involved in the reaction as such. Further experimental data are needed to understand the details of the N_2O_5-ice interaction, for example from layered films of pure N_2O_5 and H_2O which probe the interaction between condensed phases at their interface. Similarly IR spectra of thin mixed films of H_2O and N_2O_5 with a variety of composition ratios condensed on an Au substrate can be used to demonstrate the effect of water availability on product formation [90]. Slow heating of these films to successively higher temperatures allows the reaction to proceed, and spectra recorded over a range of temperatures reveal the reaction products under different degrees of hydration. The full dataset obtained from these experiments therefore contains information about the dependence of the hydrolysis reaction on temperature and water availability which can be used to interpret the real-time reaction spectra.

Various general features of N_2O_5 behaviour are evident from IR spectra of layered samples. N_2O_5 condenses in either its covalent or ionic form depending upon the temperature and the condensed covalent form is metastable with respect to the ionic form, converting rapidly above 100 K. However, neither condensed form appears to be particularly reactive towards H_2O or ice, either at the interface between condensed layered N_2O_5/ice films or for gaseous H_2O adsorbing on condensed N_2O_5 films. This suggests that the material formed in Fig. 13.4 arises from gaseous N_2O_5 interacting with condensed ice. From the deposition of mixed films of H_2O and N_2O_5 of various stoichiometry below 100 K, we identified absorption bands attributable to unreacted covalent N_2O_5 and water-ice when formed, regardless of ratio. Heating such films reveals other aspects of the mechanism of the N_2O_5/ice interaction. For less than 1:1 N_2O_5:H_2O mixed films, heating results in slower transformation of the covalent N_2O_5 to nitronium nitrate, in contrast to the behaviour of a pure N_2O_5 film which converts fully and rapidly at ca. 100K. This behaviour reflects the presence of a barrier to the motion of N_2O_5 molecules in the presence of even a small amount of H_2O, hindering the formation of the ionic solid. Above 160 K, all absorptions due to covalent N_2O_5 are lost, leaving films consisting of polycrystalline nitronium nitrate and small amounts of molecular nitric acid. At intermediate compositions (up to ca. 4:1), fewer nitronium ions are observed during film heating from the direct N_2O_5 covalent/ionic

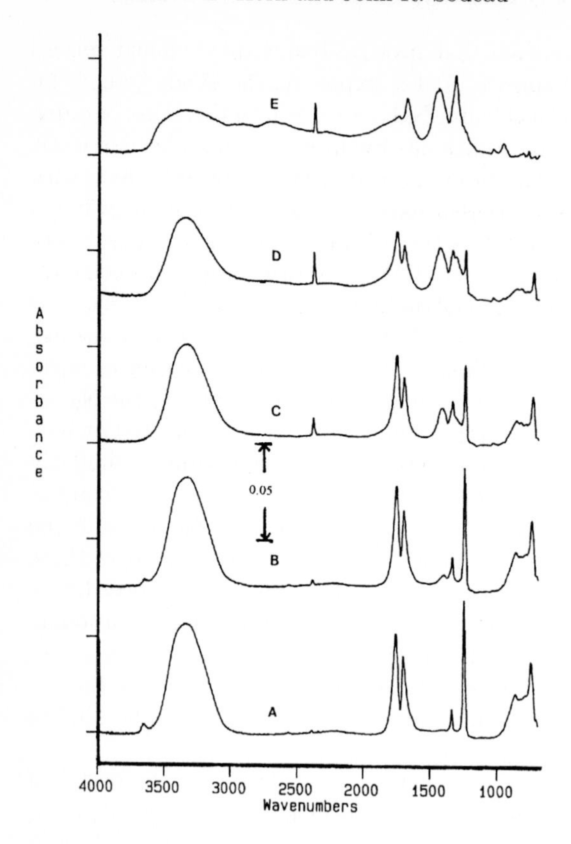

Fig. 13.5. IR spectra of a film of co-condensed H_2O and N_2O_5 in a 3:1 ratio at a variety of temperatures. (A) as deposited at 80 K; (B) 125 K; (C) 140 K; (D) 155 K; (E) 170 K. Copyright of the American Geophysical Union,1997.

transition, since the individual molecules are sufficiently far apart that they react before they can diffuse together in sufficiently large concentrations to form the ionic solid. Above 140 K, reaction occurs producing material which is mainly composed of molecular nitric acid and amorphous nitric acid hydrates with a small trace of nitronium ions as shown in Fig. 13.5. In contrast to the low water content films, the nitrate ions which are produced during reaction in these intermediate composition films show evidence of solvation compared to the symmetrically coordinated, unsolvated anion in nitronium nitrate crystals. This is demonstrated by the characteristic band splitting of the v_3 mode and the appearance of the v_1 mode of the nitrate ion, typical of coordinated nitrate in salts and the crystalline dihydrate of nitric acid. At high dilution (10:1 or higher), virtually no nitronium nitrate is formed at any stage of the heating process and instead, covalent N_2O_5 converts smoothly to amorphous NO_3^- and H_3O^+. Beyond 170 K, this material converts with the loss of some water to nitric acid trihydrate, identified by its characteristic IR spectrum. These experiments have recently been repeated by Agreiter et al.[91], who reached similar conclusions.

These matrix-type experiments show that the reaction products are dependent not only on temperature but also on the availability of water for reaction. In the presence of excess H_2O, ionised nitric acid is formed directly, whereas under reduced H_2O conditions both molecular and ionised forms are produced along with some nitronium nitrate from the covalent/ionic transition. On the basis of these observations, we previously proposed a reaction mechanism in which covalent N_2O_5 arriving from the gas-phase reacts with the ice surface in a number of ways, depending upon the number of available water molecules at the impact site. Initially, the incoming N_2O_5 molecule interacts with a surface water molecule and becomes trapped into a weak physisorption potential well. From this site, it may either desorb or react further. When a large number of other N_2O_5 molecules are nearby (i.e. at high N_2O_5 partial pressures), extended condensed covalent films are formed which rapidly ionise to form nitronium nitrate for which no further reaction can occur. Under conditions where additional water molecules are available, hydrolysis of the trapped molecule occurs. On the basis of IR evidence and some simplistic frontier orbital arguments, we proposed that the initial step occurs by nucleophilic attack of the oxygen atom of a surface H_2O molecule upon one of the electrophilic nitrogen atoms, accompanied by a lengthening of one of the central O-N bonds until separation into two ions occurs:

$$H_2O + N_2O_5 \longrightarrow [H_2ONO_2]^+ + ONO_2^- \tag{13.3}$$

This protonated nitric acid species was then predicted to produce NO_2^+ and H_2O by unimolecular decomposition (13.4), or if a small excess of water is available undergo a proton transfer reaction to give molecular nitric acid and an H_3O^+ ion (13.5):

$$[H_2ONO_2]^+ \longrightarrow H_2O + NO_2^+ \tag{13.4}$$

$$H_2O + [H_2ONO_2]^+ \longrightarrow H_3O^+ + HONO_2 \tag{13.5}$$

$$nH_2O + HONO_2 \longrightarrow H_3O^+ + ONO_2^- + (n\text{-}1)H_2O \tag{13.6}$$

In the presence of excess water, this molecular nitric acid can undergo further proton transfer to a neighbouring H_2O molecule to give a film entirely composed of ionised nitric acid (13.6). This mechanism is consistent with the relative amounts of NO_3^- (formed in the initial hydrolysis step, (13.3)) and $HONO_2$ observed spectroscopically (*i.e.* in 3:1 mixtures, approximately 1:1) and satisfactorily accounts for the species identified in the spectra shown in Fig. 13.4, (*i.e.* totally dissociated nitric acid at low exposure, mixtures of $HONO_2$ and its ionised form at intermediate exposures and additional nitronium nitrate in excess N_2O_5). However, although the formation of protonated nitric acid is known to be feasible as a result of gas-phase studies [92,93] and in clusters ([94] and chapter 4) it is unlikely to be particularly stable in condensed media and this mechanism has been questioned. A number of computational studies of varying sophistication have been applied in recent years to address this problem [95,96]. The most comprehensive study of this system has been reported by McNamara and Hillier, who used high-level electronic structure calculations to investigate the hydrolysis in small neutral water clusters containing one to six solvating water molecules [97]. This work showed that in larger clusters (5 or 6 H_2O molecules), N_2O_5 shows a strong polarization through the elongation of one of the central O-N bonds, enhancing the electrophilicity of one of the nitrogen atoms and facilitating the nucleophilic attack of one of the solvating H_2O molecules. This step is immediately followed by a proton transfer to directly produce H_3O^+ and $HONO_2$. Of particular note is the fact that the reaction is barrierless for 6 or more H_2O molecules. This mechanism is entirely consistent with existing spectroscopic data for N_2O_5 heterogeneous hydrolysis on ice, and this example nicely illustrates the synergy which accrues from the use of vibrational spectroscopy combined with electronic structure calculations for reaction mechanism determination.

In conclusion, N_2O_5 hydrolysis again relies upon the potential of a crystalline ice surface to form a weak physisorption well for a precursor state, from which subsequent reactions can occur. When adsorbed on clean ice, an N_2O_5 molecule is presented to the surface in such a way that interaction with a number of H_2O molecules in the surface serves to polarize and thereby activate one of the central O-N bonds to such an extent that even a weak nucleophile such as H_2O can approach the electrophilic centres of N_2O_5 and initiate reaction. Additional surface H_2O molecules then participate in the reaction and their availability and/or mobility determines the nature of final reaction products. These results also demonstrate why the reaction saturates under static conditions (i.e., where the ice surface is not renewed), since at stratospheric temperatures the build-up of a stable surface hydrate phase will lock up all available surface water, removing both the nucleophile and the activation mechanism. Under the dehydrated conditions of the polar winter stratosphere, there will be few adsorbing water molecules to refresh the reactivity of the surface.

Chlorine Nitrate. Of all the heterogeneous reactions of nitrogen-containing species to be found in the atmosphere, the hydrolysis of chlorine nitrate on ice surfaces has attracted by far the most detailed laboratory scrutiny, representing as it does a direct channel for the reactivation of stratospheric chlorine species. Although the direct photolysis of $ClONO_2$ to produce both Cl and ClO is known to occur at mid-latitudes [98], it is not a particularly efficient channel given the low solar flux at appropriate wavelengths in the polar winter stratosphere. Kinetic studies show that $ClONO_2$ is rapidly heterogeneously hydrolysed on ice surfaces from 140 – 185 K to produce gaseous HOCl [43,50,99,100]. Post reaction analysis reveals the presence of nitric acid in the ice film [43], leading to an overall reaction mechanism:

$$ClONO_2 + H_2O \longrightarrow HOCl + HONO_2 \tag{13.7}$$

Again, the details of the heterogeneous mechanism of this reaction cannot be reliably extracted from kinetic data and vibrational spectroscopy and electronic structure calculations are both needed to fully quantify the reaction steps. As a result of such studies, it is apparent that the nature of the reaction also depends upon temperature in a previously unsuspected manner. The combined results from studies both performed by us and from those reported in the literature are used below to describe the various stages of the interaction between $ClONO_2$ and H_2O in ice surfaces.

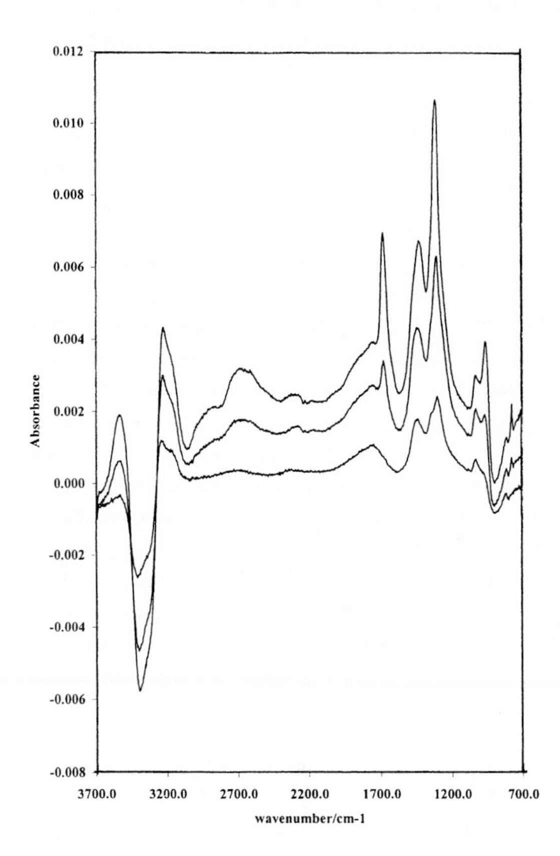

Fig. 13.6. RAIR spectra obtained from the exposure of a crystalline ice film at 140 K to gaseous $ClONO_2$. Exposure increases from bottom up. Copyright of the Royal Society of Chemistry, 1998.

Starting from low temperatures, Fig. 13.6 shows typical IR spectra obtained upon exposure of a crystalline ice film to gaseous $ClONO_2$ at 140 K [101]. Nominal exposures are given in Langmuirs (1 L = $1.333 \mathrm{x} 10^{-6}$ mbar s). Again, these are absorbance-difference spectra with positive bands for gained species and negative bands for lost species. Initially, loss bands are seen for water consumed in a reaction with the incoming $ClONO_2$ and the growth of various products. Upon increasing exposure, different products are observed. These spectroscopic features can be assigned by reference to well-defined spectra of the decomposition products of $ClONO_2$ and HNO_3 (including ions and molecular species). In the initial stages of the reaction (*i.e.* upon exposure to small amount of $ClONO_2$), the formation of absorption bands characteristic of the H_3O^+ ion and the NO_3^- ion are observed. Furthermore, these ions are present in an unstructured amorphous form as described above for the interaction of N_2O_5 with ice under similar conditions. As exposure increases, additional sharp features due to molecular nitric acid are seen to grow in. According to (R5), the production of nitric acid (whether hydrated or molecular) should also be accompanied by the formation of HOCl. Although not clear from the spectra in Fig. 13.6, there is spectroscopic and mass-spectrometric evidence that this is indeed the case and that the resultant HOCl remains adsorbed onto the surface or is lost to the gas-phase during the reaction depending upon the surface temperature [102]. This process is further corroborated by a recent study of the low temperature (110 K) reaction of $ClONO_2$ by Berland *et al.*, in which surface HOCl was detected using laser induced thermal desorption spectroscopy [103]. As the reaction continues, it is clear that molecular rather than ionized $HONO_2$ is being formed in direct relation to the reduction in the number of available surface water molecules. This process is directly comparable to the effect of water availability upon the hydrolysis of N_2O_5 when deposited upon a thin ice film as shown in Fig. 13.4. It can be concluded from this behaviour that the initial reaction product on the surface at low temperature is most likely to be the molecular acid, which subsequently ionizes in the presence of excess water. When there is insufficient H_2O (per surface $HONO_2$), ionization becomes unfavourable. This mechanism concurs with the results of a theoretical study of the $ClONO_2$/ice system by Bianco and Hynes [104], in which a concerted mechanism involving simultaneous nucleophilic attack of a surface water oxygen atom and proton transfer from a neighboring water molecule to the NO_3 moiety was observed for clusters containing one $ClONO_2$ and three H_2O molecules. Their studies showed that although an isolated H_2O molecule is too weak a nucleophile to attack $ClONO_2$ directly, in the presence of solvating water a cyclic transition state can be formed for the concerted reaction, for which a substantially lower activation barrier is obtained. From this evidence, it seems apparent that the predominant mechanism for low temperature ($\leq$ 140 K) $ClONO_2$ hydrolysis produces molecular $HONO_2$ and HOCl.

At higher temperatures, there is however strong evidence indicating that there may be additional reaction mechanisms contributing to the hydrolysis process. Experimentally, the observation of adsorption-induced polarization of $ClONO_2$ [105], its unprecedented reactivity on organic surfaces [99] and the observation of significant time delay in the ejection of reaction products in kinetically sensitive experiments [100,106] suggest that $ClONO_2$ reacts in a rather different way at higher temperatures. For example, Fig. 13.7 shows IR spectra of the condensation of $ClONO_2$ on an Au foil at 180 K [105] in both the presence and absence of H_2O. The spectrum of the adsorbed anhydrous $ClONO_2$ at this temperature looks rather different to that of the condensed

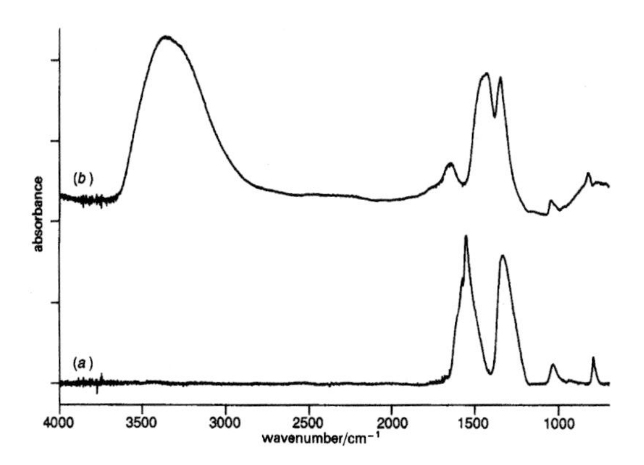

Fig. 13.7. RAIR spectra of thin films condensed at 180 K. (a) anhydrous $ClONO_2$; (b) $ClONO_2$+H_2O at a gas pressure ratio of 10:1 [NB. This does not imply that the concentration ratio in the condensed film is 10:1 as the relative sticking coefficients for $ClONO_2$ and H_2O are different at this temperature]. Copyright of the Royal Society of Chemistry, 1995.

molecular solid at lower temperatures, and predominantly exhibits features which are attributable to a strongly coordinated nitrate species. On the basis of vibrational mode assignment and comparison to literature spectra of coordinated salts, this material can be identified as strongly polarised $Cl^{\delta+}$-$O^{\delta-}NO_2$. Co-condensation of $ClONO_2$ with small amounts of H_2O (from a 10:1 $ClONO_2$:H_2O gas mixture) results in the formation of the material with a spectrum such as that shown in Fig. 13.7b. Again, this spectrum shows features attributable to a coordinated nitrate group (although less strained than in Fig. 13.7a) and broad OH features. We have previously proposed that this material consists of H_2O strongly coordinated with polarised $Cl^{\delta+}$-$O^{\delta-}NO_2$ in the form $[H_2OCl]^+ONO_2^-$, with the former ion being geometrically analogous to the H_3O^+ ion. The assignment to a full contact ion pair is considered speculative by some [107], but this behaviour nevertheless demonstrates the tendency of $ClONO_2$ to polarize when condensed upon a surface, particularly in the presence of H_2O.

On the basis of the spectroscopic observations of the tendency of $ClONO_2$ to become polarized when condensed or when associated with condensed water, we have previously proposed a two-step mechanism in which an intermediate of the form described above participates directly in the hydrolysis [90,102]:

$$ClONO_2 + H_2O \longrightarrow [H_2OCl]^+NO_3^- \tag{13.8}$$

$$[H_2OCl]^+NO_3^- + H_2O \longrightarrow H_3O^+ + HOCl + NO_3^- \tag{13.9}$$

Computational studies of this mechanism show that full ionization into a (contact) ion pair may not in fact be necessary as part of the reaction mechanism [108-110]. The initial modeling of $ClONO_2$ on ice by Bianco & Hynes indicated that in the presence of an appropriate number of solvating H_2O molecules, the nucleophilicity of the attacking H_2O could be increased in a cyclic structure generating a species similar to OH^-. As discussed above for low temperatures, subsequent proton and electron transfer events in their scenario lead to molecular nitric acid products and strongly hydrogen-bonded

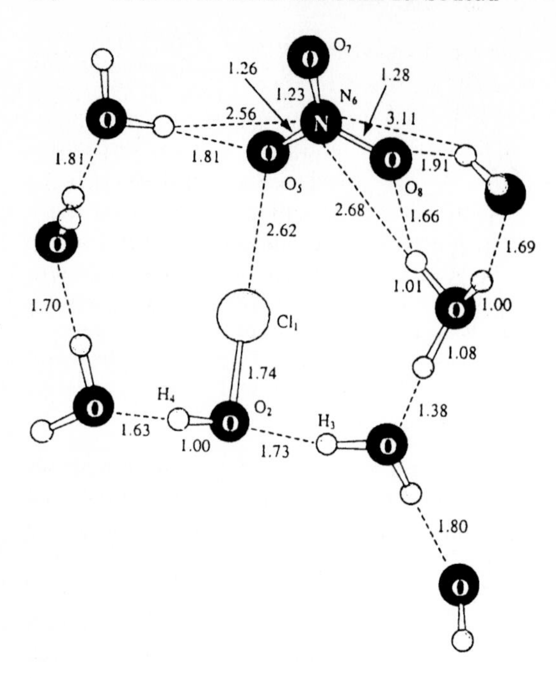

Fig. 13.8. Calculated ion-pair structure of the $ClONO_2(H_2O)_8$ cluster, showing ionized nitric acid and molecular HOCl. Taken from McNamara and Hillier [108]. Copyright of the American Chemical Society, 1999.

HOCl. These deductions seem reasonable and indeed, the latter observation may indeed account for the delayed ejection of HOCl in kinetic experiments. However, electronic structure calculations in larger clusters by McNamara & Hillier [108,109] reveal that species akin to $[H_2OCl]^+$ may be involved as a result of the solvation-dependent polarization of the Cl-O bond in $ClONO_2$. They investigated clusters of $ClONO_2$ with three to eight H_2O molecules for which spontaneous Cl-O bond lengthening was observed with increased cluster size. In the range 3:1 – 6:1, the Cl-O bond length increases from 0.179 nm to 0.190 nm and is accompanied by a commensurate shortening of the distance between the Cl atom and the nearest water O atom. For the 6:1 cluster, a stable ion pair could be identified containing HOCl, H_3O^+ and NO_3^-, again with the HOCl strongly bound. In an 8:1 cluster, a species akin to the ionic $[H_2OCl]^+$ is observed for one of its isomers, although with a structure which is considerably different from the optimised gas-phase form. From this structure, the formation of a slightly lower energy structure containing a well-defined HOCl molecule can occur by proton transfer. Figure 13.8 shows the calculated structure of this cluster. In support of this mechanism, solvation-dependent polarisation of $ClONO_2$ has been observed experimentally in static secondary ion mass spectrometric measurements [111]. The dependency of the hydrolysis process upon available H_2O on ice is replicated in uptake studies of $ClONO_2$ on the surface of nitric acid hydrates of controlled stoichiometry and water-richness. Using IR spectroscopy on ice films held in a Knudsen cell, Barone *et al.* have shown that the reaction efficiency scales with P_{H_2O} at 185

K and that ionized nitric acid hydrates formed from the reaction are readily incorporated into existing films [112]. In related experiments, Geiger *et al.* have used second harmonic generation (SHG) methods to study the specific role of the HOCl product in the hydrolysis mechanism using laser light which is resonant with the electronic absorptions of HOCl [106]. Their results show that a proportion of HOCl formed in the initial ice-based hydrolysis is retained by the surface and may in fact act (auto)catalytically to increase the rate of subsequent $ClONO_2$ reaction. Such results are not inconsistent with the observation of a significant amount of HOCl evolution during $ClONO_2$ hydrolysis above 155 K [102], since only a small amount need be retained for catalysis to occur. All of the above studies concur in their observation that competition between the nitric acid product and incoming $ClONO_2$ for surface water sites eventually stops the hydrolysis.

In summary, the reaction between crystalline ice and $ClONO_2$ once again shows a strong dependence upon the coordination of the incoming molecule with the ice surface. Through weak initial hydrogen-bonding, the incoming molecule most probably traps into a shallow physical adsorption well and becomes equilibrated with the surface. Subsequent realignment on the surface results in the formation of coordinated structures with $ClONO_2$ bound between three or more H_2O molecules. The stabilisation which results from this surface coordination leads to a spontaneous elongation of the Cl-O bond which serves to increase the electrophilicity of the Cl atom and facilitates the hydrolysis. Depending upon the number of available H_2O molecules and the temperature, further reaction occurs to liberate HOCl, H_3O^+ and NO_3^-, with molecular $HONO_2$ being formed when H_2O is scarce or its motion is restricted and ionized $H_3O^+NO_3^-$ being formed at higher temperatures and when sufficient H_2O is available. This behaviour also has obvious implications for the heterogeneous reactivity of $ClONO_2$ on other atmospherically relevant surfaces such as nitric acid hydrates [77,113], aqueous sulfuric acid and supercooled ternary solutions of H_2SO_4, HNO_3 and H_2O [114,115], and sea-salt mimics [116,117], where competition between the incoming molecules, existing adsorbates and the substrates themselves will potentially lead to rather complex reaction kinetics.

13.3.3 Mechanism of the ice-supported $ClONO_2$+ HCl reaction

The solvation and ionisation processes described above represent reaction between incoming molecules and the substrate (ice), since water molecules in the surface are clearly either activated or capable of being activated in the presence of an adsorbate and can therefore participate in chemical reactions in different ways at the surface compared to the gas phase or the bulk. Comparable activation effects in the reaction between adsorbed states and incoming gases are also observed, and the basis of such reactions are discussed below using the HCl + $ClONO_2$ system as an example.

Reaction between $ClONO_2$ and $H_3O^+(H_2O)_nCl^-$
The kinetics of the heterogeneous reaction between these important reservoir species on ice surfaces has been extensively studied in coated-wall flow tubes and Knudsen reactors [43,100]. It is known to be fast, producing gas-phase Cl_2 and condensed nitric acid under stratospheric conditions. However, as for the hydrolysis reactions, mechanistic details require the application of surface sensitive methods, which themselves require electronic structure calculations for clear interpretation. We have previously reported IR spectroscopic and mass spectrometric observations of the reaction between gaseous $ClONO_2$ and adsorbed HCl using a film with an appreciable concentration of hydrated ionic HCl as a substrate [102]. Such a film can easily be formed *in vacuo* by co-condensing HCl and water upon a metal substrate at 160 K, leading to stoichiometric or amorphous HCl hydrates depending upon deposition conditions. For example, the RAIR spectrum of such a film is shown in Fig 13.9a, and exhibits features which are readily assigned to an amorphous film of HCl hydrate of *ca.* 4:1 H_2O:HCl stoichiometry. Upon exposure of such a film to gaseous $ClONO_2$, the growth of strong features which are readily assigned to crystalline forms of ionised $HONO_2$ and HCl hydrates is observed

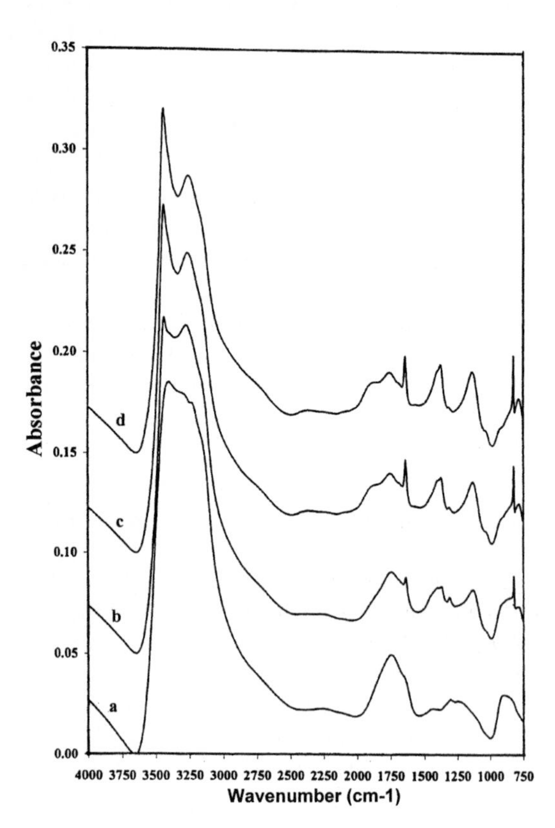

Fig. 13.9. RAIR spectra of the result of exposing an amorphous HCl/H_2O hydrate film at 160 K to gaseous $ClONO_2$ From the bottom spectrum up: unreacted HCl/H_2O film; subsequent spectra show exposure to $1x10^{-7}$ mbar $ClONO_2$ for 20 s; 40 s; 60 s. Copyright of the American Chemical Society, 1998.

(Figs. 13.9b – 13.9d). These spectra indicate therefore that the heterogeneous reaction of gas-phase $ClONO_2$ with condensed, hydrated HCl films results in the formation of condensed nitric acid hydrates as expected. Mass spectrometric monitoring of the gases present above the substrate surface during exposure to constant pressure sequential doses of $ClONO_2$ gave clear evidence for a saturating reaction. In each case, the surface was exposed to a constant pressure of $ClONO_2$ for a fixed period (20 s). For the first dose, a rapid rise in the m/e 70 signal due to $^{35}Cl_2^+$ was immediately observed, indicating that the reaction promptly produces gas-phase Cl_2. The IR spectrum of the first post-dose film is presented in Fig. 13.9b, showing the traces of nitrate deposition. Subsequent dosing showed that the $^{35}Cl_2^+$ signal rapidly decreases with cumulative exposure, as does the rate of surface nitrate hydrate formation. Eventually, no further reaction occurs. The IR spectrum of the substrate after reaction has ceased, in Fig. 13.9d, shows the presence of known, stable $HONO_2$ and HCl hydrates (specifically, nitric acid trihydrate and HCl hexahydrate). This composition is corroborated by static secondary ion mass spectrometric measurements of the surface composition after reaction, which reveal the presence of nitrate and $H_3O^+(H_2O)_n$ ions only. No further reaction is observed by any of these methods unless the film is refreshed with water.

Prompt production of Cl_2 upon exposure to $ClONO_2$ is evidence for a direct reaction with adsorbed HCl. As described in section 3.2 above, HCl is present in an ionised form, $H_3O^+Cl^-$, under these conditions and it therefore seems reasonable to conclude that the reaction occurs between Cl^- and incoming $ClONO_2$. This conclusion is supported by the recent observation of direct reaction between $ClONO_2$ and water clusters containing DCl in a flow reactor ([118] and section 2 of chapter 5). Once again, computational studies allow the details of the reaction mechanism to be explored. Bianco & Hynes [119] report studies of the reaction in clusters containing nine H_2O molecules in addition to the reactants, and observe an initial reaction complex which contains ionized HCl (as an $H_3O^+Cl^-$ contact ion pair) and molecular $ClONO_2$. They calculate a reaction path which proceeds via a proton transfer from the initial contact ion pair in the ice lattice, accompanied by a nucleophilic attack of the Cl^- on the chlorine atom of $ClONO_2$. They also estimate the net exothermicity of the removal of the product Cl_2 into the gas phase as -11.4 kcal mol^{-1} and calculate a barrier to the reaction of *ca.* 6 kcal mol^{-1}. From this evidence, they conclude that the proton transfer step from the contact ion pair is essential in facilitating the reaction, and suggest that solvent-separated ions are less able to participate in such a reaction. In contrast, more recent calculations on similar $HCl/ClONO_2/H_2O$ clusters by McNamara et al. (which also predict the ionisation of HCl in the cluster) arrive at the conclusion that it is the solvent-separated ions in larger clusters which are involved on the basis that reaction with a free Cl^- ion leads directly to the experimentally observed ionised nitrate ion product and the prompt release of a gas phase Cl_2 molecule [120]. The principal conclusion to be drawn from these theoretical treatments appears to be that the ice surface plays a lesser role in facilitating the reaction between Cl^- and $ClONO_2$ compared to the hydrolysis process, since $ClONO_2$ presentation and rearrangement is less critical in the minimisation of the energetic barriers to reaction.

13.4 Photochemistry in ices and on ice surfaces

The study of photochemical processes in environments for which molecular motion is restricted has been pursued for many years. The substrate solids and surfaces involved can range from the truly organized, *i.e.* crystalline, through to low-temperature, inert "gas matrices" in which random orientations between guests and the host are apparent. These latter interactions are often interpretable using IR spectroscopy and are generally grouped under the heading "site-effects". Monographs have been devoted to these specialised fields of research in constrained media often in the context of diverse applications such as the photo-production of organic compounds, information storage within surface thin-films and biological photosynthesis [121].

In contrast, studies of constrained photochemistry in ices and on their surfaces are far fewer in number, much more recent in publication and mainly with applications directed toward the study of the Earth's atmosphere or more distant planetary/interstellar phenomena. A review published in 1995 summarised potential reasons for the study of heterogeneous atmospheric photochemistry associated with ices as follows: "The Antarctic ozone hole caught the atmospheric science community off-guard because heterogeneous thermal reactions were overlooked; let us not disregard the potential effect of heterogeneous photochemistry"[122]. Therefore the purpose of this section is to review and link the work that has been published to date on the photochemistry of organic and inorganic compounds constrained within two main ice environments: (i) The Earth, particularly with relation to its polar atmospheric and snow-pack regions; (ii) Interstellar Space.

13.4.1 The Earth

The chemistry associated with the Earth's atmosphere, particularly in relation to the processes that control levels of ozone has been studied intensively in recent years. Effectively the critical balance of ozone in the stratosphere (*ca.* 16-40 km altitudes) has implications for the filtering of ecologically harmful UV-B light from *terra firma* and the oceans. In the troposphere (*ca.* 0-12 km) increasing the concentration levels of ozone and other reactive species causes changes to the oxidising (oxidative) capacity of the atmosphere. In this regard, the contrast between the natural toxicity of ozone and its so-called "detergent" effect in cleaning the troposphere of harmful chemical species is crucial. The complexities associated with understanding the beneficial/harmful paradox of ozone in the Earth's atmosphere has led to major advances in field-measuring analytical techniques for a variety of reactive species, the development of sophisticated, predictive computer models and the recognition of the role that heterogeneous chemistry plays in air pollution events. In the specific area of "ice photochemistry" the main contributions of the studies have been in the observations associated with the release of active species from the polar ice/snow-packs and the role of Polar Stratospheric Clouds (PSCs) in halogen-initiated stratospheric ozone depletion events over the Polar regions [1].

Ice/Snow-packs. As alluded to above, the oxidative capacity of the atmosphere is controlled by the presence of highly reactive species. One of the most important is the hydroxyl radical (OH), formed as a result of ozone photolysis. Monitoring the dramatic depletion of ozone in Arctic surface air during polar sunrise therefore offers an opportunity to improve understanding of the processes controlling ozone abundance and hence the oxidative capacity of the atmosphere. Arctic ozone destruction is catalysed, in part, by bromine atoms and terminated once bromine reacts with formaldehyde to form relatively inert hydrogen bromide. However, current computer models cannot simulate the high formaldehyde concentrations actually found in Arctic surface air. It has been suggested from field measurements that photochemical production at the air-snow interface accounts for the discrepancy between observed and predicted formaldehyde concentrations [123-125]. This work has been extended to include the measurement of acetaldehyde and acetone in ambient and snow-pack air as part of the SNOW99 study in Michigan, USA [126].

Hydrogen peroxide (H_2O_2) also contributes to the atmosphere's oxidizing capacity and measured bidirectional summertime H_2O_2 fluxes from the snow-pack at Greenland have revealed a daytime H_2O_2 release from the surface snow reservoir. These observations provided the first direct evidence of a strong net summertime H_2O_2 release from the snow-pack. Relative to calculations estimated from photochemical modelling in the absence of any snow-pack source, the measured H_2O_2 concentrations were found to be approximately seven-fold higher. The hydroxyl (OH) and hydroperoxyl (HO_2) concentrations were also greater by 70% and 50%, respectively. The total H_2O_2 release over a 12-day period was of the order of 5×10^{13} molecules m^{-2} s^{-1}, which compared well with observed concentration changes in the top snow layer. Photochemical and air-snow interaction modelling indicated that the net snow-pack release is driven by temperature-induced release of H_2O_2 from deposited snow. The conclusion was reached that the physical cycling of H_2O_2 represents a key to understanding how snow-packs act as complex physical-photochemical Earth reactors [127].

Observations of OH, NO, and actinic flux at the South Pole surface during December 1998 have also implied a surprisingly active photochemical environment, which should result in the photochemical *production* of ozone [128]. Both snow manipulation experiments and ambient measurements during the Polar Sunrise Experiment 2000 at Alert (Alert2000) indicate intensive photochemical production of nitrous acid (HONO) in the snow-pack. This process constitutes a major HONO source for the overlying atmospheric boundary layer in the Arctic during the springtime, and sustained concentrations of HONO high enough that upon photolysis they became the dominant hydroxyl radical (OH) source. Measurements of NO_x ($NO + NO_2$) and the sum of reactive nitrogen constituents, NO_y, were made near the surface at Alert, Canada during 1998. In early March when solar insolation was absent or very low, NO_x mixing ratios were frequently near zero. After polar sunrise when the sun was above the horizon for much or all of the day a diurnal variation in NO_x and NO_y was observed with amplitudes as large as 30-40 pptv. The source of active nitrogen is attributed to release from the snow surface by a process that is apparently sensitized by sunlight. It was suggested that the observed release of NO_x may have been initiated by photolysis of nitrate, present in relative abundance in surface snow at Summit [129-131].

A series of experiments has subsequently been designed to test the hypothesis that the observed NO_x production is the result of nitrate photolysis [132]. Snow produced from deionized water with and without the addition of nitrate was exposed to natural sunlight in an outdoor flow chamber. While NO_x release from snow produced without added NO_3^- was minimal, the addition of 100 μM NO_3^- resulted in the release of > 500 pptv NO_x. The rate of release was highly correlated with solar radiation. Further addition of radical trap reagents resulted in greatly increased NO_x production (to > 8 ppbv). In snow produced from deionized water plus sodium nitrate, production of NO_2 dominated that of NO. The reverse was true in the presence of radical trap reagents. An investigation of nitrogen dioxide release in the photolysis of spray-frozen aqueous nitrate solutions has also been undertaken [133]. The results suggest that NO_3^- photolysis in ice takes place in a liquid-like environment and that actual quantum yield values may depend on the morphology of the ice deposits. Further studies and interpretation of the *photolysis* mechanism for nitrate species at low-temperatures are presented in the "Polar Stratospheric Cloud" discussion below.

In addition to their more general pollutant effects, many organic compounds contribute to the atmosphere's oxidising capacity. In this regard the recent observation of the deposition of carboxylic acids in snows and ices, possibly from their *in situ* photochemical processing, may be relevant [134]. In fact, the distribution, accumulation, and chemical/photochemical transformations of persistent, bio-accumulative, and toxic compounds (PBTs) in water ice, especially in the polar regions is now an active field of study. Hence models have been proposed in which significant amounts of PBTs might be generated by the photochemistry of primary pollutants in ice [135]. Furthermore results obtained after the photolysis of chlorobenzene, o- and p-dichlorobenzene, bromobenzene and p-dibromobenzene in water ice have been reported. All the photo-transformations appeared to be based on dehalogenation followed by coupling, and rearrangement reactions in ice cavities. Many of the products were very toxic substances of high environmental risk, such as PCBs [136,137]. These laboratory results support the hypothesis that, in the atmosphere, secondary pollutants can be formed on snow and ice particles by the action of solar irradiation.

Indeed measurements at Summit, Greenland, performed from June-August 1999, showed significant enhancement in concentrations of several trace gases in the snowpack (firn) pore air relative to the atmosphere. Measurements have been reported of alkenes, halocarbons, and alkyl nitrates that are typically a factor of 2-10 higher in concentration within the firn air than in the ambient air 1-10 m above the snow. Profiles of concentration to a depth of 2 m into the firn show that maximum values of these trace gases occur between the surface and 60 cm depth. The alkenes show highest pore mixing ratios very close to the surface, with mixing ratios in the order ethene > propene > 1-butene. Mixing ratios of the alkyl iodides and alkyl nitrates peak slightly deeper in the firn, with mixing ratios in order of methyl > ethyl > propyl. These variations are likely consistent with different near-surface photochemical production mechanisms [138].

Polar Stratospheric Clouds. As discussed in sections 2 and 3 of this review, stratospheric heterogeneous chemistry catalysed by PSC (water-ice) surfaces is now well-established as playing an important role in the processes which mediate polar ozone levels. Since 1998 a number of publications have also centred on the water-ice *photochemistry* of the main surface species involved such as chlorine nitrate, hydrochloric acid, nitric acid and dinitrogen pentoxide.

Hence the heterogeneous photochemistry of chlorine nitrate ($ClONO_2$) adsorbed on HCl-doped crystals has been studied at 181 K and at 190 K. Under the experimental conditions employed the main gaseous products found at 181 K were Cl_2O and Cl_2, while at 190 K, Cl_2O was mainly observed. At both temperatures a net enhancement in the rate of gaseous product formation was observed when light of wavelength longer than 350 nm was used for photolysis [139]. The photodissociation of the closely related $ClNO_2$ embedded in argon and water low-temperature clusters has also been investigated by detecting the chlorine atom photo-fragments using resonance enhanced multiphoton ionization time-of-flight spectroscopy (REMPI-TOF). By utilizing various cluster formation conditions the measured speed distributions of the products show that, in contrast to monomer photolysis, only one decay channel is found to be active. The implication to atmospheric chemistry is that $ClNO_2$ embedded in water clusters represents a photolytic source of Cl and NO radicals [140].

Given the prevalence of condensed nitric acid hydrates in the atmosphere, reflection-absorption infrared spectroscopy (RAIRS) has been employed in

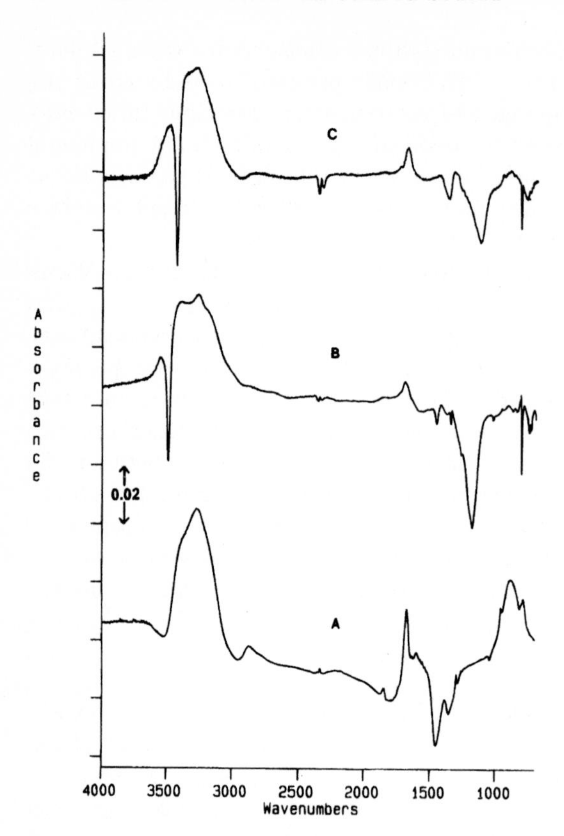

Fig. 13.10. RAIR spectra resulting from the photolysis of thin films of nitric acid hydrates (300 min at $\lambda > 230$ nm) at 120 K. (A) amorphous nitric acid dihydrate; (B) crystalline nitric acid dihydrate; (C) crystalline nitric acid trihydrate. Copyright of the American Chemical Society, 1996.

order to investigate the low-temperature photochemistry (90 - 140 K) of thin films of appropriate composition (including ammonium nitrate) grown *in vacuo*. Figure 13.10 shows that photolysis of amorphous nitric acid hydrate, the crystalline dihydrate (NAD) and trihydrate (NAT) at $\lambda > 230$ nm results in the formation of molecular nitric acid due to rapid protonation of the excited nitrate ion. Secondary photolysis of $HONO_2$ produces NO_2 and NO. Similarly, when neat films of molecular, anhydrous nitric acid are irradiated, nitrate and nitronium ions are observed. In contrast, ammonium nitrate photolysis at 140 K did not result in a proton transfer to produce NH_3 and $HONO_2$ but in the formation of the peroxynitrite ion ($ONOO^-$) as a precursor for NO_2^-. Molecular dinitrogen tetroxide and nitrous oxide were also detected in the film [141]. What these measurements show is that in favourable optical conditions, transformations of these materials into less benign (and more reactive) forms are possible, and that photochemistry may result not only in additional surface reactions but also in the modification of the reactive surface. This is an area requiring much more detailed scrutiny, since few atmospheric models include such effects.

Other NO_x and ClO_x compounds have been photolysed in ice systems over the last few years in order to determine whether the processes might impinge on the chemical composition of the atmosphere. For example, monolayer coverages of N_2O_4 have been prepared *via* the adsorption of gaseous NO_2 at *ca.* 100 K on 5-10 monolayers of condensed H_2O in a high vacuum system. Exposure to continuous near-UV irradiation resulted in the photodesorption of NO_2 from the surface. The interaction with the ice was determined to be purely physical in character and there was no reaction observed between NO_2 and the water surface. The temporal profiles of both the NO_2 photodesorption rate and of the surface-bound N_2O_4 concentration exhibited simple monotonic decays and were explained within a simple kinetic scheme for the photochemistry of adsorbed N_2O_4. The wavelength dependence of the NO_2 yield was shown to be comparable to the gas-phase absorption cross-section for N_2O_4. The experiments show that the photochemistry of N_2O_4 physisorbed on low-temperature ice is not significantly different to that of its gas-phase counterpart. The dynamics of photodissociation in physisorbed N_2O_4 have also been investigated by time-of-flight mass spectrometry. It was shown that the photodissociation dynamics are dependent on the unique properties of the ordered physisorbed system [142-144].

In a separate set of studies, the photochemistry of NO_2 adsorbed on an Au(111) surface has been investigated at < 120 K in an ultrahigh vacuum system. The adsorption states of the molecules were characterized by thermal desorption spectroscopy and IR reflection absorption spectroscopy. It was concluded that the photolysis of physisorbed N_2O_4 is inhibited significantly when the Au surface is covered with a thin water ice film, suggesting that the photolysis is enhanced by metal substrate photoexcitation or electron transfer from the substrate to the adsorbates [145].

The photochemistry of chlorine dioxide, OClO, in amorphous ice has also been investigated using IR and UV-vis spectroscopy. Exposure to ultraviolet light (360 nm) quantitatively converted the OClO to the chlorine peroxy radical, ClOO. Under the dilute conditions used, the only photoproduct appeared to be ClOO, in contrast with the gas-phase photolysis which yields predominantly ClO. It was suggested that these findings have implications for stratospheric ozone loss because the photochemical conversion of OClO to ClOO in stratospheric ice particles may represent a new ozone-depleting mechanism. Thus, under the very dilute conditions that would exist on polar stratospheric ice particles, the photochemical conversion of OClO to Cl atoms and O_2 would represent a new source of active halogen radicals for ozone depletion [146,147]. In a separate TOF-QMS/RAIR study it was shown that the predominant photoproduct retained in the ice films at coverages between 0.5 and 2 ML was $ClClO_2$ without formation of other photoproducts of formula Cl_xO_y [148]. A third research group has also investigated the OClO system using RAIR spectroscopy and mass spectrometry. It was confirmed that photolysis of a neat film of OClO ($\lambda > 300$ nm, 90-110 K) produces chloryl chloride, $ClClO_2$. However it was also shown that irradiation of a co-deposited $OClO/H_2O$ film produces chlorine superoxide, ClOO. This result suggests that OClO isomerisation is the first step in the process which produces $ClClO_2$. Photolysis of Cl_2O was also undertaken and shown to produce OClO, initially, which is subsequently converted to $ClClO_2$ [149].

Recently the photodecomposition of $CFCl_3$ on an ice surface on Ru(001) has been studied using infrared absorption spectroscopy. It appears that so-called anionic chlorine coordinated to free OH species is formed as a result. If extrapolated to the field, the mechanism would enable the highly concentrated storage of chlorine on ice surfaces [150].

The above studies are, of course, laboratory-based but, to help complete any theoretical picture of polar and mid-latitude ozone erosion by chlorine free radical catalysis, field measurements which include ice analysis are required. Hence recent simultaneous, high-resolution observations of ClO, H_2O, tropopause height, particle reactive surface area, and ice saturation occurrence frequency have been obtained using the NASA ER-2 aircraft. The objective was to test the hypothesis that the existence of cirrus clouds or cold aerosols in the first few kilometers above the tropopause at mid-latitudes is responsible for increasing the ratio of chlorine free radicals to total inorganic chlorine, thus amplifying the rate of catalytic ozone destruction. The observations revealed a sharp decrease in ice saturation frequency at the tropopause, a marked degree of under-saturation just above the tropopause and a corresponding sharp gradient in the product of cold aerosol reactive surface area. Finally, the consistent absence of enhanced concentrations of ClO immediately above the tropopause was noted. The results suggest that mid-latitude ozone loss is not controlled *in situ* by the mechanism of cirrus cloud and/or cold aerosol enhancement of chlorine radicals in the vicinity of the tropopause [151].

Of final note in this section is the report that IR spectroscopy has been used to study the vibrational spectrum of ozone trapped in amorphous ice. The photochemistry of ozone in excess ice was also investigated using laser irradiation at 266 nm. H_2O_2 was shown to be produced at low-temperatures through a hydrogen-bonded complex between ozone and free OH bonds. At higher temperatures, when a solid solution of ozone in water exists, the H_2O_2 is apparently formed by the reaction of an excited oxygen atom $O(^1D_2)$ with its nearest water molecules [152].

Amorphous ice also represents a constrained low-temperature environment observed on outer satellites of the solar system. Therefore further photochemical studies, which have been made and are relevant to interstellar regions, are summarised below(and presented in more detail in Chapter 15).

13.4.2 Interstellar space

Current knowledge on the chemical composition of interstellar ices has recently been summarized in inventory form. The listing was determined by infrared observations from the Infrared Space Observatory (ISO) and also from laboratory spectroscopic experiments. Sources of radiolysis, UV photolysis, and ice heating are also given and, by use of specific examples, it is shown how "energetic processing" explains the observed solid-state characteristics of the key chemicals CH_3OH, CO_2, and OCN^-. The solid-state photochemistry of the first organic acid detected in interstellar ices, HCOOH, is also described [153].

The genesis of molecular material in the universe, from hydrogen to complex organic matter is also of interest and therefore both experimental and theoretical studies on relevant gas/solid/gas transformations have been published [154]. Clearly, interstellar ices can be chemically processed by ultraviolet radiation to form complex products and therefore a variety of UV photolysis experiments on ices for molecules of astrophysical interest such as H_2O, NH_3, CH_4, CO, CO_2, O_2, N_2, H_2CO, and CH_3OH have been performed. New molecules produced during photolysis were identified on the basis of their characteristic infrared features. Rates of formation were also estimated for first-order products. In experiments with CH_4 and H_2CO as guest molecules products with as many as 8 carbon atoms are produced. The results may indicate how and why very complex organic molecules are formed in molecular mixtures characteristic of interstellar ice mantles [155].

The infrared spectra of CO frozen in non-polar ices containing N_2, CO_2, O_2 and H_2O have also been reported along with associated UV photochemistry. The spectra were used to test the hypothesis that the narrow 2140 cm^{-1} interstellar absorption feature attributed to solid CO might be produced by CO frozen in ices containing non-polar species such as N_2. Good matches to the interstellar band at all temperatures between 12 and 30 K both before and after photolysis were found using this approach. Ultraviolet photolysis of the ices produced a variety of photoproducts including CO_2, N_2O, O_3, H_2CO, and possibly NO plus NO_2. XCN was not observed in these experiments, placing important constraints on the origin of this enigmatic interstellar feature. Interestingly N_2O has not previously been considered as an interstellar ice component [156].

It has long been thought that the delivery of extraterrestrial organic molecules to Earth by meteorites may have been important for the origin and early evolution of life. In this context, a very recent laboratory demonstration that glycine, alanine and serine naturally form after UV photolysis of the analogues of icy interstellar grains is of particular note. Such amino acids would naturally have a deuterium excess similar to that seen in interstellar molecular clouds, and the formation process could also result in enantiomeric excesses if the incident radiation is circularly polarized. The results suggest that at least some meteoritic amino acids are the result of interstellar photochemistry rather than formation in liquid water on an early Solar System body [157]. The polycyclic aromatic hydrocarbon (PAH) naphthalene has also been exposed to uv radiation in water-ice under astrophysical conditions; the products were analyzed using infrared spectroscopy and high-performance liquid chromatography. As found in earlier studies on the photoprocessing of coronene in water-ice, aromatic alcohols and ketones (quinones) were formed. The regiochemistry of the reactions leads to specific predictions of the relative abundances of various oxidized naphthalenes that should exist in meteorites. It is suggested that since oxidized PAHs are present in carbon-rich meteorites and interplanetary dust particles, the delivery of such extraterrestrial molecules to the early Earth may have played a role in the origin and evolution of life [158].

Although the majority of this review has been centred on photo-reactions in/on water-ice, it is important to remember that on more distant solar system bodies such as Triton and Pluto *nitrogen-ices* exist and are therefore worthy of some investigation. Hence IR spectroscopy has been used to show that irradiation of solid N_2 and N_2-rich ices with 0.8 MeV protons produces the N_3 (azide) radical. In contrast, no N_3 was observed after analogous photolysis by far-UV photons. The study apparently represents the first documented difference in reaction products between the radiation chemistry and photochemistry of a nonpolar astronomical ice analogue. The difference in reaction chemistries could be used to identify ion-irradiated ices on interstellar grains and in the outer solar system [159].

Finally, the IR transmission spectra and photochemical behaviour of various organic compounds isolated in solid N_2 ices have been published. It is shown that "excess absorption" in the surface spectra of Triton and Pluto, *i.e.*, absorption not explained by present models incorporating molecules already identified on these bodies such as N_2, CH_4, CO, and CO_2 , may be due to alkanes frozen in the nitrogen. Thus the photochemistry of N_2:CH_4 and N_2:CH_4:CO ices was explored, from which it was demonstrated that the reactive molecule diazomethane, CH_2N_2 is formed. The observation is thought to be particularly important since this compound may be largely responsible for the synthesis of larger alkanes from CH_4 and other small alkanes [160].

References

1. Finlayson-Pitts, B. J.; Pitts, J. N. *Chemistry of the Upper and Lower Atmosphere*; Academic Press: San Diego, 2000.
2. McCoustra, M. R. S.; Horn, A. B. *Chemical Society Reviews* **1994**, 23, 195.
3. Winkler, A. S.; Holmes, N. S.; Crowley, J. N. *Physical Chemistry Chemical Physics* **2002**, *accepted for publication.*
4. Tolbert, M. A.; Middlebrook, A. M. *Journal of Geophysical Research-Atmospheres* **1990**, 95, 22423.
5. Tolbert, M. A.; Koehler, B. G.; Middlebrook, A. M. *Spectrochimica Acta Part a-Molecular and Biomolecular Spectroscopy* **1992**, 48, 1303.
6. Chesters, M. A.; Horn, A. B.; Kellar, E. J. C.; Parker, S. F.; Raval, R. Spectroscopy of Crystal Growth Surface Intermediates on Silicon. In *Mechanisms*

and Reactions of Organometallinc Compounds With Surfaces; Cole-Hamilton, D. J., Williams, J. O., Eds.; Plenum Press: New York, 1989.
7. Bradshaw, A. N.; Schweizer, E. K. Infrared Reflection-Absorption Spectroscopy of Adsorbed Molecules. In *Advances in Spectroscopy*; Clark, R. J. H., Hester, R. E., Eds.; John WIley and Sons: Chichester, 1988; Vol. 16; pp 413.
8. Heavens, O. S. *Optical Properties of Thin Solid Films*; Butterworths Scientific Publications: London, 1955.
9. Horn, A. B.; Banham, S. F.; McCoustra, M. R. S. *Journal of the Chemical Society-Faraday Transactions* **1995**, 91, 4005.
10. Zondlo, M. A.; Onasch, T. B.; Warshawsky, M. S.; Tolbert, M. A.; Mallick, G.; Arentz, P.; Robinson, M. S. *Journal of Physical Chemistry B* **1997**, 101, 10887.
11. Fluckiger, B.; Chaix, L.; Rossi, M. J. *Journal of Physical Chemistry A* **2000**, 104, 11739.
12. Materer, N.; Starke, U.; Barbieri, A.; Vanhove, M. A.; Somorjai, G. A.; Kroes, G. J.; Minot, C. *Journal of Physical Chemistry* **1995**, 99, 6267.
13. Materer, N.; Starke, U.; Barbieri, A.; VanHove, M. A.; Somorjai, G. A.; Kroes, G. J.; Minot, C. *Surface Science* **1997**, 381, 190.
14. Devlin, J. P. *Journal of Geophysical Research-Planets* **2001**, 106, 33333.
15. Devlin, J. P.; Buch, V. *Journal of Physical Chemistry* **1995**, 99, 16534.
16. Mantz, Y. A.; Geiger, F. M.; Molina, L. T.; Molina, M. J.; Trout, B. L. *Journal of Chemical Physics* **2000**, 113, 10733.
17. Zhang, Q.; Buch, V. *Journal of Chemical Physics* **1990**, 92, 5004.
18. Zhang, Q.; Buch, V. *Journal of Chemical Physics* **1990**, 92, 1512.
19. Trakhtenberg, S.; Naaman, R.; Cohen, S. R.; Benjamin, I. *Journal of Physical Chemistry B* **1997**, 101, 5172.
20. Dohnalek, Z.; Ciolli, R. L.; Kimmel, G. A.; Stevenson, K. P.; Smith, R. S.; Kay, B. D. *Journal of Chemical Physics* **1999**, 110, 5489.
21. Dohnalek, Z.; Kimmel, G. A.; Ciolli, R. L.; Stevenson, K. P.; Smith, R. S.; Kay, B. D. *Journal of Chemical Physics* **2000**, 112, 5932.
22. Lankau, T.; Cooper, I. L. *Journal of Physical Chemistry A* **2001**, 105, 4084.
23. Lankau, T.; Nagorny, K.; Cooper, I. L. *Langmuir* **1999**, 15, 7308.
24. Zhdanov, V. P.; Norton, P. R. *Surface Science* **2000**, 449, L228.
25. Buch, V.; Devlin, J. P. *Journal of Chemical Physics* **1991**, 94, 4091.
26. Rowland, B.; Fisher, M.; Devlin, J. P. *Journal of Physical Chemistry* **1993**, 97, 2485.
27. Keyser, L. F.; Leu, M. T. *Microscopy Research and Technique* **1993**, 25, 434.
28. Horn, A. B.; Chesters, M. A.; Mccoustra, M. R. S.; Sodeau, J. R. *Journal of the Chemical Society-Faraday Transactions* **1992**, 88, 1077.
29. Holmes, N. S.; Sodeau, J. R. *Journal of Physical Chemistry A* **1999**, 103, 4673.
30. Manca, C.; Allouche, A. *Journal of Chemical Physics* **2001**, 114, 4226.
31. Palumbo, M. E. *Journal of Physical Chemistry A* **1997**, 101, 4298.
32. Allouche, A.; Verlaque, P.; Pourcin, J. *Journal of Physical Chemistry B* **1998**, 102, 89.
33. Manca, C.; Martin, C.; Allouche, A.; Roubin, P. *Journal of Physical Chemistry B* **2001**, 105, 12861.
34. Graham, J. D.; Roberts, J. T.; Brown, L. A.; Vaida, V. *Journal of Physical Chemistry* **1996**, 100, 3115.

35. Fenter, F. F.; Rossi, M. J. *Journal of Physical Chemistry A* **1997**, 101, 4110.
36. Borget, F.; Chiavassa, T.; Allouche, A.; Aycard, J. P. *Journal of Physical Chemistry B* **2001**, 105, 449.
37. Ahmed, M.; Apps, C. J.; Hughes, C.; Watt, N. E.; Whitehead, J. C. *Journal of Physical Chemistry A* **1997**, 101, 1250.
38. Ahmed, M.; Apps, C. J.; Hughes, C.; Whitehead, J. C. *Chemical Physics Letters* **1995**, 240, 216.
39. Sokolov, O.; Abbatt, J. P. D. *Journal of Physical Chemistry A* **2002**, 106, 775.
40. Schaff, J. E.; Roberts, J. T. *Journal of Physical Chemistry* **1994**, 98, 6900.
41. Schaff, J. E.; Roberts, J. T. *Langmuir* **1998**, 14, 1478.
42. Chu, L. T.; Leu, M. T.; Keyser, L. F. *Journal of Physical Chemistry* **1993**, 97, 7779.
43. Leu, M. T.; Moore, S. B.; Keyser, L. F. *Journal of Physical Chemistry* **1991**, 95, 7763.
44. Chu, L. T.; Leu, M. T.; Keyser, L. F. *Journal of Physical Chemistry* **1993**, 97, 12798.
45. Thibert, E.; Domine, F. *Journal of Physical Chemistry B* **1997**, 101, 3554.
46. Hanson, D. R.; Mauersberger, K. *Journal of Physical Chemistry* **1990**, 94, 4700.
47. Abbatt, J. P. D.; Beyer, K. D.; Fucaloro, A. F.; Mcmahon, J. R.; Wooldridge, P. J.; Zhang, R.; Molina, M. J. *Journal of Geophysical Research-Atmospheres* **1992**, 97, 15819.
48. Barone, S. B.; Zondlo, M. A.; Tolbert, M. A. *Journal of Physical Chemistry A* **1999**, 103, 9717.
49. Foster, K. L.; Tolbert, M. A.; George, S. M. *Journal of Physical Chemistry A* **1997**, 101, 4979.
50. Hanson, D. R.; Ravishankara, A. R. *Journal of Physical Chemistry* **1992**, 96, 2682.
51. Graham, J. D.; Roberts, J. T. *Journal of Physical Chemistry* **1994**, 98, 5974.
52. Kroes, G. J.; Clary, D. C. *Geophysical Research Letters* **1992**, 19, 1355.
53. Banham, S. F.; Sodeau, J. R.; Horn, A. B.; McCoustra, M. R. S.; Chesters, M. A. *Journal of Vacuum Science & Technology a-Vacuum Surfaces and Films* **1996**, 14, 1620.
54. Koehler, B. G.; Mcneill, L. S.; Middlebrook, A. M.; Tolbert, M. A. *Journal of Geophysical Research-Atmospheres* **1993**, 98, 10563.
55. Graham, J. D.; Roberts, J. T. *Chemometrics and Intelligent Laboratory Systems* **1997**, 37, 139.
56. Donsig, H. A.; Vickerman, J. C. *Journal of the Chemical Society-Faraday Transactions* **1997**, 93, 2755.
57. Boryak, O. A.; Kosevich, M. V.; Stepanov, I. O.; Shelkovsky, V. S. *International Journal of Mass Spectrometry* **1999**, 189, LI.
58. Wang, L. C.; Clary, D. C. *Journal of Chemical Physics* **1996**, 104, 5663.
59. Gertner, B. J.; Hynes, J. T. *Faraday Discussions* **1998**, 301.
60. Svanberg, M.; Pettersson, J. B. C.; Bolton, K. *Journal of Physical Chemistry A* **2000**, 104, 5787.
61. Mantz, Y. A.; Geiger, F. M.; Molina, L. T.; Trout, B. L. *Journal of Physical Chemistry A* **2001**, 105, 7037.
62. Mantz, Y. A.; Geiger, F. M.; Molina, L. T.; Molina, M. J.; Trout, B. L. *Chemical Physics Letters* **2001**, 348, 285.

63. Toubin, C.; Picaud, S.; Hoang, P. N. M.; Girardet, C.; Demirdjian, B.; Ferry, D.; Suzanne, J. *Journal of Chemical Physics* **2002**, 116, 5150.
64. Re, S.; Osamura, Y.; Suzuki, Y.; Schaefer, H. F. *Journal of Chemical Physics* **1998**, 109, 973.
65. Smith, A.; Vincent, M. A.; Hillier, I. H. *Journal of Physical Chemistry A* **1999**, 103, 1132.
66. Bacelo, D. E.; Binning, R. C.; Ishikawa, Y. *Journal of Physical Chemistry A* **1999**, 103, 4631.
67. Bolton, K.; Pettersson, J. B. C. *Journal of the American Chemical Society* **2001**, 123, 7360.
68. Milet, A.; Struniewicz, C.; Moszynski, R.; Wormer, P. E. S. *Journal of Chemical Physics* **2001**, 115, 349.
69. Uras, N.; Rahman, M.; Devlin, J. P. *Journal of Physical Chemistry B* **1998**, 102, 9375.
70. Kang, H.; Shin, T. H.: Park, S. C.; Kim, I. K.; Han, S. J. *Journal of the American Chemical Society* **2000** ,122, 9842
71. Horn, A. B.; Sully, J. *Journal of the Chemical Society-Faraday Transactions* **1997**, 93, 2741.
72. Livingston, F. E.; George, S. M. *Journal of Physical Chemistry A* **2001**, 105, 5155.
73. Uras-Aytemiz, N.; Joyce, C.; Devlin, J. P. *Journal of Physical Chemistry A* **2001**, 105, 10497.
74. Andersson, P. U.; Nagard, M. B.; Pettersson, J. B. C. *Journal of Physical Chemistry B* **2000**, 104, 1596.
75. Haq, S.; Harnett, J.; Hodgson, A. *Journal of Physical Chemistry B* **2002**, 106, 3950.
76. WMO "Scientific Assessment of Ozone Depletion,", 1994.
77. Hanson, D. R.; Ravishankara, A. R. *Journal of Geophysical Research-Atmospheres* **1991**, 96, 5081.
78. Hisatsune, I. C.; Devlin, J. P.; Wada, Y. *Spectrochimica Acta* **1962**, 18, 1641.
79. Koch, T. G.; Horn, A. B.; Chesters, M. A.; Mccoustra, M. R. S.; Sodeau, J. R. *Journal of Physical Chemistry* **1995**, 99, 8362.
80. Bencivenni, L.; Sanna, N.; SchriverMazzuoli, L.; Schriver, A. *Journal of Chemical Physics* **1996**, 104, 7836.
81. Zondlo, M. A.; Barone, S. B.; Tolbert, M. A. *Journal of Physical Chemistry A* **1998**, 102, 5735.
82. Ritzhaupt, G.; Devlin, J. P. *Journal of Physical Chemistry* **1991**, 95, 90.
83. Koehler, B. G.; Middlebrook, A. M.; Tolbert, M. A. *Journal of Geophysical Research-Atmospheres* **1992**, 97, 8065.
84. Toon, O. B.; Tolbert, M. A.; Koehler, B. G.; Middlebrook, A. M.; Jordan, J. *Journal of Geophysical Research-Atmospheres* **1994**, 99, 25631.
85. Middlebrook, A. M.; Berland, B. S.; George, S. M.; Tolbert, M. A.; Toon, O. B. *Journal of Geophysical Research-Atmospheres* **1994**, 99, 25655.
86. Richwine, L. J.; Clapp, M. L.; Miller, R. E.; Worsnop, D. R. *Geophysical Research Letters* **1995**, 22, 2625.
87. Peil, S.; Seisel, S.; Schrems, O. *Journal of Molecular Structure* **1995**, 348, 449.
88. Nakamoto, K. *Infrared and Raman Spectra of Inorganic and Coordination Compounds*, 5th Ed. ed.; John Wiley: New York, 1997.
89. Horn, A. B.; Koch, T.; Chesters, M. A.; McCoustra, M. R. S.; Sodeau, J. R. *Journal of Physical Chemistry* **1994**, 98, 946.

90. Koch, T. G.; Banham, S. F.; Sodeau, J. R.; Horn, A. B.; McCoustra, M. R. S.; Chesters, M. A. *Journal of Geophysical Research-Atmospheres* **1997**, 102, 1513.
91. Agreiter, J.; Frankowski, M.; Bondybey, V. E. *Low Temperature Physics* **2001**, 27, 890.
92. Cacace, F.; Attina, M.; Depetris, G.; Speranza, M. *Journal of the American Chemical Society* **1990**, 112, 1014.
93. Choi, J. H.; Kuwata, K. T.; Cao, Y. B.; Haas, B. M.; Okumura, M. *Journal of Physical Chemistry A* **1997**, 101, 6753.
94. Cao, Y. B.; Choi, J. H.; Haas, B. M.; Johnson, M. S.; Okumura, M. *Journal of Chemical Physics* **1993**, 99, 9307.
95. Hanway, D.; Tao, F. M. *Chemical Physics Letters* **1998**, 285, 459.
96. Snyder, J. A.; Hanway, D.; Mendez, J.; Jamka, A. J.; Tao, F. M. *Journal of Physical Chemistry A* **1999**, 103, 9355.
97. McNamara, J. P.; Hillier, I. H. *Journal of Physical Chemistry A* **2000**, 104, 5307.
98. Minton, T. K.; Nelson, C. M.; Moore, T. A.; Okumura, M. *Science* **1992**, 258, 1342.
99. Hanson, D. R. *Journal of Physical Chemistry* **1995**, 99, 13059.
100. Oppliger, R.; Allanic, A.; Rossi, M. J. *Journal of Physical Chemistry A* **1997**, 101, 1903.
101. Horn, A. B.; Sodeau, J. R.; Roddis, T. B.; Williams, N. A. *Journal of the Chemical Society-Faraday Transactions* **1998**, 94, 1721.
102. Horn, A. B.; Sodeau, J. R.; Roddis, T. B.; Williams, N. A. *Journal of Physical Chemistry A* **1998**, 102, 6107.
103. Berland, B. S.; Tolbert, M. A.; George, S. M. *Journal of Physical Chemistry A* **1997**, 101, 9954.
104. Bianco, R.; Hynes, J. T. *Journal of Physical Chemistry A* **1998**, 102, 309.
105. Sodeau, J. R.; Horn, A. B.; Banham, S. F.; Koch, T. G. *Journal of Physical Chemistry* **1995**, 99, 6258.
106. Geiger, F. M.; Pibel, C. D.; Hicks, J. M. *Journal of Physical Chemistry A* **2001**, 105, 4940.
107. Bianco, R.; Thompson, W. H.; Morita, A.; Hynes, J. T. *Journal of Physical Chemistry A* **2001**, 105, 3132.
108. McNamara, J. P.; Hillier, I. H. *Journal of Physical Chemistry A* **1999**, 103, 7310.
109. McNamara, J. P.; Tresadern, G.; Hillier, I. H. *Chemical Physics Letters* **1999**, 310, 265.
110. McNamara, J. P.; Hillier, I. H. *Journal of Physical Chemistry A* **2001**, 105, 7011.
111. Donsig, H. A.; Herridge, D.; Vickerman, J. C. *Journal of Physical Chemistry A* **1999**, 103, 9211.
112. Barone, S. B.; Zondlo, M. A.; Tolbert, M. A. *Journal of Physical Chemistry A* **1997**, 101, 8643.
113. Hanson, D. R.; Ravishankara, A. R. *Journal of Geophysical Research-Atmospheres* **1993**, 98, 22931.
114. Hanson, D. R.; Ravishankara, A. R. *Journal of Geophysical Research-Atmospheres* **1991**, 96, 17307.
115. Hanson, D. R. *Journal of Physical Chemistry A* **1998**, 102, 4794.

116. Caloz, F.; Fenter, F. F.; Rossi, M. J. *Journal of Physical Chemistry* **1996**, 100, 7494.
117. Gebel, M. E.; Finlayson-Pitts, B. J. *Journal of Physical Chemistry A* **2001**, 105, 5178.
118. Gilligan,J.J.; Castleman,A.W. *Journal of Physical Chemistry A* **2001**, 105, 1028.
119. Bianco, R.; Hynes, J. T. *Journal of Physical Chemistry A* **1999**, 103, 3797.
120. McNamara, J. P.; Tresadern, G.; Hillier, I. H. *Journal of Physical Chemistry A* **2000**, 104, 4030.
121. Ramamurthy, V. *Photochemistry in organized and constrained media*; VCH Publishers: NY, 1991.
122. Sodeau, J. R. Atmospheric Cryochemistry. In *Spectroscopy in Environmental Science*; Clark, R. J., Hester, R,E, Ed.; John Wiley & Sons: NY, 1995; pp 349.
123. Sumner, A. L.; Shepson, P. B. *Nature* **1999**, 398, 230.
124. Yang, J.; Honrath, R. E.; Peterson, M. C.; Dibb, J. E.; Sumner, A. L.; Shepson, P. B.; Frey, M.; Jacobi, H. W.; Swanson, A.; Blake, N. *Atmospheric Environment* **2002**, 36, 2523.
125. Sumner, A. L.; Shepson, P. B.; Grannas, A. M.; Bottenheim, J. W.; Anlauf, K. G.; Worthy, D.; Schroede, r. W. H.; Steffen, A.; Domine, F.; Perrier, S.; Houdier, S. *Atmospheric Environment* **2002**, 36, 2553.
126. Couch, T. L.; Sumner, A. L.; Dassau, T. M.; Shepson, P. B.; Honrath, R. E. *Geophysical Research Letters* **2000**, 27, 2241.
127. Hutterli, M. A.; McConnell, J. R.; Stewart, R. W.; Jacobi, H. W.; Bales, R. C. *Journal Of Geophysical Research-Atmospheres* **2001**, 106, 15395.
128. Crawford, J. H.; Davis, D. D.; Chen, G.; Buhr, M.; Oltmans, S.; Weller, R.; Mauldin, L.; Eisele, F.; Shette, r. R.; Lefer, B.; Arimoto, R.; Hogan, A. *Geophysical Research Letters* **2001**, 28, 3641.
129. Zhou, X. L.; Beine, H. J.; Honrath, R. E.; Fuentes, J. D.; Simpson, W.; Shepson, P. B.; Bottenheim, J. W. *Geophysical Research Letters* **2001**, 28, 4087.
130. Ridley, B.; Walega, J.; Montzka, D.; Grahek, F.; Atlas, E.; Flocke, F.; Stroud, V.; Deary, J.; Gallant, A.; Boudries, H.; Bottenheim, J. W.; Anlauf, K.; Worthy, D.; Sumner, A. L.; Splawn, B.; Shepson, P. B. *Journal Of Atmospheric Chemistry* **2000**, 36, 1.
131. Honrath, R.; Peterson, M. C.; Guo, S.; Dibb, J. E.; Shepson, P. B.; Campbell, B. *Geophysical Research Letters* **1999**, 26, 695.
132. Honrath, R. E.; Guo, S.; Peterson, M. C.; Dziobak, M. P.; Dibb, J. E.; Arsenault, M. A. *Journal Of Geophysical Research-Atmospheres* **2000**, 105, 24183.
133. Dubowski, Y.; Colussi, A. J.; Hoffmann, M. R. *Journal Of Physical Chemistry A* **2001**, 105, 4928.
134. de Souza, S. R.; de Carvalho, L. R. F. *Quimica Nova* **2001**, 24, 60.
135. Klan, P.; Holoubek, I. *Chemosphere* **2002**, 46, 1201.
136. (134) Klan, P.; del Favero, D.; Ansorgova, A.; Klanova, J.; Holoubek, I. *Environmental Science And Pollution Research* **2001**, 8, 195.
137. Klan, P.; Ansorgova, A.; del Favero, D.; Holoubek, I. *Tetrahedron Letters* **2000**, 41, 7785.
138. Swanson, A. L.; Blake, N. J.; Dibb, J. E.; Albert, M. R.; Blake, D. R.; Rowland, F. S. *Atmospheric Environment* **2002**, 36, 2671.
139. (137) Faraudo, G.; Weibel, D. E. *Progress In Reaction Kinetics And Mechanism* **2001**, 26, 179.

140. Li, Q.; Huber, J. R. *Chemical Physics Letters* **2002**, 354, 120.
141. Koch, T. G.; Holmes, N. S.; Roddis, T. B.; Sodeau, J. R. *Journal Of Physical Chemistry* **1996**, 100, 11402.
142. Rieley, H.; McMurray, D. P.; Haq, S. *Journal Of The Chemical Society-Faraday Transactions* **1996**, 92, 933.
143. Rieley, H.; Colby, D. J.; McMurray, D. P.; Reeman, S. M. *Journal Of Physical Chemistry B* **1997**, 101, 4982.
144. Rieley, H.; Colby, D. J.; McMurray, D. P.; Reeman, S. M. *Surface Science* **1997**, 390, 243.
145. Sato, S.; Senga, T.; Kawasaki, M. *Journal Of Physical Chemistry B* **1999**, 103, 5063.
146. Pursell, C. J.; Conyers, J.; Alapat, P.; R., P. *Journal Of Physical Chemistry* **1995**, 99, 10433.
147. Pursell, C. J.; Conyers, J.; Denison, C. *Journal Of Physical Chemistry* **1996**, 100, 15450.
148. Graham, J. D.; Roberts, J. T.; Anderson, L. D.; Grassian, V. H. *Journal Of Physical Chemistry* **1996**, 100, 19551.
149. Gane, M. P.; Williams, N. A.; Sodeau, J. R. *Journal Of The Chemical Society-Faraday Transactions* **1997**, 93, 2747.
150. Ogasawara, H.; Kawai, M. *Surface Science* **2002**, 502, 285.
151. Smith, J. B.; Hintsa, E. J.; Allen, N. T.; Stimpfle, R. M.; Anderson, J. G. *Journal Of Geophysical Research-Atmospheres* **2001**, 106, 1297.
152. Chaabouni, H.; Schriver-Mazzuoli, L.; Schriver, A. *Low Temperature Physics* **2000**, 26, 712.
153. Ehrenfreund, P.; d'Hendecourt, L.; Charnley, S.; Ruiterkamp, R. *Journal Of Geophysical Research-Planets* **2001**, 106, 33291.
154. (152) Cottin, H.; Gazeau, M. C.; Chaquin, P.; Raulin, F.; Benilan, Y. *Journal Of Geophysical Research-Planets* **2001**, 106, 33325.
155. Gerakines, P. A.; Schutte, W. A.; Ehrenfreund, P. *Astronomy And Astrophysics* **1996**, 312, 289.
156. Elsila, J.; Allamandola, L. J.; Sandford, S. A. *Astrophysical Journal* **1997**, 479, 818.
157. Bernstein, M. P.; Dworkin, J. P.; Sandford, S. A.; Cooper, G. W.; Allamandola, L. J. *Nature* **2002**, 416, 401.
158. Bernstein, M. P.; Dworkin, J. P.; Sandford, S. A.; Allamandola, L. J. *Meteoritics & Planetary Science* **2001**, 36, 351.
159. Hudson, R. L.; Moore, M. H. *Astrophysical Journal* **2002**, 568, 1095.
160. Bohn, R. B.; Sandford, S. A.; Allamandola, L. J.; Cruikshank, D. P. *Icarus* **1994**, 111, 151.

Part IV

Amorphous Solid Water (ASW) Films

14 Molecular Beam Studies of Nanoscale Films of Amorphous Solid Water

R. Scott Smith, Zdenek Dohnálek, Greg A. Kimmel, Glenn Teeter, Patrick Ayotte, John L. Daschbach, and Bruce D. Kay

14.1 Introduction

What is Amorphous Solid Water? Amorphous solid water (ASW) is a solid phase of water that is metastable with respect to the crystalline phase [1,2]. It is metastable because it is "trapped" in a configuration that has a higher free energy than the equilibrium crystalline configuration [3]. Amorphous solids, also known as glasses, are often described as structurally arrested or "frozen" liquids. Amorphous solids are most often formed when a liquid is cooled fast enough that crystallization does not occur prior to the system reaching a temperature where the structural relaxation timescale is long compared to the laboratory timescale, i.e. 100 s. The temperature where this occurs is called the glass transition temperature, T_g.

Figure 14.1 shows a model phase diagram for supercooled liquid, glass, and crystal formation. If a liquid is cooled below its melting temperature, T_m, at a rate that is faster than the crystallization rate, then a material can sometimes exist as a supercooled liquid that is thermodynamically metastable with respect to the crystalline phase. Upon further cooling the structural relaxation timescale increases until the supercooled liquid no longer behaves ergodically on an experimental timescale [3,4]. Below T_g, the relaxation timescale increases rapidly because the viscosity diverges to that of a solid (infinite viscosity) and thus the liquid's structure is "frozen". The material behaves less like a liquid and more like a solid i.e. an amorphous solid or glass. The ability to supercool a liquid and the lifetime of that supercooled liquid are limited by the crystallization kinetics to the thermodynamically favored state. An alternate approach to create a supercooled liquid is to create an amorphous solid and then heat it above its T_g, whereupon it transforms into a deeply-supercooled liquid prior to crystallization. The advantage of this approach is that at temperatures near T_g the viscosity is extremely high and therefore the crystallization kinetics are slowed. This results in an increase in the lifetime of the metastable supercooled liquid. This pathway is indicated by the dashed arrow in Fig. 14.1 and is the method that we use in the experiments discussed in this chapter.

Glassy phases of water ice can be formed via several methods, including vapor deposition on a cold substrate [2], quenching of liquid water [5], and by high pressure amorphization of crystalline ice [6,7]. The original observation

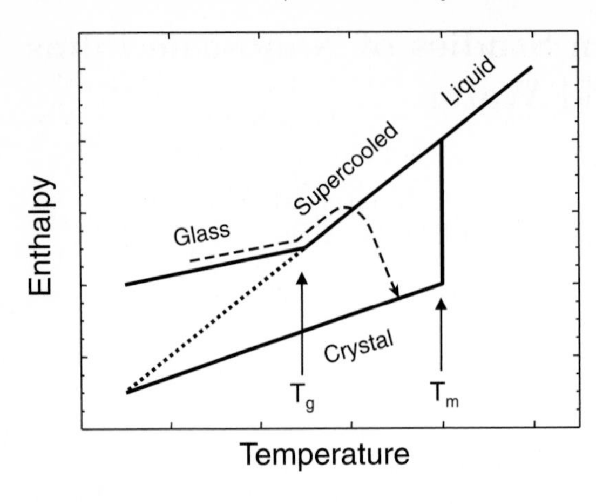

Fig. 14.1. A schematic phase diagram of a model liquid, supercooled liquid, and glass. The melting point is at T_m and the glass transition temperature is at T_g. Traditional experiments to form supercooled liquids proceed by rapid cooling from the liquid phase. An alternate pathway, heating the glass above T_g, is shown here (*dashed arrow*).

of amorphous phases of water and reports on their structural and physical properties are described in the classic book on ice by Hobbs [8]. The amorph formed by vapor deposition on a cold (<140 K) substrate is commonly referred to as amorphous solid water (ASW) [2]. More recently there has been a resurgence in the interest in amorphous water because of questions regarding its thermodynamic relationship to supercooled and liquid water and its applicability as a model for liquid water [1,2,4,9,10]. In addition, there is interest from the astrophysical community because amorphous water is believed to be a component in cometary and other astrophysical ices ([11-14] and chapters 13 and 15).

Despite the interest in ASW and its fundamental importance, a quantitative understanding of its structural, thermodynamic, and kinetic properties has been difficult. In this chapter, we discuss the use of nanoscale thin films to explore the structural and physical properties of amorphous solid water.

14.2 Growth of Nanoscale Thin ASW Films

The physical and chemical properties of ASW, which are intimately related to its morphology, are of considerable interest to physical chemists, astrophysicists, planetary scientists, and cryobiologists. Nevertheless, conflicting results have been reported for such fundamental properties as the surface area, density, and porosity of ASW. The morphology of ASW grown by vapor deposition, as with any substance, depends strongly on the growth conditions. The growth and annealing temperatures are well known to strongly influence the physical properties of ASW. Controlling the morphology of ASW by the method of deposition has important implications for experimental studies concerning the physical and chemical properties of ASW.

Molecular beam techniques are ideally suited to synthesize nanoscale thin films of amorphous solid water because they allow for the precise control of many of the important growth parameters including the deposition flux, collison energy, incident angle, and growth temperature. It is known that ASW can be grown porous, with the extent of the porosity dependent on the impingement flux and substrate temperature [11,15]. In addition to these parameters, we have recently shown that the morphology of ASW grown by vapor deposition also depends on the incident growth angle of the water molecules from the gas phase [16-18]. Films with structures from non-porous to highly porous can be grown by increasing the angle of incidence of the impinging molecules.

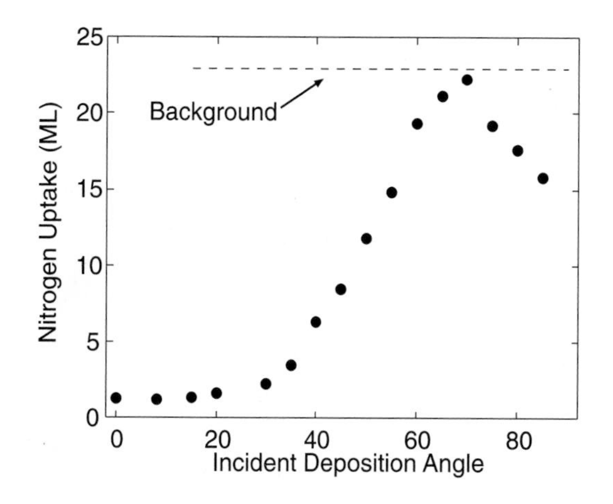

Fig. 14.2. The nitrogen uptake by 50 ML ASW films versus the incident growth angle (*circles*). Also shown is the nitrogen uptake for an ASW film grown by background deposition (*dashed line*) [16].

Nitrogen adsorption at low temperature can be used to characterize the surface area and, depending on the adsorption conditions, the porosity of the ASW films. The information obtained in these experiments is analogous to isothermal nitrogen adsorption measurements (often called BET isotherms) which are typically made at higher temperatures. The effect of the incident growth angle on the surface of an ASW film grown at 22 K is shown in Fig. 14.2. The data show the nitrogen uptake is strongly dependent on the incident growth angle. Films grown at near normal incidence grow relatively dense, whereas films grown at larger incident angles grow very porous.

We find that the films grown at large incident angles, e. g. $\theta = 70°$, have an adsorbed nitrogen to water molecule ratio of about 1:2. Assuming that the adsorbed nitrogen completely fills the pore volume, this ratio yields a density of about 0.6 g/cm^3 and an apparent surface area of about 2700 m^2/g. At larger incident angles ($\theta > 70°$), the ability to condense enough nitrogen to completely fill the pore volume may not be a valid assumption. This is because at large incident angles the average pore size in the film increases, i.e. the average pore radius is larger. A larger pore radius results in an increase in the relative vapor pressure of the condensing gas and thus a decrease in the condensation coefficient [17,18]. Therefore, at large incident angles the amount of nitrogen uptake may not be a direct measure of a film's density. Hence, the nitrogen uptake maximum in Fig. 14.2 is not necessarily a maximum for film porosity (minimum in the density).

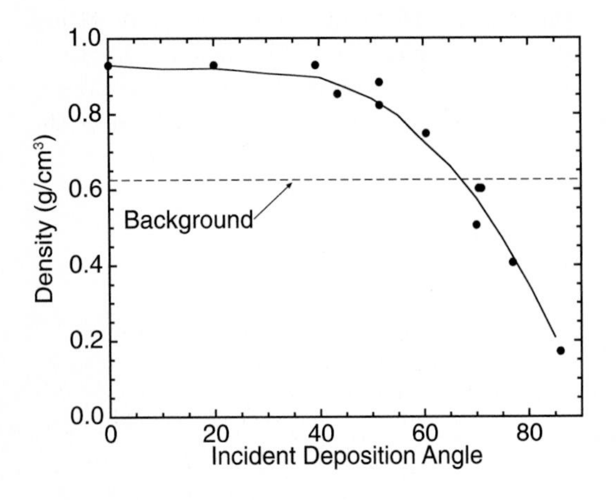

Fig. 14.3. ASW film density versus the incident growth angle (*circles*) determined using laser interferometry. Also shown is the density of an ASW film grown by background deposition (*dashed line*). The densities obtained from Monte Carlo simulations using a "hit and stick" growth model (*solid line*) are in excellent agreement with the experimental data.

Another method to determine the density of an ASW film is laser interferometry. The details of this technique have been discussed elsewhere [15,19]. In this method, the overall index of refraction of the film is measured and used to deduce the film density and porosity. Figure 14.3 is a plot of ASW density (circles) as a function of incident growth angle determined using the laser interference method [20]. The data show the ASW film density decreases with increasing incident growth angle reaching densities as low as $\sim 0.2\,\mathrm{g/cm^3}$ [20]. The data are in excellent agreement with the densities obtained from Monte Carlo simulations (solid line) using a growth model discussed below.

Figure 14.2 and Fig. 14.3 also show the results for films grown by background deposition (dashed lines). During background deposition, molecules of H_2O impinge on the substrate with a cosine distribution of incident angles. The nitrogen uptake for a film grown by background vapor deposition (dashed line in Fig. 14.2) is comparable to those grown at oblique angles and has density of about $0.6\,\mathrm{g/cm^3}$ (dashed line in Fig. 14.3). These results show that the growth method can dramatically affect the ASW morphology and explain the apparently conflicting results for the density of ASW [15,19].

The effect of the incident growth angle on the porosity of a vapor deposited film can be qualitatively understood using a simple physical picture based on ballistic deposition [21]. For ballistic-deposition simulations of film growth, randomly positioned particles are brought to a surface with straight-line trajectories, and the particles stop as soon as they encounter an occupied site ("hit and stick"). For off-normal deposition angles, regions that are initially thicker by random chance have higher growth rates and shadow the regions behind, leading to the formation of a porous film. Computer simulations have tested the above models and have related our experimental observations to some of the physical properties of ASW [18]. Figure 14.4 displays a two dimensional section from a three dimensional (3D) ballistic deposition simulation of ASW growth at an incident angle of $\theta = 70°$. The shadowing mechanism leads to the formation of porous, columnar films. Because the length of the shadow is proportional to the tangent of the incident angle, the resulting film morphology is a sensitive function of the incident angle. The shadowing mechanism is of course not unique to the growth of ASW films, and recently we have developed a reactive ballistic deposition technique for synthesizing highly porous MgO films [22].

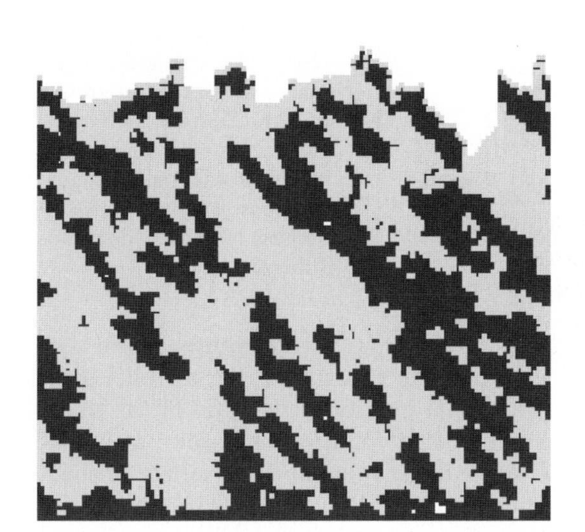

Fig. 14.4. Three dimensional ballistic deposition simulations of 50 ML ASW films deposited at an incident angle of 70° (*black*). The simulations are based on a "hit and stick" model. A shadowing effect leads to the formation of porous, columnar films. Since the length of the shadow is proportional to the growth angle the resulting film morphology is a sensitive function of the incident angle. The pore area accessible to adsorbed molecules is also shown (*gray*).

The ballistic deposition model described above requires that the incoming molecules "stick" where they "hit", i. e. the molecules have very limited or no diffusion upon adsorption. The amount of diffusion is temperature dependent and therefore the porosity of the ASW films will depend on the deposition temperature. The effect of the deposition temperature and the annealing temperature on the porosity of ASW films is shown in Fig. 14.5. The nitrogen uptake for 50 ML ASW films grown at an incident angle of 60° as a function of growth temperature is shown in Fig. 14.5 (circles). The data show that despite the high incident growth angle (60°), ASW films grow "dense" when the deposition temperature is above about 90 K. This result is consistent with the ballistic deposition model where the increased surface temperature during growth leads to formation of dense, smooth films through enhanced surface diffusion [18]. Also shown in Fig. 14.5 is the nitrogen uptake for 50 ML ASW films grown at an incident angle of 60° and at 22 K as a function of the annealing temperature (squares). The data show that an initially porous film will eventually densify when it is heated to a high enough temperature. Complete densification requires annealing to a higher temperature than that required to "grow" a dense film. This is likely because a molecule initially sticking to the surface typically has a lower coordination than the average molecule within a film and thus has a lower energetic barrier to diffusion.

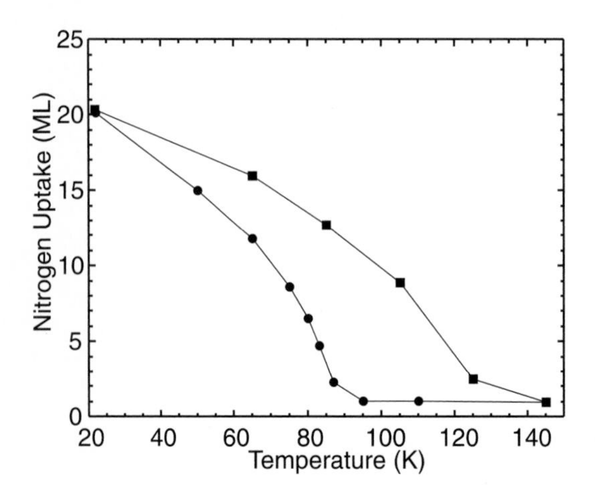

Fig. 14.5. The nitrogen uptake by 50 ML ASW films grown at 60° versus the growth temperature (*circles*) and the anneal temperature (*squares*) after growth at 22 K. Despite the large incident growth angle, at higher growth temperatures the films grow less porous because of enhanced surface diffusion.

These data show that by adjusting the growth temperature, anneal temperature, flux, and angle of deposition, we can control the morphology of the ASW film and hence the physical properties of the film. In the subsequent sections, unless explictly noted, all ASW films were grown at normal incidence thus creating "dense" ASW films. In addition, a variety of single crystal metal substrates (including Au(111), Ru(100), Pt(111)) have been used and the results for the desorption energetics, crystallization kinetics, and the diffusivity are substrate independent. The desorption kinetics, however, can depend on the hydrophobic/hydophillic nature of the substrate and this is discussed elsewhere [23,24]. The effect of other substrates on the crystallization of ASW is discussed in Section 4.

14.3 Desorption, Free Energy, and Entropy of ASW

The desorption kinetics for both amorphous and crystalline H_2O and D_2O from 145 to 160 K have been measured using temperature programmed desorption (TPD) [25]. The temperature dependent desorption rates calculated from these kinetic parameters are plotted in Fig. 14.6. Note that the temperature range plotted in Fig. 14.6 is an extrapolation from the original experimental measurement, however the equilibrium vapor pressures for crystalline ice calculated using these desorption rates are in excellent agreement with the tabulated vapor pressure measurements [24]. The desorption rate for H_2O is slightly higher (∼2 times) than the desorption rate for D_2O due to the lower zero point energy of the deuterated water. The desorption rates from the amorphous phases of H_2O and D_2O are about two times greater than from the corresponding crystalline phases. The metastable amorphous ice has a higher free energy than the crystalline ice, and as a result, the water desorption rate from the amorphous ice is higher [23,25-28].

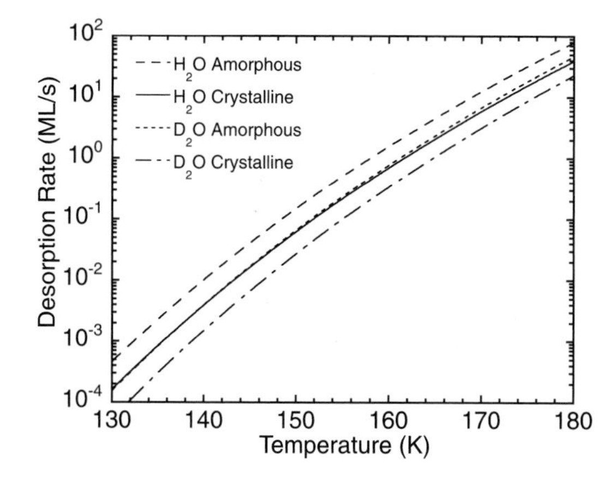

Fig. 14.6. The desorption rates (in ML/s where $1\,\mathrm{ML}=1\times10^{15}$ molecules/cm^2) for amorphous and crystalline H_2O. Also shown are the desorption rates for amorphous and crystalline D_2O. The amorphous desorption rate is about a factor of 2 greater than the crystalline desorption rate for both H_2O and D_2O.

A question has been raised as to whether amorphous water ice is a metastable extension of liquid water or a distinct thermodynamic phase [4,10,29]. The basis for the argument is that the thermodynamic free energy and entropy data do not allow for a continuous thermodynamic path connecting liquid water and ASW [29]. An accurate measurement of the free energy of amorphous ice has been difficult. Estimates of the free energy can be obtained from the available calorimetric determined enthalpy data [30-32] and theoretical calculations of the entropy [2]. These estimates give values that imply that amorphous solid water ice is not a metastable extension of liquid water [29]. The ratio of the amorphous and crystalline desorption rates, which are directly related to the equilibrium vapor pressures, can be used to calculate the relative free energy between ASW and crystalline ice [25].

The thermodynamic continuity problem is illustrated in Fig. 14.7 where the excess free energy is plotted versus temperature. Measurements of the free energy of normal supercooled liquid water extend down to 236 K. Prior to our measurement, estimates for excess free energy at 150 K were calculated using the equation $\Delta G = \Delta H - T\Delta S$, where ΔH has the calorimetrically determined value, $\Delta H(150\,K) = 1350 \pm 150\,\mathrm{J/mole}$ [30-32], and ΔS is obtained from theoretical estimates. Various estimates of the free energy using a range of values for ΔS are shown in Fig. 14.7. If a continuous thermodynamic path connecting the free energy at 150 K and the supercooled liquid data exists, then the excess free energy at 150 K must be greater than or equal to the value at 236 K [29]. This is because the heat capacity of the liquid or an amorphous

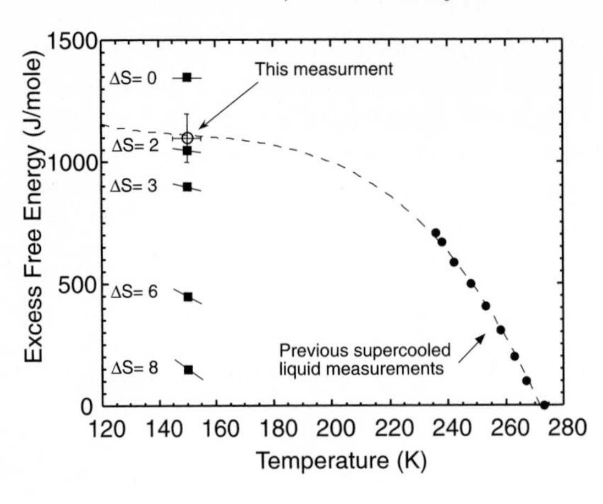

Fig. 14.7. The excess free energy for supercooled water (*closed circles*) from 273 to 236 K and for ASW (*open circle*) at 150 K. Estimates of the excess free energy at 150 K for various values of the entropy are also shown (*closed squares*). This plot shows that a reversible path (*dashed line*) to connect glassy water and the normal supercooled liquid is possible.

solid is always greater than the crystal. Previous theoretical estimates of the entropy of glassy water at 150 K have ranged from 6-9 J/(K mole) [2,29,33,34] and as is seen graphically in Fig. 14.7 these entropy estimates do not allow for a continuous reversible thermodynamic path. Our measured value of the excess free energy at 150 K shows that it is possible to connect the amorph with normal supercooled water by a thermodynamic continuous path thus resolving the continuity problem. A value for the excess entropy at 150 K of 1.7 ± 1.7 J/(K-mole) is obtained using this free energy measurment and the previously measured enthalpy [25]. The unexpectedly low value of ΔS suggests the possibility of an essentially unique ideal glass with no residual entropy [35].

14.4 Crystallization Kinetics of ASW

A nanoscale thin film of metastable ASW will eventually crystallize when heated to a high enough temperature for a sufficient amount of time. The crystallization kinetics of ASW have been studied by a number of techniques including calorimetry [36], electron diffraction [13,37-39], FTIR [40,41], inert gas physisorption [42,43], and temperature programmed desorption (TPD) [23,27,28]. Typically, crystallization is observed to occur between 140 and 170 K with crystallization times varying as a result of the experimental parameters such as the temperature ramp rate. The variation in the observed crystallization temperature further demonstrates that the crystallization of ASW is a kinetic rather than a thermodynamic phase transition.

Figure 14.8 shows a typical TPD spectrum for an initially amorphous film grown at 85 K on Au(111) where the temperature is increased at a linear rate of 0.6 K/s. As the temperature is increased, the desorption rate from the film increases until about 160 K where the rate deviates from the normally expected exponential increase [25,44]. The change in the desorption rate tem-

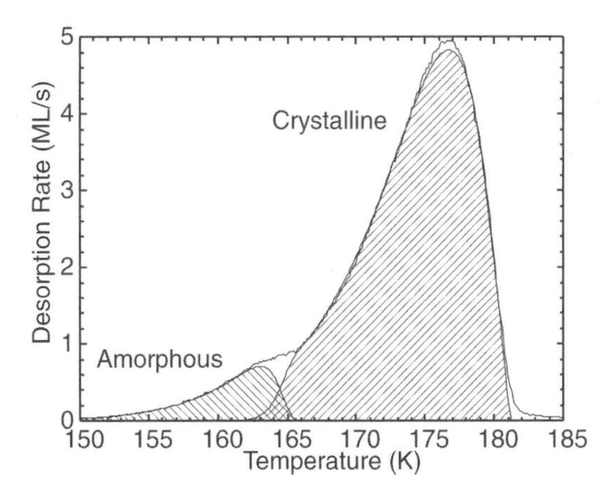

Fig. 14.8. The TPD spectrum for a 85 ML ASW film grown at normal incidence on Au(111) at 80 K (*solid line*. Kinetic model simulation results for the desorption from the component amorphous and crystalline phases are shown as hatched areas.

perature dependence results in an apparent "bump" in the spectrum. Figure 14.9 shows a typical isothermal desorption spectrum for an initially amorphous film grown at 85 K on Au(111), where the temperature is ramped and then held at a constant value of 154 K [23,24]. The desorption rate exhibits an initial rapid increase due to ramping to the isothermal temperature, but then the desorption rate decreases to a value approximately half the initial rate. As discussed below, both the apparently complex TPD (Fig. 14.8) and isothermal (Fig. 14.9) lineshapes are a consequence of the irreversible conversion of the amorphous phase to crystalline ice.

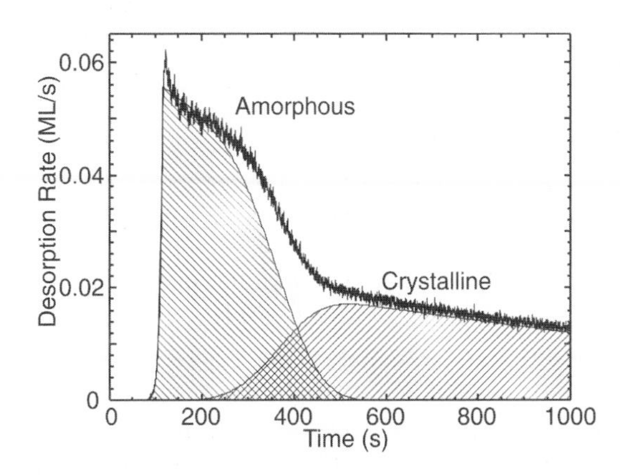

Fig. 14.9. The isothermal desorption spectrum for a 35 ML ASW film grown at normal incidence on Au(111) at 80 K (*solid line*) and then ramped to and held at 154 K. Kinetic model simulation results for the desorption from the component amorphous and crystalline phases are shown as hatched areas.

A mean field kinetic model can be used to quantify the desorption spectra in Fig. 14.8 and Fig. 14.9. In this model, the total desorption rate is a linear combination of the amorphous and crystalline desorption rates weighted by their respective mole fractions, $R(T)_{Total} = \chi(T)R(T)_{Crystalline} + (1 - \chi(T))\,R(T)_{Amorphous}$. The desorption rates for the amorphous and crystalline phases are given in Fig. 14.6. The crystallization kinetics are contained in the time dependence of χ and are best described using the classical nucleation and growth kinetic model represented by the Avrami equation, $\chi(t) = (1 - \exp(-(kt)^n)$ [45-47]. In this equation, χ is the time dependent crystalline mole fraction, k is a phenomenological rate constant, and n is a parameter dependent on the nucleation and growth mechanism of the crystalline phase [46,47].

The model simulations are in excellent agreement with the experimental data and also provide a clear explanation for the apparently complicated TPD and isothermal lineshapes. The hatched areas in Figs. 14.8 and 14.9 are the desorption rates from the component amorphous and crystalline phases. In Fig. 14.8, the observed "bump" in the overall TPD desorption rate begins when the desorption from the crystalline component begins, i. e. when crystallization starts, and ends when the desorption from the amorphous phase ends. As discussed in Section 3, the amorphous phase has a higher desorption rate because of the excess free energy of the metastable phase, hence the conversion to the crystalline phase results in the observed "bump". Similarly, the isothermal desorption spectrum in Fig. 14.9 shows an initial rapid decrease in the desorption rate as a result of the transformation of the amorphous phase into the crystalline phase. After complete crystallization, desorption occurs only from the crystalline phase which has a lower desorption rate.

Quantitative information about the crystallization kinetics and mechanism is contained in the k and n parameters in the Avrami equation. Figure 14.10 is an Arrhenius plot of the crystallization rate constants, k, obtained from desorption experiments (circles) [23,24] and physisorption experiments (squares) [43] for ASW deposited on single crystal metal substrates. The physisorption method allows for the crystallization to be measured at temperatures where there is negligible desorption and thus allows for measurements over a much wider temperature range. A fit to these data yields a crystallization activation energy of 70 kJ/mole. In both of these experiments a value for $n = 4$ was obtained indicating a crystallization mechanism consistent with a spatially random constant nucleation rate and spatially isotropic three-dimensional growth.

This apparent activation energy is a convolution of both the nucleation and growth steps in the crystallization process. In order to determine the activation energies of these individual steps, experiments where ASW was deposited on a crystalline ice thin film were conducted. The crystalline ice substrate acts as a two-dimensional template for crystal growth thereby eliminating the need for the nucleation step. The results of these experiments using the physisorption method are also shown in Fig. 14.10 (triangles) [43]. The data show that there is a dramatic enhancement of the crystallization rate for amorphous films on crystalline ice substrates. An Arrhenius fit to the data yields an activation energy of ~56 kJ/mole. The increased rate originates from the absence of the nucleation process because the crystalline ice substrate serves as a 2-dimensional nucleus for the growth of the crystalline phase

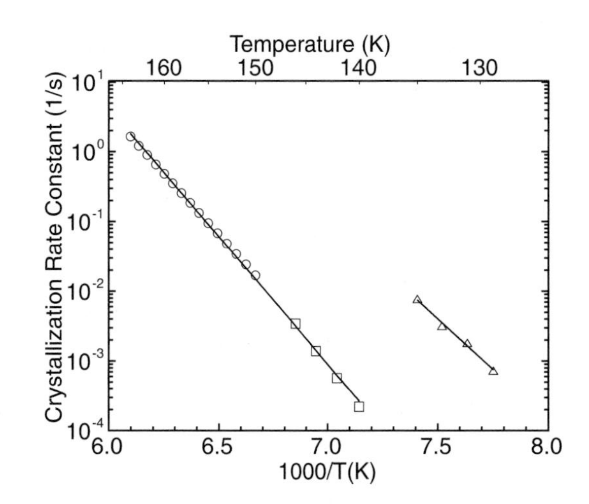

Fig. 14.10. Arrhenius plot of the crystallization rate constants for ASW films grown on metal crystal substrates determined by desorption experiments (*circles*) and by nitrogen physisorption (*squares*). The crystallization rate for ASW films grown on a crystalline ice film is also shown (*triangles*). Arrhenius fits (*lines*) yield crystallization activation energies of 70 kJ/mole and 56 kJ/mole on metal crystal and crystalline ice substrates, respectively.

and thus the highly activated process of nucleation is circumvented. This contrasts with the crystallization on a metal substrate (circles and squares in Fig. 14.10) which requires bulk nucleation before 3-dimensional growth can proceed. We can use the activation energy for nucleation and growth (metal crystal substrate) and the activation energy for growth only (crystalline ice substrate) to extract the activation energy for nucleation which is determined to be 140 kJ/mole [43].

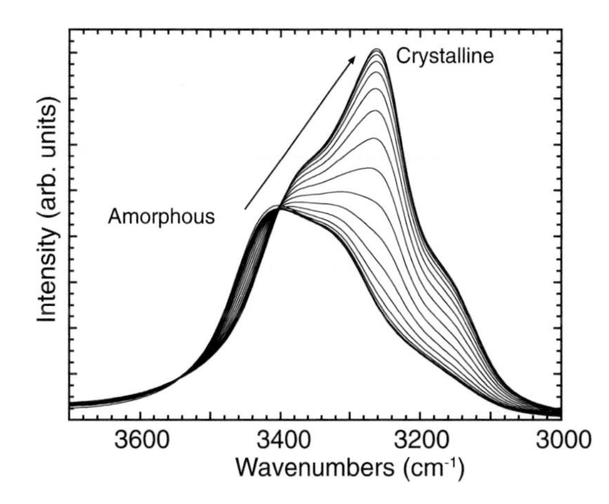

Fig. 14.11. A time series of FTIR spectra from a 50 ML ASW film grown on Pt(111) at 80 K and then annealed at 146 K. The spectra show the evolution from a pure amorphous film to a completely crystalline film (*arrow*).

One potential limitation of the TPD and inert gas physisorption methods is that they are sensitive only to the outer surface of the thin film. A probe that is sensitive to changes that occur within the entire film is FTIR and has been used in many studies of ASW [40,41,48-51]. An example of how FTIR is used to study ASW crystallization is shown in Fig. 14.11 [52]. A time series of reflection-absorption infrared spectra of the OH stretching region during the isothermal annealing at 146 K of an ASW film grown on Pt(111) is shown. The first spectrum in the series (tail of arrow) displays the broad band characteristic of an amorphous solid, while subsequent spectra show the emergence of a narrower peak centered at 3260 cm^{-1}. The last spectrum in the time sequence corresponds to the fully crystallized film. The spectra display an isosbestic point (a point where all the spectra intersect) indicating the interconversion from the metastable amorphous to the stable crystalline phase. The intermediate spectra are a combination of the pure amorphous and the pure crystalline spectra and can be analyzed to determine extent of crystallization.

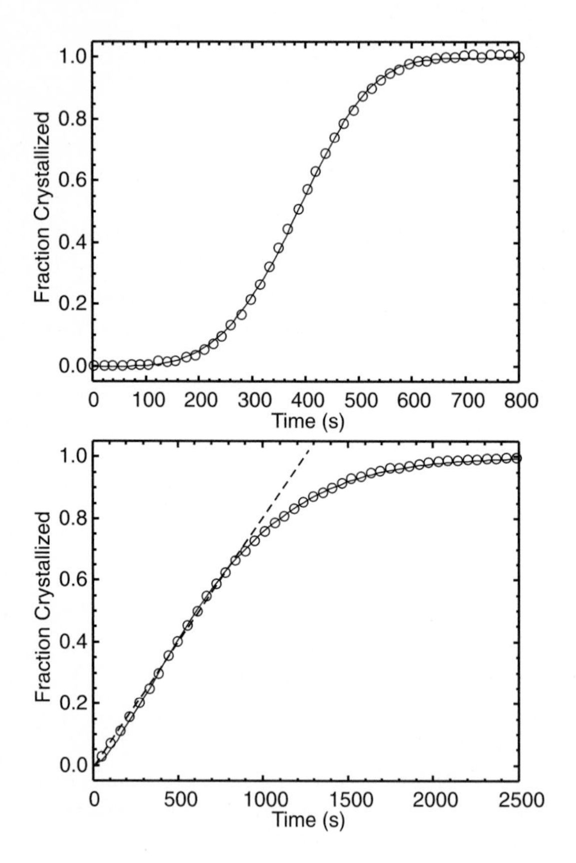

Fig. 14.12. (*Top panel*) The time evolution of the crystallized fraction for a 50 ML ASW film grown on Pt(111) at 80 K and then ramped to and held at 146 K (*circles*). A fit using the Avrami equation (*solid line*) yields a value of $n \approx 4$ indicating a nucleation and three dimensional growth crystallization mechanism. (*Bottom panel*) The time evolution of the crystallized fraction for a 30 ML ASW film grown on a crystalline ice template at 80 K and then ramped to and held at 128 K (*circles*). A fit using the Avrami equation (*solid line*) yields a value of $n \approx 1.4$ suggesting a different crystallization mechanism then that shown in the top panel. A linear crystallization time dependence is shown by the dashed line.

A comparison of the isothermal crystallization of ASW on a clean Pt(111) substrate (top) and on a crystalline ice template (bottom) is shown in Fig. 14.12. The extent of crystallization was determined using FTIR [52]. The crystallization of ASW on Pt(111) requires a higher temperature (146 K) and the kinetics are well fit (solid line) by the Avrami equation with a value of $n = 4$. This suggests a crystallization mechanism of random nucleation and three dimensional growth.

On the other hand, the crystallization of ASW on the crystalline template occurs at a much lower temperature (128 K), and the kinetics show a much different temporal behavior. While the Avrami equation (solid line) gives a good fit to the data and yields a value of $n = 1.4$, it is not reasonable to expect the Avrami model to realistically describe a crystallization mechanism in which a two dimensional crystallization front progagates through the ASW film. In this case, one might expect a linear time dependence if the two dimensional growth front maintains a constant velocity as it moves through the film. The dashed line in Fig. 14.12 (lower panel) shows that the crystallization kinetics are initially linear, but deviate from linearity when about 70% of the film has crystallized. A more complicated Monte Carlo model simulation is required to extract the rate and mechanistic details for this case.

The FTIR results agree well with the TPD [23,24] and nitrogen physisorption [42,43] experiments. Reported values for the activation energy for the homogeneous crystallization of ASW vary from 44 kJ/mole [27] to values close to 70 kJ/mole [23,28,40,43], while reported values for n vary from values ~ 1 [13,27,40,41] to values ~ 4 [23,28,43]. As shown in Fig. 14.12 and elsewhere [42,43], the substrate can have a dramatic effect on the crystallization rate and mechanism. It could be that the discrepancies in these values are due to substrate effects. Another reason for the discrepancies could be differences in ASW preparation, which as discussed in Section 14.2, could result in ASW films with different physical properties or in the preseeding of the amorphous film with crystalline nuclei. Further experiments and detailed kinetic modelling are currently being conducted to address these issues.

14.5 Diffusivity of Amorphous Solid Water

A long-standing question has been whether the melt of ASW is connected to normal supercooled liquid water or whether it is a distinct liquid phase [1,4,9]. Part of the reason for the controversy has been the anomalous thermodynamic and transport properties displayed by supercooled liquid water. Several theoretical proposals to explain these anomalies have been made; however, no consensus has emerged [4]. Some theories preclude the existence of normal supercooled liquid water below an apparent temperature singularity at 228 K [9,53], while others are consistent with a continuity of metastable states from the melting point at 273 K to the glass transition temperature at 136 K [4,32,54,55]. The requisite experimental data to distinguish between the various theoretical models has been difficult to obtain because of the inability to supercool liquid water to temperatures approaching the apparent divergence at 228 K [9].

In principle one could create a metastable deeply supercooled liquid by heating a glass above its T_g. Such an experiment is illustrated by the dashed line in Fig. 14.1. When heated above T_g, the material would be a supercooled liquid that could at some point undergo an irreversible transormation to the crystalline state (see Section 4). The lifetime of the metastable liquid prior to crystallization would determine whether or not one could measure its diffusivity. For example, at T_g, a material would have a diffusivity of 10^{-18} cm^2/s [1]. This is a value commonly used to define T_g and corresponds to a viscosity of about 10^{13} Poise (assuming that the diffusivity and viscosity obey the Stokes-Einstein relationship [56]). A diffusivity in this range would result in motion of about 1 molecular diameter (about 3 Å for water) in about 100 s. Such a diffusivity would be nearly impossible to observe with a macroscopic sample – a 1 cm thick sample would require about 10^{10} years to completely mix!

In recent experiments, our group has used nanoscale thin films to overcome the problem of observing extremely small diffusion lengths on a reasonable laboratory time scale (10 - 100 s) to determine the self-diffusivity of ASW upon heating above its T_g [57-59]. In these experiments nanoscale films of varying isotopic composition are created by sequential dosing of the isotopic vapor at low temperature (85 K). After ASW film growth, the sample temperature was raised above the glass transition (T_g ≈140 K), through the crystallization region (155-160 K), and to a temperature where the ice film had completely desorbed (170-180 K). The desorption spectra during this temperature ramp reveal the crystallization kinetics of the ASW films and also the extent of intermixing between the $H_2^{16}O$ and $H_2^{18}O$ layered interfaces.

Our initial results indicated that there was long-range translational diffusion that occurs concomitantly with the amorphous to crystalline ice phase transition [57]. The apparent diffusivity estimated from these data was a factor of $10^{6\pm1}$ greater than an Arrhenius extrapolation for diffusion in crystalline ice. These findings suggested that the amorphous material exhibits liquid-like translational diffusion prior to crystallization at temperatures near 155 K. Thus the observed diffusive behavior is more consistent with that of a liquid (albeit a very viscous liquid) than a solid. In subsequent work, we developed a kinetic model to quantify the magnitude and temperature dependence of the diffusivity [58,59]. The temperature dependent diffusivity is quantified using a mathematical model that couples our previous mean-field description of the desorption/crystallization kinetics (see Section 4) to a one-dimensional representation of the diffusive transport between layers. The diffusion rate is dependent on the phase of the material (amorphous or crystalline) and the diffusion rate is treated as a linear combination of the amorphous and crystalline diffusivity weighted by their respective mole fractions. After complete crystallization the diffusive motion is effectively "frozen" out. Details on the experimental technique, analysis, and results are discussed elsewhere, so here we present only the ASW diffusivity results [56,58,59].

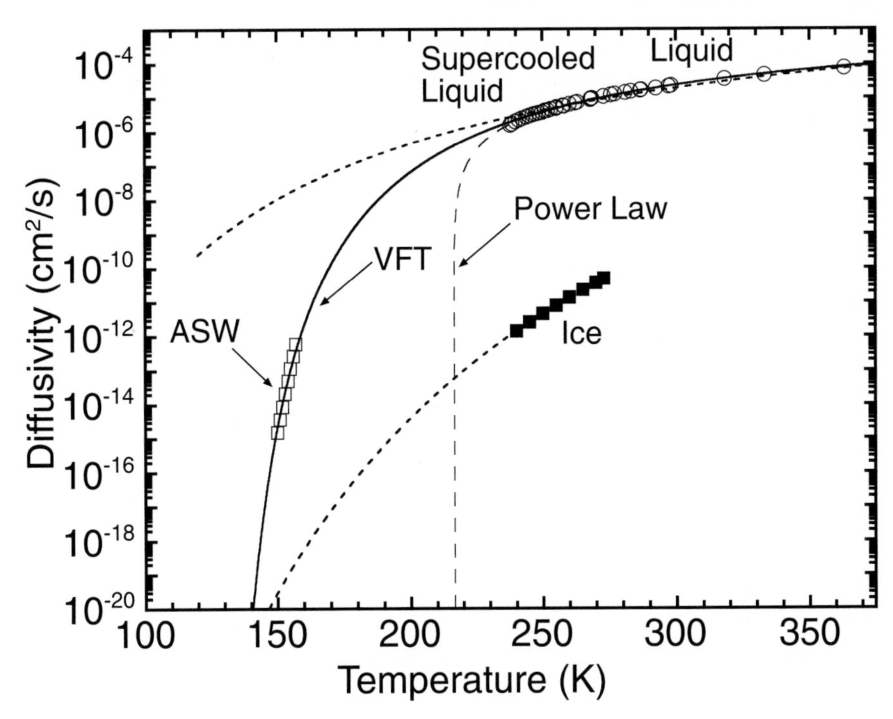

Fig. 14.13. The ASW (*open squares*), liquid, supercooled liquid (*open circles*), and crystalline ice (*solid squares*) diffusivities are plotted. The dotted lines are Arrhenius extrapolations. The solid line labeled "VFT" is a fit of the liquid/supercooled liquid and ASW data to, $D = D_o \exp(-E/(T - T_o))$. The dashed line labeled "Power Law" is a fit of liquid/supercooled liquid data to, $D = D_o T^{1/2}(T/T_s - 1)^{\gamma}$.

A plot of the ASW diffusivity [56,58,59], along with the supercooled liquid [60,61], and liquid water diffusivities [62] is shown in Fig. 14.13. The individual ASW diffusivity points (open squares) are from 150-157 K which is the temperature range where we observe significant ASW diffusion prior to crystallization. The low temperature diffusivity data show that the amorphous diffusion is highly activated. An Arrhenius extrapolation of the liquid and supercooled liquid (dotted line) data does not agree with our ASW diffusivity results that exhibit a much stronger temperature dependence. It is known that many glass-forming liquids exhibit markedly non-Arrhenius behavior as they are supercooled below their freezing point [1]. The temperature dependence of this non-Arrhenius behavior is often well represented by the empirical Vogel-Fulcher-Tammann (VFT) equation [1]. The data displayed in Fig. 14.13 are well described by a VFT equation that fits both the supercooled liquid data and the diffusivity observed in ASW. On the other hand, the power law equation, which has been used to suggest that there is singularity in the water phase diagram near 220 K, fits only the liquid and supercooled liquid data and not the ASW data [10,63]. This is not a surprise since the ASW data were not included in the fit. Obviously, any equation with a singularity above the temperature of the ASW data cannot simultaneously fit both the ASW and liquid diffusivity data.

Both the VFT and power law interpretations require assumptions about the diffusivity in temperature regions where data do not presently exist. The VFT fit requires interpolation between about 240 K and 160 K. The power law fit requires an extrapolation of about 18 K between the lowest experimental data point at 238 K and the singularity at 220 K. In general, data much closer to the proposed singularity are needed to confirm its existence. Unfortunately experiments in the 160-240 K temperature region have not been possible due to the rapid crystallization of supercooled liquid water below 240 K and of ASW above 160 K. While our new ASW diffusivity data provide support for continuity between ASW and liquid water at low pressure, an unambiguous resolution of the continuity conundrum must await further experiments in the unexplored temperature region from 160 to 240 K.

14.6 Molecular Volcanos: Structural Changes to ASW During Crystallization

The interactions between ice and an adsorbate have important applications in atmospheric and astrophysical science. In particular, the kinetics for the adsorption, desorption, trapping, and release of an adsorbate interacting with an ice substrate are needed in order to quantitatively model macroscopic processes. In the previous sections, we described how molecular beam techniques can be used to grow, characterize, and determine the physical properties of nanoscale ASW films themselves. Here we use similar techniques to study the fundamental interactions between gas molecules and an ASW or an ice substrate.

Molecular beams with high deposition thickness control are ideally suited for creating "*chemically tailored*" nanoscale (< 1000 Å) films with layered interfaces. The desorption kinetics of a species interacting with a water-ice layer can be used to probe this interaction. Water and CCl_4 are immiscible at room temperature, and as such provide a test system for studying hydrophobic interactions [44]. Figure 14.14 displays the TPD spectra from layered films where 5.4 ML CCl_4 was deposited on top of (upper panel) and underneath of (lower panel) 30 MLs of dense ASW at 85 K. When CCl_4 is deposited on top of ASW, the CCl_4 film completely desorbs prior to the onset of significant water desorption because the CCl_4 desorption rate is much greater than the water desorption rate. These TPD spectra are indistinguishable from samples in which CCl_4 and ASW are deposited in spatially separate columns. This indicates that there is very little interaction and/or no intermixing between the CCl_4 and ASW layers.

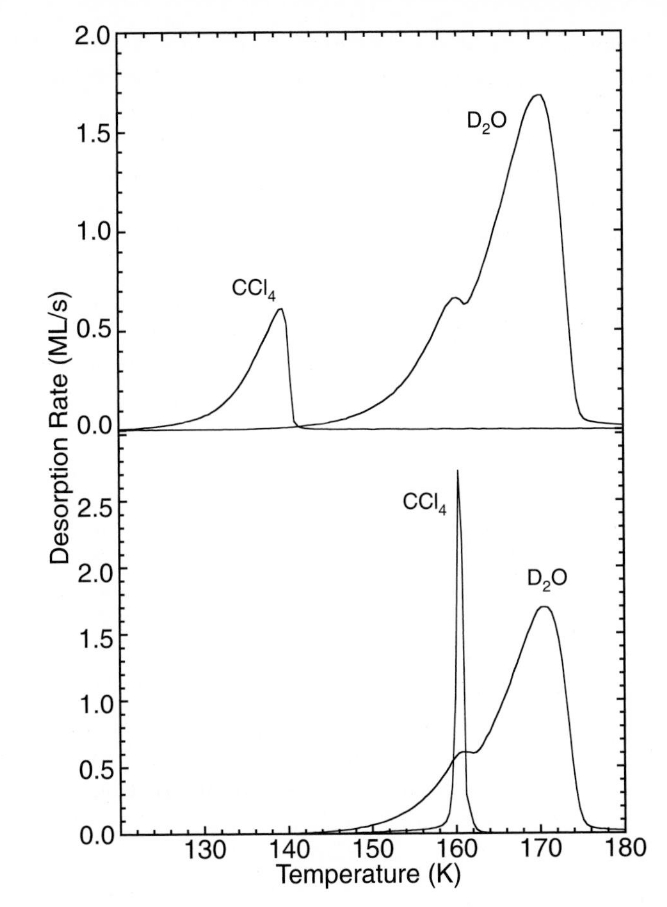

Fig. 14.14. (*Upper panel*) The TPD spectra for 5.4 ML of CCl_4 deposited on top of 30 ML of ASW at 85 K. When CCl_4 is deposited on top of ASW, it desorbs prior to H_2O desorption and with minimal or no interaction with the ASW film. (*Lower panel*) The TPD spectra for 30 ML of ASW deposited on top of 5.4 ML of CCl_4 at 85 K. When CCl_4 is deposited underneath of ASW its desorption is impeded by the ASW overlayer until the ASW film begins to crystallize. We have termed this effect the "molecular volcano".

When ASW is deposited on top of CCl_4, the desorption of CCl_4 is impeded by the amorphous water overlayer until the temperature range for amorphous water crystallization. At this higher temperature, the abrupt CCl_4 desorption occurs completely over a narrow temperature range prior to the desorption of the majority of the water film. The CCl_4 desorption peak shifts with the temperature of the phase transition "bump" for both H_2O and D_2O. This correspondence suggests that the CCl_4 desorption through the water overlayer is directly correlated with the amorphous-to-crystalline phase transition. These experiments demonstrate that the CCl_4 is trapped beneath the ice until the phase transition occurs, and that desorption occurs in a dramatic fashion in concert with the crystallization of amorphous ice. We have termed the abrupt desorption the "molecular volcano" [44].

The observed abrupt desorption is likely due to structural changes such as cracks, fissures, and/or grain boundaries that accompany the crystallization kinetics. The abrupt desorption occurs through connected pathways that are formed during the nucleation and growth of crystalline ice. The onset of the abrupt desorption corresponds to the threshold for dynamic percolation [64]. This threshold physically corresponds to the point at which the grain boundaries or microcracks develop between the growing crystallites and impinge upon each other to form a connected escape path that traverses the water overlayer.

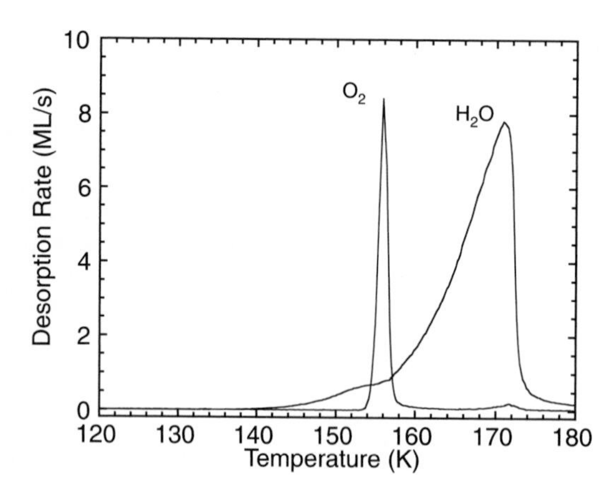

Fig. 14.15. The TPD spectra for 125 ML of ASW deposited on top of 5 ML of O_2 at 22 K. The O_2 layer that would normally desorb at much lower temperatures (30-40 K), is trapped by the ASW overlayer. The O_2 eventually desorbs when the ASW crystallizes which is consistent with the '"molecular volcano" mechanism.

More recently, experiments using a variety of underlayer species (Ar, N_2, O_2, CO, and CH_4) deposited at temperatures near 20 K confirm that the molecular volcano mechanism is a general phenomenon [65]. For example, Fig. 14.15 shows the TPD spectra for 125 ML of ASW deposited on top of 5 ML of O_2 at 22 K. The data show that the majority of the O_2 desorbs in the "volcano" peak although a small amount remains trapped and desorbs with the depletion of the water overlayer. This confirms that the "volcanic" desorption mechanism is primarly due to changes in the ASW overlayer and not with the interactions of the trapped species. An interesting question is what is the form of the molecules that remain trapped after the crystallization is complete. The structure could be a simple closed pore or a clathrate hydrate cage. Future studies using TPD and FTIR probes should help to answer this question.

The details about the dynamics of the microstructural evolution accompanying the crystallization of ASW may be important in the study of astrophysical ices. Comets occasionally release trapped gases in unexpected bursts. Furthermore, the extent to which these ices can trap and retain gases is an important parameter in determining the chemical content of water ice in comets and on planets [66-69].

14.7 Summary

Molecular beam techniques, in combination with mass spectrometric detection and in-situ FTIR spectroscopy, are powerful research tools that can be used to synthesize "chemically tailored" nanoscale films, characterize their structure, and probe their physical properties. The use of nanoscale films allows for the measurment of diffusivities between 10^{-16} and 10^{-12} cm^2/s that would be impossible to observe with a macroscopic sample – a 1 cm thick sample would require between 10^4 and 10^8 years to completely mix! Elucidation of these processes will further our understanding of solvation and reactions in multi-phase, multicomponent solutions and in determining reaction mechanisms in heterogeneous systems. For example, we have recently observed the formation of a binary supercooled liquid solution of methanol and ethanol [70].

This work was supported by the U. S. Department of Energy Office of Basic Energy Sciences, Chemical Sciences Division. Pacific Northwest National Laboratory is operated for the U. S. Department of Energy by Battelle under contract DE-AC06-76RLO 1830.

References

1. C. A. Angell, Science **267**, 1924-1934 (1995).
2. M. G. Sceats, S. A. Rice. in *Water: A Comprehensive Treatise, Volume 7* (ed. Franks, F.) 83-214 (Plenum Press, New York, 1982).
3. R. Zallen. *The Physics of Amorphous Solids* (John Wiley & Sons, New york, 1983).
4. P. G. Debenedetti. *Metastable Liquids: Concepts and Principles* (Princeton University Press, 1996).
5. G. P. Johari, A. Hallbrucker, E. Mayer, Nature **330**, 552-553 (1987).
6. O. Mishima, L. D. Calvert, E. Whalley, Nature **310**, 393-395 (1984).
7. O. Mishima, H. E. Stanley, Nature **396**, 329-335 (1998).
8. P. V. Hobbs. *Ice Physics* (Oxford University Press, London, 1974).
9. C. A. Angell. in *Water: A Comprehensive Treatise, Volume 7* (ed. Franks, F.) 1-81 (Plenum Press, New York, 1982).
10. C. A. Angell, Ann. Rev. Phys. Chem. **34**, 593-630 (1983)
11. E. Mayer, R. Pletzer, Nature **319**, 298 (1986)
12. A. H. Delsemme, J. Phys. Chem. **87**, 4214-4218 (1983)
13. P. Jenniskens, D. F. Blake, Science **265**, 753-756 (1994)
14. J. Klinger, J. Phys. Chem. **87**, 4209-4214 (1983)
15. D. E. Brown, S. M. George, C. Huang, E. K. L. Wong, K. B. Rider, R. S. Smith, B. D. Kay, J. Phys. Chem. **100**, 4988-4995 (1996)

16. K. P. Stevenson, G. A. Kimmel, Z. Dohnálek, R. S. Smith, B. D. Kay, Science **283**, 1505-1507 (1999)
17. G. A. Kimmel, K. P. Stevenson, Z. Dohnálek, R. S. Smith, B. D. Kay, J. Chem. Phys. **114**, 5284-5294 (2001)
18. G. A. Kimmel, Z. Dohnálek, K. P. Stevenson, R. S. Smith, B. D. Kay, J. Chem. Phys. **114**, 5295-5303 (2001)
19. M. S. Westley, G. A. Baratta, R. A. Baragiola, J. Chem. Phys. **108**, 3321-3326 (1998)
20. Z. Dohnálek, G. A. Kimmel, P. Ayotte, R. S. Smith, B. D. Kay, submitted to J. Chem. Phys. (2002)
21. A. L. Barabasi, H. E. Stanley. *Fractal Concepts in Surface Growth* (Cambridge Univ. Press, Cambridge, 1995).
22. Z. Dohnálek, G. A. Kimmel, D. E. McCready, J. S. Young, A. Dohnálkova, R. S. Smith, B. D. Kay, Journal of Physical Chemistry B **106**, 3526-3529 (2002)
23. R. S. Smith, C. Huang, E. K. L. Wong, B. D. Kay, Surf. Sci. Lett. **367**, L13-L17 (1996)
24. R. S. Smith, B. D. Kay, Surf. Rev. and Lett. **4**, 781-797 (1997)
25. R. J. Speedy, P. G. Debenedetti, R. S. Smith, C. Huang, B. D. Kay, J. Chem. Phys. **105**, 240-244 (1996)
26. A. Kouchi, Nature **330**, 550 (1987)
27. N. J. Sack, R. A. Baragiola, Phys. Rev. B **48**, 9973 (1993)
28. P. Löfgren, P. Ahlstrm, D. V. Chakarov, J. Lausmaa, B. Kasemo, Surf. Sci. Lett. **367**, L19 (1996)
29. R. J. Speedy, J. Phys. Chem. **96**, 2322-2325 (1992)
30. M. A. Floriano, Y. P. Handa, D. D. Klug, E. Whalley, J. Chem. Phys. **91**, 7187-7192 (1989)
31. A. Hallbrucker, E. Mayer, G. P. Johari, J. Phys. Chem. **93**, 4986-4990 (1989)
32. G. P. Johari, G. Fleissner, A. Hallbrucker, E. Mayer, J. Phys. Chem. **98**, 4719-4725 (1994)
33. G. P. Johari, Philos. Mag. **35**, 1077 (1977)
34. M. G. Sceats, S. A. Rice, Journal of Chemical Physics **72**, 3260-3262 (1980)
35. R. J. Speedy, P. G. Debenedetti, Molecular Physics **88**, 1293-1316 (1996)
36. V. P. Koverda, N. M. Bogdanov, V. P. Skripov, Journal of Non-Crystalline Solids **57**, 203-212 (1983)
37. A. Kouchi, T. Yamamoto, T. Kozasa, T. Kuroda, J. M. Greenberg, Astronomy and Astrophysics **290**, 1009-1018 (1994)
38. A. Kouchi, T. Yamamoto, Progress in Crystal Growth and Characterization of Materials **30**, 83-108 (1995)
39. P. Jenniskens, D. F. Blake, Astrophysical Journal **473**, 1104-1113 (1996)
40. W. Hage, A. Hallbrucker, E. Mayer, G. P. Johari, J. Chem. Phys. **100**, 2743-2747 (1994)
41. W. Hage, A. Hallbrucker, E. Mayer, G. P. Johari, J. Chem. Phys. **103**, 545-550 (1995)
42. Z. Dohnálek, R. L. Ciolli, G. A. Kimmel, K. P. Stevenson, R. S. Smith, B. D. Kay, J. Chem. Phys. **110**, 5489-5492 (1999)
43. Z. Dohnálek, G. A. Kimmel, R. L. Ciolli, K. P. Stevenson, R. S. Smith, B. D. Kay, J. Chem. Phys. **112**, 5932-5941 (2000)
44. R. S. Smith, C. Huang, E. K. L. Wong, B. D. Kay, Phys. Rev. Lett. **79**, 909-912 (1997)

45. M. Avrami, J. Chem. Phys. **9**, 177 (1941)
46. C. N. R. Rao, K. J. Rao. *Phase Transitions in Solids* (McGraw-Hill, New York, 1978).
47. R. H. Doremus. *Rates of Phase Transformations* (Academic Press, New York, 1985).
48. J. P. Devlin, J. Geophys. Res.-Planets **106**, 33333-33349 (2001)
49. B. Rowland, J. P. Devlin, J. Chem. Phys. **94**, 812-813 (1991)
50. J. P. Devlin, V. Buch, J. Phys. Chem. **99**, 16534-16548 (1995)
51. M. Fisher, J. P. Devlin, J. Phys. Chem. **99**, 11584-11590 (1995)
52. G. Teeter, Z. Dohnálek, P. Ayotte, J. Daschbach, G. A. Kimmel, R. S. Smith, B. D. Kay, Manuscript in preparation
53. R. J. Speedy, C. A. Angell, J. Chem. Phys. **65**, 851-858 (1976)
54. G. P. Johari, J. Chem. Phys. **98**, 7324-7329 (1993)
55. G. P. Johari, J. Chem. Phys. **107**, 10154-10165 (1997)
56. R. S. Smith, Z. Dohnálek, G. A. Kimmel, K. P. Stevenson, B. D. Kay. in *ACS Symp. Series 820: Liquid Dynamics, Experiment, Simulation. and Theory* (ed. Fourkas, J. T.) 198-211 (American Chemical Society, Washington, DC, 2002).
57. R. S. Smith, C. Huang, B. D. Kay, J. Phys. Chem. B **101**, 6123-6126 (1997)
58. R. S. Smith, B. D. Kay, Nature **398**, 788-791 (1999)
59. R. S. Smith, Z. Dohnálek, G. A. Kimmel, K. P. Stevenson, B. D. Kay, Chem. Phys. **258**, 291-305 (2000)
60. F. X. Prielmeier, E. W. Lang, R. J. Speedy, H.-D. Ldemann, Ber. Bunsenges. Phys. Chem. **92**, 1111-1117 (1988)
61. W. S. Price, H. Ide, Y. Arata, J. Phys. Chem. A **103**, 448-450 (1999)
62. H. Weingärtner, Z. Phys. Chem. **132**, 129-149 (1982)
63. F. X. Prielmeier, E. W. Lang, R. J. Speedy, H.-D. Ludemann, Phys. Rev. Lett. **59**, 1128-1131 (1987)
64. D. Laufer, E. Kochavi, A. Bar-Nun, Phys. Rev. B: Condens. Matter **36**, 9219-9227 (1987)
65. P. Ayotte, R. S. Smith, K. P. Stevenson, Z. Dohnálek, G. A. Kimmel, B. D. Kay, J. Geophys. Res.-Planets **106**, 33387-33392 (2001)
66. T. Owen, A. Bar-Nun, Icarus **116**, 215-226 (1995)
67. J. A. Nuth, H. G. M. Hill, G. Kletetschka, Nature **406**, 275-276 (2000)
68. A. Bar-Nun, I. Kleinfeld, Icarus **1989**, 243-253 (1989)
69. C. F. Chyba, Nature **343**, 129 (1990)
70. P. Ayotte, R. S. Smith, G. Teeter, Z. Dohnálek, G. A. Kimmel, B. D. Kay, Phys. Rev. Lett. **88**, art. no.-245505 (2002)

15 Microporous Amorphous Water Ice Thin Films: Properties and Their Astronomical Implications

Raul A. Baragiola

Summary. Vapor-deposited amorphous solid water (ASW) is ubiquitous in astronomical environments like interstellar grains, icy satellites, comets, and planetary rings, and artificial environments on Earth. This chapter reviews some of the extraordinary properties of ASW, especially those affected by the presence of microporosity. There is an emphasis on the most recent experiments performed under ultrahigh vacuum since enormous amounts of gas, whose effects are not well-understood, can be adsorbed by ASW from its environment. Several instances are given where the properties of ASW play a role in icy grains in interstellar space and on the icy surfaces of satellites in the outer solar system. The chapter concludes with a summary list of unsolved questions and needs for further theoretical and experimental studies.

15.1 Introduction

This chapter addresses the microporous amorphous phase of vapor deposited water ice, which is the dominant form of water in the universe [1, 2]. It exists as mantles on dust grains (mostly silicates) in interstellar clouds, usually combined with other condensed gases that are deposited (accreted) from the gas phase. These grains live in a radiation environment of ultraviolet light, ion winds, and cosmic rays which induce many effects in the ice mantles, like chemical reactions, electrostatic charging, lattice damage, desorption, and evaporation. The radiation-processed grains are thought to aggregate into comets, which are found to have a similar composition as that of star forming interstellar clouds. Vapor deposited ice is also a principal constituent of the surface of many satellites and rings in the outer solar system [3, 4], which are subject to relatively intense magnetospheric ion bombardment [5].

Deposits of water condensed from the gas phase onto a substrate cooled to temperatures higher than about 190 K, consist of hexagonal ice, Ih. If condensation occurs below 190 K but above $\sim$ 135 K, the deposit is cubic crystalline ice, Ic, which transforms into Ih if heated above 160-200 K [6, 7]. At substrate temperatures below about 130 K, a deposit is formed with a molecular structure lacking long-range order. This phase, called vitreous or amorphous solid water (ASW), can be formed in different structures with local tetrahedral ordering [8–10]. It was first observed in studies of X-ray diffraction of vapor-deposited ice at low temperatures [11, 12]; diffuse diffraction structures resembled those resulting from liquid water, leading to the conclusion that the deposit was of an amorphous or glassy phase. Nano-crystalline ice would also produce diffuse X-ray scattering and the distinction between nano-crystalline and truly amorphous ice has been based on controversial evidence of a glass transition temperature. The metastable amorphous phase converts to the more stable Ic at a rate that depends on temperature [13]. One finds in the literature often references to a crystallization temperature, but it is important to realize that crystallization depends both on temperature and time. In laboratory time scales, crystallization is observed in the range 130-160 K. On an experiment where ice is warmed up from a low temperature, the crystallization transition will occur at a temperature that is lowest for the lowest heating rate.

Possibly the most important property of the amorphous phase condensed from the vapor is its microporosity (pore dimensions down to ~2nm), which, we will argue, affects most other properties. In particular, microporosity results in a large capacity to adsorb gases (the effective surface area for adsorption can reach several hundred m^2/g), especially at very low temperatures (<100 K). Thus, freshly prepared amorphous ice films act like a high capacity cryogenic vacuum pump, capable of pumping small molecules from a room temperature source at a rate of $\sim 0.4 - 3 \times 10^{15}$ Pmolecules cm^{-2} s^{-1}, where P is the pressure of condensable gases in units of μTorr (10^{-6} Torr). In fact, gas adsorption by ASW deposited at low temperatures was proposed as the basis of a vacuum pump four decades ago [14]. This pumping ability means that thin ice films prepared in modest vacuum conditions (0.1-1 μTorr, typical of most reported experiments) may contain substantial contamination. For experiments not done in UHV, with the exception of a few careful infrared absorption measurements, reports have usually not specified the gas composition or the duration of the experiment, making it difficult or impossible to assess in most cases the degree of contamination of the ice. A vacuum environment at 0.1 μTorr may be mostly composed of water, in which case the film will grow at a rate of $\sim$ 0.1 μm/hour, but will likely contain a substantial partial pressure of CO, CO_2, H_2 and also N_2 and O_2 from leaks. Each of these gases will be incorporated into the ice film at a rate proportional to its partial pressure if the ice is at temperatures below that at which the gas might condense in multilayers. Below about 15 K, the main gas adsorbed in ice will be hydrogen. In general, if a given gas can penetrate into the pores, the fractional contamination of a film of thickness d at time t will be initially proportional to Pt/d and saturate for long times. We can expect that contamination can affect properties of the amorphous deposits, and speculate that different vacuum conditions are at the root of the large discrepancy in results often found in the literature, but there have not been systematic studies of such effects. In addition, another likely source of discrepancies between data of different investigators may have been identified from our recent finding that microporosity depends on the angle at which the molecules arrive at the surface, thus on the gas dosing technique [see also Chapter 14.1].

15.2 Amorphous Ice Phases

Since 'amorphousness' is so hard to quantify, the different forms of amorphous ice that have been reported are distinguished by particular properties, rather than by a measure of long-range disorder. One such distinction is the density. There is a low-density form, ASW (ρ=0.94 g/cm^3) and a high-density form, HD-ASW (ρ=1.1 g/cm^3), which grows at 10 K or below [9]. HD-ASW transforms into ASW at ~114 K with a heat release of ~6 meV/molecule [6]. In addition there are other amorphous forms that are produced by methods other than vapor deposition. HGW (hyperquenched glassy water) is formed by rapidly cooling liquid water [15-18]. HDA (high density amorph), formed by pressurizing hexagonal ice (Ih), has ρ= 1.31 g/cm^3 while compressed [19]. The ice stays amorphous after releasing the pressure at low temperatures, and transforms slowly into LDA (low-density amorph, ρ=1.17 g/cm^3) through a continuum of amorphous structures [20]. A recent method reported to produce amorphous ice is that of cutting crystalline ice; it works most likely by creating HDA at the high pressures in the cutting treatment [21] followed by relaxation to LDA. A promising method to produce high-purity HGW is by ultra fast (~10^6 K/s) cooling of liquid water melted in cold crystalline sample by a strong laser pulse [22]. In addition, a type of HGW, called watergel, was recently obtained by the condensation of water in helium gas at below 1.5 K, and found to intriguingly decompose upon warming to a few K.

Several authors have argued, based on gas adsorption experiments in ultrahigh vacuum (UHV), that two or three different forms of ASW exist [23-26]; some experiments could just as well be explained by the reasonable assumption of different surface adsorption sites in just one form of ice. ASW coexisting with crystalline ice, called restrained ice by Jenniskens and Blake [27], has been observed for very thin films (~50 nm) in an electron microscope under moderate vacuum conditions. It is likely that a continuum of amorphous structures can be formed, as shown by infrared absorption spectra [28], at least partially due to different levels of porosity. Evidence for restrained ice was searched for and not found using thermal desorption [29].

15.2.1 The High Density ASW Phase

Narten et al. [9] found a denser form of ASW when condensing on a single crystal Cu substrate at 10 K. The density was estimated to be 1.1 g/cm^3 from the X-ray diffraction pattern. This form could not be deposited onto fused sapphire or a polycrystalline metal [30]. Heide and Zeitler confirmed this high-density form using the electron microscope, and found that ASW, Ic, or

Ih can be converted into HD-ASW by irradiating in the electron microscope at very low temperatures (8-20K) [31]. Upon warming, this phase remains stable up to 70 K and then transforms to ASW. Other electron microscope studies by Jenniskens et al. [27,32-34] report that HD-ASW forms below 30 K at deposition rates < 1ML/s. The very interesting and extensive results might be affected by gas adsorption since the use of very thin films ($\sim$50 nm) and relatively high pressures (0.1-1 μTorr) constitute conditions that favor substantial trapping of gases in the film (see above).

The X-ray experiments on HD-ASW have not been done in ultrahigh vacuum either. X-ray or electron diffraction cannot distinguish between oxygen in water and oxygen, carbon, and nitrogen in contaminant gases. In addition, heavy loading of thin ice films with hydrogen (a typical background gas in high vacuum systems), which must occur at low temperatures, may put the ice structure under compression and result in a transformation to a high-density form. Thus, we must wait for new experiments that include compositional analysis of the films to draw definitive conclusions about this form of ice.

15.3 Electronic Structure and Optical Properties

The electronic structure of the free water molecule is [35] $(1a_1)^2$ $(2a_1)^2$ $(1b_2)^2$ $(3a_1)^2$ $(1b_1)^2$ X $^1A_1$. The $1a_1$ are the non-bonding O-1s orbitals. The $2a_1$ is mainly O-2s (74%), mixed with O-2p and H-1s, with 20% of the bond energy. The $1b_2$ is bonding, and a mix of H-1s and O-2p orbitals. The non-bonding $3a_1$ and $1b_1$ are the lone-pair orbitals: $3a_1$ is largely derived from O-2p but mixed with H-1s. The broadening and decrease of intensity of this level in ice compared to the gas, as seen by electron spectroscopy, means that this level participates in the bonding in ice [36]. The higher occupied molecular orbital, $1b_1$, formed from O-2p, is oriented perpendicular to the molecular plane [37] and is broader in ice because it contributes to the hydrogen bond. XPS shows no energy shift in the $1a_1$ (mostly O-1s) level and a slight shift in the $2a_1$ (mostly O-2s) level when ASW grown at 77K is taken to 147 K [38]. Reissner and Schulze [39] found no difference in the UPS spectra of multilayer water films grown at 120 and 143 K although they produce different shifts in the surface work function.

The description used for orbitals of free water molecules has limited validity in the condensed state, particularly for excited states, which partially overlap adjacent water molecules. Comparisons of the electronic excitations of ASW, Ic, and Ih, that have been done using photoabsorption [40] and electron energy loss spectroscopy (EELS), reveal only slight differences between the solid phases. The minimum energy to place a valence electron outside the ice is the photoelectric threshold. For ASW, it has been reported to be 8.7 eV [38,41] and 10.5 eV [42], close to that for liquid water, 10.1 eV [43]. The threshold, seen in photoabsorption [42] and in EELS [44,45] has

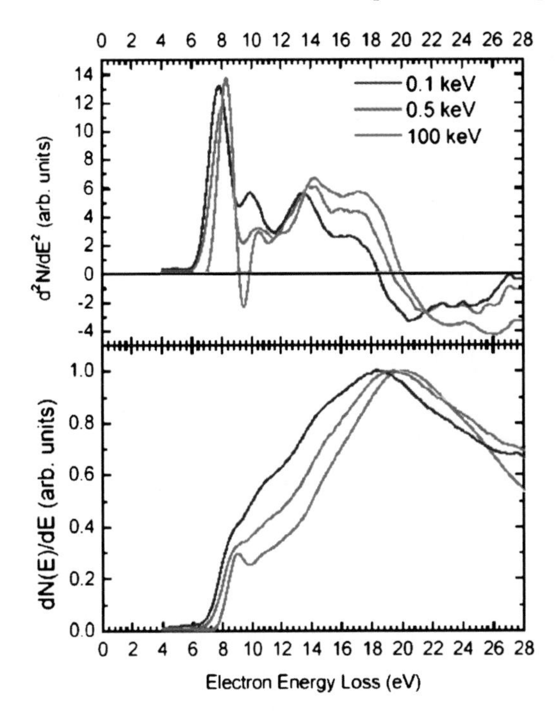

Fig. 15.1. Bottom: EELS spectrum of ASW at 100 and 500 eV [45], compared with 100 keV data from ref. [44]. The intensity of each spectrum is normalized so that the height of the main energy loss peak is 1. Top: Derivative spectra, where the main structure correlates closely in energy with those in the photoabsorption spectrum.

a Gaussian broadening without the temperature-dependence of an Urbach tail. Figure 15.1 shows the electronic excitation spectrum of ASW obtained by using EELS [44,45]. Spectra obtained at high electron energies represent bulk properties, those at progressively lower energies are more indicative of surface properties but also contain excitation to triplet states disallowed for dipolar excitations. More studies are needed to ascertain the ways in which the properties of the surface differ from those of the bulk.

15.4 Molecular Structure and Vibrational Spectroscopy

Absorption of light in the infrared is due to atomic and molecular vibrations, and carry information on the molecular structure of ASW at relatively short range. The most direct way of obtaining the structure of ice is by X-ray, electron or neutron diffraction. The local tetragonal ordering in ASW has been reviewed [18, 46], but without considering the effect of porosity in the interpretation of the measurements. Another way to determine the near neighbor structure in ASW is by examining the fine structure near the O-1s core threshold in X-ray absorption spectra [47, 48]. This technique provides information about slight changes in configuration, like those resulting from annealing, and can be extended to give information specific to surfaces.

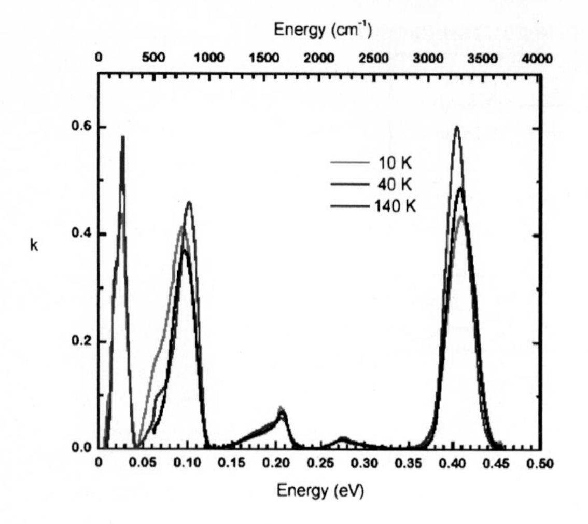

Fig. 15.2. Optical absorbance **k** of ASW deposited at 10 K and after annealing at 40 K and 140 K. Data from Hudgins et al. [50].

Comprehensive reviews of infrared spectra [18, 49, 50] have been complemented by several other measurements [51-59]; see also chapter 17. Optical constants of ASW and Ic at different temperatures are tabulated [50] and shown in Fig. 15.2. The spectra consist of intense bands with absorbance that peaks approximately at 3.1μm (O-H stretch), 6.1 μm (bending or scissoring), 12μm (libration or frustrated rotation), the lattice modes (frustrated translations) above 20 μm (peaking at 46μm), a combination band at 4.6 μm, weak overtones at 2μm, 1.5μm, 1.0μm and progressively weaker ones at lower wavelengths into the visible range. Reflection spectra (especially at glancing angles) are different from bulk transmission spectra due to structure in the index of refraction and to difference in vibrations in the surface compared to the bulk [60, 61].

The most studied infrared absorption is that corresponding to the O-H stretch which, most prominently is wider in ASW than in Ic. Besides the proton disorder (also present in Ic), ASW has two additional types of disorder that contribute to the bandwidth. One is the disorder in the oxygen positions and the other that caused by the micropores. The latter contribution, not considered in the early analysis of ASW spectra [62–64], can be inferred from the fact that compacting the ice by annealing (but maintaining the amorphous phase) narrows the bands (see below under *Annealing*). The dependence of the absorption band shape and position on temperature [10, 65, 66] and thermal history can be used to diagnose astronomical ices [10]. For this application, mixtures of water with other gases are important, and there are several studies of the corresponding IR spectra [67, 68].

Of particular importance for ASW is the finding of sharp O-H bands in the absorption spectra of microporous ice [69, 70]. These bands are assigned to dangling bonds based on their high frequency (hydrogen bonding decreases the O-H stretch frequency), on their sensitivity to adsorbed gas, and on the fact that they are more prominent in clusters, which have a large surface to volume ratio. The oscillator strength of dangling bonds on the surface of ultrathin ice films is ~30 times smaller than that of the O-H stretch for hydrogen bonded molecules [71], a fact that needs to be considered when examining published infrared spectra. The relatively strong dangling bond bands in the microporous ice shift in frequency when gas is adsorbed, and decrease in intensity during annealing. Similar findings were reported recently on ASW containing Ar [72].

Energy loss to vibrations can be measured also with low energy electrons [73], and neutrons. The latter provide freedom from the constraints of the dipolar photoabsorption selection rules. Different forms of amorphous ice show an enhanced probability of vibrations at energies < 6 meV, related to the mesoscopic structure of the ice. However, the few experiments reported on ASW [74-77] have been without impurity analysis of the ice and with unspecified vacuum conditions in long experiments, so there are uncertainties on the degree of gas loading of the ice samples and on the extent of differences with LDA ice [78, 79].

15.5 Growth Of ASW Films

The structure and properties of ASW depend on a number of experimental conditions; the importance of some of these conditions has only been realized recently. The purity of the films is determined not only by the purity of the starting water and on how it was degassed but also, very importantly, by the vacuum conditions. Since microporous water absorbs large quantities of gas, as mentioned above, the purity of a film will degrade with time after deposition. The flux of incoming impurity molecules is determined by their partial pressure; thus the importance of post-deposition contamination on bulk properties will be larger for thinner films. In experiments where dosing is done by admitting water vapor to the whole experimental chamber (back filling), impurities on the walls (like CO, CO_2, etc.) can be replaced by the water molecules, desorb into the ambient and condense on the film. The following parameters have been reported to affect the properties of thin ASW films and should, therefore, be controlled and reported in publications: substrate (composition, cleanliness, surface roughness, temperature), and the temperature, magnitude, direction and uniformity of the flux of incoming water molecules.

The growth of amorphous ice from condensation of low-pressure vapor has been addressed in several experimental [9, 64] and a few recent theoretical [80, 81] studies. A surprising finding has been that low condensation rates (less than a few monolayers/sec) produce ASW at < 110 K, and high rates produce Ic [82-84]. This is contrary to expectations based on studies on other amorphous solids, which are formed when the deposition is so fast that diffusion cannot succeed in arranging atoms in a more stable, crystalline structure. We note here that heat release from condensation has a negligible effect on surface temperature, at the condensation rates used in experiments in high vacuum (one monolayer/sec with condensation energy of 0.5 eV (surface binding) plus kT_{vapor} equals a heat input of only 84 $\mu W/cm^2$). It is conceivable that a reason for the need of low condensation rates to produce amorphous films is to allow a substantial amount of contaminants from the background gas to co-deposit on the film, distorting the lattice. For instance, in the growth of solid argon from the gas phase, 3 % impurities are needed to produce amorphous films [85].

Bruneaux-Poulle et al. [86], using X-ray diffraction, found conditions of temperature and growth rate for the growth of ASW, Ic and Ih and connected the formation of ASW with a low diffusivity. Hallbrucker et al. [87] were able to obtain films > 98% amorphous (as determined by X-ray diffraction) at 77 K and 1 μTorr background pressure, at growth rates of 100-200 μm/h, two orders of magnitude higher than the recommendations of Olander and Rice [82], which they state is possible due to better vacuum in their experiments.

A helium scattering experiment [88] showed that crystalline ice films grew on Pt(111) at 125 K if the dosing was done over hours in UHV, and resulted in disordered films when growing fast or at lower temperatures. On the other hand, reflection IR measurements by Mitlin and Leung [89] on 1.5 μm films showed non-crystalline ice growing at 128-145K on polycrystalline copper in about one hour by backfilling the chamber to a water pressure (i.e., no directional flow).

15.5.1 ASW vs. Ic Content

The ASW deposits may contain embedded crystalline grains [90, 91]. Olander and Rice state that above 77 K it is not possible to form pure ASW, and that many previous studies had ASW contaminated with crystalline ice grains. As we shall see below, others have found that the conditions are not that stringent, which suggest that unspecified parameters affected the experiments of Olander and Rice. Using UHV techniques, we found that the sublimation rate of ASW depends in a systematic way on growth temperature and suggested that this is due to a temperature dependence of the content of crystallites in ASW [92]. Trakhtenberg et al. [84] studied the effect of substrate morphology on the structure of ice films, and found that films grown on smooth surfaces are more crystalline than those grown on rough surfaces. This contrasts with the finding that crystalline ice forms on small silicate grains below 20 K [93].

On the other hand, Jenniskens et al. [1] found that very thin films of ASW exist inside Ic at temperatures higher than those quoted for crystallization. Using X-ray diffraction, Kohl et al. [94] found that HGW contains at most 20% amorphous component when annealed to 183K, quite lower than that reported by Jenniskens et al.

15.5.2 Effect of Deposition Rate

The often-quoted recipe of Olander and Rice to prepare purely amorphous films of ice is to either use growth temperatures below 55 K or limit the growth rate to less than 100 μm/h. As mentioned above, this may be related to contamination with background gas at the moderate vacuum used in those experiments. Using infrared spectroscopy, Hagen et al. [95] found, in contrast, that a more disordered ASW results at higher deposition rates. Also, Kouchi et al. [96] have argued that deposition that is very slow (astronomical rates) will produce Ic.

There has been discussion in the literature about the effect on film morphology [97] when using a directed capillary, a microcapillary array, a molecular beam, and deposition from the background or using a capillary with a baffle that impedes line of sight to the substrate. In some cases, needle-like structures [23, 98] or dendritic growth [99] have been observed. The conditions for the formation of these structures is not clear; whereas Laufer et al. [23] found them when placing a diffuser between the gas doser and the surface, Mayer and Pletzer [98] found the opposite: tiny "stalactites" formed when growing ice directly from a capillary in conditions of what they call supersonic flow while smooth surfaces were obtained with a diffuser baffle. They attributed this behavior to clustering and they warn about the presence of clustering in other experiments. However, clustering in the gas phase is not a concern in most cases when the growth rate is slow, since typically conditions in the doser tube are those of molecular flow, where the mean free path for intermolecular collisions is larger than the diameter of the tube. The difference between directed flow and the use of an obstructing baffle may be related to the fact that the direction of the incoming gas flux has an important effect in the porosity of the ice [100, 101]. The dependence of microstructure on deposition rates and type of doser has been reexamined recently using infrared spectroscopy for ASW grown at 11 K at ~0.1 μTorr [28]. There is clearly a need for systematic studies to resolve this issue.

15.5.3 Development of Porosity

The reason why microporosity develops in ASW films is not well understood. Buch [102] has suggested that incoming molecules attach preferentially to dangling bonds sticking out of the surface and thus stick out even more forming increasingly longer surface protrusions. Simpler "hit and stick" models with simplified intermolecular potentials give an open structure, with porosity increasing with the angle of incidence of the incoming water flux [103-106]. The question of the temporal and spatial extent of the dissipation of the kinetic energy and condensation energy of incoming water molecules, underlying the hit and stick assumption, has not yet been solved. Incoming molecules give the solid the heat of adsorption and their (quite smaller) kinetic and internal energy. The fact that limited microporosity results during growth even at relatively high temperatures close to the crystallization transition suggests that diffusion during deposition is negligible and that the hit and stick mechanism is accurate. The efficiency of the dissipation of the condensation energy, that enables this mechanism of micropore formation, can be rationalized by noting the different mechanisms of substantial energy loss in ice, as illustrated by the infrared absorption spectra of Fig. 15.2. The strong absorption of a single quantum of O-H stretch vibration amounts to 0.4 eV which can be later dissipated in multiphonon relaxation in the lattice (as seen in thermal He scattering [107]) without going into translational motion of water molecules.

15.5.4 Effect of Background Gas

Diffraction studies of ASW using X-ray, neutrons and high energy electrons have not been done till now in ultrahigh vacuum, and the long measurement times in some cases must have meant that the highly absorbent samples were significantly contaminated. For instance, Narten et al. [9] state that X-ray diffraction studies took about two weeks per sample. The lack of any systematic changes in the diffraction patterns during that time could signal either that there was no significant contamination or, on the contrary, that the concentration of contaminants reached saturation levels. Sivakumar et al. [83] found no changes in the Raman spectrum of ASW at 10K for two months. Although the pressure is not given in this report, it is conceivable that the ASW had absorbed a large amount of residual gases during such a long time. Even though the cold shields around the sample provide high pumping, the ice faces surfaces at room temperature, which release gases continuously. Long experimental times are also typical of neutron scattering studies; for instance Kolesnikov et al. [74] state that the deposition of ASW took 45 hours, in unreported vacuum conditions.

As shown by several studies, gases that are co-deposited with water at low temperatures cannot be removed completely by heating, even if taken to crystallization temperatures. This has to be taken into account to analyze even recent experiments in which water was co-condensed with Ar [54, 72].

15.6 Nanostructure

The porosity of ASW may be derived directly from measurements of the density of the films, but more usually through measurements of the index of refraction, related to porosity by the Lorentz-Lorenz relation. The range of values of porosity reported in the literature is very wide, from 0.05 to ~0.6 [108-111]. The morphology of the micropores is not known, but from the large effective areas for absorption of gases at low temperatures, it can be inferred that part of the pores are interconnected. The fact that annealing decreases the internal area for gas adsorption by more than an order of magnitude but the porosity remains (~0.1) shows that many pores remain in the ice closed, not accessible by molecules from the outside.

The intrinsic density of ASW (not including micropores), $\rho g \sim$ g.94 g/cm^3 is obtained from the short-range atomic oxygen distribution derived from X-ray diffraction [9]. Values of 0.94 g/cm^3 [112] and 0.91 g/cm^3 have been measured by flotation in cryogenic liquids, suggesting that the liquids penetrated the pores or that the ice films were annealed during the measurements. Optical measurements yield the lower macroscopic density, which includes pore space [3,108,113-115]. Densities can be as low as ~0.6 g/cm^3 at 20K, when the films are grown with omni directional water flux from a relatively high ambient water pressure. This low density agrees, within errors, with the porosity of 0.39 derived heuristically from internal friction [110] and gas adsorption [116] measurements but is hard to connect with the conclusion that ice grown at 5 K has no discernible porosity [117] based on the lack of dangling O-H bonds which signal micropores.

Westley et al. [100] pointed out that the different values of density reported in different studies could be explained by an increase of the porosity with the angle of incidence of the incident water molecules with respect to the surface normal. This was recently confirmed in our laboratory [118] by combined optical and microbalance techniques, and is consistent with the recent report by Stevenson et al. [101] that the gas adsorption capacity (indicative of the open porosity) depends on incidence angle.

15.6.1 Mesoscale Morphology

The nano- and mesoscopic nature of low temperature ice has not been resolved, and there are conflicting reports on the question of whether the surfaces are smooth or rough. Optical microscopy by Laufer et al. [23] showed needle-like structure in pure and gas-filled amorphous ice condensed at temperatures below 100 K. The structure, described to be similar to a "shaggy woolen carpet", persisted when the ice was warmed up to 180 K until the ice films evaporated completely. Mayer and Pletzer [97] found that tiny stalactites formed when using a gas doser in conditions where clustering of water molecules could occur. Other optical and electron microscope studies have not yet found this type of structure. Our films grown on an optically flat gold surface have very little roughness as long as the thickness is not so large to cause cracking.

Differences in the morphology of the ice affect the way it reflects light. ASW deposited at or above 100 K is transparent like a glass, whereas below 100 K, films thicker than ~10 μm, grown below 100 K, are granular, non-uniform and opaque [83]. The appearance of thin transparent ASW films changes to translucent and then to opaque white, upon warming. Gormley and Hochanadel [112] found that the reflectivity for visible light increased little up to 140 K where it started to increase more rapidly. Heating at ~10

K/min, they found a slight drop in reflectance at 152 K which they attributed to a glass transition. The large increase of reflectance at higher temperatures was thought to be due to crystallization and to the increase of the size of the crystal grains from smaller to larger than the wavelength of light. A closer examination by Westley et al. [100] revealed that the high scattering resulted from cracks. Cracks can also occur at low temperatures during deposition.

15.6.2 Gas Adsorption

Gas adsorption studies provide information on the pores that are connected to the outside of the ice, but not of the enclosed pores. In the gas adsorption technique, a fresh ASW film is exposed to a directed flow or a background pressure of a gas of small molecules. Under conditions that a single monolayer of gas is adsorbed on the walls of the pores, one can obtain the total area of the connected pores, and hence a measure of the connected porosity, via measurement of the quantity of gas released as the ice film is warmed up. In addition to the use of gas adsorption for characterization of porosity, the information on gas trapping is needed in astrophysics (where one is concerned also with the effect of photon and particle irradiation that can cause the synthesis of new molecules). The question of the interaction of hydrogen with ice mantles coating dust in interstellar clouds prompted the first measurements of adsorption of H_2 on ASW at very low temperatures by Brackman and Fite in 1961 [14]. At the lowest temperatures, two effects come into play: gases may not be able to enter the micropores but rather form overlayers. If they do penetrate the pores, they may form multilayers. For instance, Devlin [119] determined that the minimum temperatures to penetrate the ice are < 12 K for H_2, $\sim$18 K for N_2, $\sim$26 K for CO, and $\sim$55 K for CF_4. The behavior of CO was later confirmed [120]. On the other hand, it was reported that ASW does not seem to adsorb H_2 above $\sim$16 K and Ne above 25 K [121], but the former is in apparent contradiction with the observation of H_2 infrared absorption bands on ice exposed to hydrogen at 20 K [122, 123].

Gas adsorption coefficients may give an indication of the surface area; however, dramatic changes in sticking probabilities can also result from a moderate surface roughness [124]. Ghormley [125], extending the measurements of Brackman and Fite [14], established that adsorption areas can be large, hundreds of m^2/g, that they decrease with temperature, that similar behavior is seen by gas co-deposited with water or dosed on the ASW, and that the gas desorbs in distinct peaks as a function of temperature, related to phase transformations, when the ice film is heated. Subsequent work confirmed these findings but found widely varying values for the effective surface area for gas absorption [23,87,98,116,119,126-129]. Similar conclusions were obtained by thermal measurements [130]. From the lack of hysteresis of adsorption isotherms, Mayer & Pletzer [98] inferred that a small fraction of pores are wider than 2 nm. Comparing adsorption isotherms of N_2 and Arin ice, Schmitt et al. [131] also concluded that there exist a large fraction of pores only a few molecular diameters wide. The adsorption area, 258 m^2/g for a 50-μm film grown at 77 K, was found to decrease slowly with time at 77K and faster at 87 K. The analysis of the N_2 isotherm of a 900 μm film, grown fast (100 ML/s) with an adsorption area of 35 m^2/g showed that 40% of the pore volume was made up of pores < 1.8 nm in radius. Annealing the samples for 10 minutes at 190K reduced the adsorption area to less than 0.1 m^2/g.

The most thorough investigations of gas adsorption in ASW are those of Bar-Nun and colleagues [23, 116, 121, 128, 132], who were motivated by the need to understand the trapping and release of gases like CO, CO_2 and CH_4 in comets. Their research revealed that the eruptive desorption of trapped gas appears to eject small grains of ice from the needles mentioned above. This work established several distinct temperature regions for gas release, beyond the peaks occurring at crystallization and at the Ic$\rightarrow$Ih transition, and that the temperatures at gas release peaks depend on heating rate. The researchers also confirmed the decrease in surface area for adsorption upon annealing; it decreases by $\sim$90% ($\sim$99%) the ASW film is heated from 10-20 K to 122 K (151K). In a contemporary study, Schmitt et al. [131] also found that annealing of ASW with preadsorbed gases leads to the occlusion of the gases, which are then released at higher temperatures.

More recently, Stevenson et al. [101] and Chapter 4.1 demonstrated the effect of the angle of incidence θ, of the water molecules condensing on the substrate, on the N_2 gas adsorption ability of very thin ASW films, < 250 bilayers. Q, the amount of N_2 adsorbed, increases with θ, consistent with the prediction of an increase of porosity with θ[100]. Q passes through a maximum at $\theta_m \sim 65^o$ and declines slowly for very oblique incidence; at this value of θ_m, Q is as large as that for ice films grown from background (not directed) gas flow obtained by backfilling the vacuum chamber with water vapor. The linear growth of Q with film thickness d suggests that microporosity is independent of d. As expected from previous experiments [115] which showed that porosity from background deposition drops with deposition temperature T_d, it was found that Q dropped by about an order of magnitude for T_d between 20 K to 90 K, and also dropped with annealing temperature. This study reported no micropores above 90 K.

A similar N_2 adsorption experiment on ASW was performed by Vichnevetski et al. [133] who, with a parametric model, obtained an effective adsorption area of 2500 m^2/gr for a 22 monolayer film grown at 45 degree incidence on a Pt(111) surface at 20 K. A more recent UHV experiment [134] used infrared absorption and CH_4 adsorption on ASW grown at 25 K with or without annealing to 135 K. It was found that gas adsorption falls to very low values for ice grown at 60 K (assigned to cavity collapse), whereas the infrared signatures of dangling bonds (assigned to micropores) nearly disappear for growth at 120 K. However, the idea that large cavities (not micropores) are needed to trap gases are at odds with the fact that large internal surface areas, required to explain gas adsorption, can only be achieved with small micropores.

15.6.3 Annealing and Pore Collapse

When ASW grown at low temperatures is heated, but kept below the temperature where it starts crystallizing, it undergoes a number of irreversible changes that are not well understood, particularly below 115 K, where diffusion for pre-annealed ASW is insignificant [135]. Ghormley [130] detected a heat release of 8 cal/g when a film of ASW was heated from 20K and 77K for the first time; since ASW is way out of equilibrium, being amorphous and microporous, the heat release signals structural relaxation towards equilibrium. Hallbrucker et al. [87] confirmed heat release due to relaxation and found that was accompanied by the narrowing of the first broad X-ray diffraction peak.

Several authors [1,28,95,136-141] have found irreversible changes in the infrared O-H stretch band (narrowing, and shift in the frequency for maximum absorption) when annealing ice deposited at 10 - 100 K. Rowland et al. [70] found that dangling H bonds in vapor-deposited ASW, seen by IR spectroscopy, diminish upon annealing to 60 K signaling collapsing of the micropores. In these dangling bonds, H does not participate in H bonding, a situation that happens at the exterior ice surface and is expected to occur in the internal surface of the micropores. Spectral changes in IR spectra upon annealing were also observed for the libration band [136] and the lattice bands [56-59,142]. A few studies examined time dependencies; whereas Mayer & Pletzer found no time dependence by holding the sample at each temperature for several hours, Zondlo et al. [143] found a decay of 2-3 hours of the intensity of the dangling bond vibration at 118-112 K.

Eldrup et al. [144, 145] studied the formation of positronium in thin films of ASW using incident positron beams. Since cavities of average diameter of ~1.7 nm are needed to trap positronium, the results were used to characterize the presence and evolution of such cavities in the films. They interpreted a transition in positronium trapping at 100 K to be the result of the disappearance, or coalescence, of the cavities, because of the start of significant vacancy migration at this temperature. Some cavities were found to persist even above the crystallization transition.

Compaction of ASW upon annealing is also seen indirectly in dielectric experiments [146, 147], in hole-burning studies [148] and in desorption of trapped gases. In the extensive studies mentioned above of the thermal evolution of ASW containing gases, Bar-Nun and colleagues found that gas desorption has two distinct temperature behaviors, suggesting two different forms of ASW, one below 85 K and one for temperatures between 85K and 136K. When increasing the temperature a few K, the rate of gas release jumped without any apparent delay. These experiments were interpreted to mean that pores close irreversibly upon annealing at those temperatures.

Other annealing studies have used measurements of internal friction of ASW films versus temperature. Härdle et al. [149] explained the decrease in internal friction observed between 20 K and 84 K as due to a "polymerization process" in a medium with a wide and continuum range of glass transition temperatures. Hessinger et al. [150] observed this type of annealing above their lowest temperature of 48 K, and assigned partially to a reduction of porosity and mostly to a reduction of local atomic disorder. The annealing processes at low temperatures might be transitions between different metastable forms of amorphous ice. In a more detailed report, Hessinger et al. [110] showed that annealing has a logarithmic dependence on time (in the range 1 h - few days) which was not understood, and also that the films did not anneal completely even after a 12 h at 126.6 K since they showed a shear modulus half the bulk value for compact ice [151].

15.6.4 Internal Stresses and Fracture of ASW Films

There are scattered reports that ice films may crack during growth. This is usually indicated by a transition from transparency to a milky or frosty appearance [131]. Wood et al. [65, 152] found that films deposited at 20 - 7 K would shatter if thicker than a few microns. Using a microscope, we found that films will fracture after a thickness that increases with growth temperature [153].

Sivakumar et al. [83] reported vaguely that films deposited at 110 K broke up when heated or cooled. This suggests that stresses caused by a mismatch between the thermal expansion of the film and the substrate might be the mechanism for cracking. However, different studies point to an effect caused by a phase transformation. The first evidence for cracking during heating, by Ghormley and Hochanadel [112], found cracking at ~200 K, possibly related to conversion from Ic to Ih. Others report that the temperature at which there is an abrupt change in optical reflectance is 145-150 K, more in line with the conversion of ASW to cubic ice [113, 154]. Drobishev et al. [155, 156] report similar behavior but at 160–162 K, which they correlated to transition to Ih, although the temperature seems rather low for this transformation. Westley et al. [100] found that cracking starts at 150 K, but the frosty appearance develops fully at ~200 K. Smith et al. [157] suggested that cracks may lead to the abrupt gas release that has been observed to occur in coincidence with phase transformations.

15.7 Thermal Properties

15.7.1 Glass Transition

In analogy with other glasses, it can be expected that amorphous ice shows a reversible transition between a glass and a viscous liquid at a temperature

T_g, called the glass transition temperature. There is a strict thermodynamic definition and several operational indications of a glass transition. There have been no observations of a discontinuous change in coefficient of thermal expansion at T_g, or an abrupt change in viscosity that characterizes the glass transition. The evidence for the glass transition is indirect and contradictory, from a heat spike to a slight change in slope of the heat evolved in scanning calorimetry. Measurements need to be done at a fast heating rate (tens of K/second) to minimize irreversible changes due to crystallization. When the glass transition is observed, the reported T_g has ranged from 135 to 139 K [91, 158]. While Hallbrucker et al. report that the values for HGW are very similar to those of ASW [17], Handa and Klug [159] found a quite lower value for LDA, $T_g = 124$ K. The difficulties inherent in these types of measurements have been pointed out in several reports [87, 130, 160], and include relaxation of the microporosity, and trapping / desorption of gases. A recent proposal [161] that T_g is at ~165 K, and a vigorous reply [162] continue a line of conflicting results and interpretations that started decades ago [91,130,158,160,163-165].

In isotopic exchange experiments in ASW, Fisher and Devlin concluded that exchange reactions occur by defect-mediated orientational diffusion, rather than by translational motion [135]. On the other hand, Smith and Kay [166] showed that intermixing in isotopic layers of ASW deposited at 85K occurs at 150-160 K when heating at 0.6 K/s but is relatively frozen at 159 K, when crystallization is thought to occur. The data was interpreted as indicating enhanced mixing in a liquid formed above T_g. However, crystallization does not occur only at 159 K, but it starts in seconds in the range 140-150 K (see next section); thus the mixing can be explained by diffusion accentuated by crystallization and its accompanying localized heating. Evidence for consequences of liquid behavior has been obtained by electron microscopy [1]; but the effect of electron irradiation on diffusion has not been studied.

15.7.2 Crystallization

Much confusion exists in the literature about the temperature for phase transformations in ASW. Since each transformation occurs over a range of temperatures, the temperature at which, say, 50% of the transformation is complete depends upon the heating rate, if observed during a temperature ramp. The metastable ASW phase is seen to convert to the more stable Ic above 125 K with a heat release of 16 meV/molecule [6] that decreases with impurity content [167], and is accompanied by long-range translational diffusion [168]. Understanding crystallization by calorimetry is complicated because of concurrent processes like closing of micropores, annealing of point defects, changes in surface area, and desorption of gases that accompanies this transition. An important aspect of the crystallization is its duration, which includes a nucleation time and a transformation time. There are several reports

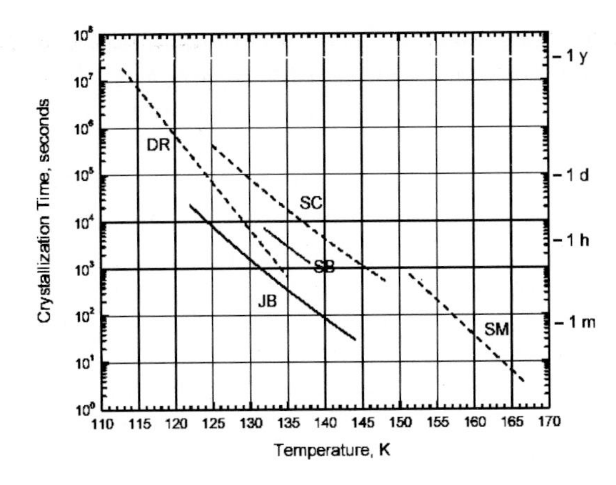

Fig. 15.3. Compilation of crystallization times, using lines as fits to the data. Dashed line, start of the transformation: Jenniskens and Blake [33]. Dotted line, 63% of the transformation: SB, Sack and Baragiola [92]. End of transformation curves are shown as dashed lines: DR, Dowell and Rinfret [13]; SC, Schmitt et al. [172]; SM, Smith et al. [183].

[13,33,92,158,169-172] of total crystallization times obtained with a variety of techniques, the most direct ones being those using X-ray or electron diffraction. As shown in the composite plot (Fig. 15.3), most of the experiments agree over the wide range from start to finish of transformation, leading to an activation energy of ~0.4 eV (close to the strength of two H bonds). Some spread is expected from different ways of preparing the ASW films that result in different contents of micropores and crystalline grains [82, 90, 92, 173] or if ASW is grown atop polycrystalline ice ([174] and Chapter 14). Such would be the condition resulting from experiments done at moderate vacuum, where films of sizable thickness would condense from residual water during cooling of the sample holder, which usually takes hours. Crystallization times measured [175, 176] for hyper-quenched amorphous ice are about an order of magnitude larger than for ASW.

There are several indications that ASW does not crystallize completely in the range 130 - 150 K, although most infrared measurements do not support this. Dowell and Rinfret [13] report that they never observe Ic in a pure state; a certain quantity of ASW was always present. Hagen et al. [10] reported that ASW annealed completely at 130 K but not into Ic. In their X-ray diffraction studies, Hallbrucker et al. [87] conclude that ASW relaxes to a different, less porous state. As mentioned above, Hessinger et al. [150] found that the internal friction of ASW films crystallized at 152 K is higher than that for a film grown at 142K (assumed to be crystalline), indicating that substantial disorder is still present, which they state as possibly due to amorphous inclusions [27].

There seems to be an inconsistency among all these studies. If amorphous ice persists when heated above 140 or even 152 K, why do sublimation rates fall to the values of crystalline ice at ~150 K [29,92]? A common aspect of the X-ray and electron diffraction experiments is that, unlike the sublimation studies, they were not done in UHV, so one is tempted to attribute the differences to the presence of trapped gases. We note that the diffraction techniques are not capable of chemical analysis, in contrast with mass spectrometric studies of the desorbed flux in sublimation experiments.

15.7.3 Sublimation

Sublimation determines the stability and transport of ice in space. The ASW found in astronomical bodies is not pristine, but fractionation due to sublimation occurs on icy satellites of Jupiter: water desorbs from dirty and therefore warm regions and condenses on bright, cold regions. The ice distilled in this way may remain clean for some mean time that depends on depth, until made dirty by debris from micrometeorite impact and ion implantation. Sublimation is one of the important parameters needed to model the evolution of comets, besides thermal conductivity, fracture stress, gas desorption and phase transformations.

Sublimation or thermal desorption of ASW has only been measured in vacuum. Therefore, references to vapor pressure, which imply evaporation and condensation in equilibrium, are incorrect. Kouchi [177] first showed that, at a given temperature, the sublimation rate of ASW is much larger (up to two orders of magnitude) than for crystalline ice, and it depends on growth conditions. The sublimation rates are higher than early estimates [66] based on the vapor pressure of crystalline ice and enthalpy differences between ASW and Ic from Ghormley [130]. Sandford and Allamandola [178] used infrared spectroscopy to measure the thermal erosion of thin films, confirmed the enhanced sublimation from ASW, obtaining sublimation enthalpies of 0.415 eV and 0.437 eV for ASW and Ic, respectively.

In Kouchi's experiments, the temperature of the ice was ramped at 1 K/min while the flux of water was measured with a mass spectrometer, a technique called temperature-programmed desorption (TPD). The sublimation rate shows a "bump" at temperatures about 135 K and decreases at higher temperatures to the rate appropriate to crystalline ice. Sack and Baragiola [92], with absolute measurements of sublimation rates, showed that under isothermal conditions sublimation decreased exponentially in time. Assuming that the decay is due to crystallization, they explained that the bump observed in TPD is a signature of the crystallization process, and that it should occur at lower temperature when the heating rate is low. This means that in the case of a new comet approaching the Sun, when the heating rate is extremely low compared with the laboratory values, the ice will have plenty of time to crystallize before sublimation can be observed as a coma. The UHV measurements of Speedy et al. [179], which are similar to those of Kouchi, confirm the results of Sack and Baragiola, and show lower sublimation rates for D_2O than for H_2O. The results are used to discuss the thermodynamic continuity between ASW and liquid water. We note that the non-equilibrium microporous nature of ASW, with its large energy content due to the internal surface energy, cannot be connected in a thermodynamic sense with liquid water. Other recent UHV studies [174,180-182] have confirmed the work mentioned above. In particular Livingston et al. [180] mentions that the failure of previous work [183, 184] to find zero-order kinetics in ultrathin films was due to the use of non-uniform deposits, such that the variation of sublimation flux when films were desorbed was due to a variation of surface area rather than intrinsic sublimation rate. Other potential experimental problems in these type of studies is the presence of spurious peaks in temperature-programmed desorption curves [24] due to sublimation of ice from parts of the apparatus (like heat shields) which are at different temperatures [29].

The implication of a higher sublimation rate for ASW than for Ic is that sublimation depends on the amorphous / crystalline content of the ice deposits [92], which in turn are affected by growth temperature and on the type of substrate [183, 184]. It also implies that the water evolution from comets will be different in the first passage of the comet around the Sun than in subsequent orbits, since heating by the Sun induces crystallization in addition to sublimation [185].

15.7.4 Thermal Conductivity

The thermal conductivity of amorphous solids like amorphous ice is much smaller, at low temperatures, than that of crystalline solids [186]. In addition, porosity, which increases phonon scattering, should decrease thermal conductivity further. Kouchi et al. [187, 188] reported a value of the thermal conductivity of ASW four or five orders of magnitude smaller than Klinger's estimate for compact amorphous ice [185]. Their approach was to determine at what thickness the surface temperature of a growing ice film, exposed to background infrared radiation from the walls of the vacuum chamber, reached that needed to crystallize the surface. To solve the thermal transport equation they assumed, incorrectly, that the infrared radiation is absorbed only in the front of the film. Thus, their analysis actually overestimates thermal conductivity. Further, the method of determining crystallinity using glancing angle diffraction of 20 keV electrons is subject to error because the electron beam can crystallize amorphous ice, especially at the relatively high temperatures used (125-135 K).

Another study [189] of the thermal conductivity of pressure amorphized LDA ice produced expected results for the thermal conductivity. Since this ice is non-porous, the statement [190] that the experiment by itself refutes the results by Kouchi et al. is incorrect. The thermal conductivity of non-porous LDA should be an upper limit to that of microporous ASW.

15.8 Interaction of Energetic Radiation with ASW: Radiolysis and Property Changes

Water ice is a good model system to study radiation induced surface chemistry due to the relatively small number of different stable species that can be formed. Irradiation of ice produces the emission of atomic and molecular particles (desorption or sputtering), the emission of electrons and luminescence photons, and chemical changes in the solid. The response of ice to photon or particle impact can be used to characterize the starting material and provide basic information about the evolution of defect states. One is interested in understanding the evolution of the solid during irradiation and the ultimate chemical and structural alterations. Different observables that can be used during irradiation are desorption or sputtering of surface atoms, and the emission of ions, light and electrons. Chemical changes can be deduced from measurements of infrared spectra and by different luminescence techniques. Line changes in XPS or Auger spectroscopy and its relation with gas-phase experiments can be used to study changes induced by particle irradiation [191, 192]

15.8.1 Basic Radiolysis of Water

Electrons and ions can induce desorption of atoms or molecules from ice either directly or through the generation of secondary electrons [193]. Unlike photons, which are totally absorbed in an inelastic interaction, electrons can create several close ionization/excitation events that can cause additional desorption. Low energy secondary electrons can also cause desorption by dissociative attachment; the resulting negative ions have very small kinetic energies (the spectra peaks at ~ 0 eV) [194]. Such low desorption energies can cause a large probability of trapping of the anion at the surface, especially if the surface is allowed to charge positively. This can occur during bombardment by positive ions and photons and also with electrons of intermediate energies, in the absence of charge compensation.

Johnson and Quickenden [195] reviewed decades of results on photolysis and radiolysis of ice; a few developments have occurred since then. Like in the gas phase, the primary molecular dissociation in water is OH + H, which is achieved directly or through ion-electron recombination. The OH radicals can be stored in ice up to temperatures of about 130 K in crystalline ice and $\sim$100 K in HGW [196]. Hydrogen atoms do not trap at 77 K in HGW [197]. Instead, they diffuse and recombine either with OH forming H_2O or with another H forming H_2, which can then escape from the solid. With continuing irradiation, OH radicals accumulate in the ice; their concentration is limited by the recombination with H and by reactions with other OH to yield H_2O_2. Fast ions produce a dense track of excitations in the solid; within this track the density of OH is high enough that hydrogen peroxide can be formed by

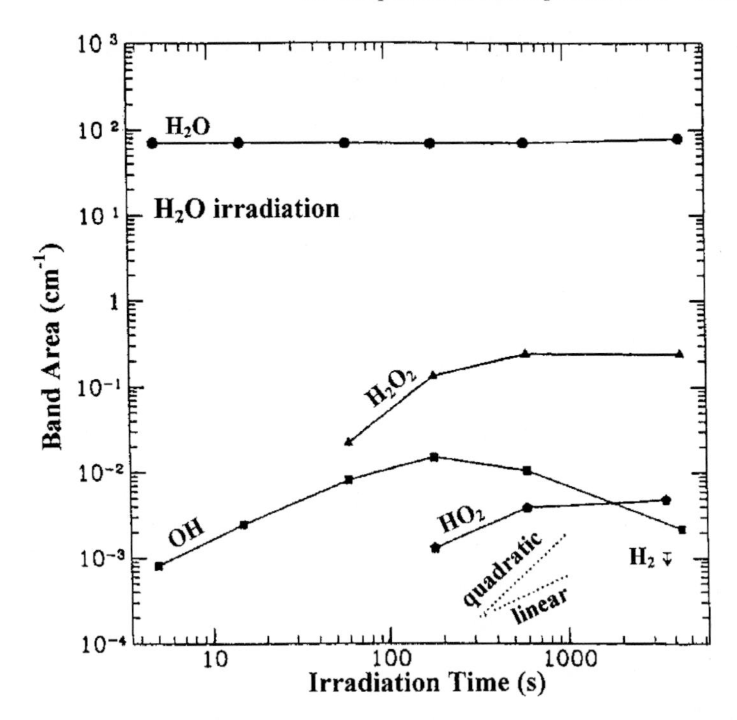

Fig. 15.4. Evolution of the area of infrared absorption bands of different radiation products in ice at 10K with the irradiation time of photons of energy > 6 eV (mostly Lyman-α) from a microwave hydrogen discharge [199]. One second of irradiation corresponds roughly to a fluence of 10^{15} photons/cm^2, or about half a year of exposure of an icy grain in a diffuse interstellar cloud. The bands used are (in cm^{-1}) at 3270 (H_2O), 3453 (OH), 2850 (H_2O_2), 1389(HO_2), and 4143 (H_2).

a single ion. Also, it was found that electrons are not trapped in HGW at 77 K [197].

Long-lived radicals and stable molecular products of radiation can be studied in ice by electron spin resonance [196, 198], infrared spectroscopy [199] and ultraviolet spectroscopy. Figure 15.4 shows the result of infrared studies [199] on the accumulation of radiation products in ice films photolyzed at low temperatures. Moore and Hudson [200] observed the H_2O_2 infrared band at 3.5μm when irradiating ice with 800 keV protons at 16K, but not at 80 K. In similar experiments using 30 keV protons, Gomis et al. [201] detected H_2O_2 both at 16K and at 77K. The importance of the discrepancy lies in that ion irradiation is thought to explain the observation of hydrogen peroxide at Europa, a satellite of Jupiter [202]. Alternative methods of producing H_2O_2 in Europa are surface reactions with atmospheric atomic oxygen, and ion irradiation of O_2 containing water ice. Mass spectrometry of gases released when heating radiolyzed ice also reveals the presence of O_2, H_2O_2, and HO_2 radiation products [203]. With this thermal desorption technique, however, there is an uncertain knowledge of the effect of temperature in inducing reactions before desorption (e.g., upon heating, trapped OH can diffuse and recombine into H_2O_2 which is subsequently desorbed at a higher temperature).

15.8.2 Sputtering and Desorption

Irradiation of ice by energetic particles and photons is a common occurrence in many astronomical bodies like ice-coated cosmic grains, icy satellites and rings, and comets. Irradiation can produce the ejection of molecules that leads to the erosion of the surface. The process is called impact desorption when it results from a single collision event at the surface and sputtering when it involves many collisions in the bulk and the surface of the solid. Here we will use the term sputtering as embracing also impact desorption. Sputtering may lead to mass loss if the ejected atoms and molecules can escape the gravitational pull of the body, or it can lead to the redistribution of molecules on the surface. The ejected species will form a transient water atmosphere that will also interact with the incoming radiation. For instance, the surfaces of the icy satellites of Jupiter or Saturn are subject to intense bombarding by magnetospheric ions [5] that have energies from eVs to MeVs, peaking at $\sim$100 keV. Water molecules, H_2 and O_2 are ejected from the surface; in addition, neutral and charged H_2, O_2, H, O, O_3, H_2O_2, and HO_2 can be formed by interactions of the incoming radiation with the water atmosphere and subsequent chemical reactions. Water molecules mostly precipitate back onto the surface, perhaps tens of kilometers away from the impact site since molecular collisions are infrequent in the tenuous atmosphere of the satellites. The temperature of the ice can reach 160 K on the sub-solar regions of the Jovian satellite Callisto and can go down to maybe a few K in permanently shielded regions of the surface. Hydrogen molecules will predominantly escape, because of their high speed. Oxygen molecules that return to the surface may be trapped in the pores of the ice, or be scattered back to the atmosphere. Trapping of O_2 will depend strongly on porosity and on surface temperature. The result is a tenuous oxygen atmosphere (10^{-3} - 10^{-8} Torr) [204–206]. Insufficient laboratory data exists currently to model the generation of atmospheres beyond the order of magnitude estimates [207] currently possible.

Sputtering is characterized by the sputtering yield, Y, the number of molecules (or atoms) ejected per incident projectile. There are two types of sputtering, depending on how energy is transferred from the projectile to the solid. Knock-on sputtering [208] is due to the momentum imparted by the projectile in collisions with target atoms; it is fairly well understood for simple metals but not for molecular insulators like water ice. In electronic sputtering, the projectile transfers energy to the electronic system that leads to the formation of repulsive interactions in the solid. Electronic sputtering is very efficient for high velocity ion impact of ice [207,209-237]; the vast majority of ejected molecules and atoms are neutral; the small ion fraction [232,238-242] is characterized by the presence of large cluster ions. Electronic sputtering of vapor-deposited ice, one of the most conspicuous examples in early studies, was discovered by Brown and co-workers [209], who readily understood that sputtering is an important phenomenon on the surface of airless astronomical bodies [243].

A summary of experimental findings at energies above 10 keV is as follows. The total sputtering yield stays essentially constant below 60-100 K and increases rapidly at higher temperatures. The main species desorbed at the lowest temperatures are intact water molecules, with near thermal energies. The most probable energies are much smaller than the surface binding energy of 0.45 eV [92] or the hydrogen bond energy (0.24 eV). They are ≤ 1 meV for H_2O ejected at 12 K by MeV He^+ and Ar^+ and ≤ 20 meV for 50 keV Ar^+ on ice at 25 K [214, 217, 220]. Mass analysis of the sputtered flux also shows the emission of H_2, and O_2 radiation products. For 50 keV He^+ and Ar^+ impact on ice at 77 K [213] emission of H_2O and O_2 appears when the ion beam hits the target, but H_2 emission is delayed. In contrast, emission of H_2 was detected as soon as the ice was hit by MeV ions, low energy electrons [244, 245], and 10.2 eV photons [246, 247]. For MeV Ne^+ projectiles, H_2 and O_2 molecules are ejected at a rate that increases with fluence (ions/cm^2) up to a saturation rate that increases roughly exponentially with ice temperature [220, 233]. Below 10 keV, the fluence dependence is not seen [224]. There is a controversy on which is the primary sputtered species at low energies: water [216, 221, 222] or molecular products (H_2 and O_2) [224].

Sputtering yields produced by ions or photons are insensitive to the conditions of growth (leading to amorphous or crystalline ice) [224,251]. Although, the ice remains approximately stoichiometric over the penetration depth of the ions [211,212,218,224,299], substantial alterations might occur in the surface layers. Auger spectroscopy, which is very sensitive to the surface, shows changes in the oxygen-KLL Auger spectrum (likely due to H_2O_2 formation) after irradiation with large fluences of keV electrons [192].

The sputtering yield is proportional to the *square* of the electronic energy deposition near the surface. This is still not understood, but suggests that two events are needed for significant electronic sputtering. Mechanisms considered by Brown *et al.* [214, 215] are thermal spikes due to overlapping events along the ionization track of the projectile and Coulomb repulsion between ionized molecules. Another possible mechanism is collision cascades initiated by the exothermic formation of H_2O_2 from OH radicals formed in the track. Below an impact velocity of $\sim 5 \times 10^7$ cm/s electronic sputtering decreases in importance as elastic sputtering becomes appreciable; this is evident in the "plateau" that appears in the energy dependence of the yields [224, 235, 237](Fig. 15.5).

Although double collisions dominate the sputtering by ions, single collisions can also lead to surface erosion. Fast (100-200 keV) electrons in electron microscopes give a sputtering yield that appears to be *linear* with the ioniza-

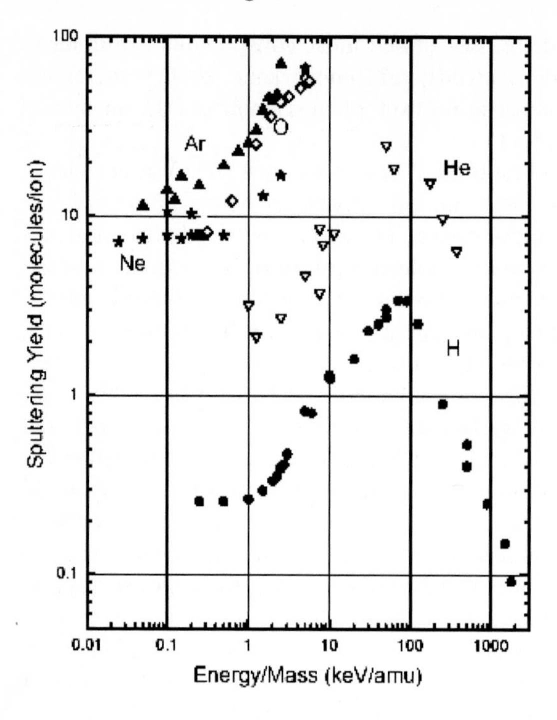

Fig. 15.5. Compilation of sputtering yields of ice vs. energy/amu for different singly charged ions [237].

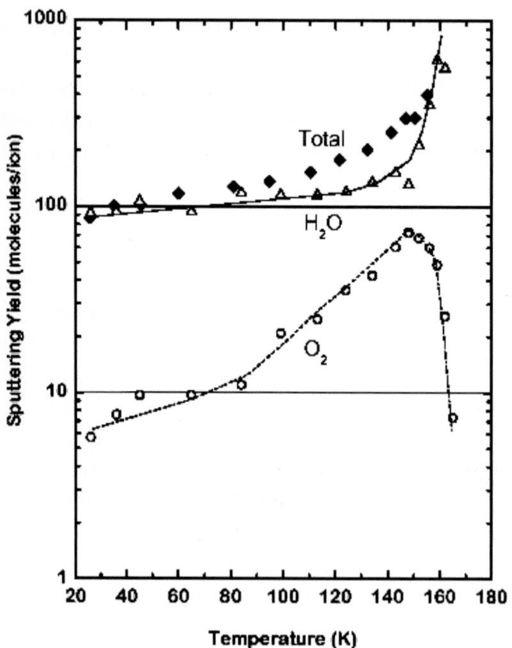

Fig. 15.6. Total and partial sputtering yields of water for 100 keV Ar^+ ions versus temperature. The dashed line shows the sputtering calculated from the mass spectrometer (MS) contributions, normalized to the yield measured by the microbalance at 17 K, taking into account that the relative contribution of O_2 and H_2O is according to their mass ratio (36/20) using $H_2^{18}O$.

tion rate [31,248-250]. Erosion by single excitation events has been demonstrated using energetic photons [246, 251, 252] and low energy electrons [244].

Figure 15.6 shows the temperature of the total and partial sputtering yields for 100 keV Ar^+ projectiles. It can be seen that the sputtering of water molecules is temperature dependent above ~120 K. The temperature-dependent component has been attributed to processes involving radicals for which trapping, diffusion, and reaction rates [217, 219, 220, 224, 249] depend strongly on temperature. The large number of processes involving electrons and multiple neutral, excited and ionized species of H, H_2, O, O_2, H_2O, HO_2, H_2O_2, makes this problem presently intractable until more information is provided by additional experimental techniques.

Sputtering by protons is independent of fluence up to $10^{16}/cm^2$ [294], consistent with results for *total* yields by MeV He^+ and Ne^+ at similar fluences [209,220,233]. Partial yields of H_2 and O_2 depend strongly on fluence at low temperatures, in the range where they don't contribute significantly to Y [211,236]. Westley et al. [246] saw a fluence dependence of photodesorption below 80 K for ice grown on polished substrate but the fluence appearance does not appear for ice grown on rough substrates [253], implying photosputtering in single photon events from defect sites.

A fluence dependence is observed in the emission of hydrogen and oxygen molecules. Reimann et al. [220] observed an increase in O_2 emission with fluence to a saturation value, for ice at 15 K irradiated by MeV neon ions. The lack of initial O_2 emission at the lowest fluences means that O_2 molecules are not formed in the dense ionization tracks of these experiments. The most likely explanation [200] is that the oxygen atoms produced in single collisions react with an adjacent water molecule to form H_2O_2 (this seems much more likely than finding another O atom to form O_2). The hydrogen peroxide would accumulate in the ice, and later be dissociated by another projectile into O_2 plus H_2.

The detection of solid oxygen on the Jovian satellite Ganymede from absorption bands in the red [254], prompted the question: how can O_2 exist at the reported high temperatures where the vapor pressure would exceed the atmospheric pressure by many orders of magnitude? There has been a controversy in the literature on the explanation of this puzzle. Johnson et al. have proposed that O_2 is formed inside the ice by radiolysis, and trapped in bubbles or inclusions [255]. Our experiments show instead that O_2 formed in the ice cannot be trapped permanently; it diffuses out at Ganymede's reported temperatures [256]. Transient trapping of O_2 in ice can be made by co-depositing O_2 and water in a film. When warmed above 70 K, the absorption bands become those of liquid oxygen, and different from those observed on Ganymede [257]. Our explanation is that very cold regions exist on Ganymede, made of segregated, bright ice patches, which are not visible to the Galileo infrared radiometer. Details of the findings and discussion of the model were published recently [258]. We note that, as expected from the measurements with ions and 10.2 eV photons, emission of O_2 has been observed under electron irradiation [259]. H_2 and O_2 emissions were recently also observed for 210 eV electrons on ice grown at ~100 K [260].

For incident electrons, besides the few works mentioned above, there have been many other studies [73,193,194,244,245,261-277], mostly giving energy and temperature dependence of the emission of ionic fragments, which are easier to detect, and neutral molecular products. The emission of secondary ions (including clusters) has also been observed under irradiation with ions [225,227,231,238-242] and with hard UV and X-ray photons [251, 273, 278, 279].

15.8.3 Reflectance Changes & Brightening

Changes in the reflectance of surfaces (albedo) due to irradiation is important in astronomy, where it can be caused by interaction with ions from the solar wind or planetary magnetospheres. Ion irradiation has also been postulated as the reason for the bright polar ice in Ganymede [280]. Several studies have shown an increase in the non-specular reflectance for visible light of *transparent* films with fluence of fast ion beams and upon bombardment with electrons and ions [209, 223, 280, 281]. In contrast, Sack et al. [282] observed a decrease in reflectance of *frost-like* deposits with 33 keV He^+ and Ar^+ ions. More recent and improved experiments in our laboratory have shown a very small effect of ion irradiation on reflectance [283, 284].

15.8.4 Phase Changes

Amorphization of crystalline ice can occur by irradiation at temperatures below 70 K. This has been reported to occur by electron [285, 286] or ion [56, 287, 288, 291] bombardment, and by illumination with UV light in the range 110-400 nm [289, 290]. The mechanism is unknown, but it is thought that the low temperatures are required to inhibit annealing of damage produced by irradiation. This damage includes molecular dissociation since it is seen that irradiation leads to the release of H_2. At higher temperatures, amorphization does not occur, possibly due to an increased mobility of the radicals created by the irradiation, which allows recombination to occur. The nature of the amorphous phase created by irradiation is not known. The recrystallization upon heating leads to the original crystalline phase that was amorphized by the radiation, suggesting a memory effect [285,286] or growth from crystallites remaining from an incomplete amorphization [6, 19, 92]. An intriguing result is the temperature dependence of the minimum dose necessary for amorphization with VUV light [290]. This dose is 1.8-3.6 x 10^{15} photons/cm^2 s^{-1} between 50 and 70 K, but below 10^{14} photons/cm^2 s^{-1} at 10 K. The later value is surprising, since it implies defect concentrations of only ~0.01% and it implies that ice can be amorphized by a dose of 0.06

meV/molecules, much less than the ~10 meV/molecule required. A recent re-examination [289] using infrared spectroscopy has shown that amorphization saturates at a dose of ~1 - 3 eV/molecule. This suggests that the amorphization seen in the earlier experiments was caused by the electron beam used to monitor the crystalline state.

Amorphization by energetic ions has been studied by monitoring changes in the infrared absorption spectra. Hudson and Moore [291] irradiated polycrystalline ice with 700 keV protons and monitored changes in the IR lattice bands. They report that, as a function of irradiation fluence, there is an oscillation between crystalline and amorphous phases, and a concurrent increase in the partial pressure of hydrogen molecules in the sample chamber. This led to a plausible model in which there is a competition between amorphization by the ions, and crystallization due to heating caused by the recombination of radicals formed and trapped in the ice. Support for the model comes from experiments which show that ice irradiated at 13K has an infrared spectrum partially resembling that of crystalline ice when warmed to temperatures from 46 to 120K, quite lower than the temperatures needed for unirradiated ice, suggesting that, when the ice is heated, trapped radicals recombine exothermically inducing crystallization. However, the hydrogen pressure oscillations were not seen by Baratta et al., [292, 293] who irradiated ice films, several tens of nm thick, with 3 keV He^+ ions at 10K, or by Baragiola et al. [294] where ice films of different thicknesses between 0.5 and 10 μm were irradiated at 7-8K with 50 keV H^+, H_2^+ or He^+ with fluences up to 10^{15} ions/cm^2.

Hudson and Moore [295] reported the decay of the spectral features of crystalline ice with dose of 700 keV protons. The fraction of crystalline ice can be represented by a function $f_c = \exp(-kD)$, where D is the dose and k an effective amorphization cross-section. At 13K, $k \sim 1.5\text{x}10^{-14}$ cm^2, huge compared with normal ionization and dissociation cross sections, indicating that amorphization is due to a cascade of secondary collisions, of the type that lead to sputtering. But k decreases fast with increasing temperature, by a factor of 1000 at 77K, suggesting that defects leading to amorphization have a very low activation energy.

Ultrathin ASW films have been crystallized with sub-bandgap ultraviolet light by a mechanism that may involve photoelectron emission from the graphite substrate into the ice [296]. Electrons can be trapped in negative ion states releasing energy that can lead to local crystallization.

Above 110 K, amorphous ice can be crystallized by high-energy ions in high-density electronic excitations [209] or in low excitation events produced by fast electrons [250,286]. Irradiation can also produce other phase transformations. It has been reported that ASW can be converted into the high-density form under the electron microscope; the required electron dose increases with temperature [250]. For T > 70 K, no conversion is possible.

15.9 Unknowns and Conclusions

We conclude with a summary of a few of the questions about ASW that need to be addressed in future investigations. Probably the main question about ASW is what is the nanoscale morphology of the microporosity, how it develops, and how it affects other properties. Other important questions are the molecular scale mechanisms involved in compaction of structural relaxation below 100 K, in the glass transition and in crystallization. There is the need of theoretical guidance to predict or interpret transformation kinetics, the effect of porosity on different spectroscopies, the thermalization of sticking of water and other molecules on the surface, the release of trapped gas over different temperature ranges, and in general the properties in the presence of radiation and co-deposited gases. The last two questions are particularly relevant to astrophysical environments, as is that of how to extrapolate results in the laboratory to astronomical time scales, especially when several mechanisms with different kinetic behaviors coexist.

In spite of nearly half a century of research on radiation effects in ice, the picture of radiolysis is very fragmentary and we still lack a model of primary processes and subsequent chemical reactions that include the effect of the density of excitations, reactant trapping and diffusion involving neutral and ionic species. We also lack characterization tools needed for a complete inventory of radiation products in the ice and for the time dependence of radiolysis in the sub-nanosecond time scale.

There has been an impressive advance in experimental methods in the last decade or so, but many conflicting results still exist. Understanding differences between experiments is hindered by the general lack of information in the literature reports of the composition of the background gas, the duration of the experiments, and the type of gas dosing. This chapter has stressed the case for the need of experiments with improved vacuum, and in-situ analysis of ice purity. In particular we look forward to new diffraction experiments with X-rays and fast electrons under ultrahigh vacuum conditions and in which the radiation damage effects are fully characterized.

On closing we acknowledge financial support from the NASA's Office of Space Science, NSF-Astronomy, and NASA's Cassini program under JPL contract 1210586.

References

1. P. Jenniskens, S. F. Barnhak, D. F. Blake, M. R. S. McCoustra, J. Chem. Phys. **107**, 1232 (1997).
2. T. L. Roush, J. Geophys. Res. **106E**, 33,315 (2001).
3. J. Klinger, D. Benest, A. Dollfus, R. Smoluchowski, Editors, *Ices in the Solar System*, Reidel Publ., Norwell, MA. (1985).
4. B. Schmitt, C. De Bergh, and M. Festou, Editors, *Solar System Ices*, Kluwer, Dordrecht, 1998.

5. R. E. Johnson, in Ref. [4] (1998) 303.
6. Y. P. Handa, O. Mishima and E. Whalley, J. Chem. Phys. **84**, 2766 (1986).
7. G. P. Johari, Phil. Mag. B **78**, 375 (1998).
8. P. Boutron and R. Alben, J. Chem. Phys. **62**, 4848 (1975).
9. A. H. Narten, C.G. Venkatesh and S.A. Rice, J. Chem. Phys. **64**, 1106 (1976).
10. W. Hagen, A. G. G. M. Tielens and J. M. Greenberg, Astron. Astrophys. Suppl. **51**, 389 (1983).
11. E. F. Burton and W. F. Oliver, *Nature* **135**, 505 (1935).
12. E. F. Burton and W. F. Oliver, Proc. R. Soc. London Ser. A **153**, 166 (1935).
13. L. G. Dowell and A. P. Rinfret, Nature **158**, 1142 (1960).
14. R. T. Brackmann and W. L. Fite, J. Chem. Phys. **34**, 1572 (1961).
15. P. Bruggeller and E. Mayer, Nature **288**, 569 (1980).
16. E. Mayer and P. Bruggeller, Nature **325**, 601 (1980).
17. G. P. Johari, A. Hallbrucker, and E. Mayer, Nature **330**, 552 (1987).
18. M. G. Sceats and S. A. Rice, in *Water, a Comprehensive Treatise*, vol. 7, ed. by F. Franks (New York: Plenum, 1982) Ch. 2.
19. O. Mishima, L. D. Calvert, E. Whalley, Nature **310**, 393 (1984).
20. C. A. Tulk, C. J. Benmore, J. Urquidi, D. D. Klug, J. Neuefeind, B. Tomberli and P. A. Egelstaff, Science **297**, 1320 (2002).
21. A. Al-Amoudi, J. Dubochet, and D. Studer, J. Microsc. **207**, 146 (2002).
22. A. J. Fowler and M. Toner, Adv. Heat and Mass Transfer in Biotechnology, ASME BED **40**, 155 (1998).
23. D. Laufer, E. Kochavi, and A. Bar-Nun, Phys. Rev. **B 36**, 9219 (1987).
24. A. Kouchi, J. Crystal Growth **99**, 1220 (1990).
25. V. Sadtchenko, K. Knutsen, C. F. Giese, and W. R. Gentry, J. Phys. Chem. **B 104**, 2511 (2000).
26. V. Sadtchenko, K. Knutsen, C. F. Giese, and W. R. Gentry, J. Phys. Chem. **B 104**, 4894 (2000).
27. P. Jenniskens and D. F. Blake, Science **265**, 753 (1994).
28. L. Schriver-Mazzuoli, A. Schriver, and A. Hallou, J. Molec. Struct. **554**, 289 (2000).
29. S. La Spisa, M. Waldheim, J. Lintemoot, T. Thomas, J. Naff, and M. Robinson, J. Geophys. Res. **106**, 33,351 (2001).
30. S. A. Rice, W. G. Madden, R. McGraw, M. G. Sceats, and M. S. Bergren, J. Glaciol. **21**, 509 (1978).
31. H. G. Heide, and Z. Zeitler, Ultramicrosc. **16**, 151 (1985).
32. P. Jenniskens, D. F. Blake, M. A. Wilson, and A. Pohorille, Astrophys. J. **455**, 389 (1995).
33. P. Jenniskens and D. F. Blake, Astrophys. J. **473**, 1104 (1995).
34. P. Jenniskens, D. F. Blake and A. Kouchi, in Ref. [4] (1998). 139.
35. N. Levine, *Quantum Chemistry*, Prentice Hall, Upper Saddle River, (2000).
36. A. J. Yencha, H. Kubota, T. Fukuyama, T. Kondow, and K. Kuchitsu, J. Electron Spec. **23**, 431 (1981).
37. W. L. Jorgensen and L. Salem, *The Organic Chemist's Book of Orbitals*, Academic Press, New York, 1973.
38. B. Baron and F. Williams, J. Chem. Phys. **64**, 3896 (1976).
39. R. Reissner and M. Schulze, Surf. Sci. **454-456**, 183 (2000).
40. M. Watanabe, H. Kitamura, and Y. Nakai, in *Vacuum Ultraviolet and Radiation Physics*, ed. by E. Koch, R. Haensel and C. Kunz (Pergamon, 1975) 70.

41. B. Baron, D. Hoover, and F. Williams, J. Chem. Phys. **68**, 1997 (1978).
42. T. Shibaguchi, H. Onuki, and R. Onaka, J. Phys. Soc. Japan **42**, 152 (1977).
43. P. Delahay and K. von Burg, Chem. Phys. Lett. **83**, 250 (1981).
44. R. D. Leapman, S. Sun, Ultramicroscopy **59**, 71 (1995).
45. C. D. Wilson, C. A. Dukes, and R. A. Baragiola, Phys. Rev. **B 63,** 12, 1101 (2001).
46. S. R. Elliott, J. Chem. Phys. **103**, 2758 (1995).
47. R. A. Rosenberg, P. R. LaRoe, V. Rehn, J. Stöhr, R. Jaeger, and C. C. Parks, Phys. Rev. **B28**, 3026 (1983).
48. J. S. Tse, K. H. Tan and J. M. Chen, Chem. Phys. Lett. **174**, 603 (1990).
49. J. P. Devlin, Int. Rev. Phys. Chem. **9**, 29 (1990).
50. D. M. Hudgins, S. A. Sandford, L. J. Allamandola, and A.G.G.M. Tielens, Astrophys. J. Supp. **86**, 713 (1993).
51. S. Tsujimoto, A. Konishi, and T. Kunitomo, Cryogenics **22**, 603 (1982).
52. E. Langenbach, A. Spitzer, and H. Lüth, Surf. Sci. **147**, 179 (1984).
53. M. J. Wojcik, V. Buch, and J. P. Devlin, J. Chem. Phys. **99**, 2332 (1993).
54. W. Langel, H. W. Fleger, and E. Knözinger, Ber. Bunsenges. Phys. Chem. **98**, 81 (1994).
55. F. Bensebaa and T. H. Ellis, Prog. Surf. Sci. **50**, 173 (1995).
56. M. H. Moore and R. L. Hudson, Astrophys. J. **401**, 353 (1992).
57. M. H. Moore and R. L. Hudson, Astron. Astrophys. Suppl. Ser. **103**, 45 (1993).
58. R. G. Smith, G. Robinson, A. R. Hyland, and G. I. Carpenter, Mon. Not. R. Astron. Soc. **271**, 481 (1994).
59. M. H. Moore, R. L. Hudson, and P. A. Gerakines, Spectrochim. Acta **A57**, 843 (2001).
60. R. G. Greenler, J. Chem. Phys. **44**, 310 (1966).
61. A. B. Horn, S. F. Banham, and M. R. S. McCoustra, J. Chem. Soc. Faraday Trans. **91**, 4005 (1995).
62. M. S. Bergren, D. Schuh, M. G. Sceats, and S.A. Rice, J. Chem. Phys. **69**, 3477 (1978).
63. W. G. Madden, M. S. Bergren, R. McGraw, S. A. Rice, and M. G. Sceats, J. Chem. Phys. **69**, 3497 (1978).
64. G. Nielsen, R. M. Townsend, and S. A. Rice, J. Chem. Phys. **81**, 5288 (1984).
65. B. E. Wood and J. A. Roux, J. Opt. Soc. Am. **72**, 720 (1982).
66. A. Léger, S. Gauthier, D. Défourneau, and D. Rouan, Astron. Astrophys. **117**, 164 (1983).
67. P. A. Gerakines, W. A. Schutte, J. M. Greenberg, and E. F. Van Dishoeck, Astron. Astrophys. **296**, 810 (1995).
68. P. Ehrenfreund, P. A. Gerakines, W. A. Schutte, M. C. van Hemert, and E. F. van Dishoeck, Astron. Astrophys. **312**, 263 (1986).
69. B. Rowland and J.P. Devlin, J. Chem. Phys. **94**, 812 (1991).
70. B. Rowland, M. Fisher and J.P. Devlin, J. Chem. Phys. **95**, 1378 (1991).
71. B. W. Cahen, K. Griffiths, and P. R. Norton, Surf. Sci. Lett. **261**, L44 (1992).
72. C. Martin, C. Manca and P. Roubin, Surf. Sci. **502-503**, 275 (2002).
73. M. Michaud and L. Sanche, Phys. Rev. Lett. **59**, 645 (1987).
74. A. I. Kolesnikov, J.-C. Li, S. Dong, I. F. Bailey, S. F. Parker, R. S. Eccleston, W. Hahn, and S. F. Parker, Phys. Rev. Lett. **79**, 1869 (1997).
75. A. I. Kolesnikov, J.-C. Li, S. F. Parker, R. S. Eccleston, W. Hahn, and C.-K. Loong, Phys. Rev. B **79**, 59 (1999).

76. O. Yamamuro, Y. Madokoro, H. Yamasaki, T. Matsuo, I. Tsukushi, and K. Takeda, J. Chem. Phys. **115**, 9808 (2001).
77. J. C. Li and A. I. Kolesnikov, Physica **B 316-317**, 493 (2002).
78. H. Schober, M. Koza, A. Tölle, F. Fujara, C. A. Angell, and R. Böhmer, Physica B **241-243**, 897 (1998).
79. D. D. Klug, C. A. Tulk, E. C. Svensson, and C.-K. Loong, Phys. Rev. Lett. **83**, 2584 (1999).
80. Q. Zhang and V. Buch, J. Chem. Phys. 92, 5004 (1990).
81. V. Buch and Q. Zhang, Astrophys. J. **379**, 647 (1991).
82. D. Olander and S. A. Rice, Proc. Natl. Acad. Sci. USA **69**, 98 (1972).
83. T. C. Sivakumar, S. A. Rice and M. G. Sceats, J. Chem. Phys. **69**, 3468 (1978).
84. S. Trakhtenberg, R. Naaman, S.R. Cohen, and I. Benjamin, J. Phys. Chem. B **101**, 5172-5176 (1997).
85. A. Kouchi and T. Kuroda, Japan. J. Appl. Phys. **29**, L807 (1990).
86. J. Bruneaux-Poulle, A. Defrain, and N. R. Linh, J. Chim. Phys. **69**, 71-75 (1977).
87. A. Hallbrucker, E. Mayer, and G.P. Johari, J. Phys. Chem. **93**, 4986 (1989).
88. J. Braun, A. Glebov, A. P. Graham, A. Menzel, and J. P. Toennies, Phys. Rev. Lett. **80**, 2638 (1998).
89. S. Mitlin and K. T. Leung, J. Phys. Chem. **106**, 6234 (2002).
90. J. A. Pryde and G. O. Jones, Nature **170**, 685 (1952).
91. M. Sugisaki, H. Suga, and S. Seki, Bull. Chem. Soc. Jpn. **41**, 2591 (1968).
92. N. J. Sack and R. A. Baragiola, Phys. Rev. **B48**, 9973 (1993).
93. M. H. Moore, R. F. Ferrante, R. L. Hudson, J. A. Nuth, III; and B. Donn, Astrophys. J. 428, L81 (1994).
94. I. Kohl, E. Mayer, and A. Hallbrucker, Phys. Chem. Chem. Phys. **2**, 1579 (2000).
95. W. Hagen, A.G.G.M. Tielens, and J.M. Greenberg, Chem. Phys. **56**, 367 (1981).
96. A. Kouchi, T. Yamamoto, T. Kozasa, T. Kuroda. and J. M. Greenberg, Astron. Astrophys. **290**, 1009 (1984).
97. E. Mayer and R. Pletzer, J. Chem. Phys. **80**, 2939 (1984).
98. E. Mayer and R. Pletzer, Nature **319**, 298 (1986).
99. T. Gonda, S. Nakahara and T. Sei, J. Crys. Growth **99**, 183 (1990).
100. M. S. Westley, G. A. Baratta and R. A. Baragiola, J. Chem. Phys. **108**, 3321 (1998).
101. K. P. Stevensson, G. A. Kimmel, Z. Dohnálek, R. S. Smith, and B. D. Kelly, Science **283**, 1505 (1999).
102. V. Buch, J. Chem. Phys. **96**, 3814 (1992).
103. P. Meakin and J. Krug, Phys. Rev. A **46**, 3390-3399 (1992).
104. U. Essmann and A. Geiger, J. Chem. Phys. **103**, 4678 (1995).
105. V. P. Zhdanov and P. R. Norton, Surf. Sci. **449**, L228 (2000).
106. G. A. Kimmel, Z. Dohnálek, K. P. Stevenson, R. S. Smith, and B. D. Kay, J. Chem. Phys. **114**, 5295 (2001).
107. J. Braun, A. Glebov, A. P. Graham, A. Menzel, and J. P. Toennies, Phys. Rev. Lett. **80**, 2638 (1998).
108. J. Kruger and W. J. Ames, J. Opt. Soc. Am. **49**, 1195 (1959).
109. B. Schmitt, S.R.J.A. Grim, J.M. Greenberg and J. Klinger, in *Physics and Chemistry of Ices*, ed. N. Maeno and T. Hondoh (Hokkaido Univ., Sapporo, 1992).

110. J. Hessinger and R. O. Pohl, J. Non-Cryst. Solids 208, 151 (1996).
111. P. Ayotte, R. S. Smith, K. P. Stevenson, Z. Dohnálek, G. A. Kimmel, and B. D. Kay, J. Geophys. Res. **106** E, 33,387 (2001).
112. J. A. Ghormley and C. J. Hochanadel, *Science* **171**, 62 (1971).
113. B. A. Seiber, B. E. Wood, A. M. Smith, and P. R. Müller, *Science* **170**, 652 (1970).
114. B. S. Berland, D. E. Brown, M. A. Tolbert, and S. M. George, Geophys. Res. Lett. **22**, 3493 (1995).
115. D. E. Brown, S. M. George, C. Huang, E.K.L. Wong, K. Rider, R. S. Smith, and B. Kay, J. Phys. Chem. 100, 4988 (1996).
116. A. Bar-Nun, J. Dror, E. Kochavi, and D. Laufer, Phys. Rev. B **35**, 1321 (1987).
117. A. Givan, A. Loewenschuss, and C. J. Nielsen, Vibrational Spectr. 12, 1 (1996).
118. K. Sekar and R. A. Baragiola, Symposium of Photolysis and Radiolysis on Solar System Ices, Maryland (1999), and to be published.
119. J. P. Devlin, J. Phys. Chem. **96**, 6185 (1992).
120. M. E. Palumbo, J. Phys. Chem. A **101**, 4298 (1997).
121. A. Bar-Nun, G. Herman, D. Laufer, and M. L. Rappaport, Icarus **63**, 317 (1985).
122. H. G. Hixson, M. J. Wojcik, M. S. Devlin, J. P. Devlin and V. Buch, J. Chem. Phys. **97**, 753 (1992).
123. J. P. Devlin, J. Geophys. Res. **106** E, 33, 333 (2001).
124. R. J. Smith, A. Kara and S. Holloway, J. Phys. Chem. **94**, 806 (1991).
125. J. A. Ghormley, J. Chem. Phys. **46**, 1321 (1967).
126. A. W. Adamson, L. M. Dormant and M. Orem, J. Colloid Interf. Sci. **25**, 206 (1967).
127. J. Ocampo and J. Klinger, J. Colloid Interf. Sci. **86**, 377 (1982).
128. A. Bar-Nun, I. Kleinfeld, and E. Kochavi, Phys. Rev. **38**, 7749 (1988).
129. E. Mayer and R. Pletzer, in Ref. [3], 81 (1985).
130. J. A. Ghormley, J. Chem. Phys. **48**, 503 (1968).
131. B. Schmitt, J. Ocampo, and J. Klinger, J. J. Phys. (Paris) **48**, C1, 519 (1987).
132. G. Notesco, I. Kleinfeld, D. Laufer and A. Bar-Nun, Icarus **89**, 411 (1991).
133. E. Vichnevetski, A. D. Bass and L. Sanche, J. Chem. Phys. **113**, 3874 (2000).
134. N. Horimoto, H. S. Kato, and M. Kawai, J. Chem. Phys. **116**, 4375 (2002).
135. M. Fisher and J.P. Devlin, J. Phys. Chem. **99**, 11584 (1995).
136. W. Hagen and A.A.G.M. Tielens, Spectrochem. Acta **38A**, 1089 (1982).
137. G. Ritzhaupt, N. Smyrl, and J. P. Devlin, J. Chem. Phys. **64**, 435 (1976).
138. A. S. Drobyshev and D.N. Garipoglyi, Low Temp. Phys. **22**, 812 (1996).
139. A. Givan, A. Loewenschuss, and C.J. Nielsen, J. Phys. Chem. B **101**, 8696-8706 (1997).
140. U. Buontempo, Phys. Lett. **42A**, 17 (1972).
141. E. Mayer and R. Pletzer, J. Chem. Phys. **83**, 6536 (1985).
142. M. M. Maldoni, G. Robinson, R. G. Smith, W. W. Duley, and A. Scott, Mon. Not. R. Astron. Soc. **309**, 305 (1999).
143. M. A. Zondlo, T. B. Oonash, M. S. Washawsky, M. A. Tolbert, G. Mallick, P. Arentz, and M. S. Robinson, J. Phys. Chem. **B**101, 19887 (1997).
144. M. Eldrup, J. Chem. Phys. **64**, 5283 (1976).
145. M. Eldrup, A. Vehanen, P.J. Schulz, and K.G. Lynn, Phys. Rev. B **32**, 7048 (1985).
146. G. P. Johari, A. Hallbrucker, and E. Mayer, J. Chem. Phys. **95**, 2955 (1991).

147. A. A. Tsekouras, M. J. Iedema, and J. P. Cowin, Phys. Rev. Lett. **80**, 5798 (1998).
148. T. Giering and D. Haarer, J. Lumin. **66&67**, 299 (1996).
149. H. Härdle, G. Weiss, S. Hunklinger, and F. Baumann, Z. Phys. B Condensed Matter **65**, 291 (1987).
150. J. Hessinger, B.E. White, Jr., and R.O. Pohl, Planet. Space Sci. **44**, 937 (1996).
151. B. E. White, J. Hessinger, and R. O. Pohl, J. Low Temp. Phys. **111**, 233 (1998).
152. B. E. Wood and A. M. Smith, in *Thermophysics and Thermal Control, vol. 65 of Progress in Astronautics and Aeronautics* (American Institute of Aeronautics and Astronautics, 1978) p. 22.
153. D. A. Bahr and R. A. Baragiola, to be published.
154. B. E. Wood, A. M. Smith, J. A. Roux, and B. A. Seiber, AIAA J. **9**, 1836 (1971).
155. A. S. Drobyshev, N. V. Atapina, D. N. Garipogly, S. L. Maksimov, and E. A. Samyshkin, Low Temp. Phys. **19**, 404 (1993).
156. A. S. Drobyshev, Low Temp. Phys **22**, 123 (1996).
157. R. S. Smith, C. Huang, E. K. L. Wong and B. D. Kay, Phys. Rev. Lett. **79**, 909 (1997).
158. J. A. McMillan and S. C. Los, Nature **206**, 806 (1965).
159. Y. P. Handa and D. D. Klug, J. Phys. Chem. **92**, 3323 (1988).
160. D. R. MacFarlane and C. A. Angell, J. Phys. Chem. **88**, 759 (1984).
161. V. Velikov, S. Borick, and C.A. Angell, Science **294**, 2335 (2001).
162. G. P. Johari, J. Chem. Phys. **116**, 8067 (2002).
163. J. A. McMillan and S. C. Los, J. Chem. Phys. **42**, 829 (1965).
164. C. A. Angell and E. J. Sare, J. Chem. Phys. **52**, 1058 (1970).
165. G. P. Johari, Phil. Mag. **35**, 1077 (1977).
166. R. S. Smith and B. D. Kay, Nature **398**, 788 (1999).
167. A. Kouchi and S. Sirono, Geophys. Res. Lett. **28**, 827 (2001).
168. R. S. Smith, C. Huang, and B.D. Kay, J. Phys. Chem. B **101**, 6123 (1997).
169. R. H. Beaumont, H. Chihara, and J. A. Morrison, J. Chem. Phys. B **34**, 1456 (1961).
170. A. H. Hardin and K. B. Harvey, Spectrochim. Acta **29A**, 1139 (1972).
171. J. Dubochet, J. J. Chang, R. Freeman, J. Lepault, and A. W. McDowall, Ultramicroscopy **10**, 55 (1982).
172. B. Schmitt, S. Espinasse, R. J. A. Grim, J. M. Greenberg, and J. Klinger, ESA SO-302, 65 (1989).
173. G. P. Johari, A. Hallbrucker, and E. Mayer, J. Chem. Phys. **97**, 5851 (1992).
174. Z. Dohnálek, R. L. Ciolli, G. A. Kimmel, K. P. Stevenson, R. S. Smith, and B. D. Kay, J. Chem. Phys. **110**, 5489 (2000).
175. W. Hage, A. Hallbrucker, E. Mayer and G. P. Johari, J. Chem. Phys. **100**, 2743 (1994).
176. W. Hage, A. Hallbrucker, E. Mayer and G. P. Johari, J. Chem. Phys. **103**, 545 (1994).
177. A. Kouchi, Nature **330**, 550 (1987).
178. S. A. Sandford and L. J. Allamandola, Icarus **76**, 201 (1988).
179. R. J. Speedy, P. G. Debenedetti, R. S. Smith, C. Huang, and B. D. Kay, J. Chem. Phys. **B105**, 240 (1996).
180. F. E. Livingston, J. A. Smith, and S. M. George, Surf. Sci. **423**, 145 (1999).

181. Z. Dohnálek, G. A. Kimmel, R. L. Ciolli, K. P. Stevenson, R. S. Smith, and B. D. Kay, J. Chem. Phys. **112**, 5932 (2000).
182. H. J. Fraser, M. P. Collings, M. R. S. McCoustra, and D. A. Williams, Mon. Not. R. Astron. Soc. **327**, 1165 (2001).
183. R. S. Smith, C. Huang, E. K. L. Wong and B. D. Kay, Surf. Sci. **367**, L13 (1996).
184. P. Löfgren, P. Ahlström, D. V. Chakarov, J. Lausmaa and B. Kasemo, Surf. Sci. **367**, L19 (1996).
185. J. Klinger, Science **209**, 271 (1980).
186. W. A. Phillips, *Amorphous Solids* (Springer, Berlin, (1981).
187. A. Kouchi, J.M. Greeenberg, T. Yamamoto, and T. Mukai, Astrophys. J. **388**, L73-L76 (1992).
188. A. Kouchi, J.M. Greeenberg, T. Yamamoto, T. Mukai, and Z.F. Xing, in *Physics and Chemistry of Ice*, edited by N. Maeno and T. Hondoh (Hokkaido Univ. Press, Sapporo, 1992) 229-236.
189. O. Andersson and H. Suga, Solid State Comm. **91**, 985 (1994).
190. R. G. Russell and J. S. Kargel, in Ref. [4], 33 (1998).
191. R. R. Rye, T. E. Madey, J. E. Houston and P. Holloway, J Chem. Phys. **69**, 1504 (1978).
192. P. H. Holloway, T. E. Madey, C. T. Campbell, R. R. Rye and J. E. Houston, Surface Sci. **88**, 121 (1979).
193. R. Stockbauer, D. M. Hanson, S. A. Flodström and T. E. Madey, Phys. Rev. B **26**, 1885 (1982).
194. P. Rowntree, L. Parenteau and L. Sanche, J. Chem. Phys. **94**, 6570 (1991).
195. R. E. Johnson and T. I. Quickenden, J. Geophys. Res. **102**, 10,985 (1997).
196. A. Plonka, W. Szajdzinska-Pietek, J. Bednarek, A. Hallbrucker, and E. Mayer, Phys. Chem. Chem. Phys. **2**, 1587 (2000).
197. J. Bednarek, A. Plonka, A. Hallbrucker, E. Mayer, and M. C. R. Symon, J. Am. Chem. Soc. **118**, 9387 (1996).
198. S. Siegel, J. M. Flournoy and L. H. Baum, J. Chem. Phys. **34**, 1782 (1961).
199. P. A. Gerakines, W. A. Schutte, amd P. Ehrenfreund, Astron. Astrophys. **312**, 289 (1995).
200. M. H. Moore and R. L. Hudson, Icarus **145**, 282 (2000).
201. O. Gomis, M. A. Satorre, G. Leto, and G. Strazzulla, to be published.
202. R. W. Carlson et al., Science **283**, 2062 (1999).
203. D. A. Bahr, M. A. Fama, R. A. Vidal, and R. A. Baragiola. J. Geophys. Res. E **106**, 33,285 (2001).
204. R. W. Carlson et al., Science **182**, 53 (1973).
205. A. L. Broadfoot, et al., Science **204**, 979 (1979).
206. D. T. Hall et al., Nature **373**, 677 (1995).
207. M. Shi, R.A. Baragiola, D.E. Grosjean, R.E. Johnson, S. Jurac, and J. Schou, J. Geophys. Res. **100** (1995). 26,387.
208. P. Sigmund in *Sputtering by Particle Bombardment I.* R. Behrisch (ed.). Springer Verlag, Berlin (1981) Chapter 2
209. W. L. Brown, L.J. Lanzerotti, J.M. Poate, and W.M. Augustyniak, Phys. Rev. Lett. **40,** 1027 (1978).
210. R. W. Evatt, M. Sc. Thesis, University of Virginia (1980).
211. W. L. Brown, W.M. Augustyniak, E. Brody, B. Cooper, L.J. Lanzerotti, A. Ramirez, R. Evatt, R.E. Johnson, Nucl. Instr. Meth. **170,** 321 (1980).

212. W. L. Brown, W.M. Augustyniak, L.J. Lanzerotti, R.E. Johnson, R. Evatt, Phys. Rev. Lett. **45,** 1632 (1980).
213. G. Ciavola, G. Foti, L. Torrisi, V. Pirronello, G. Strazzulla Rad. Eff. **65,** 167 (1982).
214. W. L. Brown, W.M. Augustyniak, E.H. Simmons, K.J. Marcantonio, L.J. Lanzerotti, R.E. Johnson, J.W. Boring, C.T. Reimann, G. Foti, and V. Pirronello, Nucl. Instr. Meth. **198,** 1 (1982).
215. R. E. Johnson, W.L. Brown, Nucl. Instr. Meth **198**, 103 (1982).
216. R. A. Haring, A. Haring, F.S. Klein, A.C. Kummel, A.E. de Vries, Nucl. Instr. Meth. **211,** 529 (1983).
217. J. W. Boring, R.E. Johnson, C.T. Reimann, J.W. Garret, W.L. Brown, K.J. Marcantonio, Nucl. Instr. Meth. **218,** 707 (1983).
218. B. H. Cooper and T.A. Tombrello, Radiat. Eff. **80,** 203 (1984).
219. W. L. Brown, W.M. Augustyniak, K.J. Marcantonio, E.H. Simmons, J.W. Boring, R.E. Johnson, C.T. Reimann, Nucl. Instr. Meth. **B1,** 307 (1984).
220. C. T. Reimann, J.W. Boring, R.E. Johnson, J.W. Garrett, K.R. Farmer, W.L. Brown, K.J. Marcantonio, and W.M. Augustyniak, Surf. Sci. **147,** 227 (1984).
221. R. A. Haring, R. Pedrys, D.J. Oostra, A. Haring, and A.E. de Vries, Nucl. Instr. Meth. B **5,** 476 (1984).
222. A. E. de Vries, R.A. Haring, A. Haring, F.S. Klein, A.C. Kummel, and F.W. Saris, J. Phys. Chem. **88,** 4510 (1984).
223. R. E. Johnson, L.A. Barton, J.W. Boring, W.A. Jesser, W.L. Brown, and L.J. Lanzerotti, in Ref. [3], 301 (1985).
224. A. Bar-Nun, G. Herman, M.L. Rappaport, and Yu. Mekler, Surf. Sci. **150** (1985).
225. A. Bar-Nun, G. Herman, M.L. Rappaport, and Yu. Mekler, in Ref. [3], 287 (1985).
226. F. Rocard, J. Bénit, J-P. Bibring, D. Ledu, and R. Meunier, Radiat. Eff. **99,** 97 (1986).
227. J. Benit, J-P. Bibring, S. Della Negra, Y. Le Beyeq, and F. Rocard, Radiat. Eff. **99,** 105 (1986).
228. D. B. Chrisey, J.W. Boring, J.A. Phipps, R.E. Johnson, and W.L. Brown, Nucl. Instr. Meth. **B13,** 360 (1986).
229. W. L.Brown, L.J. Lanzerotti, K.J. Marcantonio, R.E. Johnson, and C.T. Reimann, Nucl. Instr. Meth. **B14,** 392 (1986).
230. J. W. Christiansen, D. Delli Carpini, and I.S.T. Tsong, Nucl. Instr. Meth. **B15,** 218 (1986).
231. J. W. Christiansen, I.S.T. Tsong, and S.H. Liu, J. Chem. Phys. **86,** 4701 (1987).
232. J. Benit, These Dr. Sc., Univ. Paris-Sud, Orsay, France (1987).
233. J. Benit and W.L. Brown, Nucl. Instr. Meth. B **46,** 448 (1990).
234. F. Spinella, G.A. Baratta, G. Strazzulla, and I. Torrisi, Rad. Eff. Def. Solids **115,** 307 (1991).
235. M. Shi, D.E. Grosjean, J. Schou, R.A. Baragiola, Nucl. Instrum. Meth. B **96,** 524 (1995).
236. R. Pedrys, F. Krok, P. Leskiewicz, J. Schou, U. Podschaske, B. Cleff, Nucl. Instr. Meth. Phys. Res. B **164-165,** 861 (2000).
237. R. A. Baragiola, C.L. Atteberry, C.A. Dukes, M. Famá and B.D. Teolis, Nucl. Instr. Meth. Phys. Res. B (2002) in press.

238. E. N. Nikolaev, G. D. Tantsyrev, and V.A. Saraev, Sov. Phys. Tech. Phys. **23,** 241 (1978).
239. G. M. Lancaster, F. Honda, Y. Fukuda, and J.W. Rabalais, Int. J. Mass Spectrom. Ion Phys. **29,** 199 (1979).
240. G. M. Lancaster, F. Honda, Y. Fukuda, and J.W. Rabalais, J. Am. Chem. Soc. **101,** 1951 (1979).
241. R. F. Magnera, D.E. David, and J. Michl, Chem. Phys. Lett. **182,** 363 (1991).
242. T. Matsuo, T. Tonuma, H. Kumagai, H. Shibata, and H. Tawara, J. Chem. Phys. **101,** 5356 (1994).
243. L. J. Lanzerotti, W.L. Brown, J.M. Poate, and W.M. Augustyniak, Geophys. Res. Lett. **5,** 155 (1978).
244. G. A. Kimmel, T. M. Orlando, C. Vézina, and L. Sanche, J. Chem. Phys. **101**, 3282 (1984).
245. G. A. Kimmel and T. M. Orlando, Phys. Rev. Lett. **77**, 3983 (1996).
246. M. S. Westley, R. A. Baragiola, R. E. Johnson, and G. A. Baratta, Planet. Space Sci. **43**, 1311 (1995).
247. N. Watanabe, T. Horii, and A. Kouchi, Astrophys. J. **541**, 772 (2000).
248. Y. Talmon, H.T. Davis, L.E. Scriven, and E.L. Thomas, J. Microsc. **117** (1979) 321.
249. H. G. Heide, Ultramicroscopy **7,** 299 (1982).
250. H. G. Heide, Ultramicroscopy **14,** 271 (1984).
251. R. A. Rosenberg, V. Rehn, V. O. Jones, A. K. Green, C. C. Parks, G. Loubriel and R. H. Stulen, Chem. Phys. Lett. **80**, 488 (1981).
252. M. S.Westley, R. A. Baragiola, R. E. Johnson, and G. A. Baratta, Nature **373,** 405 (1995).
253. D. A. Bahr, M. Famá, R.A. Vidal, J. Schou, and R.A. Baragiola, to be published.
254. J.R. Spencer, W.M. Calvin, and M.J. Person, J. Geophys. Res. **100**, 19,049 (1995).
255. R.E. Johnson and W.A. Jesser, Astrophys. J. **480**, L79 (1997).
256. R.A. Vidal, D. Bahr, R.A. Baragiola, and M. Peters, Science **276,** 1839 (1997).
257. R.A. Baragiola and D.A. Bahr, J. Geophys. Res. **103,** 25865 (1998).
258. R.A. Baragiola, C.L. Atteberry, D.A. Bahr and M. Peters, J. Geophys. Res. E **104**, 14,183 (1999).
259. M. T. Sieger, W. C. Simpson, and T. M. Orlando, Nature **394**, 554 (1998).
260. A. J. Wagner, C. Vecitis, and D. H. Fairbrother, J. Phys. Chem. **B 106**, 4432 (2002).
261. G. A. Kimmel, R. G. Tonkyn, and T. M. Orlando, Nucl. Instr. Meth. B **101**, 179 (1995).
262. G. A. Kimmel, T. M. Orlando, P. Cloutier, and L. Sanche, J. Phys. Chem. B **101**, 6301 (1997).
263. G. A. Kimmel, and T. M. Orlando, Phys. Rev. Lett. **75**, 2606 (1995).
264. G. R. Floyd and R. H. Prince, Nature (Phys. Sci.) **240,** 11 (1972).
265. R. H. Prince and G. R. Floyd, Chem. Phys. Lett. **43**, 326 (1976).
266. T. E. Madey and J. T. Yates, Chem. Phys. Lett. **51**, 77 (1977).
267. T. E. Madey and F. P. Netzer, Surf. Sci. **117**, 549 (1982).
268. R. Stockbauer, D. M. Hanson, S. A. Flodstrom, E. Bertel and T. E. Madey, Phys. Scripta **T4**, 126 (1983).
269. R. H. Stulen, J. Vac. Sci. Technol. **A1**, 1163 (1983).

270. R. H. Stulen and P. A. Thiel, Surf. Sci. **157**, 99 (1985).
271. J. O. Noell, C. F. Melius, and R. H. Stulen, Surf. Sci. **157**, 119 (1985).
272. E. G. Avdiev and G. M. Bogolyubov, Geomagn. Aeron. **30**, 389 (1990).
273. D. Coulman, A. Puschmann, U. Höfer, H-P. Steinrück, W. Wurth, P. Feulner, and D. Menzel, J. Chem. Phys. **93**, 58 (1990).
274. S. L. Bennett, C. L. Greenwood, E. M. Williams, and J. L. de Segovia, Surface Sci. **251/252**, 857 (1991).
275. C. P. Safvan, U. T. Raheja, and D. Mathur, Rapid Com. Mass Spectr. **7**, 620 (1993).
276. L. Sanche, Scanning Microsc. **9**, 619 (1995).
277. M. T. Sieger, W. C. Simpson, and T. M. Orlando, Phys. Rev. B **56**, 4925 (1997).
278. R. H. Stulen and R. A. Rosenberg, J. Vac. Sci. Technol. **A2**, 1051 (1984).
279. Y. R. Wu, B. W. Yang, and D. L. Judge, Planet. Space Sci. **42**, 273 (1994).
280. R. E. Johnson, Icarus **62**, 344 (1985).
281. G. Strazzula, L. Torrisi and G. Foti, Europhys. Lett. **7**, 431 (1988).
282. N. J. Sack, J. W. Boring, R. E. Johnson, R. A. Baragiola, and M. Shi, J. Geophys. Res. **96**, 17,535 (1991).
283. C. Atteberry, Ms. Thesis, U. Virginia (1998).
284. D. A. Bahr, Ph. D. Thesis, U. Virginia (2000).
285. J. Lepault, R. Freeman and J. Dubochet, J. Microsc. **132**, RP3 (1983).
286. J. Dubohet and J. Lepault, J. Phys. (Paris) **45,** C7, 85 (1984).
287. G. A. Baratta, G. Leto, F. Spinella, G. Strazzulla, and G. Foti, Astron. Astrophys. **252**, 421 (1991).
288. G. Strazzulla, G. A. Baratta, G. Leto, and G. Foti, Europhys. Lett. **18**, 517 (1992).
289. G. Leto and G. A. Baratta, Astron. Astrophys. (in press).
290. A. Kouchi and T. Kuroda, Nature **344**, 134 (1990).
291. R. L. Hudson and M. H. Moore, J. Phys. Chem. **96**, 6500 (1992).
292. G. A. Baratta, A. C. Castorina, G. Leto, M. E. Palumbo, F. Spinella, and G. Strazzulla, Planet. Space Sci. **42**, 759 (1994).
293. G. Leto M.E. Palumbo, and G. Strazzulla, Nucl. Instr. Meth. **B 116,** 49 (1996).
294. R. A. Baragiola, R. Vidal, M. Shi, W. Svendsen, J. Schoub, and D. Bahr, Nucl. Instr. Meth. Phys. Res. (in press).
295. R. L. Hudson and M. H. Moore, Radiat. Phys. Chem. **45**, 779 (1995).
296. D. Chakarov and B. Kasemo, Phys. Rev. Lett. **81**, 5181 (1998).

Part V

Ice Particles

16 Nucleation of Ice in Large Water Clusters: Experiment and Simulation

Lawrence S. Bartell and Yaroslav G. Chushak

Summary. Experimental studies of water in greatly confined spaces carried out at the University of Michigan are reviewed. In particular, measurements of rates of homogeneous nucleation of ice in large clusters of water probed by electron diffraction are discussed. Nucleation rates were astronomically higher than any previously observed in the laboratory. Measurements of rates permit inferences to be drawn about interfacial free energies of the ice-water boundary. Diffraction patterns also show that the phase of ice formed when supercooling is deep is the metastable cubic ice. This is because the interfacial free energy for the cubic ice boundary is lower than that for the stable hexagonal phase. Moreover, it is shown that very finely divided water can be cooled substantially below the temperature at which bulk water has been proposed to freeze catastrophically. Possible reasons for small drops avoiding such a critical point are proposed. Molecular dynamics simulations of large, crystalline and deeply supercooled liquid clusters were carried out with a variety of potential functions. They indicated that, despite the disorder found in the surface layers of the crystalline clusters, this disorder was not responsible for the nonideal profiles of the Bragg reflections seen in experiments. Simulations show promise in the field of nucleation. Fully realistic simulations of the freezing of water would be much more enlightening than the traditional nucleation experiments because of the detailed accounts of the underlying cooperative molecular motions they would afford. Such simulations have proven to be elusive, partly because of the enormous demands on computer times involved. Even with advances in computer technology showing signs of overcoming that obstacle, it is not clear that a suitable interaction potential function is available for the purpose. Steps that may be necessary to resolve the problem are discussed briefly.

16.1 Introduction

Because the investigation of water in confining geometries was never a research goal at the University of Michigan before it began by chance, the experimental developments leading to the work in this chapter will be reviewed. For many years the main focus of research in our group was the structural chemistry of free molecules (i.e., those in the vapor phase) [1, 2]. Our primary research tool was electron diffraction, supplemented by quantum chemical computations. About 20 years ago it seemed that a good fraction of interesting volatile molecules had either already been studied or, because of the enormous advances in computational power, were now able to be studied quite accurately by quantum calculations. If one were to continue in structural chemistry, an area of investigation that seemed more important and more challenging to pursue was the structure of liquids. Liquids were much more poorly understood than crystals or gas molecules, making them a more opportune subject for investigation. We had devised certain schemes for the acquisition and treatment of diffraction data in determinations of gas phase structures that made us believe that we could profitably apply them to the study of liquids. Prior experimental studies of liquid structure had made use of X-ray or neutron diffraction. We believed that electrons could provide higher signal-to-noise ratios and higher resolution of diffraction detail than either of the two traditional techniques. The trouble in applying electron diffraction, however, is that electrons are scattered perhaps 100 million times more strongly by matter than are X rays or neutrons. Therefore, it is impossible to study bulk liquids in the same way as they are studied by X rays or neutrons. It is necessary to find a way to produce exceedingly thin samples to avoid multiple scattering and absorption. Samples no thicker than a few dozen nanometers, at most, are desirable. The only way it seemed possible to accomplish this goal routinely and reproducibly would be to generate large clusters by condensing vapors of the subject in supersonic flow. Two other groups had already used this technique and had even studied the structures of their large clusters by electron diffraction with notable success [3]. These groups, however, were physicists and gas dynamicists with interests very different from ours, and their diffraction units had been designed to optimize supersonic flow of the introduced samples. The subjects of these prior investigations had principally been rare gases, and all of the clusters they obtained had turned out to be solid. It wasn't obvious that condensation in supersonic flow could yield the liquid clusters we sought.

Nevertheless, we constructed a device to enable us to introduce samples, cooled by a supersonic expansion, into our diffraction unit. We designed the device to be flexible and versatile, rather than to achieve particularly high Mach numbers [4]. Since we were chemists, not physicists, and were interested in studying liquids, not solids, we examined substances that physicists would have been unlikely to have tried - materials such as benzene and butane. These gave liquid clusters right away, and the clusters were large, of just the size we had hoped for. At approximately 10,000 molecules, they appeared to be large enough to serve as useful models of bulk liquids, yet small enough to transmit electrons without degradation of the beam. Moreover, and we had failed to anticipate this advantage over X-ray and neutron studies, our clusters, having been produced in the chilling flow of a supersonic jet, were very cold, deeply supercooled, and therefore yielded sharper pair correlation functions than had been possible to acquire by the traditional structural techniques with bulk liquids. These sharper pair correlation functions provided a more discriminating test of the intermolecular potential functions governing liquid structure than the more diffuse functions derived from warm liquids.

To the best of the authors' knowledge, our study of liquid benzene yielded the most accurate structure of that well-studied material that had ever been published [5].

It is probable that our program to investigate structures of liquids would have continued if we had not stumbled onto some totally unexpected results. Some substances gave clusters that were crystalline, just as had always been the case prior to our studies. Others, as mentioned above, yielded liquid clusters. But for some substances, under certain flow conditions, we obtained clusters that were liquid, while under other conditions, we obtained crystalline clusters of the same material. Our analyses of the nonequilibrium gas dynamic processes going on during the course of condensation in supersonic flow showed convincingly that these particular clusters must have been born as liquids and then had transformed into crystals while still *inside* our supersonic nozzle. Therefore, we supposed it might be possible to choose conditions which would move the transition to a region outside the nozzle so that the transformation from liquid to solid could be observed as it happened along the flight of the cluster beam. This strategy worked [6]. Times at various stages of the transition could be inferred from the times-of-flight, and the duration of the transformation turned out to be mere microseconds. We were too inexperienced at the time to realize that the implied nucleation rates were unprecedented in studies of freezing. When we began to search the literature on the kinetics of nucleation, we found that the field was very underdeveloped, both in theory and in experiment. It was a much less explored field than the field of liquid structure, and therefore a topic riper for exploitation than the field we had set out to study. Consequently our research program changed directions [7, 8].

16.2 Utility of research on water in confined spaces

One important area involving nucleation of ice is atmospheric science. What happens in cirrus clouds is considered to play an important role in the weather [9, 10]. In these clouds, liquid aerosols consisting of exceedingly small droplets of water freeze, presumably by homogeneous nucleation. Laboratory data on the homogeneous freezing of water at the temperature of the upper atmosphere are sparse. Studies of large clusters are now able to provide new information about such systems.

On a more general level, many processes in science, technology, and nature are mediated by nucleation. This led to a Perspective in the journal *Science* [11] lamenting that our comprehension of one of the simplest and most ubiquitous transformations known, the freezing of liquid water, remains fragmentary. Moreover, deficient though the current theory of nucleation may be, it still allows inferences to be made of interfacial free energies between liquids and solids by measurements of homogeneous nucleation rates [12]. These measurements are most readily made on liquids confined to very small spaces [13]. Such interfacial information is so difficult to get by other methods that almost all the interfacial free energies published to date for liquid-solid boundaries have been derived from nucleation experiments with very small droplets. Those for water in prior work have been on emulsions of microscopic water droplets dispersed in oil [14–16].

A problem with the emulsion technique is that many water-insoluble oils actively promote nucleation, requiring careful tests to ensure that nucleation in a given oil medium is homogeneous. If measurements of homogeneous nucleation are attempted using bulk liquid instead of very fine, dispersed droplets, trace amounts of inadvertent impurities can catalyze the formation of nuclei. Heterogeneous nucleation may take place and freeze the entire sample, spoiling the experiment. If the liquid is finely divided and care is taken, presumably only a small fraction of the drops will be contaminated, and they should contribute very little to the measurements. A virtue of the cluster technique is that the submicroscopic drops are surrounded by no material medium in their flight through space.

16.3 Factors governing cluster phase

Our study of the system of water, not to mention water in confined spaces, the subject of this book, was accidental. We hadn't recognized how important the problem might be. Unlike the physicists who studied subtle and interesting variations in the structures of argon clusters brought about by variations in conditions of supersonic flow we, as chemists, studied many dozens of different materials. We sought to understand what it was about individual substances that made some of them condense into crystalline clusters and others into liquid clusters. Examining the physical properties of the materials, we recognized certain thermodynamic quantities that were correlated with the observed behavior [17]. This analysis led us to formulate a simple empirical criterion involving an index, R_c , a dimensionless quantity defined by

$$R_c = (T_b - T_m)/T_b + 0.007(\Delta S_{fus}/R)^2, \tag{16.1}$$

characterized by the boiling and melting points and entropy of fusion. Substances with an index R_c substantially greater than 0.32 invariably yielded liquid clusters while those with an index appreciably lower than 0.32 gave microcrystals. Subsequently we were able to explain the basis of this criterion [18, 19]. Substances whose phase on condensation could be controlled by choosing appropriate flow conditions were those with an index close to 0.32. Water just happened to be one of those materials. Therefore we tried water and found that it indeed could be watched as it underwent freezing [20]. Our sole motivation at the time was just to try another example of a substance which might transform if our criteria were correct.

16.4 Experimental technique

Illustrated in Fig. 16.1 is a schematic diagram of the diffraction experiment. A 40 kV electron beam with a current of a few hundred nA to 1 μA is focused on a detector by a magnetic lens, to a spot size of a few hundredths of a mm. Our earlier work used photographic plates to detect scattered electrons. This technique gave a high signal-to-noise ratio since it averaged over photographic grains by spinning the circularly symmetric diffraction patterns as they were being scanned. Currently the detector is a photodiode array which, at a cost of lowering the signal-to-noise ratio, greatly increases the ease of finding optimum conditions for cluster formation. It provides patterns in real time, during experiments. A supersonic cluster beam is generated by passing

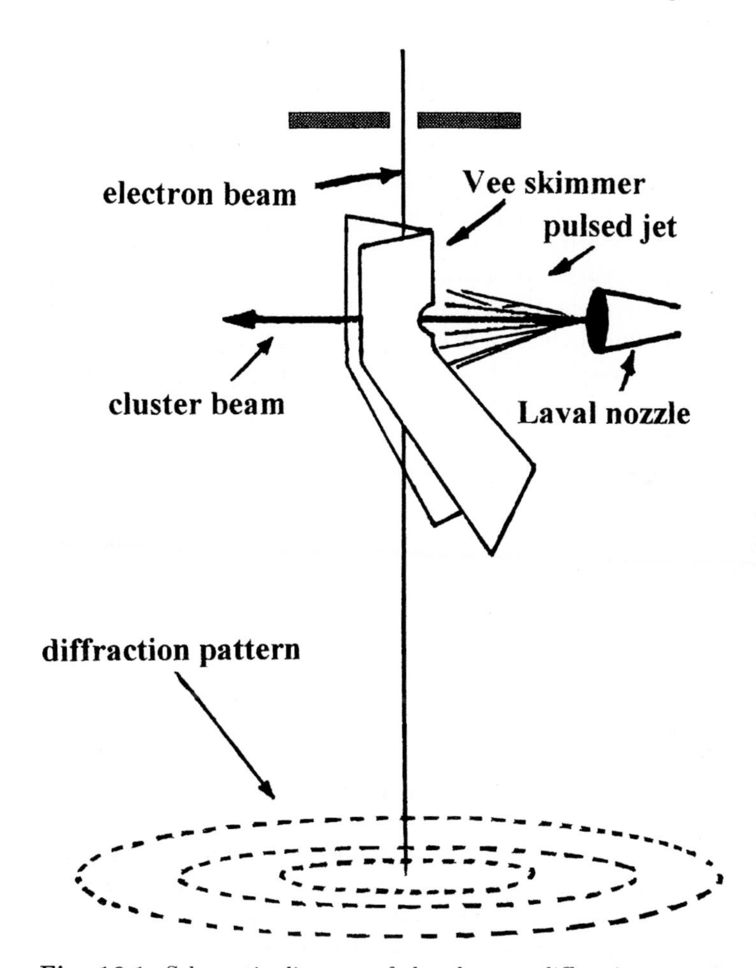

Fig. 16.1. Schematic diagram of the electron diffraction experiment monitoring the structure and composition of clusters. Increasing times of flight are obtained by withdrawing the nozzle away from the electron beam.

the subject vapor through a miniature Laval nozzle, a nozzle whose bore expands from the throat to the exit. In such a nozzle the flow goes supersonic immediately and the gas cools rapidly in the adiabatic expansion. Expansion through a Laval nozzle generates large clusters more efficiently than free-jet expansion because it slows cooling, allowing many more collisions per degree of drop in temperature. The vapor is usually seeded into a rare gas carrier to enhance the cooling and to absorb the heat of condensation. Adiabatic cooling condenses the flowing vapor to a cluster beam which is transmitted through a "Vee skimmer" [21] to intersect the electron beam at a point somewhere between the lens and the detector. Electrons are diffracted by the clusters they pass through, and the pattern of diffraction contains information about the structure of the clusters. The electron beam and cluster beam are pulsed in synchrony. Pulses are so short that gas rebounding from the walls of the apparatus does not have time to reach the electron beam and degrade the diffraction patterns or to interfere with the supersonic flow. Enough time is provided between pulses for a cold trap to scavenge the vapor and for residual gases to be pumped out of the diffraction chamber. Diffraction patterns are recorded, accumulated over a series of many pulses, then processed by a computer.

Diffraction patterns identify the phases of clusters and the relative amounts of each phase during the period of transition. Sizes of the clusters can be inferred in the usual way from the breadths of the diffraction rings of the solids. Volumes of the liquid clusters may require modest corrections for the evaporative loss in dissipating the heat evolved during freezing. Diffraction patterns are monitored at spatial intervals along the supersonic jet. Times during the transformation are calculated from times-of-flight along the jet, cluster velocities having been measured in separate experiments [22].

16.5 Inference of temperature

There is no completely direct method for measuring the temperature of the clusters produced in supersonic flow. The most direct inference to date of temperature in experimental studies of clusters generated in this manner has been that of Nibler and coworkers who recorded Raman spectra of the clusters they produced [23, 24]. Profiles of vibrational features provided information about the temperature when the spectra of clusters were compared with spectra of bulk material at known temperatures. In our opinion, the problem with this approach is that the temperatures deduced for the clusters seem to have been too high. They implied that certain phase changes took place exceedingly rapidly when there was no supercooling. Such an unlikely behavior, never seen in our experiments, was probably due to the high surface-to-volume ratio characteristic of clusters. This made the temperature seem to be higher than it actually was because the disorder in the clusters' surface layers gave the appearance of thermal agitation. This inference is supported by results of experimental studies of clusters by Torchet, et al [25]. In bulk systems the Debye-Waller factor is a measure of the amplitudes of motions of molecules and, hence, of temperature. In clusters, however, Torchet et al report that the variation of Debye-Waller factors, over a large range of cluster sizes, is mainly due to a size effect.

In our investigation it was not possible to carry out spectroscopic measurements. Therefore, it was necessary to resort to indirect inferences of cluster temperatures. We constructed a computer program to calculate the evolution with time of temperatures of clusters and of the surrounding gas, taking into account the nonequilibrium gasdynamics [26, 27]. Included in the program were the internal shape of the nozzle, the effect of the adiabatic expansion moderated slightly by friction with the walls of the nozzle, the heat evolved when material condensed, and removed by thermal accommodation by the surrounding gas and by evaporation. The program provided a reasonable estimate of the temperature of the clusters when they exited the nozzle. The interplay between the evaporation, condensation, and thermal accommodation after the jet exited the nozzle and flowed into the vacuum continued be calculated but the most important factor governing the temperature of the clusters in the free jet was the evaporative cooling. Evaporative cooling rates depend upon the size of the evaporating submicroscopic drops. Although our computer program usually yielded quite plausible sizes of clusters as they exited the nozzle, the program was designed to incorporate the experimentally determined sizes in the free flow beyond the nozzle.

The fact that evaporative cooling is the principal agent influencing the temperature gives us some confidence in the calculated results. Cooling is so rapid when the jet emerges from the nozzle that the exact temperature at the exit is not as important as the length of time experienced by a cluster after its departure from the nozzle. Initial cooling is extremely fast due to evaporation, but the cooling quickly slows the evaporation rate which, in turn, slows the cooling. Indeed, the temperature attained by clusters in a typical supersonic jet some distance from the nozzle is sufficiently systematic that it was named by Gspann the "evaporative cooling temperature." Gspann proposed that this temperature could be estimated by the relation [28]

$$T_{evap} = 0.04 \Delta E_{vap}(T_{evap})/R \tag{16.2}$$

where ΔE_{vap} is the molar energy of vaporization at the evaporative cooling temperature. Our gasdynamic program gives a detailed profile of temperature along the jet that includes the results of (16.2) in mid-range. In practice, the program yields temperature profiles not greatly different from results of rigorous calculations based solely on evaporative cooling [29], making it probable that the gasdynamic calculations, which are more detailed than those of simple evaporative cooling, cannot be greatly in error. A typical example of the calculated temperature profile of a cluster jet of water is plotted in Fig. 16.2 [20].

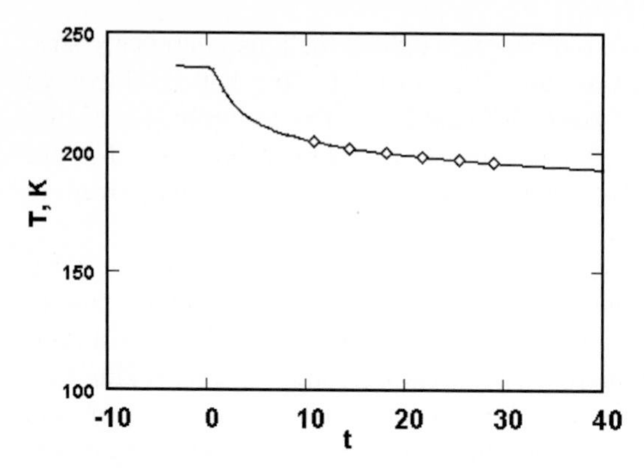

Fig. 16.2. Temperatures of water clusters at intervals along their trajectories away from the nozzle. Evaporative cooling dominates the temperature drop. Diamond marks on the curve indicate the times in microseconds after the clusters left the nozzle, at which the patterns in Fig. 16.3 were recorded.

16.6 The freezing of supercooled water clusters

Not only did water clusters freeze during their flight while being probed by electrons, but their nucleation rate was astronomically higher than any that had ever been measured in the laboratory before [20]. Lest it be supposed that it was irradiation by the electron beam which caused the rapid freezing, it can be shown that no cluster is struck by more than one electron, a particle which speeds through it at one-third the velocity of light. The electron "sees" the structure existing at the time of its impact. Any excitation the electron might have caused could not change the structure by a detectable amount before the electron had departed, carrying the information about the object it had encountered. The same conclusion can be reached more rigorously if the electron is regarded as a wave packet encountering a quantum system [30].

It should be pointed out that the calculation of nucleation rates from the diffraction experiments assumed that the fraction of water seen to have frozen by a certain time corresponds to the fraction of clusters which have frozen completely. Calculations based on Kashchiev's criterion [31] for freezing to be "mononuclear" (i.e. for only one nucleation event per cluster) indicated that the freezing of water at 200 K was indeed mononuclear and that the time for an individual cluster to freeze was short in comparison with the time it took the ensemble of clusters to freeze. Therefore, the number of critical nuclei formed per second was the same as the number of clusters frozen per second.

Another remarkable result was the temperature to which the liquid water clusters could be cooled before they froze [32]; this temperature was much lower for the verifiably liquid state than had ever been attained in the laboratory in the many previous studies. This degree of supercooling was partly a consequence of the extreme confinement of water in a small volume, and partly due to the rapid cooling rate. Water clusters survived in the liquid state until they reached 200 K. They passed without freezing through the temperature region ($\sim$ 227 K) previously considered to be the region of catastrophic freezing, and so denoted as T_H, the temperature of inevitable freezing by homogeneous nucleation [33, 34]. This temperature is about 8 K lower than the liquid had been cooled before it froze in prior experiments. Clusters were able to avoid freezing at T_H because the nucleation rate in this vicinity was not high enough to produce significant nucleation in such small volumes during the time available before the temperature dropped well below T_H. This limiting temperature below which the liquid was believed not to be able to exist had been inferred from the extrapolation of the diverging heat capacity, coefficient of expansion, and compressibility of bulk water characteristic of an approach to a critical point. Whether the apparent instability of water near T_H is due to a true spinodal is still a matter of controversy. To what extent the properties of water in confined geometries might be responsible for the avoidance of the postulated instability will be discussed subsequently.

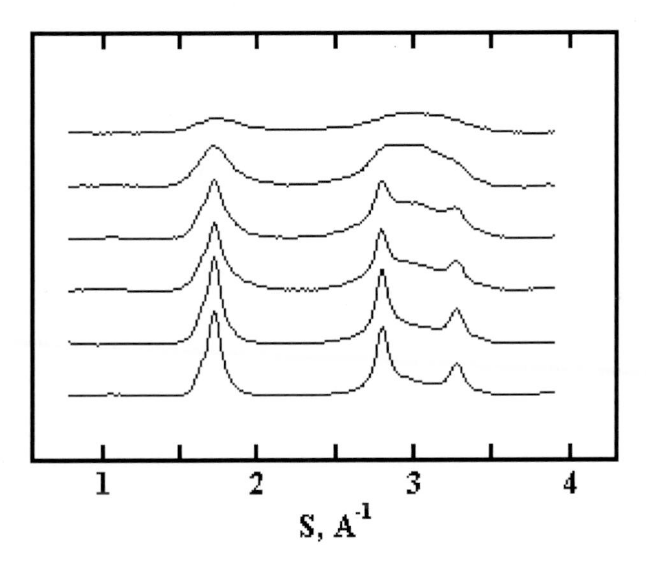

Fig. 16.3. Electron diffraction patterns of water clusters, containing $\sim$ 6000 molecules, taken at intervals of 3.6 μs as clusters froze to cubic ice.

In view of the widespread belief that water suffers a catastrophic solidification at temperatures lower than T_H, another concern about the interpretation of our results might be raised. Is it possible in our experiments that water clusters solidified, solidifying not to a crystal but to an *amorphous* phase, a phase virtually indistinguishable in diffraction experiments from that of a true liquid phase, and then froze? Might this have happened instead of the clusters having remained liquid until nucleation began? After all, Skripov [35] has described the "explosive" crystallization of amorphous solid water at low temperatures. The clear answer is that Skripov's amorphous solid water nucleates 5 orders of magnitude more slowly than the water in the cluster experiments at the same temperature, implying that the clusters were true liquids when they froze, not a glassy form with much reduced mobility of the molecules.

The extremely high nucleation rates we encountered were, of course, due to the deep supercooling attained. That is not to say that it was inevitable that water could achieve a rate as high as the result we found, namely 10^{30} nuclei per cubic meter per second [20]. What can be said is that water in such a small volume as possessed by our clusters could not have frozen during its residence time in the supersonic jet unless its nucleation rate had been in the vicinity of 10^{30} m^{-3} s^{-1} or higher. Evidence for the nucleation rate is recorded in the electron diffraction patterns. Figure 16.3 shows the time evolution of the diffraction pattern of clusters in a typical run as they transformed from the disordered liquid state to an ordered crystal.

Another advantage of our diffraction technique of probing clusters is that the diffraction features reveal the structure at all times, as is implicit in Fig. 16.3. When the water clusters froze, the positions of the Debye-Scherrer peaks in Fig. 16.3 confirmed that the product was cubic ice, I_c, a metastable form, rather than the thermodynamically stable hexagonal ice, I_h. Prior nucleation experiments [14–16] had been blind to the form of ice produced, although some of the experimentalists took it for granted that the ice they generated was the stable form and this caused them to draw incorrect inferences [14]. For reasons to be discussed, there is little doubt that the product in the prior studies was cubic, not hexagonal ice, just as in our experiment. In such highly nonequilibrium processes as transitions in deeply supercooled substances, the product is governed by kinetics, not by thermodynamics. Before our work, some atmospheric scientists had also suggested that highly supercooled liquid aerosols in the upper atmosphere freeze to cubic ice and then, as additional water condenses on the faces of the cubic microcrystals, it grows as hexagonal ice [36, 37]. This inference was partly based on estimated energetic properties of the faces of ice crystals, and partly on the angles at which observed growth features protruded from the particles generated.

16.7 Derivation of interfacial properties

As mentioned in a previous section, one reason for studying nucleation rates is to derive interfacial properties that are difficult to measure by other techniques. This section sketches how it is done. For homogenous nucleation the rate (critical nuclei per unit time per unit volume) can be expressed by [12, 38]

$$J(T) = A \exp\left(\frac{-\Delta G^*}{k_B T}\right) \tag{16.3}$$

where A is a temperature-dependent prefactor, and ΔG^* is the free energy barrier to the formation of a critical nucleus from the liquid. Two general treatments of nucleation exist which are simple to apply to experimental rate data, namely the classical nucleation theory [12, 38] (CNT), and diffuse interface theory [39, 40] (DIT). Neither is fully rigorous but the next step in complexity is the density functional theory [41] which, itself, is neither easily applied to experimental rate measurements nor entirely rigorous. For the purposes of this chapter it will suffice to consider only the application of the CNT.

The conventional formula for the CNT prefactor is

$$A_{CNT} = 16\left(\frac{3}{4\pi}\right)^{1/3}\left(\frac{\sigma_{sl}}{k_B T}\right)^{1/2}\frac{D}{v_m^{2/3}}\Delta r^2 \tag{16.4}$$

where σ_{sl} is the interfacial free energy for the solid-liquid boundary, D is the coefficient of diffusion in the liquid, v_m is the volume of a molecule in the solid, and Δr is the molecular jump distance from the liquid to solid usually taken to be $v_m^{1/3}$. Also, according to this theory, the nucleation barrier ΔG^* for a spherical nucleus is given by

$$\Delta G^* = \frac{16\pi\sigma_{sl}^3}{3(\Delta G_v + w')^2} \tag{16.5}$$

in which ΔG_v is the free energy change in freezing per unit volume of the solid, and w' , the work per unit volume of changing the surface area of the liquid phase during the formation of the nucleus. The latter quantity for a drop of radius R is

$$w' = P_L(\rho_l - \rho_s)/\rho_l \tag{16.6}$$

where P_L is the Laplace pressure $2\sigma_{lv}/R$ exerted by the surface tension upon the interior of the cluster and the ρ's are the densities of liquid and solid.

Provided that the thermodynamic information entering the prefactor is available or can be estimated, it is evident that the nucleation barrier ΔG^* can be calculated from the measured rate, $J(T)$. From ΔG^* the interfacial free energy σ_{sl} can be inferred according to (16.5) if the free energy of freezing can be estimated. If bulk-like values are assumed to apply to critical nuclei, this quantity can be calculated from the standard relation involving the difference between the heat capacities of the solid and liquid. A procedure to extrapolate measured values to deep supercooling is briefly sketched in the next section. Plotted in Figs. 16.4 and 16.5 are the observed temperature-dependent nucleation rates, $J(T)$, and the corresponding derived interfacial free energies per unit area.

16.8 Extrapolation of thermodynamic data to deep supercooling

When applying classical nucleation theory in order to derive the interfacial free energy, it is necessary to have an estimate of thermodynamic properties of the liquid down to low temperatures. Since experimental values of these properties are not available at deep supercooling, it is necessary to propose some procedure to guide the extrapolation of properties from the temperatures of available measurements to the temperature of nucleation. No universally accepted method of extrapolation exists – particularly in view of the divergent behavior of the properties. What seemed to be a plausible method was to adopt a model – the so-called two-state model [42] – which had been applied by a number of investigators. A well known review of the subject by Kauzmann a quarter of a century ago concluded that the two-state model has fatal flaws [43]. Nevertheless, a recent resurrection of the two-state model by Robinson and co-workers provided provocative evidence that Kauzmann's conclusion was too harsh [44, 45]. Robinson's model was very promising but marred by a few thermodynamic inconsistencies. Stimulated by Robinson's work we reanalyzed the two-state model. The result gave us a means of extrapolating the heat capacity, the coefficient of thermal expansion, and the compressibility to low temperatures. It should be pointed out that the

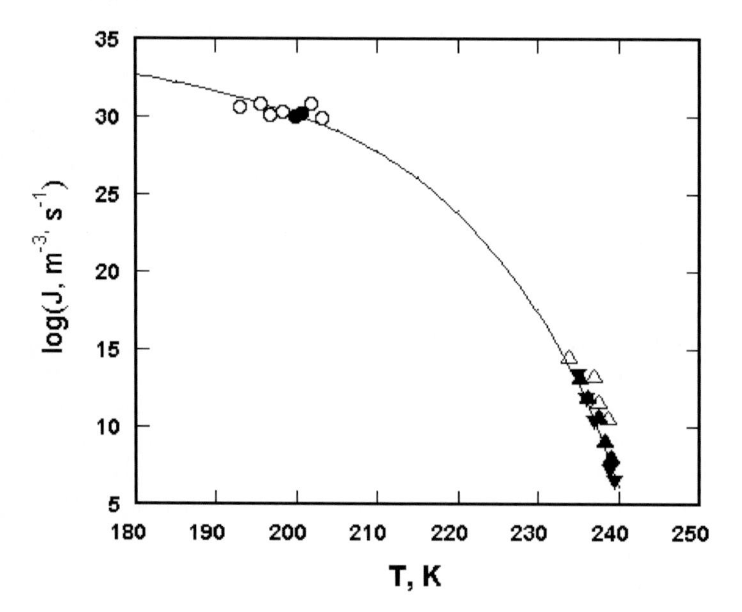

Fig. 16.4. Temperature dependent nucleation rates ($m^{-3}s^{-1}$) in the freezing of water: Clusters in supersonic beams (• , [20]; ○ , present chapter), water-in-oil emulsions (Δ , ref. [14]; solid triangles, ref. [15]; inverted solid triangles, ref. [16]). The solid curve was calculated via classical nucleation theory.

computer power available to us was orders of magnitude greater than that Kauzmann had at his disposal, enabling us to find a way around his presumed fatal flaws. Details of the model are given in reference [42]. Moderate errors in the model would affect the most important property sought, the free energy of freezing in deeply supercooled water, by only a modest amount. Although the model is consistent with the known anomalous properties of water, it leaves unexplained many peculiarities of water at very low temperatures, including the low-density and high-density amorphous phases [46].

A few words should be said about an alternative model proposed by Pruppacher [10], a recognized authority in atmospheric science and author of a definitive treatise on the subject [9]. According to his model, the heat of fusion of water vanishes at T_H, the temperature of hypothesized catastrophic freezing. An argument against Pruppacher's model is that it makes the entropy of liquid water lower than the entropy of ice at a temperature above T_H.

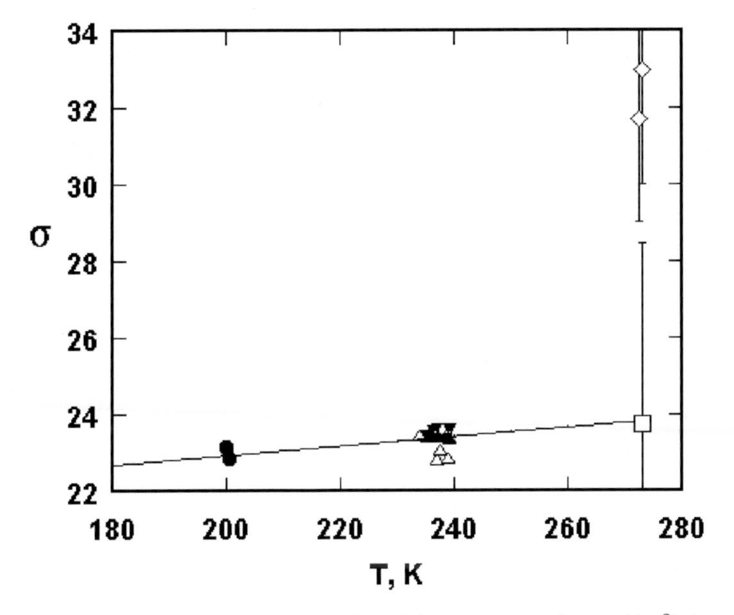

Fig. 16.5. Experimental interfacial free energies (in mJ/m^2) between cubic ice and liquid water, calculated from the nucleation rates in Figure 16.4. Symbols correspond to those of Figure 16.4. Diamonds with error bars: the measured interfacial free energies between hexagonal ice and water. Uppermost value, ref. [47], the value available to Wood and Walton. Intermediate, more recent value, ref. [48]. Square, inferred for the dominant [111] planes of cubic ice by Granasy [49] from the result for the edge plane of hexagonal ice, a plane with the same packing. Results from nucleation experiments represent a mean over the exposed planes of the crystals.

16.9 Ice formed when supercooling is deep

Why do we conclude that the ice produced in prior studies of nucleation was cubic and not hexagonal? There are several lines of evidence. First of all, from our diffraction patterns it is indisputable that the ice obtained in our experiments was cubic. In Fig. 16.4 it can be seen that the classical nucleation theory connects our own data at $\sim$ 200K quite smoothly with the prior experimental data on emulsions acquired at about 235 K and above. Even more convincing is the agreement shown in Fig. 16.5 between the interfacial free energies inferred from the nucleation rates for the emulsions in the 235 K range and the interfacial free energy derived from the cluster data. Significantly, these interfacial free energies are much lower than that measured between hexagonal ice and liquid water at the freezing point [47, 48]. When it is considered that the free energy barrier to forming a nucleus depends on the cube of the interfacial free energy, it is clear that the free energy barrier is considerably lower if the nucleus is cubic than if it is hexagonal. That is why the ice formed at deep supercooling is not the stable hexagonal phase. All of this evidence compellingly indicates that in all of the nucleation experiments plotted in Figs. 16.4 and 16.5, the ice produced was *cubic.*

One of the classic nucleation experiments is that of Wood and Walton [14] who had mistakenly supposed their ice to be hexagonal. One of the occasionally quoted results of these authors is the *extremely* sharp rise in interfacial free energy between ice and water as the temperature increases from 235 K to the freezing point, a *much* sharper rise than seen in Fig. 16.5. There are two problems with this extrapolation. One is that if the original raw data points are examined, it is found that the scattering of the data is too chaotic to justify the authors' extrapolation to the freezing point. The other is that the interfacial free energy obtained by the Wood-Walton extrapolation of the liquid-*cubic ice* result to the melting point just happens to agree with the considerably greater interfacial free energy reported at the time for the liquid-*hexagonal ice* interface, [47] i.e., where the interface is to the wrong phase. Such a result is not credible, particularly in view of Fig. 16.5.

A word of caution is in order. It must be admitted that the above argument is supported by just two points, the validity of which can be questioned. Yet, either point, by itself, would settle the argument if correct, and both appear to be plausible. The first is the supposition that the interfacial free energy of large clusters derived via the CNT (which is known to be crude, and cruder, the deeper the supercooling) can be directly compared, legitimately, with that derived from the warmer and enormously larger droplets in emulsions. The second is that the interfacial free energy of the basal plane of hexagonal ice at the freezing point can be derived via the Wulff relations from the shapes of small water inclusions in ice, knowing the interfacial free energy of the edge planes. These points, of course, relate to the interfacial free energies plotted in Fig. 16.5. At present, it is not possible to rule out rigorously the hypothesis that neither of the two points is correct. Nevertheless, it seems likely that kinetic control of the freezing of deeply supercooled water favors the cubic phase.

16.10 The proposed critical point at T_H

In 1976, Speedy and Angell proposed that the apparent instability of liquid water at T_H corresponded to a spinodal decomposition [33, 34]. These authors showed that the measured divergences of water's thermodynamic properties are consistent with an approach to a critical point at T_H. The fact that our liquid clusters were able to pass through this temperature without freezing does not prove that there is no instability of the *bulk* liquid at T_H because the critical temperature would probably depend strongly on the pressure and degree of confinement. The Laplace pressure exerted upon the interiors of our liquid clusters by the surface tension is hundreds to thousands of atmospheres, depending upon the size of the clusters. This pressure, the small dimensions, and the very short times the clusters exist at very deep supercooling may very well inhibit the spatial and temporal fluctuations characteristic of critical phenomena. Therefore, our results for water under the condition of extreme confinement cannot disprove the hypothesis of Speedy and Angell. On the other hand, the divergence of the thermodynamic quantities can be accounted for, with some success, by the two-state model of water's structural arrangement. Water is considered to be a mixture of a denser component favored at higher temperatures and a more ice-like (or clathrate-like) low-density structure. In this model, which is presented in detail in reference [42], the steep temperature dependence of the composition at low temperatures accounts for the rapid change in properties. At sufficiently low temperatures the higher-density form has disappeared and the behavior of water becomes "normal" until the inadequately understood amorphous phases are encountered.

16.11 More recent nucleation experiments

Our only published study of nucleation rates in water is that which appeared in 1995 [20]. It was a somewhat quick-and-dirty experiment initiated solely to test our criterion to identify substances whose clusters could be seen both in liquid and in solid form, as mentioned above. Once it was recognized that the results were of great potential interest in atmospheric science, further experiments were carried out to check the reliability of the first experiments. Because our original nozzle had been destroyed in an accident and we began to have trouble with replacement nozzles, the clusters we were able to generate were substantially smaller than our original clusters, and gave diffraction patterns of lower quality than the original patterns. Therefore, hoping to obtain better results, we did not publish our findings. Since that time we have been able to generate water clusters considerably larger than the original ones, but nucleation rates for these have yet to be analyzed.

Despite the small size of the more recent clusters, ranging from about 800 to 4,000 molecules, the nucleation rates in these clusters essentially corroborated our original results. This is shown in Fig 16.4, where the more recent nucleation rates are plotted along with the original cluster rates and the rates determined by others at higher temperatures for water in water-in-oil emulsions [14–16]. In our newer results there is marginal evidence that the smaller the cluster, the larger the rate, a tendency found definitively in molecular dynamics (MD) simulations of clusters of other substances. In the MD simulations three roughly comparable factors were shown to be responsible for the size dependence [50]. These were the effect of Laplace pressure, the larger coefficients of diffusion in smaller clusters, and the higher surface-to-volume ratio in smaller clusters. In the case of water, where the difference between the density of liquid and solid at deep supercooling appears to be small, the larger Laplace pressure for smaller clusters would influence nucleation rates only modestly. Coefficients of diffusion enter the rate prefactor A (16.4) directly. That diffusion, averaged over all molecules, is more rapid in smaller clusters is due to the fact that the surface layers make up a relatively larger proportion of the clusters and also tend to be more diffuse, the smaller the cluster.The surface-to-volume ratio enters inasmuch as nucleation rates are expressed as the number of critical nuclei per time per unit volume. Because of the observed tendency in MD simulations of the freezing of other materials for nucleation to occur at or near a cluster surface rather than in the interior, smaller clusters have a higher rate per unit volume than larger clusters. Although a universally accepted account of why nucleation tends to occur at the surface has not yet appeared, Pluis et al. have proposed a criterion which suggests thermodynamic conditions which would favor surface nucleation [51].In our results, however, surface nucleation occurs whether or not the thermodynamic criterion is met.

16.12 Puzzling aspects of the diffraction data for clusters

A determination of nucleation rates from the time-dependence of freezing requires a knowledge of cluster size. As stated in the foregoing, this information is provided by the breadths of the diffraction peaks of the solid. But it is not entirely clear that the theory of peak breadths for completely crystalline particles applies accurately to clusters. Debye-Scherrer diffraction peaks encountered in the 1995 experimental study were somewhat different in shape from the ideal peaks expected for quasispherical crystals [20]. Observed peaks were broader at the base relative to the full width at half height. It is possible that this was simply a consequence of a certain proportion of unfrozen clusters distorting the peaks, but two other interesting possibilities had to be considered. Tanaka, in computations of structures of small, cold clusters, had found that a pancake shape was more stable than a quasispherical shape, at least in classical simulations at low temperatures [52]. Computations of diffraction peaks based on his molecular coordinates, however, ruled out the possibility that our clusters froze to pancake-shaped solids. The pancake-shaped array of molecules predicts a diffraction pattern bearing no resemblance to ours.

A second possibility was that the surface layers of molecules in our clusters are disordered and sufficiently different from the interior in structure that their contribution to the diffraction intensity distorted the peaks. A significant difference between the structure of the surface layers and the interior had been found in spectroscopic studies of ice surfaces by Devlin and Buch ([53] and chapter 17) and in molecular dynamics (MD) simulations of the transition bilayer by Buch [54]. If such a transition layer constituted an appreciable fraction of cluster volumes and accounted for our peak profiles, it was important that we find out and make corrections to our cluster volumes.

16.13 Molecular dynamics simulations of water clusters

16.13.1 Simulations to assess surface disorder of clusters

To check the possibility that disorder at the surface of clusters compromised our determinations of cluster sizes, Buch began an MD simulation of a very large quasispherical ice I_c cluster; at 6,000 molecules, this cluster was the size of the clusters found in our original experiments. In these simulations the TIP3 potential function [55] was used. We continued the MD runs to provide a long annealing, using the TIP4 function [55]. Results did yield a transition layer of the sort seen before by Buch but it did not account for our peak shapes. The existence of a fairly thick transition layer had raised concerns that our inferences of cluster size from the Debye-Scherrer peak breadths might be invalidated because the volume of the transition layer was a fairly large proportion of the total volume of our crystals. It was found, however, that the cluster size implied by the diffraction peak breadths calculated from the molecular coordinates of the simulation agreed very well with the known physical diameter of the entire simulated cluster. Therefore, it is probable that the nonideal peak shape yielded by our clusters was due to the presence of unfrozen clusters in our sample.

16.13.2 Investigations of kinetic properties of supercooled water

Once our program of research on nucleation began, rates of freezing of various liquids were extrapolated via the classical nucleation theory to much deeper supercooling than attained in our experiments. These crude computations suggested that freezing rates in some cases might become so high that the transformation might be accessible to MD simulations. This proved to be true [6, 7]. In the case of water, however, such extrapolations gave us no optimism that MD simulations would show spontaneous freezing in clusters. Nevertheless, it still seemed worthwhile to investigate the behavior of clusters of water in MD simulations.

To begin the simulations it was necessary to generate liquid clusters. The most reliable way to do this, with hydrogen bonds connected reasonably instead of dangling haphazardly, was to start with an ideal crystal, then melt it. This was accomplished with the aid of the MD program MOLDY [56]. Initial simulations to be outlined were performed on clusters comprised of 496 and 1023 water molecules represented by the TIP4P potential function with a cutoff of 10.0 Å. In later simulations, cluster sizes up to 4096 molecules were examined and several other potential functions were applied. Simulations were carried out at constant temperature by rescaling velocities every 10 steps to keep the system at the desired temperature. Equations of motion were integrated by a modified Beeman algorithm [57], with a time step of 0.5 fs.

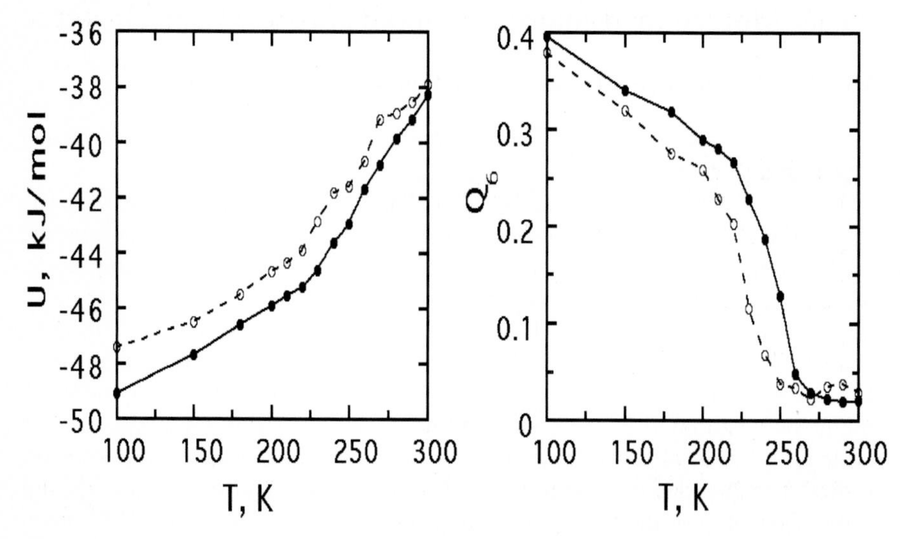

Fig. 16.6. Temperature dependence of the potential energy per molecule U and the global order parameter Q_6 for water clusters during heating. Open circles, 496-molecule; solid circles, 1023-molecule clusters.

For all of the cluster sizes, an initial, approximately spherical cluster was constructed to possess an ideal ice I_c cubic structure with a density of 1.0 g/cm^3. Each cluster was heated from 100 K to 300 K with temperature steps of 10 K in the 200 - 300 K region, and equilibrated at each temperature for 50 ps. The temperature of a melting transition can be identified in several ways. We have used the changes with temperature of the configurational energy U and the global order parameter Q_6 [58]. Results of the heating runs for the initial two clusters are presented in Fig. 16.6. The global order parameter Q_6 can be obtained by averaging of the local order parameter $q_6(i)$ around a molecule i over all molecules in cluster. In an isotropic liquid state, the global order parameter goes to zero while in the crystalline state it has some nonzero value [59]. For both of the initial cluster sizes the melting transition took place only gradually over a wide range of temperature, a range approximately 30-40 K in breadth. The change in configurational energy accompanying the transition was much less sharp than in prior MD runs with metallic or ionic clusters. Melting began at the surface. From the plot of the order parameter Q_6, it appears that the 469-molecule cluster melted completely at 250 K whereas the larger cluster melted at a temperature higher by approximately 10 K. Figure 16.7 shows images of a 496-molecule cluster at different temperatures during the heating. Despite the excessively high initial density the cluster retained its well-ordered crystalline structure after equilibration at 100 K. At 200 K the structure is still crystalline though somewhat disordered, while at 220 K only the core remains crystalline, the surface layer being severely disordered. As mentioned above, the cluster melted completely at 250 K.

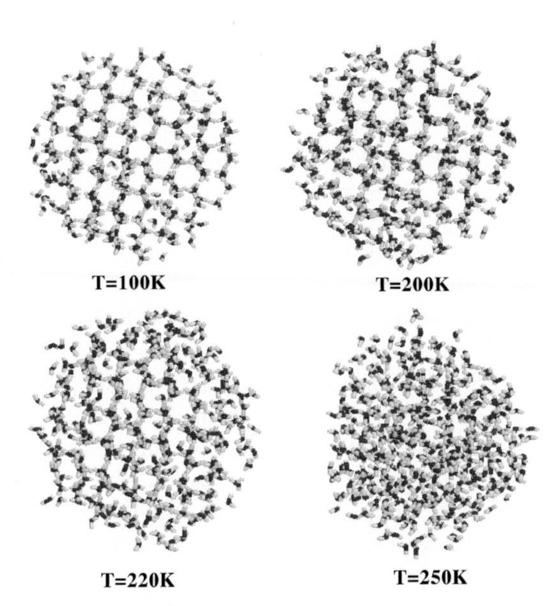

Fig. 16.7. Images of a 496-molecule cluster at different temperatures during heating stages.

In the heating runs we did not observe a transition from the metastable cubic ice to hexagonal ice, a transition found in the bulk system in the range

160-210 K. Nor in cooling runs did we observe the liquid to undergo a glass transition, a transition seen in the bulk at ∼136 K [60, 61]. While this difference from bulk behavior might be ascribed to a size effect or to imperfections in the potential function adopted, it may simply reflect the fact that the times in the simulations were many orders of magnitude shorter than those in experiments.

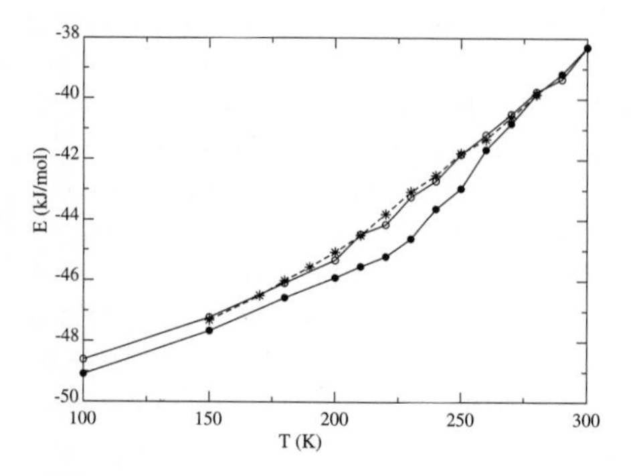

Fig. 16.8. Temperature dependence of the potential energy of a 1023-molecule cluster: a) during heating, starting with a cubic structure (solid circles); b) during cooling (asterisks); c) during reheating of the cooled cluster (open circles).

In simulations to find whether any evidence of spontaneous nucleation could be detected, liquid clusters were cooled to 220 K, then equilibrated at each of three temperatures (200K, 190K, 150K) for 1 ns in order to observe what happened. As can be seen in Fig. 16.8, the configurational energy of this cluster during the cooling runs, and also in subsequent heating runs, was appreciably higher than the energy of the crystalline cluster, as expected. Times of simulations during the cooling runs were far too short to achieve nucleation but it was possible to observe how the coefficient of diffusion fell as the degree of supercooling increased. Converting diffusion to viscosity for the warmer clusters via a modified Stokes-Einstein relation [62, 63] indicated that although the clusters became quite viscous when cooled, their effective viscosity was lower than that projected for bulk water at the same temperature and certainly far too low to be indicative of the formation of a glass. The fact that the coefficients of diffusion for the clusters were higher than expected for bulk water is associated less with the surface diffusion than to the TIP4P potential function used. The extended Simple Point Charge (SPC/E) potential [64] has been reported to give a more realistic diffusivity down to low temperatures than TIP4P. On the other hand, the SPC/E function must be used with care in studies of nucleation because, unless the electrostatic interactions are included for *all* interactions (in the bulk by incorporating the Ewald sum) it yields a *higher* free energy for the stable crystalline phase of water than for the liquid throughout the temperature range of interest [65].

Another reason the SPC/E function would be an unsatisfactory choice for studies of nucleation is the following. When a quasispherical cluster of 1023 molecules, with a starting structure of organized oxygens in the diamond lattice, was annealed at 100 K under the influence of this potential, it relaxed to an oblate ellipsoid of revolution with a minor axis 72 % that of the major axes. By contrast, when a cluster with an identical starting structure was annealed using the TIP4P interaction potential function, it remained cubic, as would be expected for real ice, and retained its quasispherical external shape. The cluster based on the SPC/E function was about 10 % more compact than that for the TIP4P function, and, in the lowest energy configuration we have so far observed, had a tetragonal structure not related to the known tetragonal structures for ice at high pressure or to the novel structures found by Baez and Clancy [66]. With a different disposition of hydrogen bonds in the starting structure, the cluster still relaxed to a prolate ellipsoid, but with a different internal structure. Clearly, as reported by Baez and Clancy, the SPC/E potential yields extraneous solid structures for water at normal pressure. It is unnaturally unspecific in the way it directs molecules to pack together into ordered arrays.

The crystallization of water in computer simulations is a long-standing problem. Even at the deepest supercooling of water measured experimentally, nucleation takes place over microseconds, not the nanoseconds of typical simulations. Shortly advances in computer science will make it feasible to carry out considerably longer trajectories.

Only by applying a static electric field had it been possible, until recently, to crystallize water in MD simulations using the TIP4P potential in systems of small numbers of molecules, systems imposing periodic boundary conditions [67, 68]. The crucial step in these simulations was the rearrangement of liquid water's hydrogen bonding network [68]. Recently, however, crystallization of water into hexagonal ice in an MD simulation with the De Leeuw & Parker potential function [69], has been reported [70]. However this transition occurred at 300 K at a rate which would require very deep supercooling. Hence, if a physically reasonable interaction function had been applied, a temperature far below 273 K would have been needed. Moreover, since the freezing was to the wrong structure for a kinetically controlled event, this potential function does not appear to yield realistic results.

After the initial submission of this chapter another report appeared describing freezing in an extremely long simulation of supercooled water [71]. In this investigation by Matsumoto, Saito, and Ohmine, systems ranging from 64 to 4096 water molecules were examined at several temperatures, each run with imposed periodic boundary conditions and a constant temperature. In runs at constant volume of the order of microseconds of trajectories (corresponding to months of cpu time on a supercomputer), one run of 512 molecules at 230 K was observed to freeze. In five other parallel simulations for systems of this size no freezing was observed, however. Although the authors did not choose to identify the frozen phase that spontaneously formed, Sastry, in his comment [72] on the report, stated that the phase was hexagonal. Images of the crystalline aggregate presented by Matsumoto, et al., did indeed show hexagonal channels of the sort corresponding to hexagonal ice but the frozen aggregate was appreciably disordered. The fact that the solid was disordered was inevitable inasmuch as a perfect structure of hexagonal ice cannot form in a cubic cell with periodic boundary conditions containing 512 molecules. The question isn't trivial because, in a physically realistic case, *cubic* ice, not hexagonal ice, is the phase expected at deep supercooling where kinetics dictates what happens, as explained earlier in this chapter. Ordered cubic ice could have fit into such a volume.

Exactly how to interpret the remarkable results of Matsumoto, et al., is unclear, quite apart from the unexpected phase apparently produced. The freezing of systems with as few as 512 molecules has been claimed by Andersen and coworkers [73, 74] to be seriously perturbed by the imposition of periodic boundary conditions. Furthermore, in a system maintained at constant volume, pressure fluctuations as large as thousands of atmospheres can be expected. Moon and Pawley [75] reported that the freezing of liquid SF_6 in molecular dynamic simulations was greatly enhanced by imposing pressure fluctuations. Therefore, it would be desirable to carry out further investigations under different conditions to gain a better understanding of the process of the freezing of water.

Whether it is possible to simulate the kinetic behavior of water realistically in molecular dynamics computations with classical pairwise additive potentials is not entirely clear. A doubt was expressed by A. Brodsky several years ago when he wrote, [76] "Is there predictive value in water computer simulations?" Brodsky's answer was "No." It would not be surprising if empirical potential functions with parameters adjusted to describe thermodynamic and some kinetic properties of water over a limited range of density, temperature and pressure failed to capture the dynamics of nucleation at deep supercooling.

Despite the considerable effort to develop a suitable intermolecular potential function for water for MD simulations and despite a perfectly enormous investment of computer time on MD simulations of water, a fully realistic potential function for water has been elusive. To what extent this has to do with the commonly neglected explicit effects of polarization or of the famous double-minimum problem [77] in water's hydrogen bonds is uncertain. It is not clear whether it is possible in the fluctuating structure associated with the thermal agitation of molecules, for momentary imbalances of electrostatic forces to initiate a modern version of the concerted Grotthus-like transfers of hydrogens from one oxygen to another [78]. Such a mechanism might conceivably be relevant under the influence of a shear, by helping to preserve the fluidity of very deeply supercooled water or, possibly, by altering probabilities of attachment of liquid molecules to solid nuclei. Whether the ability of hydrogens to jump from one oxygen to another, a jump denied them by the potential functions universally adopted in MD simulations, alters dynamic properties of deeply supercooled water is not known. Another factor neglected in classical MD simulations is the effect of the zero-point motions present in real systems of water. Large amplitudes of vibration facilitate relaxation. At low temperatures zero-point vibrations make amplitudes of molecular motions larger than those calculated by Newtonian mechanics until kT approaches the quantum energy gap separating excited states from the ground vibrational states of the individual modes. For ice this classical limit does not occur until the system is much warmer than 200 K [79]. One can only speculate about the properties of supercooled liquid at such low temperatures. Since there is some evidence that the densities of the liquid and solid approach each other at such low temperatures [42], their local vibrational modes may be similar. In any event, the effects on the dynamic properties of water at low temperatures of the double-minimum in the potential function and zero-point vibrations have largely been ignored insofar as we are aware.

16.14 Concluding remarks

The most important conclusions of our work are as follows. In our experiments we found that water in highly confined volumes can be cooled well below the temperature T_H at which bulk water has been postulated to freeze catastrophically. It is possible that the Laplace pressure inside our exceedingly small drops may have lowered the postulated critical temperature. Moreover, the small volume and brief time of passage of water clusters through T_H may inhibit the spatial and temporal fluctuations associated with a critical transition. These factors might have allowed our large clusters (10^3 – 10^4 molecules) to avoid freezing until the temperature dropped far below T_H. Therefore, our observations do not necessarily contradict the existence of the postulated critical point at T_H for bulk water. On the other hand, this critical point has not yet been unequivocally demonstrated. When deeply supercooled, closely confined water does freeze, the rate is astronomically higher than observed in conventional experiments with water-in-oil emulsions. Moreover, it is clear that when deeply supercooled water spontaneously freezes, the product is not the thermodynamically stable phase. Instead, it is the metastable cubic phase, I_c, because the laws of kinetics, not thermodynamics, govern the transition. The foregoing observations illustrate some of the effects of extreme confinement on the properties of water. As to the future, it will be interesting to find whether classical molecular dynamics simulations applying conventional potential functions can provide a realistic account of the behavior of deeply supercooled water. Much more detailed simulations made feasible by advances in computer technology may be able to determine whether zero-point vibrations and effects of the double minimum potential describing $O-H-O$ interactions play a significant role in water's dynamic properties at low temperatures and in confined spaces.

Acknowledgments. The research outlined in this chapter was supported by a grant from the National Science Foundation. We thank Dr. J. Huang and Paul Lennon for their roles in the unpublished measurements of nucleation rates in water clusters. We are indebted to Professor Buch for initiating molecular dynamics simulations of very large crystalline clusters of water and to Professor Tanaka for the molecular coordinates in his flat clusters.

References

1. See, for example, L. S. Bartell: In: *50 Years of Electron Diffraction*, International Union of Crystallography, ed. by P. Goodman (Reidel, Dordrecht, 1981), pp. 235-242.
2. L. S. Bartell: In: *Stereochemical Applications of Gas-Phase Electron Diffraction*, ed. by I. Hargittai (VCH Publishers, New York, 1988), pp. 55-83.
3. L. S. Bartell: Chem. Rev. **86**, 492 (1986).
4. L. S. Bartell, R. K. Heenan, M. Nagashima: J. Chem. Phys. **78**, 236 (1983).
5. L. S. Bartell, L. R. Sharkey, X. Shi: J. Amer. Chem. Soc. **110**, 7006 (1988).
6. L. S. Bartell, T. S. Dibble: J. Amer. Chem. Soc. **112**, 890 (1990).
7. See, for example, L. S. Bartell: J. Phys. Chem. **99**, 1080 (1995).
8. L. S. Bartell: Ann. Rev. Phys. Chem. **49**, 43 (1998).
9. H, R. Pruppacher, J. D. Klett: *Microphysics of Clouds* (D. Reidel, Dordrecht, 1978).
10. H, R. Pruppacher: J. Atm. Sci. **52**, 1924 (1995).
11. J. M. McBride: Science **256**, 814 (1992).
12. D, Turnbull, J. C. Fisher: J. Chem. Phys. **17**, 71 (1949).
13. D, Turnbull, B. Vonnegut: Industr. and Eng. Chem. **44**, 1292 (1952).
14. G. R. Wood, A. G. Walton: J. Appl. Phys. **41**, 3027 (1970).
15. G. T. Butorin, V. P. Skripov: Krystallografiya **17**, 379 (1972).
16. P. Taborek: Phys. Rev. **B32**, 5903 (1985).
17. L. S. Bartell, L. Harsanyi, E. J. Valente: J. Phys. Chem. **93**, 6201 (1989).
18. L. S. Bartell: J. Phys. Chem. **96**, 108 (1992).
19. L. S. Bartell: J. Phys. Chem. **100**, 8197 (1996).
20. J.Huang, L. S. Bartell: J. Phys. Chem. **99**, 3924 (1995).
21. L. S. Bartell, R. J. French: Rev. Sci. Instrum. **60**, 1223 (1989).
22. J. W. Hovick, R. J. French, L. S. Bartell: J. Mol. Struct. **376**, 59 (1995).
23. R. D. Beck, J. W. Nibler: Chem. Phys. Lett. **148**, 271 (1988).
24. R. D. Beck, M. F. Hineman, J. W. Nibler: J. Chem. Phys. **92**, 7068 (1990).
25. G. Torchet, M. F. de Feraudy, B. Raoult: J. Chem. Phys. **103**, 3074 (1995).
26. L. S. Bartell: J. Phys. Chem. **94**, 5102 (1990).
27. L. S. Bartell and R. A. Machonkin.: J. Phys. Chem. **94**, 6468 (1990).
28. J. Gspann: In: *Physics of Electronic and Atomic Collisions*, ed. by S. Datz (Hemisphere, Washington, DC, 1976), pp. 29-96.
29. C. E. Klots: Nature **322**, 222 (1987).

30. L. S. Bartell: In *Enciclopedia della Chemica*, Vol. IV, ed by S. Califano (USES, Firenze, Italy 1975). pp. 448-453.
31. D. Kashchiev, D. Verdoes, G. M. van Rosmalen: J. Cryst. Growth **110**, 373 (1991).
32. L. S. Bartell, J. Huang: J. Phys. Chem. **98**, 7455 (1994).
33. R. Speedy, C. A. Angell: J. Chem. Phys. **65**, 851, (1976).
34. C. A. Angell: J. Phys. Chem. **97**, 6339 (1993).
35. V. P. Koverda, N. M. Bogdanov, V. P. Skripov: J. Non-Cryst. Solids **57**, 203 (1983).
36. T. Kobayashi, K. Furukawa, T. Takahashi: J. Cryst. Growth **35**, 262 (1976).
37. T. Takahashi, T. Kobayashi: J. Cryst. Growth **64**, 593 (1983).
38. E. R. Buckle: Proc. Roy. Soc. London. **A261**, 189 (1961).
39. L. Granasy: J. Non-Cryst. Solids **162**, 301 (1993).
40. L. Granasy: Mater. Sci. Eng. **A178**, 121 (1994).
41. D. Oxtoby: In: *Liquids, Freezing and Glass Transition*, ed. by J. P. Hansen, D. Levesque and J. Zinn-Justin (Elsevier, Amsterdam, The Netherlands, 1991), pp. 147-191.
42. L. S. Bartell: J. Phys. Chem. **B101**, 7573 (1997).
43. W. Kauzmann: L'Eau Syst. Biol. Colloq. Int. C.N.R.S. **246**, 63 (1975).
44. M. Vedamuthu, S. Singh, G. W. Robinson: J. Phys. Chem. **98**, 2222 (1994); **99**, 9263 (1995).
45. G. W. Robinson, S. Zhu, S. Singh, M. W. Evans: *Water in Biology, Chemistry and Physics*, (World Scientific, Singapore, 1996).
46. R. J. Speedy: Nature **380**, 289 (1996).
47. P. V. Hobbs, W. M. Ketcham: In *Physics of Ice*, ed. by N. Riehl (Plenum, New York, 1969).
48. W. B. Hillig: J. Cryst. Growth **183**, 463 (1998).
49. L. Granasy: private communications.
50. J. Huang, L. S. Bartell: J. Phys. Chem. (in press).
51. B. Pluis, D. Frenkel, J. F. van der Veen: Surf. Sci. **239**, 282 (1990).
52. H. Tanaka, R. Yamaoto, K. Koga, X. C. Zeng: Chem. Phys. Lett. **304**, 378 (1999).
53. J. P. Devlin, V. Buch: J. Phys. Chem. **99**, 16534 (1995).
54. B. Rowland, N. S. Kadagathur, J. P. Devlin, V. Buch, T Feldmann, M. J. Wojcik: J. Chem. Phys. **102**, 8328 (1995).
55. M. W. Mahoney, W. L. Jorgensen: J. Chem. Phys. **112**, 9810 (2000).
56. K. Refson: Comput. Phys. Commun. **126**, 310 (2000).
57. K. Refson: Physica **B 131**, 256 (1985).
58. J. S. van Duijneveldt, D. Frenkel: J. Chem. Phys. **96**, 4655 (1992).
59. Y. Chushak, L. S. Bartell: J. Phys. Chem. **A 104**, 9328 (2000).
60. M. Sugisaki, H. Suga, S. Seki: Bull. Chem. Soc. Japan, **41**, 2591 (1968).
61. K. Ito, C. T. Moynihan, C. A. Angell: Nature, **398**, 492 (1999).
62. J. C. M. Li, P. J. Chang: J. Chem. Phys. **23**, 518 (1955).
63. T. V. Lokotosh, S. Magazni, G. Maisano, N. P. Malomuzh: Phys. Rev. **E 62**, 3572 (2000).
64. H. J. C. Berendsen, J. R. Grigera, T. P. J. Straatsma: Phys. Chem. **91**, 6269 (1987).
65. B. W. Arbuckle, P. Clancy: J. Chem. Phys. **116**, 5090 (2002).
66. L. A. Baez, P. Clancy: J. Chem. Phys. **103**, 9744 (1995).

67. I. M. SBvishchev, P. G. Kusalik: Phys. Rev. Lett. **73**, 975 (1994).
68. I. Borzak, P.T. Cummings: Phys. Rev. **E 56**, R6279 (1997).
69. N. H. de Leeuw, S. C. Parker: Phys. Rev. **B 58**, 13901 (1998).
70. P. J. van Maaren, D. van der Spoel: J. Phys. Chem. **105**, 2618 (2001)
71. M. Matsumoto, S. Saito, I. Ohmine: Nature, **414** 409 (2002).
72. S. Sastry: Nature, **414**, 376 (2002).
73. J. D. Honeycutt, H. C. Andersen: Chem. Phys. Lett. **108**, 535 (1984).
74. W. C. Swope, H. C. Andersen: Phys. Rev. **B 41** 7042 (1990).
75. C. Moon, S. G. Pawley: J. Mol. Struct. **485**, 479 (1999).
76. A. Brodsky: Chem. Phys. Lett. **261**, 563 (1996).
77. L. Pauling: J. Amer. Chem. Soc. **57**, 2680 (1935).
78. B. E. Conway: In *Physical Chemistry. An Advanced Treatise IXA*. Ed by H. Eyring (Academic Press, New York 1970).
79. Conclusion drawn from inspection of the heat capacity of ice reported by W. F. Giauque, J. W. Stout: J. Amer. Chem. Soc. **58**, 1144 (1936).

17 Ice Nanoparticles and Ice Adsorbate Interactions: FTIR Spectroscopy and Computer Simulations

J. Paul Devlin and Victoria Buch

Summary. Studies of ice particles can be designed to obtain basic knowledge of ice nanocrystals and large water clusters, or to provide insight to specific properties of macroscopic condensed phase water systems. The latter properties include the nature of the surface of ice and the H-bond chemistry of ice, examined in this chapter, as well as the homogeneous nucleation and growth of crystalline ice, central concerns of Chapter 16. Important basic properties of ice particles include structure, structural dynamics, spectra and the interactions of an abundant surface with adsorbate molecules. With the results of early electron-diffraction studies as a basis, our recent combined spectroscopic and simulation efforts have led to a simple self-consistent view of the structure, dynamics and 'reactivity" of ice particles at temperatures below ~140 K.

Cold ice particles divide into ice nanocrystals and large water clusters, the latter being amorphous particles smaller than ~2.8 nm which lack a crystalline subsurface or core. Much remains to be learned of these smaller particles of 20 – 300 molecules. The overall structure of ice nanocrystals (formed in a collisional cooling cell) consists of a crystalline core and a disordered surface joined by a transition layer of "strained crystalline" subsurface. The surface is disordered with a significant decrease in population of 3-coordinated water molecules compared to a full-bilayer-terminated flat crystalline surface. A difference between surface and subsurface relaxation rates is the basis of a method used to determine separate surface and subsurface spectra. The disordered surface retains elements of mobility (i.e., simple molecular rotation) down to ~60 K, while the subsurface shows orientational relaxation only above ~110 K. Nevertheless, the *subsurface* orientational mobility exceeds that of interior bulk ice by more than an order of magnitude. We argue, somewhat tentatively, that the disordered surface is a better model for the surface of (most) bulk ice than is a full-bilayer-terminated surface.

The reduced population of active (unsaturated) surface sites is a critical factor in the H-bond chemistry of ice particles. The reduced density of such sites limits the ability of the surface to solvate H-bonding adsorbates such as NH_3 and HCl; a necessary step in the nucleation of new "hydrate" phases at the ice surface. However, above some temperature, the stronger H-bonding adsorbates can "de-reconstruct" the ice surface by insertion into the weak surface H-bonds formed during disordering of the particle surface. Above this temperature (~110 K), strong adsorbates are more fully solvated and reaction with the particle may proceed. However, conversion of an ice particle to a hydrate particle requires an abundance of adsorbate molecules at the particle surface, as heterogeneous nucleation of a new phase will not proceed unless the low energy surface sites are fully occupied. When the conversion to a hydrate phase does proceed, the conversion rate is often limited by the rate of reactant diffusion through the shell of hydrate product that immediately encases the particle. For particles, and presumably for macroscopic ice samples as well, the diffusion of reactant molecules through this hydrate crust, rather than diffusion within the ice itself, is an important aspect of the transformation.

17.1 Introduction

Ice and ice-like substances are very common forms of water known to exist abundantly on earth, within the earth's atmosphere, on the satellites of planets of our solar system [1], within comets [2-4] and in the dark clouds of interstellar space [5,6]. In several important instances, such as the polar stratospheric clouds that play a critical role in generating the ozone hole [7], the icy substance is present in the form of particles or large clusters. However, whether particulate or bulk in form the impact on an environment is primarily through the surface of an icy substance. Though bulk icy matter has been the subject of almost innumerable studies dating back centuries, intense efforts to deduce the nature of the ice surface and its interaction with other substances at the molecular level have a relatively short history. There is a familiar reason; for icy substances, as most matter, the surface normally represents a very small fraction of the total, so that probe signals of the surface are buried within relatively massive signals from the bulk. To overcome the difficulty this presents in studying the surface properties, two approaches are most obvious. Either a technique can be chosen that gives a highly selective surface signal, or a form of matter must be used that optimizes the amount of surface relative to the volume of ice.

There have been several powerful methods developed to advance the molecular science of surfaces in recent years, some of which are applicable to icy matter. Those that are applicable include the diffractive scattering of helium [8,9], low energy electron diffraction (LEED) [10], reactive ion scattering (RIS) [11], infrared reflection-absorption spectroscopy (IRAS) [12,13], sum-frequency-generation spectroscopy (SFGS) [14 and 2.6.2] and, perhaps to a lesser degree, glancing-angle X-ray scattering [15]. Although each of these techniques has natural limitations, their application has expanded our knowledge of the ice surface and its interaction with other substances in ways that will be noted in later sections.

Since the surface-selective techniques have limitations, methods based on optimization of the amount of surface, relative to bulk ice, have also been pursued. Except for porous substances, such as microporous amorphous solid water (ASW) [16,17], the amount of surface exposed by a given amount of matter that is subdivided into units of quasi-spherical geometry is proportional to the inverse of the diameter of the units. Applying this view rigorously, it might be concluded that cold small water clusters represent a form of water ideal for investigation of the ice surface. Certainly, much valuable related information, including water-water intermolecular potentials and mode frequencies characteristic of water coordination geometries, has been deduced from studies of small water clusters [18-21], but small clusters differ in major respects from either the ice surface or the interior of ice. These differences include the number and strength of hydrogen bonds (H-bonds) per water molecule as well as the angles between H-bonds formed by the molecules. More directly useful insights are available, if the size of the water particles investigated is not so limited that the form of matter is no longer legitimately crystalline ice. Though ice particles are themselves of both fundamental and practical interest, one important aspect of their study concerns the relationship of their structural, dynamic and surface properties to those of bulk ice. An appreciation of certain properties of bulk ice is therefore useful before we turn our attention to ice nanoparticles.

17.1.1 Properties of Ice I

Throughout, we will generally not distinguish between ice Ic (cubic ice) and ice Ih (hexagonal ice), the two very similar crystalline forms of ice that occur at low pressures. Hexagonal ice is the stable bulk phase of water below 273 K, but cubic ice commonly forms and exists as a metastable structure at temperatures below ~180 K, or in a finely divided state over a greater temperature range. Whether in a perfect cubic or hexagonal ice crystal, the oxygen atoms form a periodic pattern which can be viewed as the result of stacking bilayers of water molecules, with each bilayer constructed from puckered hexagonal rings [22,23]. Cubic ice differs from hexagonal ice only in the stacking sequence of the hexagonal bilayers. The structure, shown in Fig. 17.1a, follows the ice rules (also called Bernal-Fowler rules), which means that each oxygen atom has four nearest-neighbor oxygen atoms arranged in a tetrahedral pattern, with a single H atom between each near-neighbor oxygen pair. Each of these H atoms is chemically bonded to one oxygen atom and H-bonded to another. Further, the water molecular unit is retained, as each oxygen is chemically bonded to two H atoms and H-bonded to two others. Within a macroscopic piece of ice, a great many arrangements of the hydrogen atoms, consistent with the ice rules, have essentially the same energy (i.e., near degenerate structures). One, the proton-ordered structure of ice XI, has the minimum energy [23]. However, within most arrangements, there is no long-range correlation of the orientation of the water molecules; so the protons are disordered within structures that have a high degree of oxygen order.

Movement between the nearly degenerate ice structures at temperatures below 150 K occurs through orientational and/or ion-defect mobility. Restructuring through molecular diffusion does not contribute at such low temperatures [24]. A transition between structures can be imagined to require the making and breaking of several hydrogen bonds, which would suggest a high energy of activation. However, the reorientational activation energy is quite low as determined from either dielectric relaxation or isotopic scrambling rates. The low activation energy (~50 kJ/mol) can be understood in

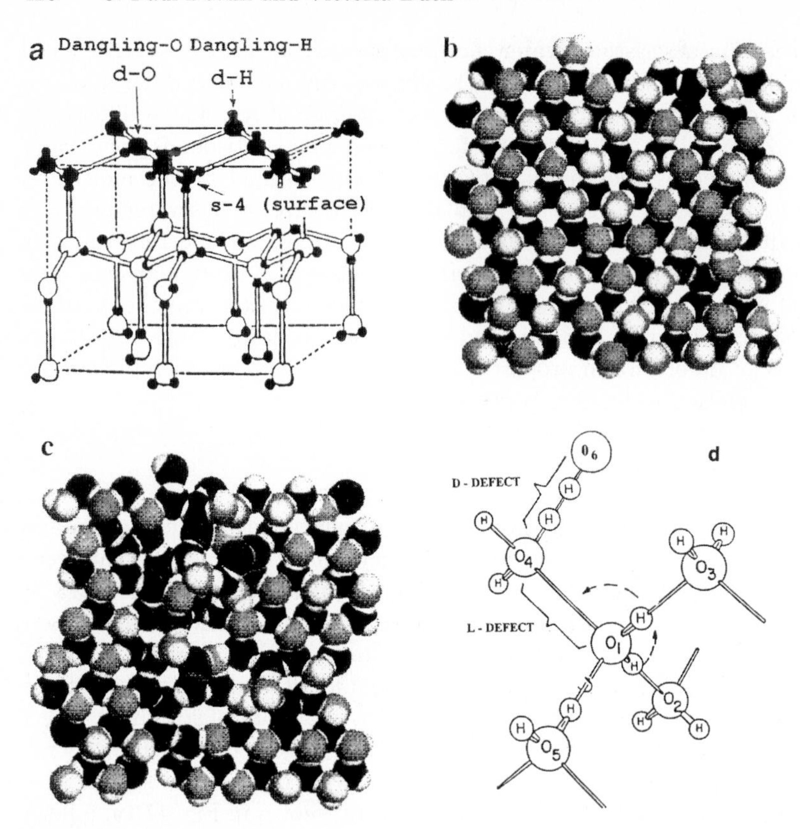

Fig. 17.1. Models of the surface region of crystalline ice (**a**) representation of the surface (filled-in circles) and near surface molecules of crystalline ice. The three types of surface-bilayer molecular sites are labeled d-H, d-O and s-4 (**b**) top view of un-relaxed (111) bilayer of cubic ice with oxygen order and proton disorder: d-H (white), d-O (gray) and s-4 (black) and (**c**) as in (**b**) except model was relaxed by heating to 200 K, running a 104 ps trajectory and re-cooling to 46 K. Note the decrease in molecules with dangling coordination sites. Classical Bjerrum point defects are shown in (**d**).

the context of orientational (Bjerrum) point defects. In pure crystalline ice, the Bjerrum defects are formed in pairs with structures that were originally imagined as either a missing hydrogen (L defect) or an extra hydrogen (D defect) along a near-neighbor oxygen-oxygen axis (Fig. 17.1d) [22,23,25]. Recent computer simulations [27] have given an updated molecular-level view of the structure and dynamics of relaxed orientational defects (Fig. 17.2); essentially, the extra OH-bond in the D-defect points into the neighboring cavity rather than towards another OH and the electrostatic repulsion increases substantially the O...O distance in the L defect. Either defect structure violates the ice rules near a particular lattice site; thus the term "point defect".

An individual defect, once formed, may reorient many water molecules while moving from one site to another, as marked by the arrows in the figure.

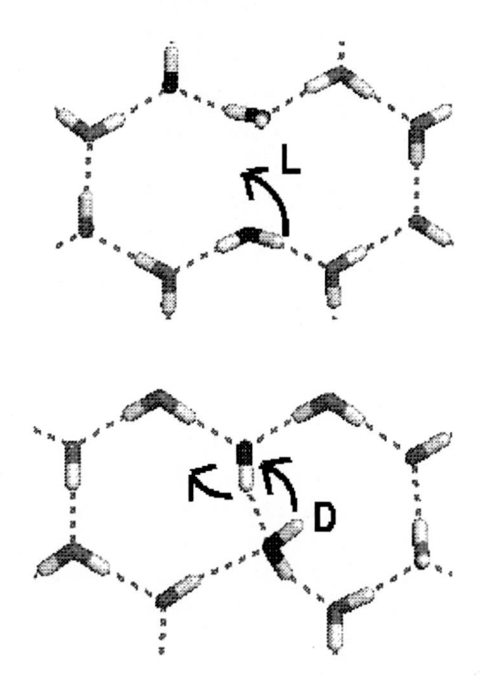

Fig. 17.2. Typical relaxed orientational defect structures, as obtained in molecular dynamics simulations [27]. Molecules within a single ice bi-layer containing a defect are shown. Arrows mark OH bond displacements associated with defect migration.

Just as for the ionic point defects (H_3O^+ and OH^-), an equilibrium population of L and D defects reflects equal rates of formation and recombination. Near the melting point, the equilibrium population of point defects enables a high orientational mobility that allows rapid motion between structures and, thus, rapid dielectric relaxation or isotopic scrambling. However, the formation and mobilization energies of point defects are sufficiently high that both the number densities and mobility decrease rapidly with decreasing temperature. Consequently, orientational relaxation of ice I occurs on an hour time scale at 140 K [28], while the relaxation of *pure* ice to the stable form below 72 K, i.e., ordered ice XI, is not observable over a period of years [23]. Above 150 K other defects, such as molecular vacancies, become active, contributing to the mobility of molecules and transitions between structures of ice [24]. However, bulk ice has a significant vapor pressure above 140 K [29], so our studies of ice particles have generally been restricted to lower temperatures to avoid unrestricted growth of the particle size by Ostwald ripening (i.e., the growth of large particles at the expense of smaller ones by vapor transport). Thus, the protonic and L defects (which are apparently orders of magnitude more mobile than the hydroxide and D defects [22]) are of most interest in current ice-particle research.

The infrared spectra of water ices are often viewed as made up of four different regions: the near infrared region of overtone and combination bands; the 3-micron (3200 cm^{-1}) O-H stretch-mode region; the region, near 6.5 microns (1650 cm^{-1}), of the water bending fundamental; the lattice-mode region, extending through the far infrared but also including the librational modes near 12 microns (800 cm^{-1}). In this chapter, we focus on the stretch and bend fundamentals. Although the stretch modes of gas-phase water absorb quite weakly, the stretch modes of ice have a high infrared intensity [30,31]. In addition to a downshift of frequency of 400 -500 cm^{-1}, these modes experience a combined $\sim$ 25-fold enhancement of the oscillator strength as a consequence of the formation of the strong hydrogen bonds of water ice. As is common for solid-state molecular systems, there is a resultant strong transition dipole-transition dipole intermolecular coupling of the stretch vibrations. Because of this strong intermolecular coupling, bands of the stretch modes range from below 3100 to near 3400 cm^{-1} at 90 K [32-36].

This dipole coupling largely determines the form of the observed infrared (and Raman) bands shown in Fig. 17.3 [35]. Here, as noted in the figure caption, five different aspects of the O-H stretch mode bands of H_2O ice are presented. The five distinctly different experimental spectra are fit, in their general form, using a numerical model that includes both intra and intermolecular coupling. In each instance a fairly complete assignment of the different spectral features was possible using an analogy to a periodic system with a tetrahedral unit cell of four oscillating dipoles surrounding each oxygen atom. Most interesting, within the present discussion of ice particles, was the demonstration of the importance of shape-effect differences between ice film and ice nanocrystal spectra (Fig. 17.3a,b), even for particles too small to exhibit reflective scatter of infrared radiation.

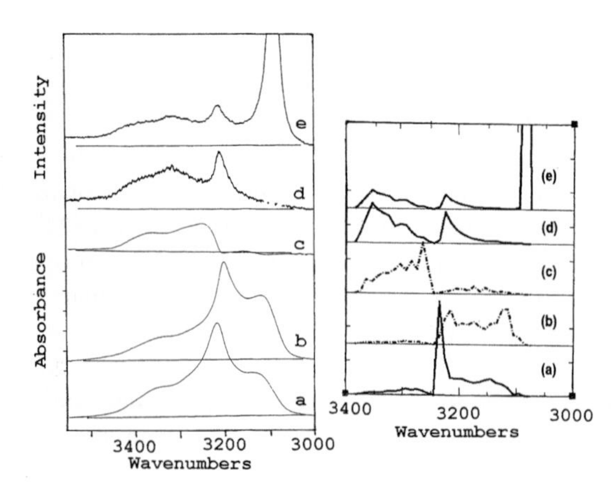

Fig. 17.3. Experimental (left) and calculated (right) O-H stretch spectra of H_2O ice (**a**) infrared absorption spectrum of $\sim$60 nm (ellipsoidal) particles at 38 K (**b**) normal to (100 nm) ice film infrared spectrum at 20 K (**c**) difference infrared spectrum between 15° off normal and normal to the film (**d**) perpendicular Raman scatter of 3 micron film (**e**) parallel Raman scatter of 3 micron film. The spectra of (c) and (d) are enhanced by a factor of $\sim$3.

Ice molecular stretch modes also exhibit an exceptional sensitivity of band positions to temperature of $\sim$ -0.2 cm^{-1}/K, which reflects the increasing strength of the H-bonds with decreasing temperature [37,38]. Recognition of this behavior is critical in studies based on difference spectra, such as those discussed in later sections. The subtracted spectra must be obtained at the same temperature (within $\sim$0.5 K) or difference spectra may contain artifacts arising from the shifting of bands.

By contrast to the intense stretch-mode absorption bands, ice I at low temperature lacks an infrared feature that can be assigned to the molecular bending mode per se [30]. Rather, below 100 K, a very broad weak band near 1550 $cm^{-1}$is the only feature in the 6-micron range. This band is caused by excitation of states that are a thorough mix of the bending mode and the overtone of the librational mode. In essence, there is no bending-mode first excited state for ice I at low temperatures. Ice I at $T < 100$ K represents a unique limiting case of condensed water systems wherein a) an inherently weak bend-mode oscillator strength and b) a maximum overlap of the states of the librational overtone with those of the bending mode result in a complete diffusion of the bending-mode band intensity [39].

17.1.2 Concepts of the Surface of Ice

Traditional discussions of crystalline ice have largely ignored the ice surface, with the exception of the question of the existence of a pseudo-liquid layer at temperatures near the melting point [23]. However, low-temperature molecular-level studies of the ice surface have multiplied rapidly in the past decade. The forms of ice analyzed in these recent investigations have been either thin crystalline films on metallic substrates [8,10,40,41] or nanocrystals having diameters in the range of 3 to $\sim$100 nm [39, 42-46]. As one can imagine from the top surface layer of Fig. 17.1A, termination of ice at a surface results in water molecules lacking the neighbors necessary for maintenance of the ice rules [47]. Cleavage of ice parallel to the hexagonal bilayers, for example, produces the three subsets of surface-bilayer molecules: single-proton-donor water molecules with a dangling O-H or O-D (d-H or d-D molecules), double-donor molecules with a dangling lone-pair of electrons (d-O molecules) and 4-coordinated (saturated) molecules (s-4 molecules).

The d-H and d-O molecules, being 3-coordinated, constitute the $\sim$50 % of the surface sites at which the ice rules are violated (i.e., defects). These are the more energetic/reactive sites of the surface and are of particular interest in studies of adsorbate-ice interactions. They are also central to questions concerning the nature of the stable structure of the thermally equilibrated surface of ice [46]. Considerations of the stable surface structure have focused on two possibilities: a) an oxygen-ordered surface with the ice rules in effect except for the topmost layer of water molecules, as in Fig. 17.1A, often referred to as a full-bilayer-terminated surface, and b) a reconstructed oxygen-disordered surface. Such reconstruction, driven by entropy and an increase in the coordination of some fraction of the 3-coordinated surface water molecules, can strongly reduce the density of dangling groups and, thereby, decrease significantly the ice surface reactivity.

It has been demonstrated that ice films prepared using carefully controlled low-temperature epitaxial growth (140 K), on metal surfaces with lattice parameters that match closely those of ice, have a predominantly full-bilayer-terminated surface. This has been well argued from both LEED data supplemented by simulated structures [10] and helium diffraction [8] measurements. However, computational studies [42,48] suggest that the oxygen-ordered surface (Fig. 17.1b) of free- standing ice is nearly isoenergetic with a disordered structure characterized by a broad distribution of ring sizes (as opposed to hexagonal rings only), and a reduced population of 3-coordinated molecules (see Fig. 17.1c). These computations suggest that entropy will drive the ice surface into oxygen disorder at some modest but not well-circumscribed temperature.

It is likely that a low-temperature epitaxially grown ordered ice surface is not representative of the surface of most ice even at quite low temperatures. The ordered structure appears to be stabilized by the metal substrate and to require special and careful deposition procedures [8,9]. Further, the LEED data did not reveal diffractive features for the outermost layer of the ice surface; a behavior attributed to large mean amplitudes of motion [10], but which may be indicative of static oxygen disorder. Our studies, which will be described in some detail in later sections, indicate that the surfaces of ice nanocrystals are oxygen-disordered [46,49]. Although the surface curvature of such nanoparticles is expected to further destabilize the "crystalline" surface, this disordered form may be more representative of "generic" ice surfaces that are not formed under special conditions that favor surface crystallinity.

17.2 Ice Nanocrystals and Large Water Clusters:

The defining distinction between a molecular cluster and a nanoparticle has not always been clear. For water particles the distinction can be determined by following the lead of El Sayed, et al. [50]. They have noted that a molecular cluster should be viewed as differing from a nanoparticle by the lack of a distinct core. In their particular example, CdSe particles < 2 nm in diameter, so small as to lack a core, do not show near-band-edge emission and are, therefore, molecular clusters. We will show that the crystalline core of an ice particle decreases smoothly with size, disappearing at $\sim$3 nm. By analogy with the CdSe particles, smaller amorphous water particles ($<$2.5 nm) composed only of surface and disordered interior are considered molecular clusters. It is useful to recognize that a transition from ice nanocrystals to large water clusters occurs in the 2-3-nm range. We will refer to particles containing more than 20 water molecules, but too small to possess an ice-like component, as large water clusters. This does not mean that large water clusters, e.g., of 32 molecules, cannot assume a structure of high symmetry [51,52]. Since our experimental methods of study of large water clusters and ice nanocrystals are very similar, the preparation and sizing of all water particles is considered below. However, much of this section will be dedicated to the behavior of ice nanocrystals of 3-60 nm size, with only a final segment on properties of large water clusters.

17.2.1 Preparation and Sizing of Water Particles

There have been two distinct approaches to preparation of cold water particles for studies at the molecular level. The initial electron-diffraction structural studies of Torchet et al. [53] were with particles in the range 2-5 nm produced in a free-jet expansion of water vapor. Bartell et al. have used a similar approach for preparation of particles for structural and kinetic studies [54], as described in Chapter 16. Other laboratories, including ours, have investigated ice particles prepared by rapidly loading inert gas mixtures, containing 0.1 to 1.0 mole % water, to pressures of hundredths or tenths of a bar into the cold inner cell of a double-walled FTIR sample cell [55]. Such cylindrical cells, referred to as collisional-cooling cells, are normally no larger than 6 cm in diameter and 20 cm in length but have origins in a large liquid-nitrogen-cooled cell first described by Ewing and Sheng [56]. Considerable versatility is gained with these and even smaller cells by using closed-cycle helium refrigerators with which the temperature can be set between ~ 20 K and room temperature [42]. An elegant variation on these cells, recently described by Baurerecker et al. [57], provides for multi-pass of the infrared beam and cooling to liquid helium temperatures.

The collisional-cooling cells can be used in two quite different modes. Excellent infrared spectra with high signal-to-noise ratios are obtained for particles with average diameter ranging from 2 to 60 nm by immediately sampling the aerosols as formed at a given choice of temperature, inert gas pressure and mole % of water vapor in the gas mixtures. However, because of particle settling, the aerosol samples normally have a size-dependent effective lifetime of tens of minutes. This seriously limits the variety of particle properties that can be conveniently probed. Limitations from particle lifetimes/optical thickness can be sidestepped, and a great increase in sampling versatility achieved, by working with 3-D arrays of particles [42]. Depending on the cell design, each loading and pumping of the cell can result in the collection of a few percent of the particles on the infrared transparent cell windows. Thus, ice particles assemble into 3-D arrays of thickness determined by the sampling parameters and the number of load-pump cycles. When held at temperatures below 130 K such arrays retain a near constant total number of water molecules and the surface is unchanged with time, or by adsorption /desorption cycles with non-reactive small molecules, over periods of weeks.

As with free-jet expansions, sampling with a collisional-cooling cell is not limited to water particles, but may be used for any molecular substance with adequate vapor pressure, i.e., most conveniently greater than a few tenths mbar at room temperature. Even more refractory substances can be studied when heated sampling ports are used. This does not necessarily imply that surface-localized vibrational modes can be usefully studied, unless the substance has modes exceptionally sensitive to the molecular environment. The spectrum of an array of nanoparticles will typically vary substantially from that of the corresponding thin film, particularly for vibrational modes with large dipole oscillator strength. However, these differences are usually related to dielectric constant and particle size/shape effects [35,58] rather than surface-localized molecular vibrations that are highlighted by the unusual surface-to-volume sample ratio. In fact, surface-localized modes of molecular nanoparticles have been reported only for strongly H-bonded substances for which variations of band frequencies with the extent of H-bonding are notoriously large. Ice is unquestionably unique in the richness of surface information accessible through coordinated spectroscopic/simulation studies and has therefore been the model system for the study of surface properties of molecular solids.

Whether for expansion beams, aerosols or arrays, the average particle size is an important parameter that can be difficult to determine. However, spectroscopic/simulation and diffraction measurements provide several complementary approaches to size determination of ice particles. Perhaps the most direct size measurements have been by transmission electron microscopy (TEM) of ice particles prepared by pulsing water-inert gas mixtures through a small collisional-cooling cell [59]. The resulting roughly spherical particles, prepared using $\sim$1% water in inert gas mixtures, are typically single crystals of cubic ice. They were observed to vary in average diameter between 15 and 100 nm, depending on cell temperature (120-150 K), loading pressure and water vapor content. The size of ice particles in expansion beams can be estimated from the breadth of the diffraction peaks [54], as described in Chapter 16.

The size of ice nanocrystals prepared in our cells have been determined by two spectroscopic methods, although a third method, using a combination of spectroscopic and simulation results, becomes more reliable for the smallest particles ($<$ 4 nm; section 2.3). The integrated intensity of the surface d-H-molecule stretch-mode band, relative to the band intensity of the interior OH-stretch vibrations, was the initial probe of ice particle size [60]. This probe is convenient but has a high uncertainty, as a conjecture must be made about the density of O-H groups on the surface and the relative oscillator strength of the surface mode vs that of the interior OH bonds. Using the gas-phase H_2O oscillator strength for the d-H mode, it was estimated that the particle size, from a 0.5% water-N_2(g) mixture at 80 K, was $\sim$25 nm [60], in reasonable agreement with the later TEM measurements [59]. Similarly, the size of particles prepared under varying conditions, or subjected to Ostwald ripening at various temperatures, can be readily estimated.

A better estimate of ice particle size can be obtained from the infrared band intensity of a monolayer of an appropriate adsorbate. The substance CF_4 is an excellent choice for several reasons. It equilibrates with a low but measurable vapor pressure (0.1 Torr) at a convenient temperature (80 K) to give an adsorbed monolayer. Further, the monolayer is readily recognized through an unusual band pattern of the asymmetric stretch mode. This triply degenerate vibration is characterized by a very large direction-independent dipole derivative that strongly couples neighbor oscillators. This produces collective in-phase modes that dominate the infrared spectrum. Computations have confirmed that a CF_4 *monolayer* on an ice particle gives a unique $\sim$77 cm^{-1} splitting of the resulting transverse (parallel)-longitudinal (perpendicular) doublet [49]. Because of the unusually large oscillator strength, the integrated intensity is insensitive to the phase of CF_4 [61], and the band intensities for the adsorbed and gas phase are easily observed. As a result, the measurement of the band intensity of monolayer adsorbed CF_4 permits a reliable estimate of the total surface area of a sample of particulate ice. From the ice band intensity the total amount of ice is known; so, knowing both the volume of ice and the surface area, the average particle size can be deduced. This application of adsorbed CF_4 is demonstrated in Fig. 17.4 where spectra for particles ranging from $\sim$2 to 20 nm are compared, with the intensity normalized to a fixed value for the adsorbed CF_4 [39]. This corresponds to a fixed surface area, so, through the implied total ice volume, the integrated intensities of the ice bands become a direct measure of the (relative) average sizes of the particles for different samples. The measurement loses accuracy with smaller particles for which surface water molecules, having reduced oscillator strengths of unknown magnitude, become a significant fraction of the total.

17.2.2 Surface and Subsurface Structure/Spectra of Ice Nanocrystals

From the earliest spectroscopic studies of nanocrystals of ice, it has been clear that the position of O-H stretch surface-localized modes can be revealed by weak adsorbates such as CF_4, H_2, Ar, N_2,O_2, CH_4, NO, CO, O_3 and ethylene (listed in order of increasing shift of the ice-surface d-H band). Such adsorbates affect only the ice surface, typically shifting the surface modes 10 –70 cm^{-1}. As a result, a difference spectrum, from subtraction of the adsorbate-coated ice spectrum from the spectrum of bare ice, show several well-defined maxima for the bare-ice surface-mode positions, plus minima at band positions assumed following shifting by the weak adsorbate [42,43]. Arguments have been made that the magnitudes of analogous shifts of the d-H mode of the pore surface of microporous amorphous ice contain information

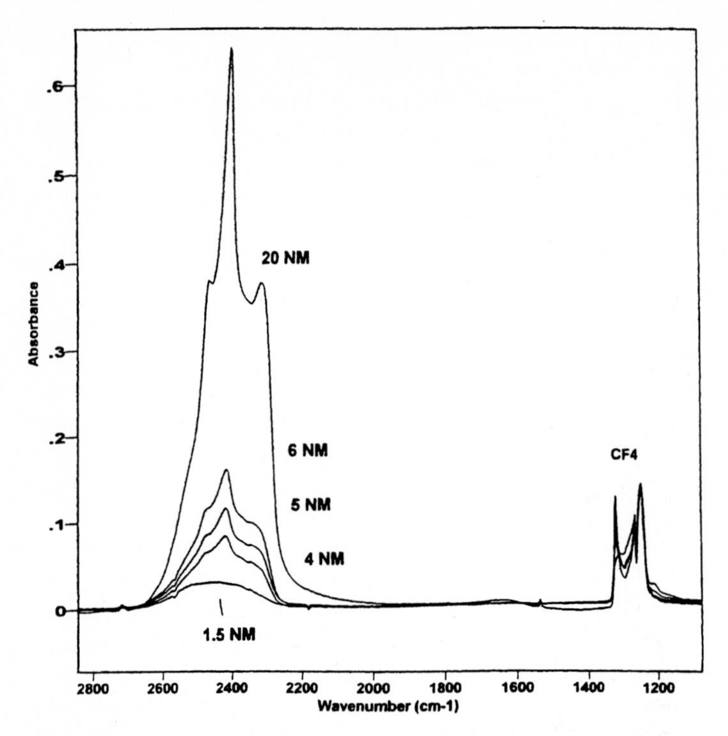

Fig. 17.4. Infrared spectra of aerosols of D_2O nanoparticles, of the indicated average diameter, coated with a near monolayer of CF_4. The cluster surface areas are normalized to the same value through the area of the CF_4 asymmetric stretch-mode bands, making the area of the O-D stretch-mode absorption proportional to the particle size.

about polarizabilty and ionization potentials of adsorbate molecules ([62] and Chapter 13).

The difference spectra from weak adsorbates are informative, but give little indication of the complete form of the spectrum of the ice surface. However, methods have been developed that reveal the surface spectrum for larger (>20 nm) ice nanocrystals and a 2nd approach is particularly useful with smaller ice particles. Both approaches are based on the recognition that ice particles have three structural parts as has been revealed most clearly through molecular-dynamic (MD) simulations. Computations for thin ice slabs, based on standard water-water intermolecular potentials, have established that an initially oxygen-ordered ice surface relaxes to a disordered form having a reduced number of 3-coordinated surface water molecules and a range of ring structures [43,44,48]. The relaxation is driven in part by the formation of additional H-bonds between 3-coordinated surface molecules. Though the original computations were used to check the presence of a liquid-like surface layer at T>200 K [48], surface disorder is indicated at significantly lower temperatures [43]. This disorder is also reflected in a bilayer (or two) of subsurface ice that joins the surface to the ice interior. It follows that ice nanocrystals can be viewed as a crystalline core surrounded by a distorted subsurface that connects the core to the disordered surface. This view has been further extended in MD simulations of a 4 nm ice particle [46], as will be described in Section 17.2.3. (In fact, all ice particles, whether with an ordered or disordered surface, contain 3 parts. Even a full bilayer-terminated surface, being different than the ice interior, places at least one subsurface bilayer under significant strain.)

The existence of a bilayer (or more) of subsurface complicates the determination of the surface spectrum, and, if ignored, would fully invalidate any so-called spectrum. Experimentally, the first indication of a significant subsurface region was from difference spectra associated with Ostwald ripening. In the absence of a subsurface, ripening should have the net effect of converting surface ice to crystalline core ice as the average particle size increases. Then, the difference between spectra taken before and after ripening would reflect only the loss of surface and the gain of crystalline ice. Since the crystalline spectrum is known (e.g., for a sample of especially large particles) it could be added out to give a unique surface spectrum. However, such "surface" spectra are markedly different for different temperatures or periods of ripening, which suggests the presence of a significant subsurface structure [45].

The response of the ice spectrum, to the presence of adsorbates that bond with 3-coordinated water molecules of the ice surface, also gives striking evidence for the presence of a significant ice *subsurface* region. This response was particularly striking for the Lewis-acid SO_2 [45], as demonstrated in Fig. 17.5 for D_2O nanocrystals. Like other "intermediate" adsorbates, such as H_2S, HCN and acetylene, adsorbed SO_2 has a qualitatively different and much greater effect on the ice particle spectrum than that described earlier for weak adsorbates. The bottom spectrum of Fig. 17.5 shows that during the initial uptake of SO_2 at 120 K three bands that match the peak positions of crystalline ice, emerge in the difference spectrum vs that of the original bare ice. During a second phase, shown in (b), the peaks of the core ice continue to grow, but, even though the d-H sites of the ice surface were saturated with SO_2 during the first phase, a sizeable band appears at the position of the SO_2-shifted d-H band (2670 cm^{-1} for d-D of D_2O ice). Then, in a third phase, the core peaks continue to strengthen but the d-D region is nearly flat. This sequence of changes, which was observed for many different SO_2-ice samples, can be understood in the context of the results of the computer simulations.

Most critical is the emergence of the extra intensity in the SO_2-shifted d-D band during the second phase (Fig. 17.5b). With the surface already saturated there must be a new source of d-D sites on the ice surface to bind with the SO_2. Their origin can be anticipated by recalling that the surface

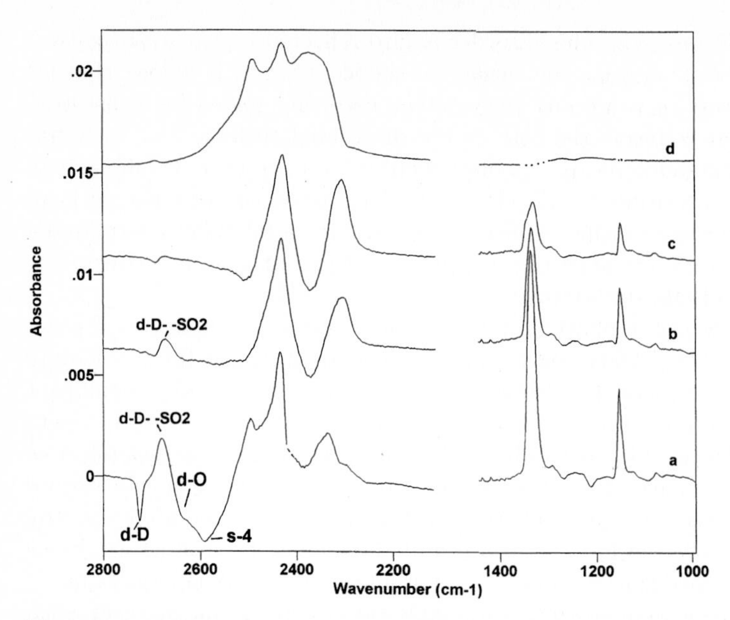

Fig. 17.5. FTIR difference spectra, showing the sequential phases of saturation, SO_2 insertion and ice relaxation at 120 K (a) SO_2 saturation of the nano-particle surface (b) insertion of SO_2 into the strained H-bonds of the surface (c) subsurface to interior ice relaxation and (d) the ice subsurface spectrum obtained from (c) by removal of the interior ice component generated in the relaxation of the subsurface. The variation of the stretch-mode bands of adsorbed SO_2 for each stage is on the right.

reconstruction, as revealed in the computer simulations, is accompanied by a pronounced decrease in 3-coordinated surface molecules through bonding of neighbor d-H and d-O groups. Further, the simulations show that the new bonds are weak and distorted [43,44,46]. It follows that a good adsorbate, such as SO_2, can cleave the strained bonds to enable adsorbate bonding with the 3-coordinated water molecules that are generated. Extensive insertion of NH_3 into weak strained bonds of the ice surface has been demonstrated in a recent simulation [63]. Adsorbate insertion transforms the ice surface structure towards that of ice, and therefore reduces the strain in the subsurface region. From this view, the third stage in the SO_2 interaction with the ice surface, in which there is no significant change in the surface spectrum (Fig. 17.5c), corresponds to the relaxation of the subsurface towards interior ice.

Thus subsurface relaxation, which also accompanies the insertion that occurs during the 2nd to 4th hour of the SO_2-ice interaction at 120 K, is the sole source of the spectroscopic change observed during the 4th to the 6th hour. Two valuable insights follow from this result: a) subsurface structural relaxation in this temperature range (120 K) is on an hour time scale and b) a method is presented for obtaining the subsurface spectrum. When the subsurface relaxes, it produces crystalline interior ice. The difference spectrum, comparing before and after that relaxation, reflects this conversion; so removal of the interior ice spectrum from the difference spectrum of Fig. 17.5c gives the subsurface spectrum of Fig. 17.5d.

With the subsurface spectrum known from the relaxation induced by a good adsorbate, the difference spectra, from Ostwald ripening of the ice nanocrystals, can be fully analyzed, the form of the subsurface spectrum further tested and the *surface* spectrum determined. It is clear that ripening must ultimately convert both surface and subsurface molecules to crystalline core molecules since the amount of surface and subsurface both decrease with increasing particle size of an array. However, as noted above, the relaxation of the subsurface to a steady state value is much slower than that of the more mobile surface. A brief ripening period, at some elevated temperature, will thus give primarily surface-to-interior conversion, whereas longer periods of ripening at the same temperature will give relatively more subsurface loss [45]. This can be seen in Fig. 17.6 where the difference spectra for two greatly different periods of ripening at 138 K are shown in curves (a) and (c) determined at 120 K. Adding out the increase in interior crystalline ice gives curves (b) and (d) of which (d), from the long ripening period, resembles that of the subsurface of Fig. 17.5d, but clearly contains a small surface component as well. Since the two spectra contain different portions of the surface and subsurface components, subtracting one from the other reveals each spectrum in turn. If (b) is subtracted from (d), using the factor required to null the 2726 cm^{-1} d-D band of the surface, the subsurface spectrum, which matches well the subsurface spectrum from adsorbate-induced relaxation (Fig. 17.5d), appears. If this spectrum is then subtracted from either (b) or (d), or, if the right amount of (d) is subtracted from (b) the surface spectrum is obtained. The resulting surface and subsurface spectra are included at the bottom of Fig. 17.6.

The surface spectrum can also be revealed in a more straightforward approach that works particularly well for small nanocrystals for which the core spectrum is not so dominant. With both the subsurface and core ice spectra known, subtraction of the "correct" amounts from a single spectrum of ice nanoparticles leaves only the surface spectrum. Here, as in the other procedure, there is subjectivity in recognizing the "correct" subtraction factors. This particular procedure, applied to 4-nm water particles, is examined in the next section where it is shown that the subjective factor is not serious.

17.2.3 A Closer Look at Ice Nanocrystals – the 4 nm Particle

The above spectroscopic/simulation study of ice nanocrystals focused on particles in the 20 nm or larger size range, with a particular goal of establishing the nature of the ice surface for systems that are likely to relate well to bulk ice. That the results may be applicable to bulk ice can be argued from their

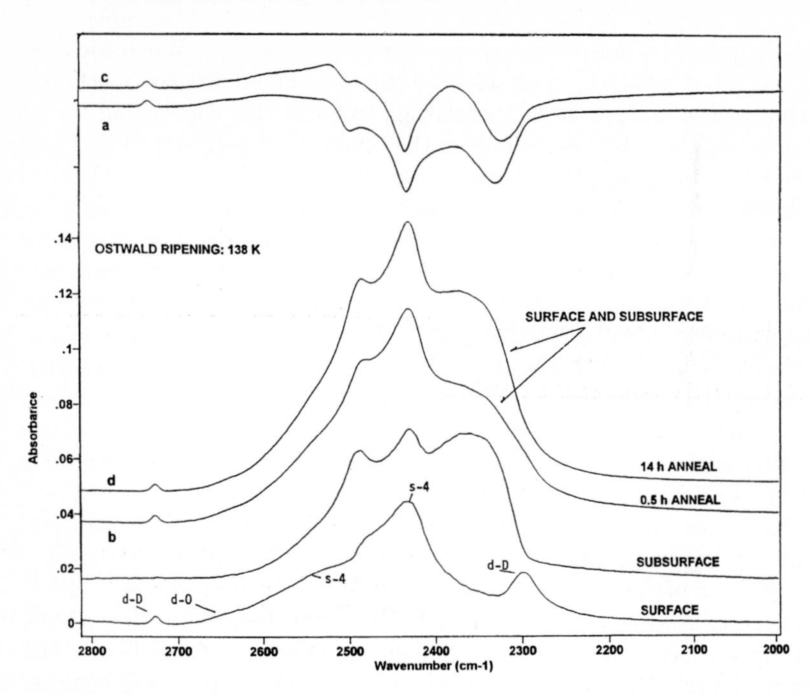

Fig. 17.6. FTIR difference spectra for sequential annealing periods for ice nanocrystals at 138 K (a) and (c) compare before and after annealing for 0.5 and 14.0 h, respectively, (b) and (d) are the same two spectra with the interior ice absorbance removed. The surface and subsurface spectra are from appropriate differences between (b) and (d) as described in the text. Spectra (c) and (d) have been expanded by 2X, relative to (a) and (b).

apparent consistency for particles over a considerable size range (20 – 60 nm diameters or 10^5 to 4 x 10^6 molecules). Further, several aspects of the infrared spectroscopic results were successfully modeled by "infinite" slabs of ice with flat surfaces. Following that study, attention was directed to the manner in which ice particles change with diminishing size. The 4 nm crystalline ice particle was chosen for particular study since a) initial spectroscopic results (as well as the earlier electron diffraction measurements [53]) indicated that such particles are sufficiently large to have a crystalline core and b) a 4 nm particle is sufficiently small to permit thorough modeling by MD/MC methods using empirical potential functions. The structure of a relaxed 4-nm particle can be judged from the space-filling and stick models of Fig. 17.7.

Though 4 nm particles were chosen for the direct comparison of experimental and computational results, the experimental study included the spectra of arrays of nanocrystals with average diameters ranging from 16 nm to 3 nm. The effect of the decreasing particle size on the stretch-mode infrared spectrum is apparent from Fig. 17.8 as determined for ice aerosols

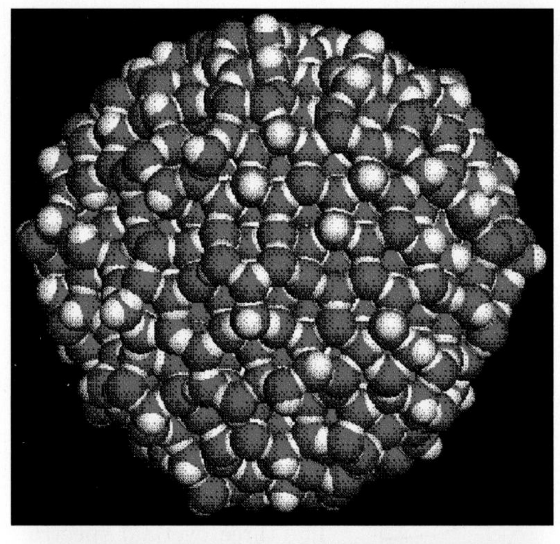

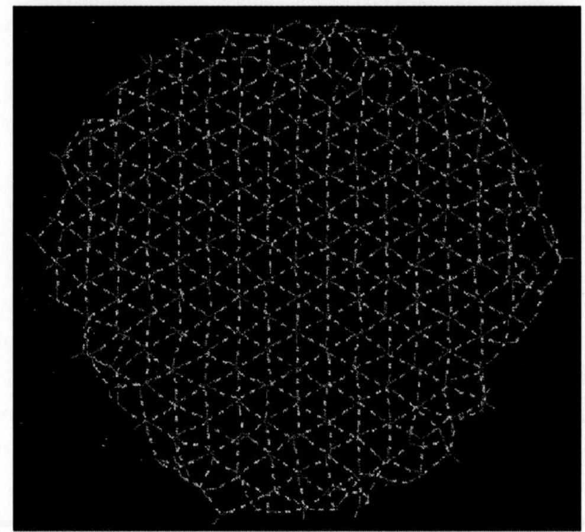

Fig. 17.7. Structure of the 4 nm ice nanocrystal as displayed using a space-filling model (top) and a stick model of a 1.1 nm middle slice of the particle that demonstrates the largely ordered nature of the interior and subsurface.

formed at 100 K from 0.1% mixtures of H_2O in He(g). For both the H_2O [39] and the D_2O [46] series it was established that, for particles < 10 nm and > 3 nm in diameter, subtraction of the core spectrum generated an invariant residual spectrum (next to bottom of Fig. 17.8). From the known changes of larger nanocrystals induced by ripening and adsorbate effects (see above), and consistent with the computational results for a 4 nm particle, this residual spectrum was recognized as the composite spectrum of the surface and subsurface components of the particles. Thus, subtraction of the known subsurface spectrum (Fig. 17.6) from the combined surface and subsurface spectrum gave the (H_2O and D_2O) ice surface spectrum. Since, together with the known subsurface and interior ice spectra, the surface spectra were found to fit experimental spectra with minimal residual errors for particles ranging

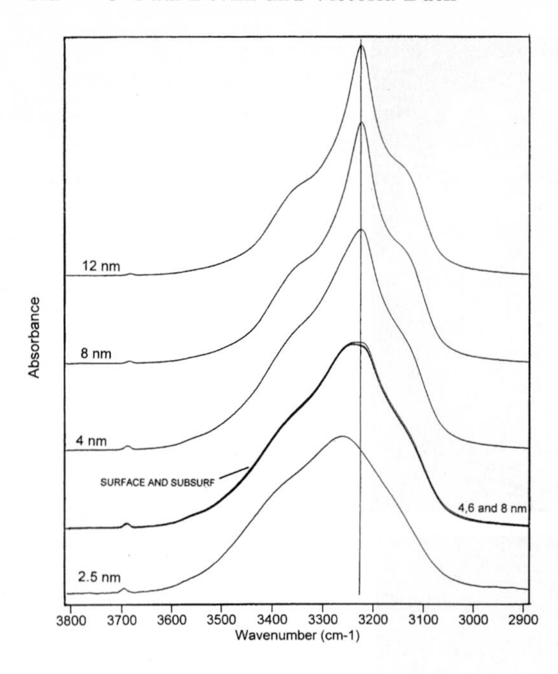

Fig. 17.8. FTIR spectra of the O-H stretch mode region of H_2O particles in the 12 to 2.5 nm range as aerosols in He(g) at 100 K. The three overlaid spectra are of the combined surface and subsurface after removal of the core component of the 4, 6 and 8 nm particles.

in size from 3 to 12 nm, subjective aspects of the analysis are apparently minor.

The experimental spectra of the surface, subsurface and interior of 4 nm H_2O nanocrystals are compared in Fig. 17.9 with the simulated spectra of the 4 nm particle of Fig. 17.7 [46]. The simulation employed a spherical particle, "cut" from a crystalline (but proton-disordered) cubic ice structure. The particle was subjected to simulated annealing, which resulted in surface reconstruction. The reconstruction was associated with substantial energy lowering, concurrent with an increase in the coordination of surface molecules. The spectrum was calculated using an exciton model for OH bond vibrations, which included intra-molecular and inter-molecular dipole-dipole coupling. The bond frequency was calibrated as a function of the electric field at the H-atom of OH, using a scheme developed in a study of $n = 8$ - 10 cage water clusters [21].

The relaxed cluster structure is displayed in Fig. 17.7b using a stick model of a 1.1 nm middle slice of the cluster. Upon relaxation, the originally "oxygen-ordered" near spherical particle of 979 water molecules, retained a crystalline core, but developed a more distorted subsurface region and a strongly disordered surface while experiencing an $\sim$50% reduction in 3-coordinated surface molecules. The simulated spectra of each of the three components of the ice nanocrystal (Fig. 17.9b) matched the experimental component spectra in a semi-quantitative manner in terms of band positions, intensity and width. This permitted the credible assignment of the various sub-bands of the surface spectrum, indicated by labels in Fig. 17.9a, and confirmed the general concept of the structure of an ice nanoparticle, i.e., an ordered core, a distorted subsurface and a disordered-surface. The results also advanced the identification of particle sizes through the use of the relative intensities of the surface, subsurface and core components of the 4-nm case as a reference state.

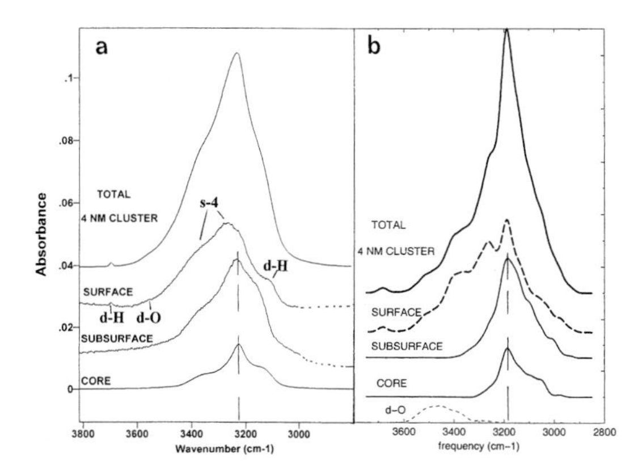

Fig. 17.9. The experimental (a) and simulated (b) absorbance spectra, to scale, of the three components, surface, subsurface and core of 4 nm nanocrystals. Labels, defined in the text, identify sub-bands of the surface spectrum. The dashed curve, labeled d-O is the contribution of d-O (or 3-coordinated double-donor water molecules) to the computed surface spectrum.

The complexity of the various spectra of H_2O (or D_2O) ice particles is enhanced by the strong dipole coupling of the molecular stretch modes common to the condensed phases of water [32-35]. This diminishes the value of the position of the sub-bands of the component spectra as a measure of the strength of individual H-bonds. This problem was overcome by a computational approach in which a 4 nm particle was composed of only HDO molecules with no coupling between bonds. The calculational results, using the scheme described above, were then compared with the "decoupled" experimental spectra for ~18% HDO molecules isolated in H_2O nanocrystals [46]. As for the H_2O particles in Fig. 17.9, the computed spectra matched the experimental spectra of 4 nm particles in a semi-quantitative manner, giving a credible assignment for the features of the surface spectrum while strengthening the tie between band positions and specific H-bond strengths.

In this respect, the nature of the ice *subsurface*, as judged from the peak positions, bandwidths and distribution of O···O···O angles and H-bond lengths (Fig. 17.9 and the simulated structure), is worth noting. Most important is the central peak position of the subsurface spectrum. Whether for H_2O or 18% HDO particles, the subsurface peak is coincident with that of the crystalline core ice in both the experimental and simulated spectra. This strongly suggests that the average *subsurface* H-bond is very similar to that of the core ice; unlike amorphous solid water for which the central peak is shifted some 50 cm^{-1} to higher frequency [37]. Further, although the computed standard deviations were at least twice those of the core ice, the values of angles and bond lengths were distributed around the same maxima. It follows that the subsurface structure is similar to that of the crystalline core and, like the core, must be nucleated. (This point gains significance from the observation that ~3 nm particles have a "crystalline" subsurface spectrum but no detectable core ice.) The subsurface appears to be "nearly crystalline ice", distorted by interaction with the disordered surface.

17.2.4 Surface-Bending Modes of Ice Nanocrystals

The spectra of the bending modes of the water molecules on the surface of ice particles have also been investigated by combined FTIR and simulation techniques [39,66]. Because the intensity of the molecular bending mode of the core and subsurface ice is very low, the bands for the bending modes of the three types of surface water molecules (d-H, d-O and s-4) stand out more clearly than those of the corresponding stretch modes. In fact, the argument is made that the bending modes of water molecules in the tetrahedral environments of the core and subsurface are vanishingly weak because of a combination of an inherently weak oscillator strength for the symmetric environments, and a mixing of the modes with the overtone of the librational modes [39]. Consequently, a relatively sharp structured feature in the 1600 cm^{-1} range of the particle spectra has been attributed to the bending modes of surface molecules superimposed on the very broad librational overtone band of the interior molecules. The broad band can be removed, using the spectrum of large nanoparticles (60 – 80 nm) or thick ice films, leaving just the surface bending mode bands (Fig. 17.10).

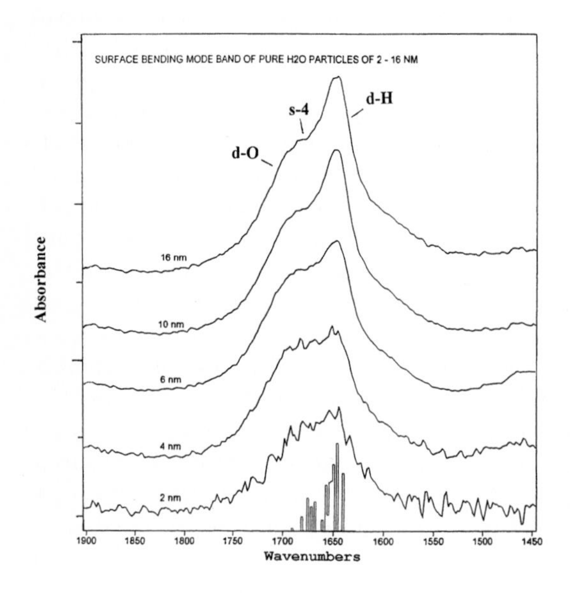

Fig. 17.10. Infrared spectra of the surface bending mode region of ice particles in the 2 to 16 nm range from difference spectra vs a bulk ice spectrum. The histogram at the bottom of the figure represents a distribution of mode oscillator strength vs frequency as calculated for surface and interior ice configurations.

Initially it was presumed that the components for each category of surface molecule in the bands of Fig. 17.10 would have frequencies that increase in the sequence d-H, d-O and s-4, corresponding to increased bonding of the surface molecules [66]. However, *ab initio* computations computations based on reduced-dimensionality models of the 3 categories of surface molecules indicate that a) the s-4 frequency falls below that of the d-O but above that of the d-H molecules and b) the s-4 intensity contribution is disproportionately low because of its greater tetrahedrality [39]. A general conclusion of the study was that, contradictory to earlier reports [31], there is a strong but irregular dependence of the bend-mode frequency and band intensity on the strength and tetrahedrality of H-bonded water structures.

17.2.5 Relaxation Rates Near the Ice Surface

There are reasons to anticipate that, for the same temperature, the structural relaxation rate within the ice particle subsurface region is greater than for bulk ice. Though the average bond is not significantly different for the subsurface structure [46], the greater distribution of bond lengths and angles makes certain that proportionately more "weak" bonds exist in the subsurface region. Also, the subsurface borders a surface that is rich with 3-coordinated sites each of which can be viewed as a point defect [67], i.e., a potential source of traveling defects capable of enhancing the subsurface mobility. For these reasons, it is not surprising that the data of Fig. 17.5 are indicative of subsurface structural relaxation times on a scale of an hour at 125 K, while similar rates for "bulk" crystalline ice are not achieved below 140 K. For example, isotopic exchange rates, dependent on mobile orientational defects, indicate that defects visit half of the lattice sites of a thick ice film on a time scale of an hour at 140 K [28] (suggesting a similar time scale for orientational relaxation).

We have reported two studies directed to the question of water molecule mobility near the surface of ice nanocrystals. Orientational mobility within the interior of small nanocrystals containing D_2O molecules isolated intact within an H_2O lattice has been probed via isotopic exchange rates [68], while the rotational mobility of HDO molecules, at the single-donor 3-coordinated (d-H) sites on the surface of H_2O nanocrystals, has been determined by following the temperature dependent relaxation to preferential deuterium-bonded configurations [69]. The latter observations were based on the lower energy ($\sim$52 cm^{-1} or 0.62 kJ/mole; from zero-point energy effects) of 3-coordinated surface HDO molecules configured with a d-H rather than a d-D. This energy difference is sufficient to give a greater d-H than d-D population at 120 K. The d-H population increases further with cooling, but relaxation ceases, on a multiple hour time scale, at $\sim$60 K, i.e., simple rotational motion of the HDO molecule at a d-H site on the nanocrystal surface onsets near 60 K.

An ability to isolate D_2O molecules intact at lattice sites within H_2O nanocrystals allows monitoring of orientational mobility within the particles. Such aerosol samples can be prepared by releasing dilute D_2O in He(g) simultaneously with more concentrated H_2O in He(g) near the same point in a collisional cooling cell held below 140 K [68]. Surprisingly, the water forms nanodroplets that quickly crystallize to nanoparticles without extensive isotopic exchange. However, the isolated D_2O eventually converts to HDO through a sequence of proton hopping and molecular turn steps, provided both mobile protons and mobile orientational defects are present [28]. Isolated D_2O is detected from a symmetric-antisymmetric OD stretch doublet at 2367 - 2445 cm^{-1}. Proton hopping through the D_2O site is associated with formation of an $(HDO)_2$ pair that is easily differentiated spectroscopically from D_2O by two OD bands at 2395-2440 cm^{-1}, split by intermolecular coupling. The latter coupling is quite strong, since the two OD bonds in $(HDO)_2$ are connected in tandem OD..OD. Further passage of an orientational defect results in two much more weakly coupled OD bonds separated by a hydrogen bond; the observable result is the collapse of the doublet to a single HDO band.

The spectra of Fig. 17.11 show stages in the isotopic scrambling within acid-doped nanocrystals. Since the difference spectra on the right indicate that isolated HDO is a major product, it follows that orientational defects are active at 125 K. Similar observations over a range of temperatures indicate that orientational mobility onsets at ~115 K with an activity corresponding to a half-life for defect passage through the lattice sites of less than an hour.

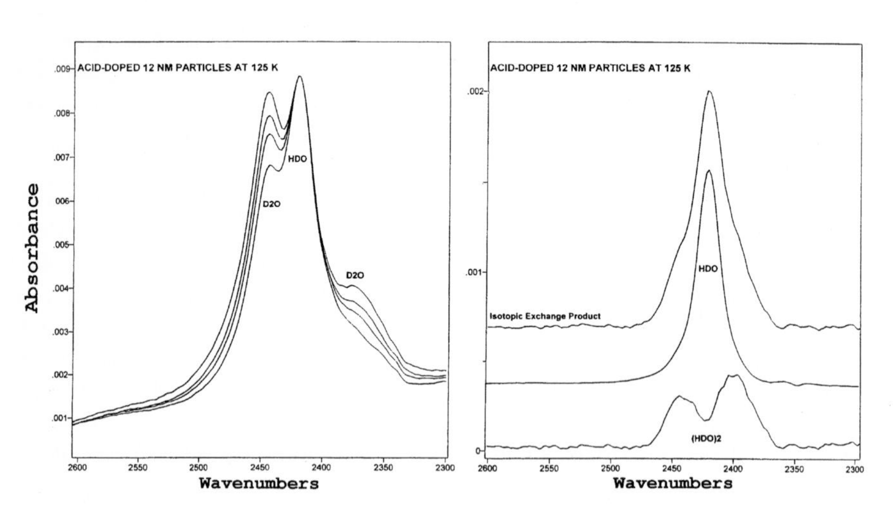

Fig. 17.11. Time variation of the O-D stretch spectra of isotopomers isolated in lightly HCl-doped 12 nm ice particles at 125 K (**a**) variation of the composite spectra for times ranging up to 22 min (**b**) isotopomer component bands of the isotopic-exchange product from a difference spectrum vs zero time; to scale, after 20 min at 125 K.

This represents an enhancement of more than an order of magnitude in the orientational relaxation rates for nanocrystals relative to bulk ice, and supports the view that the ice surface region is a source of orientational defects.

17.2.6 Non-crystalline Large Water Clusters

As noted earlier, 3 nm ($\approx$ 400 molecule)ice particles are "strained crystalline" but without a significant "perfectly crystalline" core. Below 3 nm the particle spectra quickly assume an appearance resembling that of amorphous ice. This quick change in structure to a fully disordered large cluster is apparent from the bottom spectrum of Fig. 17.8 for which the peak of the OH stretch mode band is blue shifted $\sim$50 cm^{-1} from that of the larger particles. This blue shift continues with further reduction in particle size, reaching a peak position of $\sim$3300 cm^{-1} for an average cluster size of $\sim$2.2 nm (150 molecules; Fig. 17.12). Measurements of spectra for still smaller water clusters (<100 molecules or <1.8 nm) are becoming feasible through the recent development of a liquid-helium-cooled multi-pass cluster cell [57] and from a sodium-ion-labeling large-cluster technique developed by Buck [70], and described in Chapter 3.

Simulated spectra (together with the new data) suggest that the blue shift of the band maximum, ultimately to a value near 3350 cm^{-1} for the smallest large clusters (<50 molecules), is the result of depletion of molecules within the 4-coordinated amorphous core and the emerging dominance of a

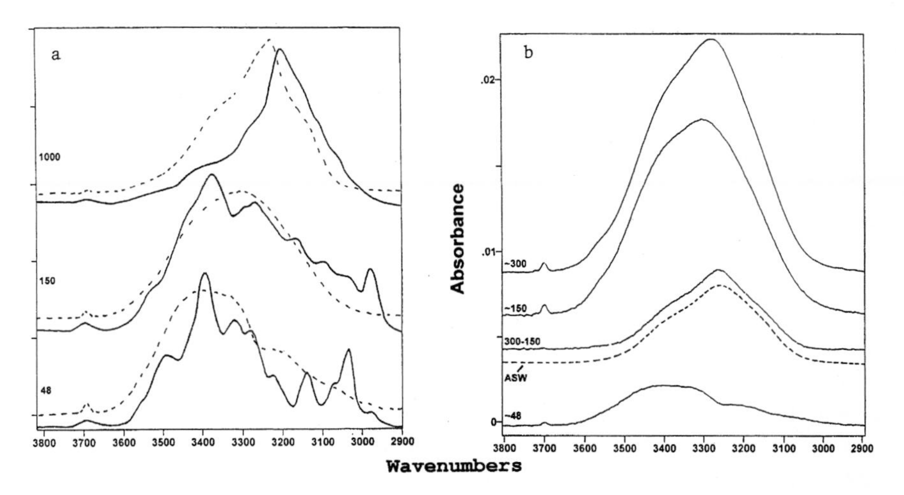

Fig. 17.12. Infrared spectra of ice particles in the range 1.5 - 4 nm (**a**) comparison of simulated and experimental (dashed) spectra of bare ice particles ranging from 1000 to 48 water molecules (4 to 1.5 nm) and (**b**) the difference spectrum for particles of 300 vs 150 water molecules compared with the spectrum (dashed) of ASW. The approximated 48-molecule spectrum is from removal of the amorphous-core component from the 150-molecule-particle spectrum.

surface spectrum quite different from that of the crystalline ice particles (for which the band maximum occurs at 3260 cm^{-1} (Fig. 17.9)). This continued evolution of the spectra with decreasing cluster size is apparent from the simulated spectra for 1000, 147 and 48 molecule spheroidal particles given in Fig. 17.12a. Ignoring fine structure, (a result of the use of a single model rather than a distribution of sizes and structures) the computed band shifts from near 3200 to about 3300 and finally to ~3350 cm^{-1}. The main loss of intensity covers a frequency range identified in the simulations as dominated by absorption of interior 4-coordinated molecules. The subtraction of the experimental spectrum of ~150 molecule (2.2 nm) particles from that of an array of clusters of ~300 molecules each (2.8 nm) confirms this view. If the 3700 cm^{-1} band of the d-H surface-molecules is nulled, the residual spectrum of the 300 molecule array is that of amorphous solid water (ASW) of a similar temperature (Fig. 17.12b). Thus, both the simulations and the experimental spectra identify the core of the large water clusters as amorphous ice.

The amorphous-core spectrum was then used to eliminate the absorption intensity of the amorphous-like interior from the experimental ~150 molecule spectrum. This yields the difference spectrum at the bottom of Fig. 17.12b that matches well the computed 48-molecule cluster spectrum (as well as spectra for clusters smaller than 50 molecules observed a) using the multi-pass cell [71] and b) the sodium-ion-labeling method of Chapter 3). By contrast, dramatic differences were revealed between a computed spectrum of a 48-molecule *sandwich* structure and the observed spectrum; an indication that large water clusters, within collisional cooling cells, adopt spheroidal shapes.

17.3 Adsorbates on the Surface of Ice Nanocrystals

Adsorbates on the ice surface can be categorized operationally as weak, intermediate or strong, based on the effect of the adsorbate on the ice surface, subsurface and interior [44,72]. A list of weak adsorbates, that influence the infrared spectrum of the ice surface only, was given in Section 2.2. A similar list of intermediate adsorbates, capable of cleaving weak bonds on the ice surface, which initiates reordering of the surface and subsurface regions of the nanocrystals, was also given in that section. Strong adsorbates can similarly cleave the strained surface bonds and initiate structural changes in the surface and subsurface of the nanocrystals. However, such substances are better characterized by an ability to cleave normal H-bonds of ice and penetrate the ice interior while converting ice nanoparticles to particles of a hydrate, even at temperatures below 120 K [72,73]. Here we examine the behavior of six interesting adsorbates; CO, ozone, acetylene, ammonia, dimethyl ether and HCl (the latter three being examples of strong adsorbates), considering their impact on ice particles as well as the effect of ice on the spectra of the adsorbed molecules.

It is axiomatic that there can be no realistic hope of understanding the interaction of adsorbates with the ice surface without a viable concept of the ice surface structure in terms of the strength, number and type of H-bonds formed by the bare-ice surface water molecules, and the density of unsaturated coordination sites available for interaction with an adsorbate. Much of the previous section was directed to an analysis of the current understanding of that structure for ice nanocrystals as derived from combined spectroscopic and simulation results. This analysis comes into play as we examine the behavior of adsorbates on the ice surface. A basis is provided for presumptions about the types of sites available for interaction with a particular adsorbate and the vulnerability of the surface to adsorbate-induced reconstruction. Equally important, the observed interplay of an adsorbate with the surface can serve to test the validity of these presumptions, and thus our knowledge of the surface.

There is, of course, a second aspect to the ice-adsorbate interaction; namely the characteristics of the adsorbate that determine a preference for and strength of interaction with different surface sites. At one level, these intermolecular interactions reduce to those between water and the adsorbate molecule in a hetero-dimer. Hetero-dimers have been the subject of numerous experimental and computational studies that provide insight to ice-adsorbate structures. Such guidance, particularly from matrix isolation, microwave and ab initio studies, is invaluable; but it is well to bear in mind that H-bond chemistry of the water molecule may be radically altered by cooperative effects within groups of water molecules, e.g., the polarity of a d-H bond of the ice surface is apparently much greater than for an individual H_2O molecule [82].

One of the interesting points that emerges in molecule-surface bonding studies is the tradeoff that often develops between molecular preference for a limited number of strong specific interactions (here, H-bonds) and more numerous physical interactions. In so far as one can state a rule, simulations indicate that an adsorbate on ice adjusts its configuration to maximize the number of attractive interactions with the surface while retaining (weakened) "H-bonds" to unsaturated water-coordination sites. Binding sites at centers of the ubiquitous H_2O rings are especially favorable for optimizing multiple interactions. An adsorbate molecule within a ring orients itself so as to take advantage of hydrogen electrostatic bonding to dangling atoms of the ring water molecules. As a result, the bonding configurations/strengths change significantly with increased surface coverage, since neighbor adsorbate molecules facilitate multiple attractive interactions and thereby enhance specific interaction(s) of an adsorbate at the H-bonding site(s). Such behavior has been revealed in computational results ranging from the very weak adsorbate H_2 [74] to the strong adsorbate NH_3 [63]. Data that support this view will be noted below.

There are also "rules" specific to the behavior of strong adsorbates on the ice surface. Observations on ice particles with "monolayer" coverage by strong adsorbates have shown no evidence of penetration of the particles by the adsorbate over long periods of exposure. Only when the abundance of adsorbate available at the surface exceeds that for "monolayer" coverage does conversion of the ice to a hydrate commence. This can be readily argued as a consequence of the need for the free energy of the adsorbate at the surface to exceed some minimum value before the initiation of a new (hydrate) phase is possible. That is, a new phase must be nucleated before it can grow, with nucleation requiring "available" molecules that are not comfortably attached to favored surface sites [63,73].

17.3.1 Carbon Monoxide and Ozone - Weak Adsorbates

Systems with moderately weak ice adsorbates at temperatures in the 30 – 90 K range play a special role. The interactions can be attributed to an ice surface that has not been transformed through the breaking/making of new H-bonds induced by either the adsorbate or Ostwald ripening of the particles. Still, the magnitude of adsorbate-induced shifts of the ice surface-molecule vibrational frequencies are significant and easily observed, and the interactions with the surface are sufficiently specific to permit identification of the sites favored by a given adsorbate. These points are nicely demonstrated by the adsorbates CO and O_3; important terrestrial/astrophysical molecules that behave in a parallel manner on the ice surface.

Original spectra of CO on the surface of microporous amorphous solid water (ASW) were interpreted in terms of the occupation of two different sites with CO frequencies of 2152 and 2136 cm^{-1} [75]. Based on computed hetero-dimer frequencies and the response of the band intensities to the presence of competitive adsorbates, these bands were assigned to d-H and oxygen surface sites, respectively [76]. Included is the notion of a significant role of multiple interactions (MI) of adsorbate with water surface molecules, particularly for the second site. This type of site will be denoted below as a MI/d-O site.

Recent interpretations support the original view of the two sites of interaction with atmospheric ice [77]. The spectra of CO on D_2O-ice nanocrystals, Fig. 17.13, mimic closely those on the amorphous surface. At lowest coverage, the absorption is broad with ill-defined maxima, but two peaks (ultimately at 2155 and 2140 cm^{-1}) move apart somewhat and grow with increasing coverage. Nevertheless, the sequence of bands shows that the initial CO favors the d-D(H) sites as the 20% cover case has a band maximum near 2150 cm^{-1}. That the interaction with the d-D group is more specific than at the second class of sites is further shown by the top (dashed) low-CO-coverage band measured in the presence of an excess of N_2 adsorbate. As observed previously using CF_4 co-adsorbed with CO on ASW [75], nitrogen, when present in abundance, takes over the less polar fraction of the surface while isolating CO on the favored d-D sites.

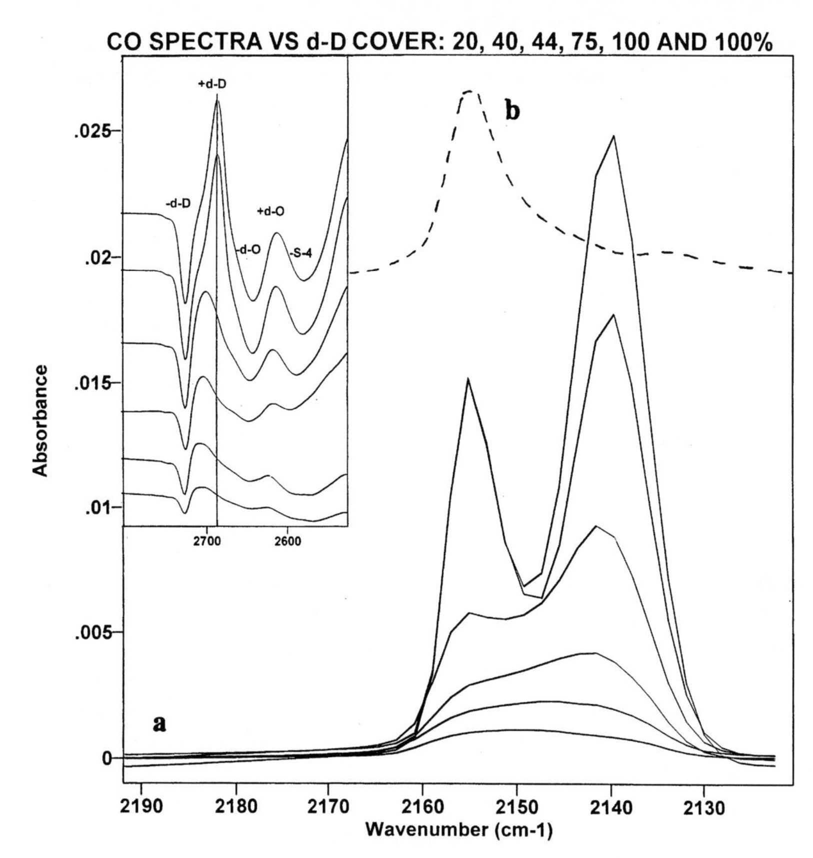

Fig. 17.13. Infrared spectra of CO adsorbed on 16 nm ice particles at 40 K as a function of % d-D sites occupied (a) CO doublet spectrum for d-D cover ranging from 20 to 100% (b) $\sim$ 20% CO cover but with remaining surface saturated with adsorbed N_2. The inset shows the variation of the surface mode bands with CO cover for the sequence of (a).

The inset of Fig. 17.13 shows the response of the ice-surface bands, for the same sequence as the main figure, through the gain and loss of intensity at the surface sites (d-D, d-O and s-4) as a function of adsorbed CO. For the top spectrum corresponding to a full monolayer, the amplitudes of loss at the d-D and d-O sites are similar, but, consistent with the evolution of the CO bands, the loss amplitude for the d-D is much greater than that of the d-O for cases of $< 50\%$ d-D cover. Also, the variation of the strength of the specific interaction with the d-D group with coverage is more obvious than implied by the minor shift of the CO bands. The magnitude of the ultimate shift of the d-D band position (to +d-D; 2690 cm^{-1}) is 36 cm^{-1}, but at 40 % cover the shift is only $\sim$22 cm^{-1}. As noted above, this variation of shift magnitude with surface cover reflects an increase in interaction strength with specific ice surface sites; an increase made possible because the more crowded CO molecules can take full advantage of specific sites without diminished generic interactions.

Results for ozone adsorption on the ice *nanocrystal* surface parallel those of CO, with occupation of two distinct types of surface sites indicated. Though there have been no previous reports of spectra of ozone adsorbed on the surface of crystalline ice, there have been recent studies of ozone adsorbed on ASW [78,79]. These studies used FTIR spectroscopy to help deduce the nature of the O_3-ice interaction, with observations made as a function of exposure and temperature. Extensive *ab initio* calculations based on an ordered ice surface, when combined with the spectral data, led to the conclusion that O_3 favors a single type of site on the walls of the ASW pores [79]. Binding of one end of the O_3 with a d-H surface site was identified with a 60 cm^{-1} downshift of the d-H frequency. The O_3, present on the surface from 50 to ~75 K, was shown to desorb into a vacuum at higher temperatures.

However our measurements for ozone adsorbate on ice nanocrystals suggest two sites, although the corresponding spectral features are not as well separated as in the case of CO. Ozone occupation of two distinct types of sites on the surface of ice *nanocrystals* is apparent from the spectra of Fig. 17.14. As for CO, a stretch mode (the asymmetric stretch; near 1037 cm^{-1}) appears as a single nearly featureless band for coverage much less than a monolayer (e.g., 20%). However, for higher coverage, it is again clear that there are two sub-bands with positions that diverge as the monolayer is completed at 75 K: a dominant band near 1035 and a shoulder at ~1040 cm^{-1}. Convincingly analogous with the CO behavior, a small amount of adsorbed ozone, on a surface otherwise saturated with oxygen, produces a single sharper band (bottom; Fig. 17.14) near the position of the high frequency shoulder of the O_3 monolayer.

The position of the d-D band (inset of Fig. 17.14) indicates the relative strength of O_3 interactions with d-D sites for the three cases: 1) partial coverage, 2) monolayer coverage and 3) dilute ozone in O_2. For both a full monolayer, and dilute ozone in adsorbed O_2, a fully shifted d-D band appears at 2691 cm^{-1}, while a weaker interaction is reflected in the smaller shift (to ~2700) for the case of a partial ozone cover. (In the mixed-adsorbate case, the band of d-D sites not occupied by ozone but rather interacting directly with O_2 shifts weakly to 2717 cm^{-1}.) The coincidence of the bands at 2691 cm^{-1} is indicative of similar interaction strengths with the d-D sites and is consistent with the 1040 cm^{-1} shoulder of monolayer ozone being caused by ozone molecules bound specifically at the d-D sites. As with CO, an increased interaction of ozone with d-D is attributed to the establishment of a more favorable alignment with the site while retaining multiple interactions with the surrounding O_2 (or other ozone molecules of the full monolayer). The result is consistent with the view [79] that ozone binds most effectively at the d-D positions. The 36 cm^{-1} shift of the d-D produced by adsorbed ozone

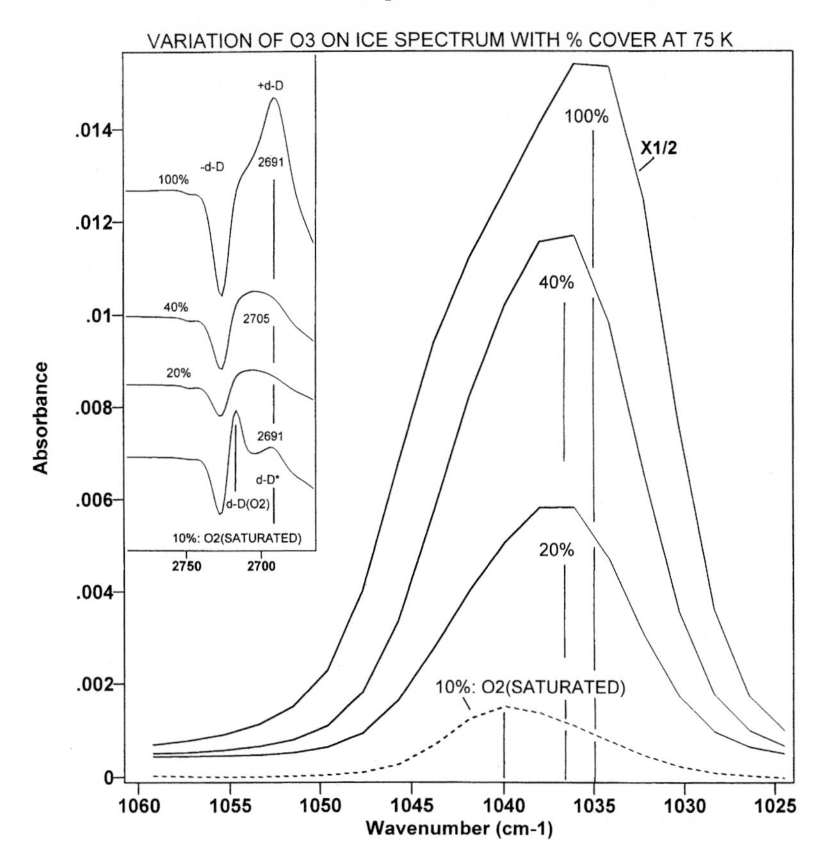

Fig. 17.14. Infrared spectra of ozone adsorbed on 16 nm ice particles at 75 K as a function of % d-D sites occupied. Solid lines show O_3 asymmetric stretch band for d-D coverage of 20, 40 and 100%, while the dashed spectrum is for 10% O_3 coverage with the remaining surface saturated with adsorbed oxygen. The variation of the intensity and position of the shifted d-D band for the same sequence is shown in the inset.

corresponds to a d-H shift of 48 cm^{-1}. As is typical, this value is significantly less than the 60 cm^{-1} reported for ozone on the d-H site of ASW micropores. (The larger shift for ASW is attributed to multi-layering of the adsorbate within the pores).

17.3.2 Acetylene - an Intermediate Adsorbate

The list of known intermediate adsorbates (i.e., substances that are able to de-reconstruct the ice surface at temperatures in the 120 K range, but are incapable of converting ice nanocrystals to a new hydrate phase) is limited to acetylene, H_2S, SO_2 and HCN. Of these adsorbates, SO_2 and HCN bind with sufficient strength to the surface that complete removal at temperatures considered here (< 140 K) is not possible. Thus, the reversal of the ordering of the surface and subsurface by SO_2 described in Section 2.2 (or by NH_3; Section 3.3) is not possible. However, both acetylene and H_2S, which bind somewhat less strongly with the ice surface groups, desorb into a vacuum at ~120 K. Data for adsorbed acetylene have therefore been used to show that the surface/subsurface ordering induced by intermediate adsorbates is, in fact, reversible.

Adsorbed acetylene has also played the interesting role of highlighting the large increase in the polarization of the ice surface dangling O-H groups from cooperative effects among water molecules (compared to the O-H bond of isolated water molecules). Matrix isolation data for the mixed acetylene-water dimer have been interpreted as showing stability only for a structure in which acetylene acts as the proton donor in an H-bonded complex [80]; and *ab initio* calculations have confirmed that proton donation from water to acetylene gives a relatively minor energy of stabilization (~2.5 kJ/mol) [81]. However, acetylene interacts with the more polar surface O-H to shift the d-H band by ~80 cm^{-1} (or 95 cm^{-1} for amorphous ice), which corresponds to a binding energy of ~12.9 kJ/mol (or 15.4 kJ/mol) [82]. This result suggests that the extra polarization of the surface O-H group causes an increase of nearly a factor of five in bonding with the π-electron system of acetylene. Further, infrared and Raman data clearly indicate that adsorbed acetylene also assumes a second configuration as a proton donor to the d-O surface sites. Acetylene is bi-functional in H-bonding with the surface of ice [83].

17.3.3 Ammonia – a Prototypical Strong Ice Adsorbate

Because ammonia is an excellent proton acceptor that interacts preferentially with the d-H surface sites of ice, a monolayer of ammonia on ice can be defined at two stages during adsorption at 120 K. Saturation of the d-H sites, corresponding to a first "monolayer", occurs with a very low ammonia ambient pressure and a relatively minor uptake at the MI/d-O sites of the surface. In fact, MC simulation of a nanocrystal surface with ~50% of an ammonia "monolayer" at 110 K indicates that ~95 % of the ammonia molecules are bound to d-H sites [63]. However, with a significant increase in the equilibrium vapor pressure, a full monolayer of adsorbed NH_3 forms as MI/d-O sites become occupied. This second saturation level is accompanied by a strong shift of the d-O surface modes and the emergence near 3000 cm^{-1} of the intense broad band of surface d-H bound to NH_3 [72].

So ammonia represents yet another example of the increased binding strength of an adsorbate with a *specific* site as a result of the presence of more generically bound molecules. Otherwise, ammonia bends over to the surface to form an optimum number of multiple weak bonds, weakening the normally strong N–H-O bond so that the band of the fully shifted d-H is not readily observed. The crowding from increased adsorbate concentration permits the H-bond to become linear and stronger. This shifts the d-H band to a much lower frequency, more normal for the ammonia - water H-bond, so that it is visible below the region of the intense ice bands. However, despite the huge shift of the d-H band (700 cm^{-1}), full monolayer adsorption is not accompanied by penetration of the ammonia towards formation of a hydrate.

The MC calculations also show that, if the ammonia concentration exceeds the density of d-H sites on the disordered ice surface, *ammonia cleaves existing weak surface H-bonds* to double the number of favorable d-H adsorption sites [63]. This insertion acts to reduce strain in the ice subsurface. For this reason it is not surprising that the FTIR spectra show a sharp increase in interior core ice; of the order of 3-5% for 20 nm particles. The increase in crystalline-core ice reflects subsurface relaxation that follows surface de-reconstruction (e.g., as observed with SO_2: Section 2.2). Consistent with this view, simulated relaxation of a system with adsorbed ammonia on an *oxygen-ordered* unrelaxed nanoparticle results in much less surface/subsurface disordering than occurs otherwise in the absence of the ammonia. The calculations show clearly that ammonia can stabilize an ice surface that is relatively rich with d-H groups. However, penetration of the ammonia into the ice was observed when an excessive amount of ammonia, beyond that required by the optimum number of d-H groups, was present. This parallels the experimental observation that conversion of ice to an ammonia hydrate requires the ammonia pressure to be increased significantly above the level for surface saturation.

17.3.4 Dimethyl Ether – a Typical Small-Ether Adsorbate

Like ammonia, ethers are much better proton acceptors than proton donors. They initially bind through oxygen to the d-H sites of the ice surface, but, at higher coverage at 120 K, also occupy MI/d-O sites. This choice of surface sites by adsorbed dimethyl ether is particularly clear from bands near 900 and 1100 cm^{-1} in the series of spectra in Fig. 17.15. The top spectrum, obtained for a sample with the d-H sites nearly saturated (and downshifted strongly to 3484 cm^{-1}; i.e., by 210 cm^{-1}), reflects primarily absorption by ether molecules attached to the d-H sites (910 and 1087 cm^{-1}). Complete saturation of the surface, as in (b) causes higher frequency bands of ether molecules on MI/d-O sites (922 and 1093 cm^{-1}) to become dominant. This change reflects two factors: there are more MI/d-O sites on the surface and the C-O stretch modes of ether have higher frequencies at such sites (i. e., C-O ether bonds are weakened by the H-bonding with the d-H sites).

The assignment of the ether bands to particular surface sites is affirmed by the effect of co-adsorption of acetylene (Fig. 17.15c). Acetylene is bifunctional, as described in Section 17.3.2, but is capable of only moderately strong bonding to the d-H sites (d-H shift of 80 vs 210 cm^{-1} by dimethyl ether). However, as a good proton donor, acetylene readily displaces ether that bonds weakly at the surface MI/d-O sites, as shown by loss of the bands at 922 and 1093 cm^{-1}, while leaving the more strongly bound ether at the

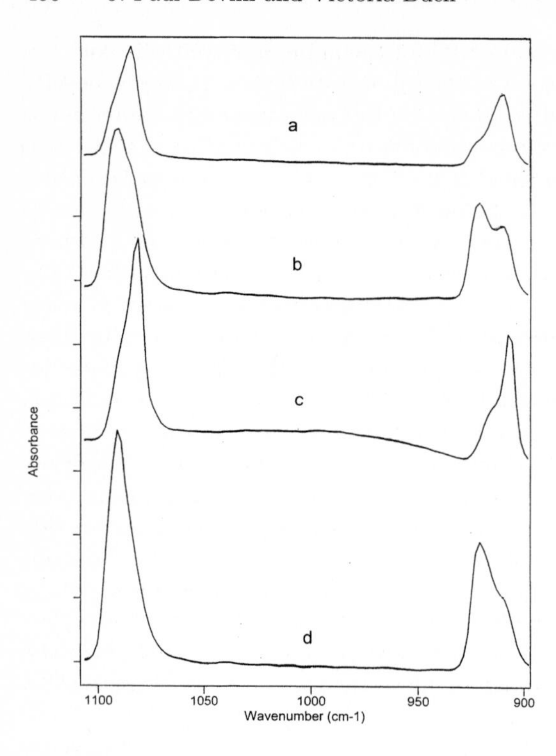

Fig. 17.15. Infrared spectrum of dimethyl ether adsorbed on 40 nm ice crystals at 120 K (a) ~80% cover of the d-H sites (b) saturation of the ice surface (c) after exposure to abundance of acetylene vapor and (d) with the acetylene pumped from the cell.

d-H sites. The reversibility of this process is clear from (d), as the 922 and 1023 bands regain dominance following removal of acetylene from the system. In fact, the ether can be quickly removed from the MI/d-O surface sites by merely warming to 130 K in a vacuum, while the d-H-bound ether remains until particle vaporization occurs.

Like ammonia, a monolayer of dimethyl ether on ice at ~120 K is not accompanied by evidence of penetration into the ice, although ethers are capable of H-bond chemistry on ice. When the abundance of ether available at the ice surface exceeds a monolayer at temperatures above 120 K, an ether hydrate is nucleated and the ice particle converts to a type II clathrate hydrate. Similar behavior has been observed for the small-ring ethers ethylene oxide (oxirane) and tetrahydrofuran, which form type I and type II hydrates, respectively [84].

17.3.5 HCl – a Representative Strong Acid Adsorbate

The strong acids HCl, HBr and HNO_3 have each been shown to meet the operational definition of strong adsorbates on ice nanocrystals. Particular interest has been directed to HCl on ice, in part because of insights available from numerous *ab initio* studies of HCl in small water clusters [85-88], and because HCl is an important actor in chemistry that leads to the seasonal depletion of ozone in the stratosphere over Antarctica [7]. By combining FTIR transmission spectroscopic studies and MC simulations of various levels of HCl coverage of ice nanocrystals with *ab-initio* computations of small water clusters (with HCl in configurations suggested by the MC calculations), we have found that cold-ice-nanoparticle surfaces provide a convenient micro-laboratory for observation of progressive solvation stages of HCl [89].

Control of the experimental dosing of acid to the ice surface was achieved by constructing 3-D arrays with alternating ice (12 nm) and acid (~1 nm) particles such that the acid moves to the ice-particle surfaces in the 40–70 K range. In such a system, the nature of the ice particle surface is critical to the adsorbate behavior. Our *ab-initio* results, in an MP2 basis, support earlier conclusions that HCl ionization requires water solvation leading to one acid proton-donor bond and 2 acceptor bonds to the chlorine [90]. The disordered nanocrystal surfaces, with a reduced number of unsaturated coordination sites as described in Sections 2.2 and 2.3, leaves an insufficient density of d-O and d-H atoms for effective acid solvation on the ice surface. As a result, at low temperatures (<60 K) and dilute HCl levels, the acid molecules are adsorbed in molecular states. However, ionization progresses with heating, but to a degree strongly influenced by the population of HCl on the surface (as self-solvation can provide the needed acceptor bond(s); [89] and chapter 4). Using the combination of tools, we identified and assigned two molecular and two ion-solvation stages in the 50 – 110 K range.

At low temperature and with dosage levels in the 15 –30% of a monolayer range, two broad molecular HCl bands were observed experimentally at 2500 and 1710 cm^{-1} with frequency downshifts of ~360 and 1175 cm^{-1} with respect to the gas phase (Fig. 17.16). The first corresponds to slightly stretched HCl while the much larger shift indicates strongly stretched HCl on the verge of ionization. The MC simulations of HCl on an ice particle suggest that the 2 bands originate from 1- and 2-coordinated HCl. At low coverage the two most common configurations of the MC results were d-O–HCl and the bridged structure d-O–HCl–d-H, with an increased population of d-O–HCl–HCl as the HCl concentration was increased. The relatively low coordination of the molecular adsorbate can be ascribed to the paucity of d-O and d-H atoms on the ice surface while *ab-initio* calculations of cluster models with 1- and 2-coordinated HCl confirm that HCl remains molecular. Similar pairs of bands, shifted in position by mass/bond-strength effects, have also been observed for HBr on H_2O and DCl on D_2O nanocrystals.

The *ab-initio* frequency shift, calculated for 1-coordinated HCl bonded to d-O in a cluster, places it quite close to the first measured band. Further, a seven-water cluster with HCl in a 2-coordinated bridge configuration, d-O–HCl–d-H, yielded a large *ab-initio* frequency shift similar to that of the second measured band (Fig. 17.16). However, such strong stretching in the bridge configuration occurs only if the d-O molecule has both H atoms bonded to other water molecules, *and* there is no additional acceptor bond from water

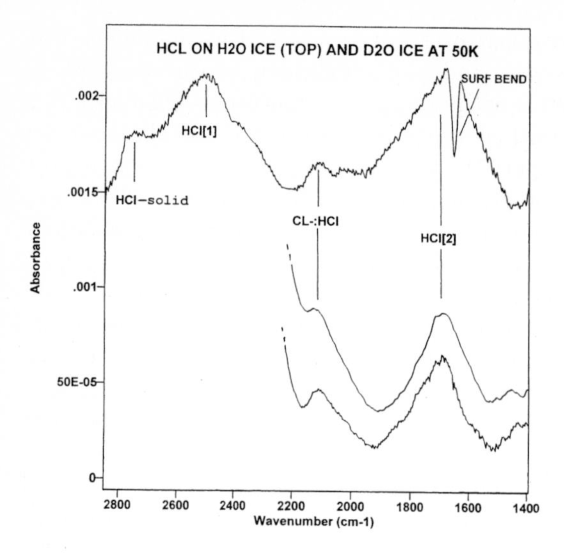

Fig. 17.16. Infrared spectra of HCl on 12 nm ice crystals at 50 K minus the spectra of bare-ice arrays (top) showing two strong bands of molecular adsorbates HCl(1) and HCl(2) on H_2O ice and (bottom) difference spectra of a single HCl-on-D_2O ice sample vs two different ice reference spectra. Possible interferences with the HCl(2) band, from surface bending and oxonium ion bands of the H_2O ice array (top), are absent for the D_2O ice array.

to the d-O atom. It has been shown in the past for liquid water that such an acceptor bond interferes with proton transfer to the d-O [91,92]. The 2-coordinated HCl structures from the MC simulations each have such an acceptor bond, a condition incompatible with identification as the strongly stretched HCl that gives the observed 1710 cm^{-1} band. It seems that the 2-coordinated HCl adsorbate modifies the ice surface beyond expectations from the MC results, which were based on rigid HCl molecules. In other words, enhanced polarization, anticipated for the strongly stretched HCl, is thought to promote the strengthening of the remaining H-bonds as part of a low activation-energy process leading to the elimination of the interfering acceptor bond.

For ∼20% monolayer acid dosage, heating above 60 K is associated with the gradual growth of ion bands in the FTIR spectra. From comparisons with amorphous hydrate/deuterate spectra [93], and ab initio calculations for ionized clusters, features have been identified for the hydronium ion in contact with Cl^- and for molecular HCl solvating Cl^-. The latter is anticipated from ionization of the bridge structures d-O–HCl–d-H and d-O–HCl–HCl. However, HCl ionization to a contact ion pair apparently requires a minimum of three H-bonds to the HCl, one to H and two to Cl. The latter provide, in turn, 2 out of 3 H-bonds to the product Cl^- (the third is provided by the cation in contact). Since extensive ionization for samples with significantly less than 20% HCl was not observed below 100 K, the evidence is strong that, for higher concentrations, the more weakly adsorbed HCl becomes sufficiently abundant and sufficiently mobile, above ∼60 K, to participate as the third coordinating unit to HCl at "bridged" d-O sites.

At a temperature near 95 K and with acid coverage of ∼ 20 %, there is an abrupt transition from a predominantly molecular to an ionic surface state [94]. During this transition the original bands of the hydronium ion are *relatively* weakened and the spectrum becomes dominated by an ionic phase rich in Zundel cations ($H_5O_2^+$). The abundance of Zundel ions is evidenced by the Zundel effect; a strong continuum underlying the band spectrum of the ions that results from an exceptionally high polarizability of the cation [95]. This formation of Zundel cations corresponds to the 4th stage in the solvation of adsorbed HCl, as the proton moves away from the chloride to a position between two water molecules. The ionized state remains a surface phase at this stage as was determined by adsorbing HCl on D_2O ice and noting that the proper H/D ratio is retained for isotopic scrambling restricted to the top surface layer. The sharp change from a largely molecular to an ionic system suggests a cooperative process involving nucleation of a 2-dimensional surface layer that resembles the amorphous dihydrate with cation structures ranging from hydronium to Zundel. The surface spectrum in fact matches that of the amorphous dihydrate solid. One may note that the structure of the crystal analog of this solid was determined by X-ray diffraction studies as $(H_5O_2{}^+)(Cl^-)$[96].

It seems apparent from the experimental and *ab-initio* results that, in the presence of an abundance of adsorbed acid, ionization of HCl molecules adsorbed at d-O sites is promoted through the formation of 3-coordinated HCl. The ionization occurs aggressively, even at 60 K, for samples with ~40% of an acid cover. This 60 K ionization is evidence that the activation energy required to remove the hindering acceptor bond of the d-O molecule is very low. It follows that the observed more highly activated process leading to extensive ionization in the 95 K range has a different source. This source is most probably either the mobilization of adsorbed HCl on the surface, allowing more weakly bound HCl to join with the stretched HCl at the d-O sites, or a reorganization of the ice surface through the influence of the acid adsorbate which leads to many new d-H and d-O surface sites (as described for other intermediate and strong adsorbates). This, in turn, stimulates ionization. Though the latter is undoubtedly a factor at some temperature, the strong correlation between ionization and HCl concentration, even at 95 K, is evidence that, for the conditions of this study, the mobility of weakly bound HCl was the critical factor.

It has been stressed for other strong adsorbates (ammonia and ethers) that penetration of a hydrate phase into ice occurs only subsequent to establishing a monolayer of adsorbate followed by a nucleation event. The nucleation requires the presence of an abundance of adsorbate molecules not comfortably bound at favorable surface sites. Extension of this view to the strong acids may require a slight twist of perception. The data, for temperatures such that the ice is not volatile (i.e., < 110 K for 12 nm particles), suggest that *isolated* molecular HCl at d-O sites on the ice surface is a stable state; so if a monolayer of adsorbed HCl is defined accordingly, HCl fits the pattern of other strong adsorbates.(The experimental coverage of 20% was determined from the observed decrease in the free d-H intensity. The ∼1:2 ratio of doubly- and singly-coordinated molecules obtained in MC suggests that the percentage of HCl-bonded d-O atoms on the nanoparticle surface is much higher than that of d-H, i.e., closer to 80%.) However, HCl is made different by the apparent nucleation and formation of an ionized *surface phase* above some level of acid dosing. The surface layer that forms from the burst of ionization at 95 K, from 20% of a monolayer of acid, apparently develops separately from any subsequent nucleation of a 3-dimensional hydrate phase, which requires yet a higher level of HCl activity [97]. It seems to follow that HCl, with two defined levels of "monolayer" coverage, behaves analogously to ammonia, with the caveat that the second level corresponds to a 2-D ionized rather than a molecular state.

17.4 Acknowledgments

Our research described in this chapter has been supported by the National Science Foundation and the Binational Science Foundation. The work represents the efforts of several graduate students, most particularly Nevin Uras-Aytemiz, to whom we are indebted. Contributions of Professor Joanna Sadlej of Warsaw University are also noted with special thanks.

References

1. M. L. Delitsky, A. L. Lane: J. Geophys. Res. **103**, 31391 (1998).
2. W. R. Thompson, B. G. J. P. T. Murray, B. N. Kare, C. Sagan: J. Geophys. Res. 92, 14933 (1987).
3. D. Blake, L. Allamandola, S. Sanford, D. Hudgins and F. Freund: Science **254**, 548 (1991).
4. J. Klinger: J. Phys. Chem. **87**, 4209 (1983).
5. F. C. Gillett, W. J. Forrest: Astrophys. J. **179**, 483 (1973)
6. W. Hagens, A. G. G. M. Tielens, J. M. Greenberg: Astron. Astrophys. **117**, 132 (1983).
7. M. J. Molina, T. L. Tso, L. T. Molina, F. C. Y. Wang: Science **238**, 1253 (1987).
8. A, Glebov, A. P. Graham, A. Menzel, J. P. Toennies, P. Senet: J. Chem. Phys. **112**, 11,011 (2000).
9. K. D. Gibson, M. Viste, S. J. Sibener: J. Chem. Phys. **112**, 9582 (2000).
10. N. Materer, U. Starke, A. Barbieri, M. A. Van Hove, G. A. Somorjai, G. J. Kroes, C. Minot: J. Phys. Chem. **99**, 6267 (1995).
11. H. Kang, T. H. Shin, S. C. Park, I. K. Kim, S. J. Han: J. Am. Chem. Soc. **122**, 9842 (2000).
12. A. B. Horn, T. G. Koch, M. A. Chesters, M. R. S. McCoustra, J. R. Sodeau: J. Phys. Chem. **98**, 946 (1994).
13. J. E. Schaff, J. T. Roberts: J. Phys. Chem. **98**, 6900 (1994).

14. W. Xing, P. B. Miranda, Y. R. Shen: Phys. Rev. Lett. **86**, 1554 (2001).
15. H. Dosch, A. Lied and J. H. Bilgram: Surf. Science **327**, 145 (1995).
16. E. Mayer, R. Pletzer: Nature **319**, 298(1986).
17. K. P. Stevenson, G. A. Kimmel, Z. Dohnalek, R. S. Scott, B. D. Kay: Science **283**, 1505 (1999).
18. C. J. Burnham, S. S. Xantheas: J. Chem. Phys. **116**, 1500 (2002).
19. R. S. Fellers, C. Leforestier, L. B. Braly, M. G. Brown, R. J. Saykally: Science **284**, 945 (1999).
20. E. M. Mas, R. Bukowski, K. Szalewicz, G. C. Groenenboom, P. E. S. Wormer and A. van der Avoird, J. Chem. Phys. **113**, 6687 (2000).
21. U. Buck, I. Ettischer, M. Melzer, V. Buch, J. Sadlej: Phys. Rev. Lett. **80**, 2578 (1998).
22. P. V. Hobbs: *Ice Physics* (Clarendon Press, Oxford, 1974).
23. V. F. Petrenko, R. W. Whitworth: *Physics of Ice* (Oxford University Press, Oxford, 1999).
24. D. E. Brown, S. M. George: J. Phys. Chem. **100**, 15460 (1996).
25. M. D. Newton: J. Phys. Chem. **87**, 4288 (1983).
26. M. Kunst, J. M. Warman: J. Phys. Chem. **87**, 4093 (1983).
27. R. Podeszwa, V. Buch: Phys. Rev. Lett. **83**, 4570 (1999).
28. P. J. Wooldridge, J. P. Devlin: J. Chem. Phys. 88, 3086 (1988).
29. N. J. Sack, R. A. Baragiola: Phys. Rev. B **48**, 9973 (1993).
30. J. E. Bertie, H. J. Labbe, E. Whalley: J. Chem. Phys. **50**, 4501 (1969).
31. J. E. Bertie, M. K. Ahmed, H. H. Eysel: J. Phys. Chem. **93**, 2210 (1989).
32. E. Whalley: Can. J. Chem. **55**, 3429 (1977).
33. S. A. Rice, M. S. Bergren, A. C. Belch, G. Nielson: J. Phys. Chem. **87**, 4295 (1983).
34. M. J. Wojcik, V. Buch, J. P. Devlin: J. Chem. Phys. 99, 2332, 1993.
35. V. Buch, J. P. Devlin: J. Chem. Phys. **110**, 3437 (1999).
36. J. P. Devlin, P. J. Wooldridge, G. Ritzhaupt: J. Chem. Phys. **84**, 6095 (1986).
37. W. Hagen, A. G. G. M. Tielens, J. M. Greenberg: Chem. Phys. **56**, 367 (1981).
38. T. C. Sivakumar, S. A. Rice, M. G. Sceats: J. Chem. Phys. **69**, 3468 (1978).
39. J. P. Devlin, J. Sadlej, V. Buch: J. Phys. Chem. A **105**, 974 (2001).
40. X. Su, L. Lianos, Y. R. Shen, G. A. Somorjai: Phys. Rev. Lett. **80**, 1533 (1998).
41. H. Witek, V. Buch: J. Chem. Phys. 110, 3168 (1999).
42. J. P. Devlin, V. Buch: J. Phys. Chem. **99**, 16534 (1995).
43. B. Rowland, N. S. Kadagathur, J. P. Devlin, V. Buch, T. Feldmann, M. J. Wojcik: J. Chem. Phys. **102**, 8328 (1995).
44. J. P. Devlin, V. Buch: J. Phys. Chem. B **101**, 6095 (1997).
45. L. Delzeit, J. P. Devlin, V. Buch: J. Chem. Phys. **107**, 3726 (1997).
46. J. P. Devlin, C. Joyce, V. Buch: J. Phys. Chem. A **104**, 1974 (2000).
47. N. H. Fletcher: Philosophical Magazine **66**, 109 (1992).
48. G.-J. Kroes: Surf. Sci. **275**, 365 (1992).
49. V. Buch, L. Delzeit, C. Blackledge, J. P. Devlin: J. Phys. Chem. **100**, 3732 (1996).
50. C. F. Landes, M. Braun, M. A. El-Sayed: J. Phys. Chem. B **105**, 10554 (2001).
51. A. Khan: J. Phys. Chem. A **103**, 1260 (1999).
52. R. Ludwig, F. Weinhold: J. Chem. Phys. **110**, 508 (1999).
53. G. Torchet, P. Schwartz, J. Farges, M. F. de Feraudy, B. Raoult: J. Chem. Phys. **79**, 6196 (1983).

54. J. Huang, L. S. Bartell: J. Chem. Phys. **99**, 3924 (1995).
55. M. L. Clapp, R. E. Miller, D. R. Worsnop: J. Phys. Chem. **99**, 6326 (1995).
56. G. E. Ewing, D. T. Sheng: J. Phys. Chem. **92**, 4063 (1988).
57. S. Bauerecker, M. Taraschewski, C. Weitkamp, H. K. Cammenga: Rev. Sci. Instrum. **72**, 3946 (2001).
58. M. L. Clapp, R. E. Miller: Icarus **105**, 529 (1993).
59. L. Delzeit, D. Blake: J. Geophys. Res. – Planets **106**(E12), 33371 (2001).
60. B. Rowland, J. P. Devlin: J. Chem. Phys. **94**, 812 (1991).
61. L. Jones, B. I. Swanson: J. Phys. Chem. **95**, 2701 (1991).
62. N. S. Holmes, J. R. Sodeau: J. Phys. Chem. A **103**, 4673 (1999).
63. N. Uras, V. Buch, J. P. Devlin: J. Phys. Chem. B **104**, 9203 (2000).
64. V. Buch, P. Sandler, J. Sadlej: J. Phys. Chem. B **102**, 8641 (1998).
65. S. Kuwajima, A. Warshel: J. Phys. Chem. **94**, 460 (1990).
66. J. Hernandez, N. Uras, J. P. Devlin: J. Chem. Phys. **108**, 4525 (1998).
67. J. P. Cowin, A. A. Tsekouras, M. J. Iedema, K. Wu and B. B. Ellison: Nature **398**, 405 (1999).
68. N. Uras-Aytemiz, C. Joyce, J. P. Devlin: J. Chem. Phys. **115**, 9835 (2001).
69. J. P. Devlin: J. Chem. Phys. **112**, 5527 (2000).
70. U. Buck: private communication.
71. S. Bauerecker, V. Buch, J. Kazimirski , and J. P. Devlin: to be submitted.
72. L. Delzeit, K. Powell, N. Uras, J. P. Devlin: J. Phys. Chem. B **101**, 2327 (1997).
73. N. Uras, J. P. Devlin: J. Phys. Chem. A **104**, 5770 (2000).
74. H. G. Hixson, M. J. Wojcik, M. S. Devlin, J. P. Devlin, V. Buch: J. Chem. Phys **97**, 753 (1992).
75. J. P. Devlin: J. Phys. Chem. **96**, 6185 (1992).
76. J. Sadlej, B. Rowland, J. P. Devlin, V. Buch: J. Chem. Phys. **102**, 4804 (1995).
77. M. E. Palumbo, J. Phys. Chem. A **101**, 4298 (1997).
78. H. Chaabouni, L. Schriver-Mazzuoli, A. Schriver: J. Phys. Chem. A **104**, 6962 (2000).
79. F. Borget, T. Chiavassa, A. Allouche, J. P. Aycard: J. Phys. Chem. B **105**, 449 (2001).
80. A. Engdahl, B. Nelander: Chem. Phys. Lett. **100**, 129 (1983).
81. M. J. Frisch, J. E. Del Bene, J. A. Pople: J. Chem. Phys. **78**, 4063 (1983).
82. S. C. Silva, J. P. Devlin: J. Phys. Chem. **98**, 10847 (1994).
83. A. Allouche: J. Phys. Chem. A **103**, 9150 (1999).
84. J. Hernandez, N. Uras, J. P. Devlin: J. Phys. Chem. B **102**, 4526 (1998)
85. M. J. Packer, D. C. Clary: J. Phys. Chem. **99**, 14323 (1995).
86. S. Re, Y. Osamura, Y. Suzuki, H. F. Schaeffer III: J. Chem. Phys. **109**, 973 (1998).
87. D. E. Bacelo, R.C. Binning, Y. Ishikawa: J. Phys. Chem. A **103**, 4631 (1999).
88. G. M. Chaban, R. B. Gerber, K. C. Janda: J. Phys. Chem. A **105**, 8323 (2000).
89. J. P. Devlin, N. Uras, J. Sadlej, V. Buch: Nature **417**, 269 (2002).
90. M. Svanberg, J. B. C. Pettersson, K. J. Bolton: J. Phys. Chem. A **104**, 5787 (2000).
91. N. Agmon: Chem. Phys. Lett. **244**, 456 (1995).
92. K. Ando, J. T. Hynes: J. Phys. Chem. B **101**, 10464 (1997).
93. L. Delzeit, B. Rowland, J. P. Devlin: J. Phys. Chem. **97**, 10312 (1993).
94. V. Buch, J. Sadlej, N. Uras-Aytemiz, J. P. Devlin: J. Phys. Chem. A, in press (November,2002).
95. G. Zundel: Adv. Chem. Phys. **111**, 1 (2000).
96. J. D. Lundgren, I. Olovsson: Acta Cryst. **12**, 17 (1959)
97. N. Uras-Aytemiz, C. Joyce, J. P. Devlin: J. Phys. Chem. A **105**, 10497 (2001).

Index

Springer Series in
CLUSTER PHYSICS

Printing: Druckhaus Berlin-Mitte
Binding: Buchbinderei Stein & Lehmann, Berlin